Ecological Studies, Vol. 154

Analysis and Synthesis

Ecological Studies

Volumes published since 1996 are listed at the end of this book.

Springer

Berlin
Heidelberg
New York
Barcelona
Hong Kong
London
Milan
Paris
Tokyo

L. Beyer M. Bölter (Eds.)

Geoecology of Antarctic Ice-Free Coastal Landscapes

With 54 Figures and 59 Tables

Springer

Prof. Dr. Lothar Beyer
Prof. Dr. Manfred Bölter
Institut für Polarökologie
Christian-Albrechts-Universität zu Kiel
Wischhofstr. 1–3
24148 Kiel
Germany

Cover illustration: Graph: T. F. Claassen, Photo: A. Tatur

ISSN 0070-8356
ISBN 3-540-42268-4 Springer-Verlag Berlin Heidelberg New York

Library of Congress Cataloging-in-Publication Data.

Geoecology of Antarctic ice-free coastal landscapes / L. Beyer, M. Bölter (eds.)
p. cm. – (Ecological studies ; v. 154)
Includes bibliographical references (p.).
ISBN 3540422684 (alk. paper)
1. Soil ecology–Antarctica. 2. Coastal ecology–Antarctica. I. Beyer, Lothar, 1959- II. Bölter, M. III. Series.

QH84.2 .G46 2002
577.5'7'09989–dc21 2001049872

Springer-Verlag Berlin Heidelberg New York
a company of BertelsmannSpringer Science+Business Media GmbH

http://www.springer.de

Printed in Germany

Cover design: *design & production* GmbH, Heidelberg
Typesetting: Kröner, Heidelberg

SPIN 10687804 31/3130 YK – 5 4 3 2 1 0 – Printed on acid free paper

Foreword

Research in Antarctica in the past two decades has fundamentally changed our perceptions of the southern continent. The new information coming in makes it increasingly evident that the ecology of Antarctica is much more distinctive, much more complex, and much more diverse than any one could have expected – witness the long list of books and reviews in the Preface that summarize, from varying points of view, the growing body of information. One could therefore ask whether a new book on the geoecology of Antarctic soils is really necessary. The answer is a definite yes, for several reasons. The editors and authors of the present volume approached the vast field of Antarctic research from a refreshingly new and original angle, not considered by earlier compilations. They also cover many new subjects and include information on geographic areas of the vast Antarctic continent not treated in earlier books. Nearly all of the authors are unrepresented in earlier reviews, and their fresh perspective alone assures that the present work offers new insights and new points of view. Last, but not least, the present work covers the most recent results of the fast-growing literature and gives the reader an excellent picture of the present state of Antarctic science.

This well-produced and authoritative book fills a gap and will become indispensable for all researchers and students interested in Antarctic ecology.

Tallahassee, Florida
Autumn 2001

E. Imre Friedmann

Preface

Polar regions have become a centre of interest during discussions on global change and global climate warming in the last decade. In the Antarctic, climate change is not only important for glaciology and atmospheric science (e.g. Weller 1992), but it also affects ecological aspects (e.g. Smith and Steenkamp 1990; Lewis Smith 1994; Kennedy 1995). Further, although many aspects of the terrestrial ecosystem, such as biodiversity, physiological and biochemical adaptations to habitats, general environmental conditions and related effects on life and habitat formations have been given much attention, overviews on the ecosystem scale are still rare; this holds especially true for links between the abiotic and biotic parts.

Antarctic terrestrial ecosystems have only limited extent, and only a small share of earth's living biomass. Therefore they play only a minor role in the global carbon cycle and in driving global climatic trends. So, what is the purpose of Antarctic terrestrial ecosystem research? Steven Chown addressed this question facing Antarctic researchers during the 7th SCAR Biology Conference held in Christchurch, New Zealand, in September 1998 (Chown et al. 2000). He summarised that Antarctic research is undoubtedly expensive, the results obtained are either specific to this region (hence, not transferable), or of a general nature (hence, the studies could be performed in any region outside Antarctica). However, on the other hand, there are several good reasons for carrying out ongoing research in Antarctica: the remoteness from human civilisation means that the Antarctic is a vast and unique laboratory for many scientific purposes, for example, understanding processes of cold adaptations in plants and insects, the spread and development of microbes, plants, and micro-arthropods, colonisation of deglaciated terrains in regions remote from sources of propagules, the global spread of pollutants, the fate of soil phosphorus and processes on initial soil formation under extreme cold conditions, as well as the effects of human disturbance in pristine environments. Individual aspects of these themes have been summarised in books or review articles, e.g. Campbell and Claridge (1987), Vincent (1988), Wynn-Williams (1990), Friedmann (1993), Lewis Smith (1993a,b), Priscu (1998).

Another initiative for this compilation is that most studies on Antarctic issues focus on local aspects of the continent or have tried to understand single processes or individual groups of organisms. We have attempted here to collate a variety of studies covering ecological aspects from landscape to habitats, from groups of organisms to individual plants and animals. A main feature is the development and genesis of soil in a real primary state.

The term geoecology points out the close interactions between soils, the actual state and processes of soil development, and their role in a polar environment with respect to its use as the basis of life processes for organisms. Geoecology means that we use soil ecology in its widest sense: soils are viewed at the centre of processes acting at scales of microbes, plants, and animals, but they are also viewed in relation to impacts from climate, geology and geography. This means that we look for a wide range of scales, as defined by necessity for geoecology by Blumenstein et al. (2000), from habitats for microbes and other soil-dwelling organisms across plant associations to landscape evolution via glacier retreat. We describe such features for three areas of the Antarctic, but have to mention that the broad sense of geoecology, namely human impact on nature, is not the main aim.

All three places; King George Island in the maritime Antarctic and the two continental sites, Ongul Islands and the Windmill Islands, share many features in common, despite being located far from each other. The places have carried permanently maintained stations for a long time and thus provide long records of environmental variables. Nevertheless, it was difficult to find directly comparable issues of research in these environments, due to national and local research interests, expertise and preferences. Biological issues can be found at all sites. Monitoring of meteorological conditions has been carried out at all stations since their inception. Many sites in Antarctica, and all these three, also have well-described vegetation. However, physical, biological and chemical processes in the soils, the soil biological component, processes central to the visible expression of the terrestrial biota, have been widely neglected, despite good geological background information. Thus, it seemed worthwhile to bring such data and information together to link these ecological components. Further, to complete soil scientific studies in the Antarctic, comparable extensive studies on soils have been carried out only in the areas of the McMurdo Dry Valleys, in some widely dispersed localities of southern and northern Victoria Land, and in the vicinity of the McMurdo Station with special emphasis on deserts (Campbell and Claridge 1987; Priscu 1998).

Hence, we have taken the chance to collate our studies from the maritime and coastal continental Antarctic and present a comprehensive view on landscape and biological issues of mainly three Antarctic places. Further locations visited during expeditions are mentioned, and relations to data from several other points are drawn and discussed. Obviously, many points of general interest can be used to describe processes in other regions of the continent and other polar or alpine environments.

We have also included features of human impact on a continent which is often referred to as a pristine environment, although this point cannot be the main focus of this book. However, despite the remote locality of this continent, increasing tourist activity, field activities of scientific programmes, erection and maintenance of national research stations, local and large-scale dispersal of pollutants, the introduction of alien organisms, all have significant effects on the Antarctic environment, individually or in combination. They cannot be neglected when describing environmental and biological processes in an area which functions as an indicator for the whole planet. There is much still to be learned from scientific approaches and results and how to protect and preserve a great wilderness.

There are enormous amounts of quantitative and qualitative materials gathered in journals, books, and many local reports. We have tried to compile some of them regarding the soils of three Antarctic coastal environments. However, there still remain many questions which cannot be answered by this contribution. We hope that it may serve as an incentive for scientists looking for new tasks in Antarctic research.

We have organised the chapters into four main parts, from specific descriptions of individual locations, groups of organisms and processes to general ecological aspects.

I General descriptions of sites and the local environmental constraints,
II Processes in the soil environment mainly driven by environmental factors,
III Organisms and biological processes on the soils (plants) and in the soils (microorganisms and soil-dwelling animals),
IV A set of summarising chapters which put together information about the soil habitat and its function on life under the constraints of Antarctic environmental conditions. The final chapter focuses on past and (anticipated) future aspects of environmental reactions on increasing anthropogenic effects on the terrestrial environment.

Two further points need to be mentioned. Parts III and IV, which deal with autotrophic and heterotrophic organisms of different (changing) taxonomic positions, follow the traditional taxonomy, i.e. lichens are still regarded as plants, cyanobacteria are combined with other autotrophs, namely algae.

All data which are used as a basis for our ideas, discussions and conclusions derive from long traditions of work in these environments; previous contributions in books, journals or reports with many other co-authors form the basis of this book. Our thanks are extended to all those who contributed directly or indirectly by numerous discussions, evaluations of proposals, revisions of papers or long talks at long evenings during our visits to Antarctic bases or at international symposia.

We would like to thank Prof. O.L. Lange and Pof. E. I. Friedmann. Both stimulated and encouraged the process of compiling this issue of *Ecological Studies*, their kind advice was very welcome and promoted our editorial efforts. The Springer-Verlag in Heidelberg, namely Mrs. A. Schlitzberger and Mr. F. Kröner, showed great patience – another main point to be greatly acknowloedged. We also thank our home institution for its help and support.

References

Blumenstein O, Schachtzabel H, Barsch H, Bork H-R, Küppers U (2000) Grundlagen der Geoökologie: Erscheinungen und Prozesse in unserer Umwelt. Springer, Berlin Heidelberg New York

Campbell IB, Claridge GGC (1987) Antarctica: soils, weathering processes and environment. Elsevier, Amsterdam

Chown SL, Gaston KJ, Gremmen NJM (2000) Including the Antarctic: insights for ecologists everywhere. In: Davison W, Howard-Williams C, Broady P (eds) Antarctic ecosystems: models for wider ecological understanding. New Zealand Natural Sciences, Christchurch, pp 1–15

Friedmann EI (1993) Antarctic microbiology. Wiley-Liss, New York

Kennedy AD (1995) Antarctic terrestrial ecosystem response to global environmental change. Annu Rev Ecol Syst 26:683–704

Lewis Smith RI (1993a) Dry coastal ecosystems of Antarctica. In: v d Maarel E (ed) Ecosystems of the world, vol 2A. Dry coastal ecosystem. Elsevier, Amsterdam, pp 52–71

Lewis Smith RI (1993b) Dry coastal ecosystems on sub-Antarctic islands. In: Maarel Evd (ed) Ecosystems of the world, vol 2A. Dry coastal ecosystem. Elsevier, Amsterdam, pp 73–93

Lewis Smith RI (1994) Vascular plants as bioindicators of regional global warming in Antarctica. Oecologia 99:322–328

Priscu JC (ed) (1998) Ecosystem dynamics in a polar desert. Antarctic Res Ser 72, Am Geophys Union, Washington, DC

Smith VR, Steenkamp M (1990) Climatic change and its ecological implications at a Sub-antarctic island. Oecologia 85:14–24

Vincent WF (1988) Microbial ecosystem of Antarctica. Cambridge University Press, Cambridge

Weller G (1992) Antarctica and the detection of environmental change. Philos Trans R Soc Lond B 338:201–208

Wynn-Williams DD (1990) Ecological aspects of Antarctic microbiology. Adv Microb Ecol 11:71–146

Kiel
Autumn 2001

Lothar Beyer, Manfred Bölter

Contents

Contributors

BEYER, LOTHAR (lbeyer@ipoe.uni-kiel.de)

Institut für Polarökologie der Christian-Albrechts-Universität zu Kiel, Wischhofstraße 1–3, 24148 Kiel, Germany

BLUME, HANS-PETER (hblume@soils.uni-kiel.de)

Institut für Pflanzenernährung und Bodenkunde, Christian-Albrechts-Universität zu Kiel, Olshausenstraße 40, 24098 Kiel, Germany

BÖLTER, MANFRED (mboelter@ipoe.uni-kiel.de)

Institut für Polarökologie der Christian-Albrechts-Universität zu Kiel, Wischhofstraße 1–3, 24148 Kiel, Germany

BRAUN, MATTHIAS (matthias.braun@uni-bonn.de)

Zentrum für Fernerkundung der Landoberfläche (ZFL), Rheinische Friedrich-Wilhelms-Universität Bonn, Walter-Flex-Str. 3, 53113 Bonn, Germany

ELSTER, JOSEF (jelster@butbn.cas.cz)

Botanical Institute, Academy of Sciences, Dukelska 1365, 37982 Trebon, Czech Republic

GOSSMANN, HERMANN (hego@ipg.uni-freiburg.de)

Institut für Physische Geographie, Universität Freiburg, Werderring 4, 79085 Freiburg, Germany

HOGG, IAN D. (i.hogg@waikato.ac.nz)

Centre for Biodiversity and Ecology Research, Department of Biological Sciences, University of Waikato, Private Bag 3105, Hamilton, New Zealand

IMURA, SATOSHI (imura@nipr.ac.jp)

National Institute of Polar Research, 9-10, kaga 1-chome, Itabashi-ku, Tokyo 173-8515, Japan

Kalk, Erich

Institut für Pflanzenernährung und Bodenkunde der Christian-Albrechts-Universität Kiel, Olshausenstraße, 24098 Kiel, Germany

Kanda, Hiroshi (hkanda@nipr.ac.jp)

National Institute of Polar Research, 9-10, kaga 1-chome, Itabashi-ku, Tokyo 173-8515, Japan

Kandeler, Ellen (kandeler@uni-hohenheim.de)

Institut für Standortslehre, Universität Hohenheim, Emil-Wolff-Str. 27, 70599 Stuttgart, Germany

Kappen, Ludger (lkappen@bot.uni-kiel.de)

Botanisches Institut, Christian-Albrechts-Universität zu Kiel, Olshausenstr. 40, 24098 Kiel, Germany

Kuhn, Dieter

Institut für Pflanzenernährung und Bodenkunde, Christian-Albrechts-Universität zu Kiel, Olshausenstr. 40, 24098 Kiel, Germany

Ohtani, Shuji (ohtanish@edu.shimane-u.ac.jp)

Shimane University, Dept. of Biology, Nishikawastu, Matsue 690, Japan

Olech, Maria (olech@ib.uj.edu.pl)

Institute of Botany, Jagellonian University, Lubicz 46, 31–512 Krakow, Poland

Pietr, Stanislaw J. (pietr@ozi.ar.wroc.pl)

Department of Agricultural Microbiology, University Wroclaw, ul. Grunwaldzka 53, 50–357 Wroclaw, Poland

Pingpank, Kristine

Institut für Pflanzenernährung und Bodenkunde, Christian-Albrechts-Universität zu Kiel, Olshausenstr. 40, 24098 Kiel, Germany

Rakusa-Suszczewski, Stanislav (profesor@dab.waw.pl)

Department of Antarctic Biology, Polish Academy of Sciences, ul. Ustrzycka 10, 02–141 Warsaw, Poland

Schroeter, Burkhard (bschroeter@bot.uni-kiel.de)

Botanisches Institut, Christian-Albrechts-Universität zu Kiel, Olshausenstr. 40, 24098 Kiel, Germany

SEPPELT, RODNEY D. (rod.seppelt@antdiv.gov.au)

Australian Antarctic Division, Cannel Highway, Kingston 7050, Tas, Australia

STEVENS, M.I. (mis1@waikato.ac.nz)

Centre for Biodiversity and Ecology Research, Department of Biological Sciences, University of Waikato, Private Bag 3105, Hamilton, New Zealand

STONEHOUSE, BERNARD (bs111@cus.cam.ac.uk)

Scott Polar Research Institute, Lensfield Road, Cambridge CB2 1ER, UK

TATUR, ANDRZEJ (ekolog@warman.com.pl)

Institute of Ecology, Polish Academy of Sciences, Dziekanow Lesny, 05-092 Łomianki and Department of Antarctic Biology, Polish Academy of Sciences, ul. Ustrzycka 10, 02-141 Warszawa, Poland

VOGT, BIRGIT

Institut für Pflanzenernährung und Bodenkunde, Christian-Albrechts-Universität zu Kiel, Olshausenstr. 40, 24098 Kiel, Germany

WHITE, DAVID (ffdmw@uaf.edu)

College of Science, Engineering and Mathematics, University of Alaska, Fairbanks, 248 Duckering Bldg., Fairbanks, Alaska 99775-5900, USA

Part I

The Environment

Preface

This first part presents descriptions of the three sites under evaluation, an overview of Antarctic coastal landscapes (Bölter et al., Chap. 1), and some overall aspects of the sites (Bölter and Beyer, Chap. 2). The sites are spread over a wide region of the Antarctic realm, and represent different areas of interest. King George Island (Rakusa-Suszczweski, Chap. 3) is a centre for research on many fields for the maritime Antarctic. Many nations have settled on this island with seasonal or permanent stations. Henryk Arctowski Station in the Admiralty Bay region has been open since 1971 and is run by the Polish Academy of Science. Many nations can visit this place and conduct scientific programmes. Casey Station (Seppelt, Chap. 4) is maintained by the Australians in the Windmill Island Region. It is a successor of the former American Wilkes Station which was founded during the International Geophysical Year 1957/1958 at Clark Peninsula, but shifted in1969 to Bailey Peninsula as Casey Station, where it was known as 'The Tunnel'. This place is closed today and fully cleared, the new station was opened in 1988 some kilometres south of Old Casey. Its location is nearly opposite Henryk Arctowski Station across the South Pole. The Japanese Station Syowa (Ohtani and Kanda, Chap. 5) was opened in 1957 on the Ongul Islands, located about halfway between Casey and Arctowski (see Fig. 1.1). Climatic conditions, geology, marine influence, and many detailed features could be mentioned, e.g. bridges or trenches between these locations by biological properties, but good knowledge of both plants and soils was a main reason that they were chosen for this compilation.

It can be seen that the individual stories on these sites also present different priorities of research conducted at those stations or in their vicinities, and that there were shifts in such priorities during time. As such, intensive geological studies have been carried out at King George Island most recently by Birkenmajer and others, whereas only few studies of this discipline can be found for the Windmill Islands or for the Ongul Islands. It is also worth mentioning that soil science was not a priority for these places and only a few papers on this matter can be noted for these three places in the literature, while many continuing soil science studies in the Dry Valley region are still going on. Intensive soil scientific studies have been conducted during the past 10 years in the Windmill Islands and at King George Island, giving new insights into this discipline for the Antarctic (see Part II), while only few comparable data from Ongul Island are available. Appropriate data for this area are summarised by Ohtani and Kanda (Chap. 5).

On the other hand, botanical studies are in good hands at all three places as shown by the large scope of record files. This has also resulted in studies not only for individual species, taxonomic issues or pattern analyses, but also for many views on physiology and ecophysiology. Environmental factors, especially the temperature regimes, have to be considered carefully for all long-term studies. Depending on the plant distribution and local interests, different aspects of this matter have gained attention. Hence, the individual chapters mirror both the scientific state-of-the-art, as well as a kind of scientific history of the places, and thus they do not all have an analogous structure.

Although many general as well as specific aspects of the sites can be found elsewhere in the literature, the authors present updated knowledge with special regard to soil ecology. Moreover, other reviews and summary reports are under evaluation, preparation, or in press, such as a recent study published for the Admiralty Bay Region at King George Island (Pol Polar Res, Vol. 20, Issue 3, 1999).

1 Antarctic Coastal Landscapes: Characteristics, Ecology and Research

M. Bölter, L. Beyer, and B. Stonehouse

1.1 Antarctic Environments and Ecosystems

The Antarctic region is divisible ecologically into three circumpolar zonal subregions (Fig. 1.1, Table 1.1), each offering limited opportunities for settlement to plants and animals. The innermost high-latitude zone includes the polar continent, an ice-covered dome rising to over 4250 m, which imposes the most rigorous living conditions. The maritime zone, widest and most clearly defined in the sector containing the Antarctic Peninsula, features winters chilled by the presence of sea ice, and longer, markedly warmer summers. The periantarctic zone, including scattered island groups in a wide expanse of ocean, is far enough north to be free of pack ice in winter, and is correspondingly warmer throughout the year.

Polar terrestrial environments generally impose special demands on indigenous plant and animal communities (Table 1.2), which constrain the extent and biotic diversity and abundance of particular habitats. In the Antarctic region the most rigorous challenge is the presence of land ice. In the form of ice caps, glaciers, ice streams and ice shelves, this presents a virtually sterile substrate covering over 98 % of the continent, and between 60 and 80 % of land in the maritime zone. It is present on all but one of the periantarctic island groups: only the small and relatively low-lying Macdonald Islands are ice-free. Land ice supports only a limited flora of bacteria and cryophilic snow algae, in particular areas close to sea level.

Where the ice has retreated, conditions for living remain harsh, but bare rock and the finer materials resulting from weathering, followed by early stages of soil formation, provide a range of opportunities for plant and animal communities to develop. Ice-free areas are common both on continental Antarctica and in the maritime and periantarctic zones, mostly, but not exclusively, close to sea level. Papers in this volume are concerned especially with three areas, two on the continental coast and one in the maritime zone (Table 1.3), that are associated with long-term research stations.

Why have we chosen these three comparatively small ice-free exposures from such a vast area? The three associated stations have long records of bio-

Ecological Studies, Vol. 154
L. Beyer and M. Bölter (eds.) Geoecology of Antarctic Ice-Free Coastal Landscapes

Table 1.1. Zonation of climate and vegetation within the circumpolar Antarctic region. (After Stonehouse 1982, 1989)

Region	Subregion	Subdivisions	Characteristics	Boundary
Antarctic	Continental Antarctica	East and West Antarctica, Alexander Land, Charcot Land, islands close to the continent. Three ecological zones: high plateau, inland slopes, coast	Mostly ice-covered; mean temperatures of warmest month below 0 °C. Annual precipitation <20 cm; clear skies inland. Soils ahumic, Protoranker, Ranker. Vegetation algae, lichens, mosses. Limited soil fauna of nematodes, mites, collemboles etc.	Extends to South Pole
	Maritime Antarctica	Antarctic Peninsula south to 69° S on west coast, 64° S on east coast, and neighbouring islands: South Orkney, South Shetland and Sandwich Is., Peter I Øy also (probably) Scott I. and Balleny Is.	Mostly ice-covered; mean temperatures of warmest months −10 to +2 °C, winter means rarely below-15 °C. Annual precipitation 20–100 cm; cloudy. Soils ahumic, Protoranker, Ranker. Vegetation: algae, lichens, mosses, 2 spp. of angiosperms; limited soil fauna of nematodes, mites, collemboles etc.	0 °C July isotherm
	Periantarctic islands	South Georgia, Bouvetøya, Heard I., Macdonald Is.	Glaciated uplands (not Macdonald Is.); mean temperature of warmest months 0 to +6 °C, winter means rarely below −2 °C. Soils mostly brown. Vegetation mosses, ferns, many angiosperm species; soil fauna includes earthworms, beetles Northern limit of pack ice	Antarctic Convergence

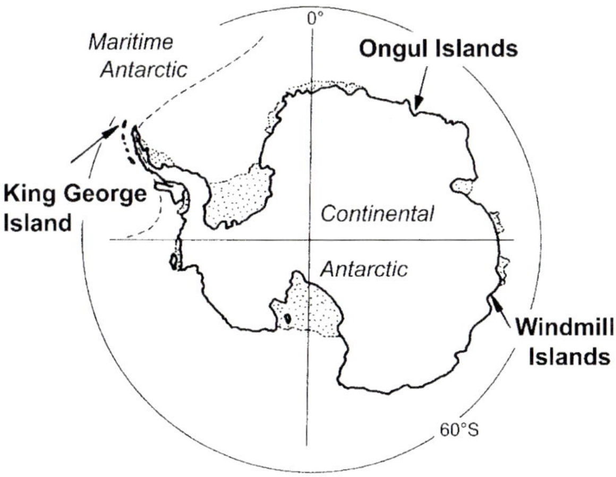

Fig. 1.1. Map of the Antarctic region 60°S, showing the locations of the three places which are the special focus of this book (cf. Chaps. 3–5). The boundary between the Maritime and the Continental Antarctic is marked by a *dashed line*. (After Seppelt 1986)

Table 1.2. Factors restricting plant settlement and growth in Antarctic ecosystems

Climatic factors	Soil factors	Recruitment factors
Low mean monthly temperatures	Prevalence of ice and bare rock	Isolation from other continents
Short growing season	Immature Lithosols	Few sources of propagules
Wide daily temperature ranges	Lack of available water	Inhospitable substrates
Frequent freeze-thaw cycles	Immature drainage systems	
Strong winds	Lack of nutrients in Lithosols	
Low precipitation	High nutrients in ornithogenic soils	
	High permafrost, thin active layer	

Table 1.3. Coastal areas and associated research stations (see Chaps. 3–5)

Landscape area	Admiralty Bay	Windmill Is.	Ongul Is.
Research station	Henryk Arctowski	Casey	Syowa
Co-ordinates	62°10' S, 58°28' W	66°17' S, 110°31' E	69°00' S, 39°35' E
Locality	King George Is., South Shetland Is.	Budd Coast, Wilkes Land	Kronprins Olav Kyst, Dronning Maud Land
Antarctic zone	Maritime	Continental	Continental
Altitude	10 m	20 m	15 m
Air temp. – annual mean	–2.7 °C	–9.1 °C	–13.4 °C
Precipitation – annual mean	510 mm	195 mm	146–192 mm

logical and other scientific research, including synoptic meteorological records. Collaborative research projects, especially since 1992, have resulted in directly comparable data sets for soils, vegetation, plants and microbes. Climatologically, Syowa Station is more like Casey than Arctowski, but its vegetation contains maritime elements that link it more closely to Arctowski than to Casey. Soils, although often poorly understood in their ecological role, have been studied at all three stations, creating a sound basis for all other aspects of terrestrial research.

1.2 Characteristics of Antarctic Coastal Landscapes

1.2.1 The Former Use of 'Oasis' as a Synonym for Ice-Free Landscapes

The term 'oasis' basically originated in Egypt to describe fertile areas of the Libyan desert. A literary note worth reading on the term 'oasis' is given by J. Pickard in his introductory chapter 'Antarctic Oases, Davis Station and the Vestfold Hills' (in: Pickard J (ed) Antarctic Oases. Academic Press, Sydney, 1986, pp. 1–19): *It gained common usage for desert areas where, because of the anomalous presence of groundwater, vegetation grew and persisted in surroundings of aridity and near-sterility*. Antarctic surveyors (e.g. Stephenson 1938, Klebelsberg 1942, Byrd 1947, quoted in Pickard (1986) misappropriated the term to describe areas of continental Antarctica where local paucity of precipitation, sometimes combined with anomalous topography, results in the local starvation of ice sheets and the consequent appearance of bedrock, gravel and soil.

Early users of the term in an Antarctic context referred simply to anomalous ice-free areas on the continent. Shumskiy (1957), reviewing contemporary discoveries along the continental coast of Antarctica, defined oases as substantial ice-free areas separated from neighbouring ice sheets by an ablation zone, and kept free from snow by ablation due to low albedo and radiation. In a much later review and catalogue of Antarctic oases, Pickard (1986a) suggested that a minimal area of at least 10 km^2 was needed to distinguish oases from smaller ice-free areas, for example nunataks, beaches and moraines. He catalogued one very large 'mountain oasis' (the Victoria Land Dry Valleys system) and six 'low coastal oases' (Bunger Hills, Vestfold Hills, Windmill Islands, Schirmacher Oasis, Sôya Coast and Thala Hills). With colleagues (Pickard 1986), he summarised aspects of the Vestfold Hills Oasis, providing a general introduction to continental coastal oases so defined.

Used in this sense, the term 'oasis' expressed the curious fact that a substantially ice-covered continent displays small, discrete ice-free areas. It bore no implications of that most striking feature of hot desert oases – abundant vegetation. Antarctic oases as originally defined are so arid as to be conspicuously lacking in plant life.

In its original Antarctic context, and in Pickard's review, 'Antarctic oasis' was restricted to continental localities and did not include the many extensive ice-free areas of the maritime Antarctic, i.e. the western flank of Antarctic Peninsula and the southern islands of the Scotia Arc. The term appeared to be broadened to include ice-free areas throughout more, possibly all, of the Antarctic region. The extension may be confusing, although expressing some truth common to all, its use is not quite correct and cannot be applied to the comparatively well-vegetated coasts of periantarctic islands. Svoboda and Freedman (1994) used the term in a similar sense for ice-free areas of Ellesmere Island in the Canadian Arctic.

Ice-free areas throughout the region have emerged relatively recently (i.e. probably within the past few thousand years) from under receding land ice, to be subject to colonisation by plants, animals and man. Ice-free areas of the maritime region, like those of the continent, are bounded at least in part by edges of the ice sheet from which they have emerged. Several differ from their continental counterparts by having substrates of thin organic soils and patches of continuous vegetation (see below). However, these do not invalidate comparison: they indicate either longer periods of exposure, or more favourable conditions for soil formation, plant recruitment and growth, or possibly both.

Glaciers currently bound several coastal areas of South Georgia and Heard Island, while periglacial evidence suggests that ice sheets previously covered most or all of the islands, down to and including the coasts. Again, the mature soils and wider diversity of vegetation on these islands contrast with those of more southerly regions. Again, these are likely to indicate only longer histories of exposure and colonisation under comparatively mild climatic conditions, in particular the absence of sea ice that ensures warmer and shorter winters.

1.3 Characteristics of Antarctic Coastal Landscapes

Ice sheets covering Antarctica and its neighbouring islands persist where solid precipitation, usually of snow or ice spicules, exceeds melting and ablation, either locally or in nearby catchment areas from which the ice has travelled. Their transient and mobile surfaces are biologically inhospitable, colonised only by a limited microflora of bacteria and snow algae. For different reasons they are inhospitable to man. Research stations built on the relatively dry interior plateau sink slowly into the ice. Those built on coastal ice shelves, where precipitation is higher, tend first to disappear under snow, then to move seaward with the general flow and be lost when the ice shelf calves.

1.3.1 Localities and Topography

Coastal landscapes occur where edges of ice sheets fluctuate and retreat locally. They appear to be stable over periods of centuries or millennia. Those of the continental coast occur between the paths of invading depressions, which bring heavy snow to neighbouring localities. Those of the maritime and peri-antarctic regions are subject to heavy winter snowfall, followed by summer melting that clears the snow and provides abundant spring melt-water.

Topography exposed by retreating ice varies widely, from broad, extensive glaciated valley systems to smooth coastal plains with superimposed moraines. Some are crossed by glacial torrents, or channels cut by former melt streams, terminating in outwash fans. Most coastal areas show evidence of secondary isostatic uplift, resulting in inland cliffs, raised beaches and benches, and of post-glacial coastal processes, including deep embayments and eroded headlands. Several include freshwater or brackish lake systems and lagoons. There is usually evidence of progressive colonisation, with the newest areas closest to the ice-cap edge. Many at present show evidence of recent expansion, due to accelerated late twentieth-century glacial retreat.

1.3.2 Soils of Antarctic Coasts

The rock surfaces exposed by glacial retreat consist initially of smooth or scoriated bedrock with a thin covering of glacial debris, mostly unsorted rock rubble. From the moment of exposure both bedrock and superficial deposits are subject to physical and chemical weathering. Fine materials are moved and re-distributed by wind. Boulders and rubble are fragmented by (1) thermal shattering, due to frost action and radical diurnal shifts in temperature, (2) mechanical weathering of wind and water and (3) chemical weathering due to rain and salt spray. Mechanical processes, which are accelerated by cold, produce particles ranging from the finest rock flours to coarse sands and gravels. Chemical processes, which require the presence of water and are inhibited by cold and aridity, produce free radicals in solution, plus solid residues including silicate and oxide clays with colloidal surfaces.

Fine particles developing from weathering accumulate to form basic ahumic soils – unsorted Lithosols in situ, interspersed with pockets of Regosols resulting from mechanical redistribution and sorting processes. Details of these processes are reviewed in Stonehouse (1989, 1999) and Williams and Smith (1989). For a comprehensive account of Antarctic soils see Claridge and Campbell (1985).

Lithosols or Regosols have little or no organic content, and lack the internal layering or structure of more mature soils. Only the driest remain completely abiotic, though many have very little organic content, with correspondingly low levels of biological activity. Ahumic soils by definition contain less than 1 % by dry weight of organic material. Their main value to living organisms rests in their capacity to hold moisture in interstices between their particles, and on the colloidal surfaces of clay minerals. Though frozen and unavailable for much of the year, its presence during the short growing season is critical both to the flora and to the further development and maturation of soils.

1.3.3 Conditions for Colonisation of Soils

While several forms of life that colonise coastal regions do not depend on the presence of a moist substrate, for most vegetation and microfauna a reservoir of moisture is essential. Interstitial moisture encourages colonisation by wind-blown spores and propagules, which in dying contribute the soluble products of their own biodegradation to the substrate, and initiate biological weathering, an advanced soil-forming process. This results in the conversion of ahumic to primitive humic soils, i.e. raising the dry weight of organic content above the critical 1 %.

The degree to which ecological maturation is possible depends on local conditions, notably climatic severity. Ice-free areas in the harshest continental conditions, combining intense cold with chronic aridity, support little or no vegetation apart from encrusting lichens. Their soils remain virtually ahumic, including only a meagre, seasonally-active microflora and fauna, often dominated by primary colonising bacteria and cyanobacteria that fix atmospheric nitrogen. Once established, the colonies are open to invasion by other organisms, making use of this and other essential elements.

Seasonal or perennial freezing is a feature of Antarctic soils that strongly affects their development. 'Permanently-frozen ground' or permafrost is ground that is permanently frozen, with a superficial 'active' layer that thaws in summer. It occurs patchily or continuously where mean annual air temperatures near the ground fall below –8 °C, an area including continental Antarctica, the maritime Antarctic and upland areas of the periantarctic islands. 'Seasonally-frozen ground', a superficial layer underlain by unfrozen subsoil, characterises periantarctic islands at sea level (Washburn 1973; Williams and Smith 1989).

In permafrost, an active layer 0.5 to 1 m deep provides mobile groundwater and extensive areas of ponds and waterlogged soils. Practically all soil movement and development processes are confined within this layer, stopping at the underlying barrier of permanent ice. Moist soils are subject to frost-heaving and frost-creep – vertical and horizontal movements that produce characteristic surface patterns of stripes and polygons. Wetter soils on slopes are liable to slumping and avalanching. Both kinds of movement inhibit soil maturation and plant colonisation, especially for slow growing forms like lichens. Soils under active cryoturbation can be well colonised by algae (including cyanobacteria) and to some extent by mosses.

1.3.4 Development and Maturation

Where seasonal melt streams provide a trickle of water in summer, local patches of algae and moss testify to the significance of even transient seasonal water. While summer aridity is an obvious factor that limits growth, also lacking is a protective cover of snow in winter, which provides a shield against severe wind blasting and erosion, and a short, transient supply of moisture in spring.

More permanent summer moisture encourages surface proliferation of algae, mosses and lichens, forming mats of vegetation from which nutrient-loaded solutions leach into the substrate, causing visible mineralisation, or chemical changes in a layer a few centimetres below the surface. 'Protoranker' soils formed in this way occur patchily in the maritime Antarctic. Continuing growth in the absence of browsers or decomposers results in accumulations of peaty material, which overlie but do not mix with the soils beneath. True peat is rare but not unknown in polar environments (Klimowicz et al. 1997).

Still higher organic content accumulates where summer moisture and warmth encourage growth of ferns and phanerogams. Their roots penetrate and mingle intimately with underlying soil particles, providing a rich substrate for soil bacteria. The resulting 'brown soils', promoted by Antarctica's two species of flowering plants – hair grass *Deschampsia antarctica* and pearlwort *Colobanthus quitensis* – occur sparsely in the maritime Antarctic region (Allen and Heal 1970; Stonehouse 1989). A wider variety of ferns and flowering plants produce more abundant brown soils on the periantarctic islands.

Both advanced soil development and proliferation of vegetation in maritime Antarctic areas, compared with those of the continent, may be attributed primarily to more favourable growing conditions, including longer, warmer and damper summers and milder winters, often with thick snow cover. An important additional factor is the presence of nesting seabirds, which introduce both nutrients and plant propagules. Penguin colonies and their immediate environs in current use support little or no vegetation. However, abandoned or disused colonies, and the less concentrated nesting areas of gulls, skuas, terns, petrels and other flying birds, provide a zonal pattern of ornithogenic soils, with leached nutrients that are acceptable especially to nitrophilous plants (Lewis Smith 1984).

Though time is clearly a key constituent of environmental maturation, we have as yet very little evidence to show how long particular landscapes have been ice-free. Circumstantial evidence suggests that maritime areas have existed longer than continental ones, and that periantarctic island coastal vegetation has existed longest of all. However, more precise dating, and more measurements of rates at which soil maturation processes occur under different climatic regimes, would be welcome from all Antarctic coastal regions.

1.4 Research Programmes and Objectives

The very limited extent of Antarctic coastal landscapes contrasts with expansive ice-free ground in the Arctic region, including wide zones of polar desert in the far north, and tundra extending south to the margins of the boreal forest. Insofar as interpolar comparisons are valid, the floras of the richest and most productive Antarctic regions are comparable with those of the harshest and least-favoured areas of Arctic polar desert, with a more limited range of species, especially among the higher plants. Of the three areas considered, only one (Arctowski) supports flowering plants, and only two species. Devon Island, in a similar environment at a higher northern latitude (75° N), supports some 90 species of flowering plants (Bliss 1971). For further comparisons see Stonehouse (1989).

Each landscape includes several terrestrial ecosystems developed during a relatively recent history of deglaciation, depending on such local parameters as temperature, moisture availability and soil nutrient contents. Ecosystems are defined by controlling ecological characteristics, for example different pedological and geomorphological features. Relationships of these controlling characteristics on biological systems became issues of comparative and descriptive ecosystem research, resulting in analyses of phenomena and processes of individual organisms or communities (e.g. Llano 1977; Kushner 1978; Horner 1985; Longton 1988; Kennedy 1995). Results from these studies have been used to describe organism adaptations and long-term evolutionary processes in the relatively undisturbed Antarctic environment.

In the region of the maritime Antarctica we find the largest areas free of glaciation under a humid climate. It is also scenically attractive, providing many places of interest for tourists. Broad-scale atmospheric influences, such as increasing UV-radiation due to diminution of ozone levels, and pollutant transport, also have an impact on the local environment. Cumulative or even exponential effects must now be taken into consideration. These new impacts on Antarctic terrestrial ecosystems may be comparable in overall effects with the massive impacts of early twentieth-century whaling and sealing on marine ecosystems.

Overview studies of terrestrial, limnological, and marine research have been published by several authors: e.g. Laws (1984a,b) for terrestrial and marine ecology; Campbell and Claridge (1987) for soils; Longton (1988) for bryophytes and lichens; Vincent and Ellis-Evans (1989) for limnology; Friedmann (1993) for microbiology in general. Distribution patterns, dispersal, colonisation, adaptations and food webs have been themes of several seminars and symposia (e.g. Llano 1977; Siegfried et al. 1985; Kerry and Hempel 1990; Battaglia et al. 1997).

The advent of new global issues, such as climate change (e.g. Weller 1992), created many new scientific programmes and initiatives. Perceptions of the Antarctic region have changed accordingly, particularly with regard to its sensitivity to global effects of feedback mechanisms on climate, biogeochemical cycles and increasing tourist activities. As a consequence, anthropogenic pollution and environmental protection issues have received increasing attention, although compilations as detailed and comprehensive as some recently published for the Arctic (e.g. AMAP 1997) have still to be attempted. Both nationally and internationally through the mediation of SCAR, the focus of scientific enquiry has changed from descriptive studies of specific areas and processes to issues of global concern.

Only a few years ago, the direct input of man could be regarded as having an effect restricted to the area closely surrounding the Antarctic research bases. There is now a much more wide-ranging scientific presence which, coupled with increasing levels of tourist operations, affects much wider areas. This has become an important issue in the maritime Antarctic Peninsula region, adjacent islands and other periantarctic islands close to South America.

References

Allen SE, Heal OW (1970) Soils of the maritime Antarctic zone. In: Holdgate MW (ed) Antarctic ecology, vol 2. Academic Press, London, pp 693–96

AMAP (1998) Arctic pollution issues: A state of the Arctic environment report. AMAP, Oslo

Battaglia B, Valencia J, Walton DWH (eds) (1997) Antarctic communities. Species, structure and survival. Cambridge University Press, Cambridge

Bliss LC (1971) North American and Scandinavian tundras and polar deserts. In: Bliss LC, Heal OW, Moore JJ (eds) Tundra ecosystems: a comparative analysis. Cambridge University Press, Cambridge, pp 8–24

Byrd RE (1947) Our navy explores Antarctica. Natl Geogr Mag 92:429–522

Campbell IB, Claridge GGC (1987) Antarctica: soils, weathering processes and environment. Elsevier, Amsterdam

Claridge GGC, Campbell IB (1985) Physical geography – soils. In: Bonner WN, Walton DWH (eds) Key environments: Antarctica. Pergamon Press, Oxford, pp 62–70

Ferris JM, Burton HR, Johnstone GW, Bayly IAE (eds) (1988) Biology of the Vestfold Hills. Kluwer, Dordrecht

Fogg GE (1998) The biology of polar habitats. Oxford University Press, Oxford

Friedmann EI (ed) (1993) Antarctic microbiology. Wiley-Liss, New York

Holdgate MW (ed) (1970). Antarctic ecology. Academic Press, London
Horner R (ed) (1985) Sea ice biota. CRC Press, Boca Raton
Kerry KR, Hempel G (eds) (1990) Antarctic ecosystems. Ecological change and conservation. Springer, Berlin Heidelberg New York
Kennedy AD (1995) Antarctic terrestrial ecosystem response to global environmental change. Annu Rev Ecol Syst 26:683–704
Klebelsberg RV (1942) Formen- und gletscherkundliche Auswertung der Lichtbildaufnamen. Wissenschaftliche und fliegerische Ergebnisse der Deutschen Antarktischen Expedition 1938–39, Bd 14, p 154
Klimowicz Z., Melke J, Uziak S (1997) Peat soils in the Bellsund region, Spitsbergen. Pol Polar Res 18:25–39
Kushner DJ (ed) (1978) Microbial life in extreme environments. Academic Press, London
Laws RM (ed) (1984a) Antarctic Ecology, vol 1. Academic Press, London
Laws RM (ed) (1984b) Antarctic Ecology, vol 2. Academic Press, London
Lewis Smith RI (1984) Terrestrial plant biology of the sub-Antarctic and Antarctic. In: Laws RM (ed) Antarctic ecology, vol 1. Academic Press, London, pp 61–162
Llano GA (ed) (1977) Adaptations within Antarctic ecosystems. Smithsonian Institution, Washington, DC
Longton RE (1988) Biology of polar bryophytes and lichens. Cambridge University Press, Cambridge
Pickard J (ed) (1986) Antarctic oasis, terrestrial environment and history of the Vestfold Hills. Academic Press, New York
Priscu JC (ed) (1998) Ecosystem dynamics in a polar desert. Antarctic Res Ser 72, Am Geophys Union, Washington, DC
Seppelt RD (1986) Lichens of the Vestfold Hills. In: Pickard J (ed) Antarctic oasis. Terrestrial environments and history of the Vestfold Hills. Academic Press, Sydney, pp 247–274
Shumskiy PA (1957) Glaciological and geomorphological reconnaissance in the Antarctic in 1956. J Glaciol 3:56–61
Siegfried WR, Condy PR, Laws RM (eds) (1985) Antarctic nutrient cycles and food webs. Springer, Berlin Heidelberg New York
Stephenson A (1938) The southern journey. In: Rymill J (ed) Southern lights. The official account of the British Graham Land Expedition 1934–37. Chatto and Windus, London, pp 177–201
Stonehouse B (1982) La zonation écologique sous les hautes latitudes australes. In: Jouventin P, Massé L, Tréhen P (eds) Colloque sur les ecosystèmes subantarctiques, Comité national français des recherches antarctiques Paris, pp 531–536
Stonehouse B (1989) Polar ecology. Blackie, Glasgow
Stonehouse B (1999) Biological processes in cold soils. Polar Rec 35:5–10
Svoboda J, Freedman B (1994) Ecology of a polar oasis, Alexandra Fiord, Ellesmere Island, Canada. Captus University Publications, Toronto
Vincent WF (1988) Microbial ecosystems of Antarctica. Cambridge University Press, Cambridge
Washburn AL (1973) Periglacial processes and environments. Edward Arnold, London
Weller G (1992) Antarctica and the detection of environmental change. Philos Trans R Soc Lond B 338:201–208
Williams PJ, Smith MW (1989) The frozen earth: fundamentals of geocryology. Cambridge University Press, Cambridge

2 Maritime and Continental Antarctic: The Sites of Investigation in Context

M. Bölter and L. Beyer

2.1 Geography and Geology

The sites used in this comparison, Casey Station (Windmill Islands) and Syowa Station (Ongul Island) on the East Antarctic continent, show several features of plant cover, soils and soil related properties which are similar to those found in the maritime Antarctic (Henryk Arctowski Station, King George Island, South Shetland Islands). Thus, it seemed worthwhile to discuss this kind of similarity by an overall approach and to present them as well-described Antarctic habitats. These locations represent different kinds of habitats on both a large and small scale.

The Polish base (Henryk Arctowski Station) can be considered as representative of the maritime Antarctic with a maritime climate and widespread floristic regions of different types; fjells, lichen heaths, grasslands and moss beds. The Japanese continental Antarctic Station (Syowa Station) is situated in a region of small, scattered, ice-free coastal islands, rocky exposures in front of glaciers, and some isolated nunataks. The Windmill Islands, with the Australian Station (Casey Station) is a region with relatively large ice-free areas on peninsulas and islands. Rocky outcrops, moraines and former marine terraces are interspersed with permanent snow and ice fields and the landscape supports small moss beds and dry fjell-like ecosystems which are covered, in part, by extensive lichen-dominated vegetation.

Geologically, the provinces of the Antarctic Peninsula and the South Shetland Islands are quite different to those locations at the East Antarctic continental shield. Consequently, the basic material from which soils are derived by weathering processes differ. At King George Island, tertiary effusive (volcanic), basalt-andesite and related pyroclastic rocks dominate, whereas lithified and loose sedimentary rocks are of less importance (Barsch et al. 1985). In the Windmill Islands, the basement rocks are primarily gneissic, in the

Ecological Studies, Vol. 154
L. Beyer and M. Bölter (eds.) Geoecology of Antarctic Ice-Free Coastal Landscapes

north, with granitic intrusives in the southern part (Robertson 1961; Blight and Oliver 1977, 1982).

2.2 Climate

Generally, the climate of continental Antarctica can be regarded as more severe than that of the Antarctic maritime islands (Holdgate 1977; Betts 1983). In between these two contrasting regions, Syowa Station (East Ongul Island) may serve as a biogeographical bridge, at least with regard to sharing several botanical features. Both the Windmill Islands and Ongul Islands share similar climatological features, particularly with regard to their mean annual wind speed and temperatures (Walton 1984; Jacka et al. 1984). King George Island has greater precipitation, lower mean wind speeds, and higher mean temperatures (Fig. 2.1). The evidently higher wind speeds at Casey and Syowa indicate much higher stress in chilling and dryness. Hence, living conditions for plants and soil animals get much more severe.

The climatic variables of mean wind speed, mean summer and winter temperatures, precipitation and annual snow balance clearly differentiate continental from maritime Antarctic localities (Phillpot 1985). The three regions being considered are all coastal landscapes, but the ameliorating effect of a maritime influence on the climate is only observed at King George Island. An important differentiation can be made by the precipitation; Arctowski Station (King George Island) reports a mean value of 510 mm, Casey Station (Windmill Islands) and Syowa Station (Ongul Islands) report values below 200 mm (cf. Chaps. 3–5). The three localities do share, however, a number of physical and biotic characteristics. This offers the opportunity of carrying out comparative analyses of plant life and several stages of soil development (Beyer et al. 1999).

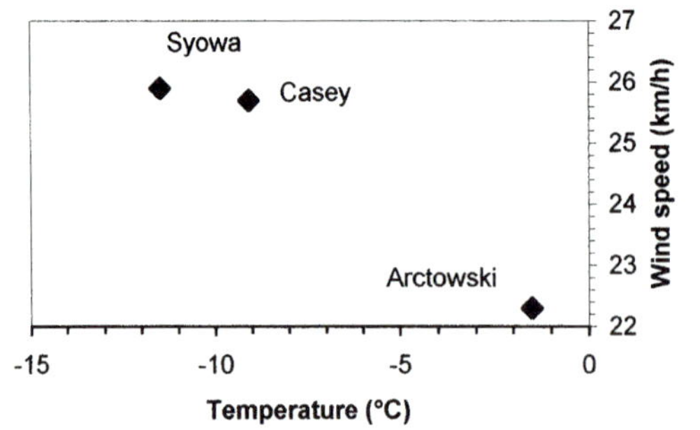

Fig. 2.1. Stations Syowa, Casey and Arctowski in relation to the meteorological features of temperature and wind speed (mean values of the years 1990–1999)

2.3 The Effects on Soil Ecosystems

Due to the pattern of snow accumulation, this set of stations can be described as significantly different from many other Antarctic places with open soils which can be described as arid (Campbell and Claridge 1987; Campbell et al. 1998). Obvious differences, however, can be found between Arctowski and Syowa Station: for example, the much lower cover and abundance of macrolichens at Syowa, where there is much less precipitation in summer with its lesser influence in diluting sea spray aerosols (Inoue 1991). The extensive lichen-dominated vegetation with, in wetter areas, moss beds, found in the northern parts of the Windmill Islands region is somewhat anomalous in the continental Antarctic context. However, recent studies on the distribution and abundance of subfossil penguin colonies and their attendant nutrient supply may show that the answer really does lie in the soil.

As in other polar and high alpine environments, microclimatic conditions are of greater importance than the macroclimate in determining the distribution and abundance of the terrestrial biota. The well-structured relief, especially in coastal regions, further enhances the microclimatic influence. Even in the milder Admiralty Bay region (King George Island) there are features of deserts, e.g. in the vicinity of glaciers and on dry fjell sites. There are several reports in the literature of soil surfaces and the surface of dry moss cushions from continental localities warming to temperatures greater than 30 °C on fine summer days, although such extremes occur only for short time periods (Lewis Smith 1986; Bölter 1992; Bölter et al. 1989, 1995). The relatively high soil density, because of the low organic matter content, may act as a good storage and transporter of energy, thus giving soils a long-term positive energy budget (Thompson 1973).

In summer, positive air temperatures combined with frequent freeze-thaw cycles become important factors for soils, vegetation and other organisms living in the soil and vegetation. These are hard conditions for life, but they provide good conditions for soil processes, together with increased weathering processes and the chemical actions within free water, thereby enhancing biological activity due to the release of soluble organic matter (Wynn-Williams 1982; Tearle 1987; Roser et al. 1993; Melick and Seppelt 1992). Nevertheless, for most of the summer season the soils and soil organisms are strongly affected by aridity, ensuring that large areas of suitable habitat support only desiccation-tolerant cryptogams and other organisms, or are effectively barren.

In many places of the Antarctic coastal areas, birds play an important role in soil development, the presence or development of plant communities, and related biogeochemical cycles. These influences modify soils over short timescales and leave their signatures in distinct ways. All sites described in this study show influences of penguins to a greater or lesser extent, particularly the Admiralty Bay region. Nests of other birds are dispersed all over the ice-free coastal zone.

The role of permanent low manuring by them and also by inland-moving seals is essential for the creation of plant communities. The continental sites are not influenced that much by active colonies. Abandoned bird colonies can still be identified by surface features as well as by typical features in depth profiles. Shifts in the location of colonies have taken place throughout the Holocene period and abandoned colonies have become valuable sources of nutrients.

New ecosystems have developed during the formation of ornithogenic soils (Tatur and Myrcha 1989). The impacts of active colonies can also be monitored over relatively long distances due to transport of volatile nitrogen compounds and particles like feathers. In an environment where a significant feature is the lack of nutrients, such inputs of allochthonous material are of special importance. Wind-blown nutrients from sea spray add another component to the soil and its living world (Greenfield 1992; Bölter et al. 1995). The distribution patterns of these aerosols can be easily monitored in freshly deglaciated areas, especially in the region of the Antarctic Peninsula.

Local variations in climatic conditions and secondary effects, such as salt crusts and patterned ground from active cryoturbation, impose a strong influence on the small-scale pattern of plant distribution, microfauna and microflora. Their responses to environmental impacts are significant for the study of further effects on the establishment of small food webs, soil development and feedback mechanisms which might enhance or stimulate other factors. One such factor is the establishment of a dense plant cover, which leads to a smoothing of soil temperatures, production of litter, and provision of a favourable habitat for soil arthropods. Short-term changes in physical properties, however, eliminate those organisms which cannot cope with such conditions.

Non-glaciated areas have permafrost layers in great depth. The milder climate in the maritime Antarctic, however, leads to a strong thaw and during summer the permafrost table is often found below 1 m depth. Higher production rates than concomitant decomposition rates, during spring to autumn, lead at many places to peat formation and brownish humic accumulations underneath mosses or lichens or – in the maritime Antarctic – meadows established by *Deschampsia antarctica* and *Colobanthus quitensis*. Such areas become favourite habitats for mites, collemboles, and other small arthropods as there is an amelioration of microhabitat temperatures.

The effect of short time windows for biological activity in combination with high degrees of patchiness can lead to further peculiarities of the Antarctic ecosystem. microorganisms, plants and animals have few opportunities for genetic exchange between separated locations. This leads to a special autochthonous system (Bornkamm 1987). The only natural disturbance to the local ecosystem autonomy arises from wind-borne sources or, to a minor extent, transport by higher animals. Such a situation is quite different from other large-scale environments, where organic and inorganic matter as well as microorganisms are transported by various vectors, e.g. earthworms, from one place to another. The lack of soil-dwelling animals which are able to perform an active bioturbation is another feature of Antarctic terrestrial ecosystems. Only in wet organic soils or underneath moss cushions do nematodes become an important factor for soil life (Block 1984). In this comparative treatment we hope to outline some of the special characteristics and to discuss their common features in soils, vegetation, soil animals and microbes.

References

Barsch D, Blümel W-D, Flügel W-A, Mäusbacher R, Stäblein G, Zick W (1985) Untersuchungen zum Periglazial auf der König-Georg-Insel, Südshetlandinseln/Antarktika. Ber Polarforsch 24:1–75

Betts M (1983) The climate of Antarctica. Antarctic Division, Dept Sci Technol, Pamphlet 2, pp 1–3

Beyer L, Bockheim JG, Campbell IB, Claridge GGC (1999) Genesis, properties and sensitivity of Antarctic Gelisols. Antarct Sci 11:387–398

Blight DL, Oliver RL (1977) The metamorphic geology of the Windmill Islands, Antarctica: a preliminary account. J Geol Soc Aust 24:239–262

Blight DL, Oliver RL (1982) Aspects of the geological history of the Windmill Islands, Antarctica. In: Craddock C (ed) Antarctic Geoscience – Symposium on Antarctic Geology and Geophysics, Madison, Wisconsin, 22–27 August, 1977, University of Wisconsin Press, Madison, pp 445–454

Block W (1984) Terrestrial microbiology, invertebrates and ecosystems. In: Laws RM (ed) Antarctic ecology. Academic Press, London, pp 163–236

Bölter M (1992) Environmental conditions and microbiological properties from soils and lichens from Antarctica (Casey Station, Wilkes Land). Polar Biol 11:591–599

Bölter M, Kappen L, Meyer M (1989) The influence of microclimatic conditions on potential photosynthesis of *Usnea sphacelata*: a model. Ecol Res 4:297–307

Bölter M, Blume H-P, Kappen L (1995) Bodenbiologische Untersuchungen in der maritimen und kontinentalen Antarktis (King George Island und Windmill Islands). Teil 1. Umweltparameter und anorganische Nährstoffe. Polarforsch 65:41–61

Bornkamm R (1987) Allochthonous ecosystems. Landscape Ecol 1:119–122

Campbell IB, Claridge GGC (1987) Antarctica: soils, weathering processes and environment. Elsevier, Amsterdam

Campbell IB, Claridge GGC, Campbell DI, Balks MR (1998) The soil environment of the McMurdo Dry Valleys, Antarctica. In: Priscu JC (ed) Ecosystem dynamics in a polar desert. Antarct Res Ser 72, Am Geophys Union, Washington, DC, pp 297–322

Greenfield L (1992) Precipitation nitrogen at maritime Signy Island and continental Cape Byrd, Antarctica. Polar Biol 11:649–653

Holdgate MW (1977) Terrestrial ecosystems in the Antarctic. Philos Trans R Soc Lond Ser B 279:5–25

Inoue M (1991) Ecological notes on the differences in flora and habitat of lichens between the Syowa station area in continental Antarctic and King George Island in maritime Antarctic. Proc NIPR Symp Polar Biol 4:91–106

Jacka TH, Christou L, Cook BJ (1984) A data bank of mean monthly and annual surface temperatures for Antarctica, the southern ocean and south Pacific Ocean. ANARE Res Notes 22:1–84

Lewis Smith RI (1986) Plant ecological studies in the fellfield ecosystem near Casey Station, Australian Antarctic Territory, 1985–86. Br Antarct Surv Bull 72:81–91

Melick DR, Seppelt RD (1992) Loss of soluble carbohydrates and changes in freezing point of Antarctic bryophytes after leaching and repeated freeze-thaw cycles. Antarct Sci 4:399–404

Phillpot HR (1985) Physical geography – climate. In: Bonner WN, Walton DWH (eds) Key environments: Antarctica. Pergamon Press, Oxford, pp 23–38

Robertson R (1961) Preliminary report on the bedrock geology of the Windmill Islands. IGY Glaciol Rep 4:1–26

Roser DJ, Seppelt RD, Ashbolt N (1993) Microbiology of ornithogenic soils from the Windmill Islands, Budd Coast, Continental Antarctica: microbial biomass distribution. Soil Biol Biochem 25:165–175

Tatur A, Myrcha A (1989) Soils and vegetation in abandoned penguin rookeries (maritime Antarctic). Proc NIPR Symp Polar Biol 2:181–189

Tearle PV (1987) Cryptogamic carbohydrate release and microbial response during spring freeze-thaw cycles in Antarctic fellfield fines. Soil Biol Biochem 19:381–390

Thompson DC (1973) Surface heat balance and climate in an ice-free area of Antarctica. World Meteorol Organ Publ 361:167–180

Walton DWH (1984) The terrestrial environment. In: Laws RM (ed) Antarctic ecology, vol 1. Academic Press, London, pp 1–60

Wynn-Williams DD (1982) Simulation of seasonal changes in microbial activity of maritime Antarctic peat. Soil Biol Biochem 14:1–12

3 King George Island – South Shetland Islands, Maritime Antarctic

S. Rakusa-Suszczewski

3.1 Introduction

Admiralty Bay is a large, complex inlet in the southern coast of King George Island, South Shetland Islands (Fig. 3.1). The bay and the land area around it have been for over two decades focal points of terrestrial and marine research, covering a wide range of disciplines. Research has been based mainly on Henryk Arctowski Station, a facility established in 1977 and operated continuously by the Department of Antarctic Biology, Polish Academy of Sciences. The base is located on an ice-free area at the western coast of Admiralty Bay (Fig. 3.2) and covers different kinds of landscapes, coastlines with penguin rookeries and seal wallows, big glacier forelands, lichen heaths, swamps, grasslands and barren areas in the upper lands.

The efforts of many scientists, both Polish and guests from other nations, have made the Admiralty Bay area one of the most intensively studied marine and terrestrial localities in the Antarctic region. It is also one of a very few Antarctic environments in which multidisciplinary studies have extended over a period long enough to demonstrate effects of natural and man-made changes. This chapter outlines key features of the area, and some of the important changes detected in the geoecosystem.

3.2 Geographical Features

3.2.1 Cartographic and Topographic Features

The name Admiralty Bay first appeared on English maps in 1822. The names Demay Point, Martin's Head and Florence Nunatak arise from this time on. The first detailed cartographic work is that of the Second French expedition

Ecological Studies, Vol. 154
L. Beyer and M. Bölter (eds.) Geoecology of Antarctic Ice-Free Coastal Landscapes

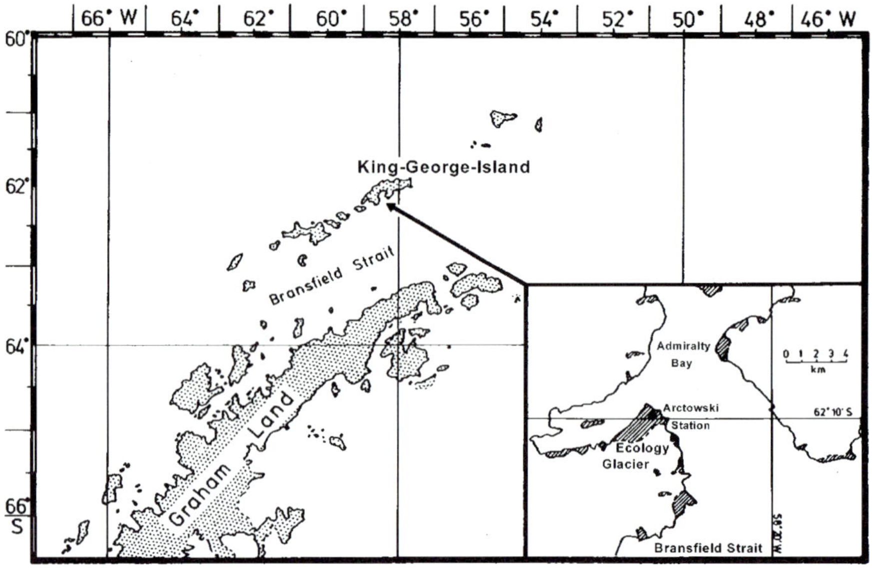

Fig. 3.1. Map of South Shetland Islands and the location of Arctowski Station and its surroundings. Redrawn after Kejna (1999)

of J.B. Charcot (Bongrain 1914; Alberts 1981). Since then, the Admiralty Bay region has appeared on several hydrographic charts, e.g. UK Admiralty Bay 1774, printed in London (cf. The Antarctic Pilot 1974), Chilean chart 1403, Argentinean chart 136 (Bahia Lasserre 1956–1957, scale 1:30,000) and US Hydrographic Office chart 6661 (1965, scale 1:50,000). Henryk Arctowski Station is located in the north-western part of the centre of Admiralty Bay, "Thomas Point". It stands on 500-year-old terraces which are 2–5 m above sea level. The thickness of the sediments is about 1.5 m: they are composed mostly of gravel, sand and cobbles on a foundation of bedrock. The geodetic point at Arctowski Station, measured accurately in 1988 using the Doppler method, has the co-ordinates: –62 09 41 56, 301 31 49 9.

The first Polish map of Admiralty Bay with the surrounding area contours (scale 1:25,000) has been elaborated by Furmańczyk and Marsz (1980). This map is based on US chart 6661 and the Polish chart of bathymetric measurements in Ezcurra Inlet (Marsz 1984). The chart of Admiralty Bay number 25121 (1984, scale 1:40,000) has been published by Brazilian cartographers. Special attention to detail has been given to the areas surrounding Ferraz Station, Martel Inlet and the west coast of Mackellar Peninsula. The second Polish chart on a scale 1:50,000 was elaborated by Battke (1980) but without isobaths. The third edition of the map was prepared to the scale of 1:50,000 (Battke 1990). This map has been used for 3-D images. The wealth of Polish names in Admiralty Bay (Birkenmajer 1980a) is the result of multi-disciplinary work by Polish and other scientists from Henryk Arctowski Station, on the western shore of the bay.

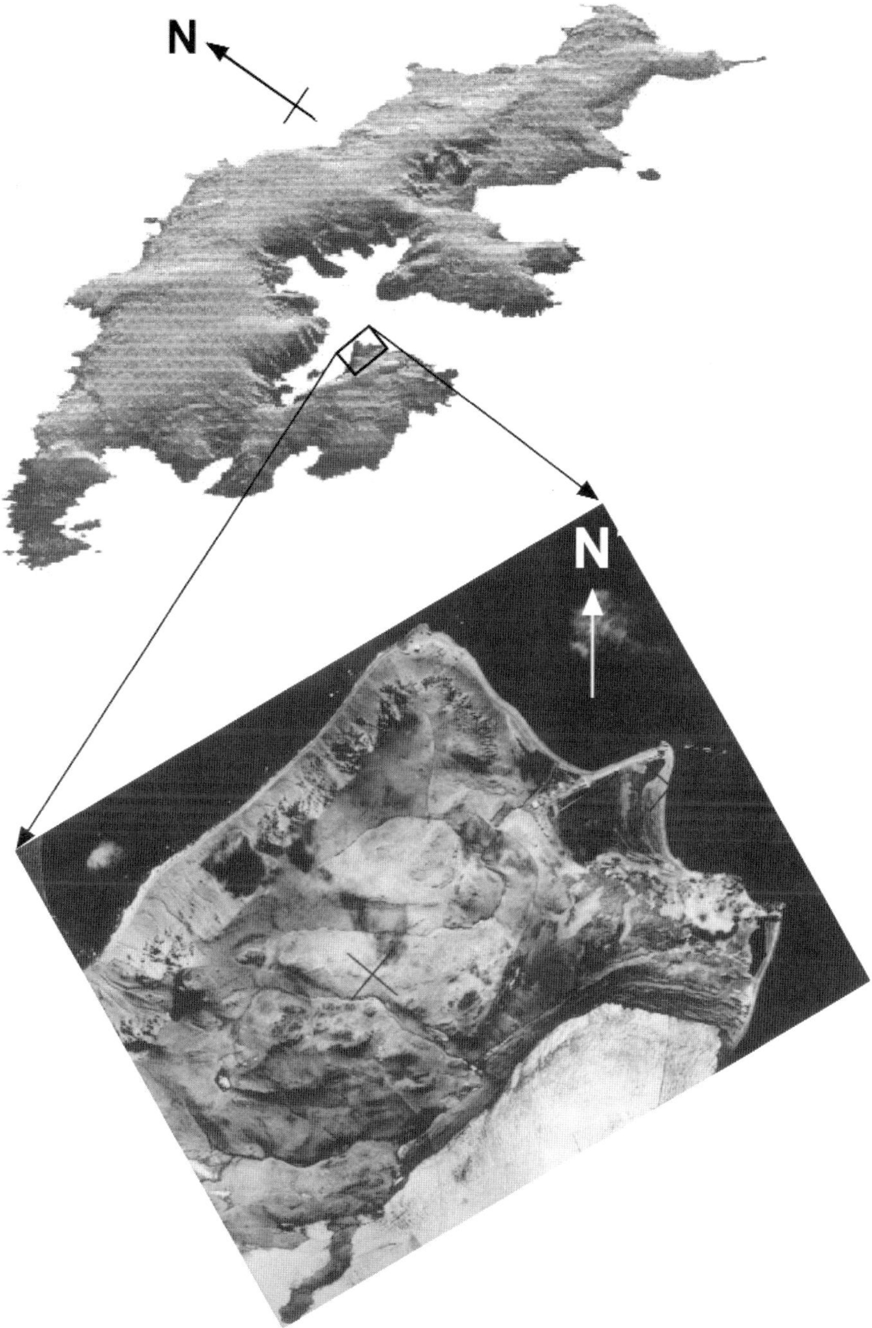

Fig. 3.2. Plot of King George Island and aerial photograph of Arctowski region, the photo corresponds to Fig. 3.1 (courtesy to M. Braun)

Table 3.1. The Admiralty Bay area

Total area of iceshed included in Admiralty Bay area	361 km^2
Total land area	239 km^2
Total sea area	122 km^2
Land area covered by ice	218 km^2 (91 %)
Land area that is ice-free	21 km^2 (9 %)
Coastal areas	19 km^2 (91 %)
Nunataks	2 km^2 (9 %)

3.2.2 Surroundings

Areas of the bay and the surrounding watershed (iceshed) are given in Table 3.1 (cf. Marsz and Rakusa-Suszczewski 1987). The rocks, which are visible through the ice, are mostly volcanic in origin, basaltic and andesitic with sedimentary intrusions (Birkenmajer 1980). This geology corresponds with two main tectonic periods, from the Cretaceous to the Miocene and from the Miocene to recent times (Groeneweg and Beunk 1992).The dominant features of the area are the ice dome and the surface of the bay. The proportion of the ice-covered area to the area covered by water is 1.78. The proportion of ice-free land surface area to the surface of water 0.17; and the proportion of ice-free land to ice-covered land is 0.09.

Land covered by the ice dome rises in the north up to 579 m, in the east to 476 m and in the west to 400 m above sea level. The north-eastern shore of the bay is a continuous ice cliff, underlain by ice-free beaches, rocks and nunataks in the near-shore area. The western and central parts of Admiralty Bay and the southern shore of Ezcurra Inlet are less icy. Ice-free areas are mostly coastal areas or nunataks. These coastal zones of ice-free land of varying widths are divided by ice-covered areas. Most of the 22 areas along the bay include beaches and sections of shore. They have shallow ponds, some permanent, others periodic, and small streams fed by snow and meltwaters from the ice cap. All streams pour into Admiralty Bay.

3.2.3 Shores

The shore of Admiralty Bay is about 84 km long and is very varied, created mostly by erosion processes, glacial accumulation and abrasion processes occur in the area of contact with the sea (Marsz and Rakusa-Suszczewski 1987). There are several different types of shores. About half the total length of shore is ice cliff, which is submerged. The southern part of Ezcurra Inlet and the area of Arctowski is abrasion accumulated. In front of some retreating glaciers, young lagoons occur, which are separated from the main part of the bay by shallow underwater ridges or slightly emergent peninsulas of sand and rocks, representing frontal moraines. Fresh water from the glacier flows into the lagoons and overflows into the bay when these ridges are destroyed, creating an exchange of water between bay and lagoon. The lagoon bottom is deepest just in front of the glaciers, and includes ridges and valleys (Rakusa-Suszczewski 1995a).

Relatively high temperatures and low pack ice on the surface of the bay during the early 1980s caused prolongation of the terminal abrasion process. Warm water melted the glacier, resulting in the retreat of the ice cliff. As a result of this abrasion the ice cliff has retreated onto the land.

3.2.4 The Sea Basin

Admiralty Bay is the largest bay of King George Island (The Antarctic Pilot 1974; Sailing Directions for Antarctica 1985). It is wide and opens to the south into Bransfield Strait. To the north it divides into three fjords, Ezcurra, Martel and Mackellar Inlet. The southern boundary is a line connecting Demay Point and Syrezol Rock. The surface of the bay in 1978 was 119.2 km^2 and in 1989 about 122.1 km^2 (Battke 1990). The difference is caused by a retreat of the fringing glaciers, amounting to 0.25 km/year. Muser (1995) established that for the period 1956–1992 the surrounding ice-covered area decreased by 12 km^2, giving a mean value of 0.3 km^2/year. Thus the rate of decrease of land ice coverage appears to have slowed during the past 20 years.

The volume of sea water in Admiralty Bay is 24 km^3. The total volume of Admiralty Bay equals about 0.29 % of the volume of Bransfield Strait (Szeliga and Rakusa-Suszczewski 1994). The central basin of Admiralty Bay is a relatively young trough-shaped valley with moraines and ridges corresponding to earlier periods of glacial maximum. The depth of this trough increases to the south. Approximately 6 km south of the entrance to Bransfield Strait, it cuts deeply into the shelf, descending to the floor at 1000 m. Ezcurra, Mackellar and Martel Inlets are older valleys, shallower than most of the central part of the basin. The mean depth of the central part of Admiralty Bay is 198 m, the depth of Ezcurra Inlet is 98 m and that of Mackellar and Martel Inlets is 115 m.

The basins and fjords have a tectonic origin (Birkenmajer 1983; Birkenmajer et al. 1988), but the Holocene glaciation had a profound influence on their shapes and sediment compositions. This influence is the effect of glacial processes, the transport of minerals and aeolian transport. In the Ezcurra inlet sediments are made up of gravel, sand, and mud. Organic substances make up only 1 % of the dry weight of the sediment (Rakusa-Suszczewski 1995a).

3.3 Changes of Climate

3.3.1 Temperature

Admiralty Bay is situated in the area of maritime climate with low temperatures which traverse 0 °C in each month of the year, large diurnal variation (Cygan 1981), high humidity and strong winds (Kowalski 1985; Tables 3.2, 3.3, 3.4).

During the period 1948–1987 mean annual surface temperatures for each 10-year period rose (Martianov and Rakusa-Suszczewski 1990). Increase of temperature in the area of South Shetland Islands reached 0.04 °C per year as calculated by Rodriquez et al. (1996). According to Kejna, between 1944 and 1996, air temperature of the South Shetland Islands increased by 1.5 °C (Kejna 1999). He also showed relations to further data sampling points at King George Island and presented related statistical analyses. Similar observations were made at Faraday Station, located on the Antarctic Peninsula, where air temperature increased by 2.5 °C from the late 1940s to the present (Ackley et al. 1996).

Data from stations located on the South Shetland Islands show a distinct decreasing trend in the range of air temperature variability, which has been

Table 3.2. Mean month and mean year air temperature during 1977–1998

Year	01	02	03	04	05	06	07	08	09	10	11	12	Mean
1977				0.0	-2.3	-4.7	-8.1	-8.0	-6.5	-2.7	0.5	1.2	
1978	2.1	2.5	0.9	-0.4	0.0	-5.1	-11.6	-6.9	-2.3	-1.2	-1.1	-1.0	-1.8
1979	3.3	2.4	2.0	0.6	-2.8	-4.9	-3.8	-4.9	-2.4	-2.0	-0.4	1.7	-0.9
1980	2.3	1.9	0.2	-4.7	-2.9	-6.5	-12.2	-11.2	-7.5	-2.2	-2.2	1.5	-3.6
1981	1.7	2.5	0.4	0.0	-3.5	-3.6	-4.2	-9.3	-4.7	-3.7	-0.5	1.3	-2.0
1982	3.6	2.6	1.5	0.7	-0.4	-2.2	-7.3	-5.4	-3.4	-2.7	-1.3	1.5	-1.1
1983	2.0	1.3	1.1	0.3	-3.1	-4.8	-4.5	-2.1	-3.7	-1.6	0.7	1.8	-1.1
1984	1.2	2.4	1.8	-1.4	-6.3	-4.5	-3.7	-2.7	-0.4	-2.6	-0.4	2.0	-1.2
1985	2.9	2.3	-0.4	-0.3	-3.8	-4.0	-2.1	-2.9	-0.7	-0.4	1.0	1.8	-0.6
1986	1.7	1.5	1.8	-1.4	-7.8	-7.7	-8.1	-6.7	-5.0	-1.4	-1.7	0.8	-2.8
1987	1.6	1.8	1.3	-2.6	-5.3	-7.6	-13.2	-4.1	-5.6	-1.4	1.0	2.1	-2.7
1988	1.5	3.4	0.8	-0.7	-1.4	-7.1	-5.4	-8.3	-3.1	-2.8	-1.3	0.5	-2.0
1989	2.2	3.2	3.2	-4.7	0.1	-1.5	0.5	-1.5	-2.4	0.0	0.7	1.5	0.1
1990	2.7			-5.3	-6.0	-5.2	-6.5	-6.1	-2.9	-2.0	-1.2	0.3	
1991	2.8	1.2	0.2	-1.4	-6.8	-8.4	-6.0	-6.4	-1.9	-2.3	0.7	-0.6	-2.4
1992	3.0	1.5	-1.0	-0.6	-8.5	-8.5	-6.7	-3.6	-1.9	-2.6	0.8	3.4	-2.1
1993	1.9	2.9	1.0	-0.6	-1.5	-3.4	-2.7	-2.9	-2.8	-1.0	1.1	1.3	-0.6
1994												2.3	
1995	3.8	3.2	0.5	-1.1	-2.0	-6.4	-13.0	-11.0	-5.8	-0.4	0.1	2.1	-2.5
1996	2.4	3.4	2.0	-1.1	-2.3	-4.9	-3.1	-3.4	-1.0	-0.9	0.3	1.7	-0.6
1997	3.4	2.6	2.2	-1.3	-0.7	-4.1	-6.8	-4.9	-6.7	-2.6	-1.9	1.7	-1.6
1998	3.2	3.4	2.1	2.1	-0.8	-0.1	-5.1	-6.9	-7.6	-1.2	0.7	1.3	-0.7

caused by an increase in the minimum temperatures in winter and the decrease in maximum temperatures in summer and autumn, so that amplitude of temperature has decreased in consecutive years (Rodriquez et al. 1996). According to Rodriquez et al. (1996), temperature data from the South Shetland Islands region display a distinct cyclical pattern with a period of 5 years. Data from the Antarctic Peninsula and South Shetland Islands are poorly correlated with observations from continental Antarctica (Ackley et al. 1996).

Table 3.3. Mean monthly and annual cloudiness, humidity, precipitation, and wind speed at Arctowski Station during the period 1978–1987

Month	Cloudiness (0–10)	Humidity (%)	Precipitation (mm)	Wind speed ($m\ s^{-1}$) Mean	Maximum
I	8.2	82	33.9	6.0	26
II	8.2	82	62.3	6.3	32
III	7.9	81	68.7	7.3	30
IV	7.9	81	58.3	7.5	60
V	7.7	82	31.5	6.4	28
VI	7.5	81	37.0	7.0	28
VII	7.0	81	31.1	6.9	30
VIII	7.1	80	25.6	7.4	36
IX	7.5	81	45.5	7.7	45
X	7.6	80	32.9	8.1	36
XI	8.1	81	42.0	7.5	35
XII	8.1	81	41.7	6.2	38
Year	7.7	81	510.4	7.0	60

Table 3.4. Seasonal frequency (in %) and direction of wind at Arctowski Station during 1978–1987

Month	N	NE	E	SE	S	SW	W	NW	Variable
I	10.8	5.8	12.7	14.6	7.4	15.1	14.2	11.2	9.6
II	12.8	6.0	10.8	11.9	5.6	15.0	13.0	14.6	10.3
III	10.5	6.4	8.9	10.1	6.1	21.6	13.6	14.2	8.5
IV	8.1	6.3	14.5	6.3	6.6	24.9	13.2	11.1	8.5
V	8.9	6.5	17.2	10.2	7.5	17.8	9.4	10.4	12.3
VI	8.8	6.0	12.7	9.2	8.3	20.8	12.5	9.6	12.0
VII	7.5	6.2	8.4	7.7	6.5	19.4	13.4	11.8	18.8
VIII	9.7	3.5	7.5	6.1	6.8	25.7	13.2	13.7	13.6
IX	12.5	5.3	7.0	6.4	5.2	24.6	11.9	15.8	10.8
X	9.3	2.7	8.2	7.8	4.4	28.8	16.3	14.4	7.7
XI	9.7	5.4	9.9	7.7	5.4	26.5	15.4	13.4	6.6
XII	10.1	4.6	8.7	10.7	6.0	19.7	18.0	12.4	9.5
Year	9.8	5.2	10.6	9.1	6.3	21.7	13.5	12.9	10.7

Rising air temperatures in the area of the Antarctic Peninsula during 1957–1982 have been recorded by Limbert (1984). In the Antarctic Peninsula, winter temperatures have risen 5.5 °C during the period 1981–1991, 4.1 °C for autumn and 1.5 °C for summer.

In the Earth's atmosphere, an increase in CO_2 and methane has been observed, reaching 360 ppm and 1720 ppb, respectively. Over Antarctica, an ozone concentration decrease in about 40% reached 100 DU in September-November 1993, which is the lowest value measured anywhere in earth's atmosphere (WMO 1995).

3.3.2 Radiation

Skies over Admiralty Bay are relatively clear. The total annual radiation in 1980 was 80.9 kcal cm^{-2} ($=3.39 \cdot 10^{-5}$ J cm^{-2}) (Kejna 1992). In the central part of the bay, near Arctowski Station, cloudiness is lower because of katabatic wind. Also, humidity and precipitation are lower than in the surroundings. Consequently, solar radiation is higher (Styszyńska 1992). During the summer, radiation varied between 165 and 834 μWh cm^{-2} (= 0.59–3.0 J cm^{-2}) (Krężel and Pęcherzewski 1981), or as mean monthly values varied from 74,000 J cm^{-2} in December to 1.6 J cm^{-2} in June. The many components of total radiation were measured by Prosek et al. (1996).

The correlation between diurnal global radiation (GL in MJ m^{-2}) and diurnal radiation of UV-B [MED day^{-1}; 1 MED=$5.83 \cdot 10^{-6}$ W cm^{-2} (= Js^{-1} cm^{-2})] were expressed for the Arctowski area by Prosek and Janouch (1996) in the equation:

$$UV=0.7482\ GL-0.0048\ (GL^2)$$

According to these authors, in January 1995, UV-B radiation varied from $1.749^{-0.5}$ W cm^{-2} (= 0.76 J s^{-1} cm^{-2}) during low cloudiness, and clean sky to $2.915^{-0.6}$ W cm^{-2} (= 0.53 J s^{-1} cm^{-2}) during high cloudiness. Changes in UV-B at Arctowski Station are in accordance with those at Ferraz Station on the other side of the bay. At the continental Australian station Mawson, UV-B radiation varied from $1.46^{-0.5}$ to $4.5^{-0.5}$ W cm^{-2} (= 177 DU; = 0.83–0.47 J s^{-1} cm^{-2}) during the ozone hole when ozone is 410 DU (Roy et al. 1994) at Palmer Station. Highest UV-B was 382.7 μW cm^{-2} ($=3.83 \cdot 10–4$ J s^{-1} cm^{-2}) in December (Gautier et al. 1994).

Solar radiation at 280–320 nm related to UV-B 320–400 nm to UV-A and 400–700 nm to PAR. In Bransfield Strait near to the South Shetland Islands 1% of energy penetrated the water to a depth of 34–48 m and, further south, to 64–95 m. Measurements show that 1% of PAR radiation penetrated to 35 m (Vosjan and Pauptit 1992), UV-A to 29 m and UV-B up to 18 and 17 m. Depending on the time of the day, length of day and atmosphere clarity, absence of yellow humus substances in the water of Admiralty Bay allows deeper penetration of radiation than in any other area of the word. As a result of the depletion of ozone, exposure of Antarctic organisms to UV-B increases. This applies to both marine organisms – phytoplankton, zooplankton and pelagic larvae of benthic animals – as well as to land, and here to cyanobacteria, algae, birds and mammals, including people. Polar organisms are likely to be more sensitive to solar radiation because they are living in areas with restricted annual radiation.

3.3.3 Climatic Differentiations

Air temperatures throughout the year oscillate around 0 °C, which is the change of physical phase for freshwater bodies. This threshold for seawater is slightly lower, and oscillates around −1.85 °C, depending on salinity. Thawing and freezing processes are important physical factors for transport of mineral matter (Rakusa-Suszczewski 1995b), nutrients and organic substances (Juchnowicz-Bierbasz 1999, Bölter and Blume, Chap. 16). The presence of snow, ice and water also have many effects on chemical and biological conditions in the nearshore geoecosystem. High variability of weather is a normal feature of the Admiralty Bay area (Marsz and Rakusa-Suszczewski 1987). On the basis of 12 years' data (1978–1989) from Arctowski Station, Marsz (1992) distinguished types and frequency (in %) of weather:

Transitional (ground frost-thaw, thawing)	44.71
Temporarily warm (without ground frost)	21.46
Temporarily frosty (without thawing)	12.44
Frosty without thawing	11.32
Very frosty	7.73
Warm without ground frost	2.09
Extremely frosty	0.25

As an adiabatic process, air masses approaching Admiralty Bay mainly from the NW, W and SW (Table 3.4) are transformed and chilled above the cold surface of the bay. Winds falling from 600 m a.s.l. are highly turbulent, with speeds 3–5 times higher than the mean wind speed at the surface. Winds from the SE and NE have a more laminar flow. The strongest gust observed at Arctowski Station registered 74 m s^{-1} (Kowalski and Wielbińska 1989).

Mean monthly and annual temperatures at Arctowski Station are a few tenths of a degree Celsius higher than at Bellingshausen Station situated further west. During winter, air temperature in Admiralty Bay can be lower than at Bellingshausen (Martianov and Rakusa-Suszczewski 1990).

Within the bay, lower temperatures were observed at Mackellar Peninsula than at Arctowski (Zubek 1980). At a smaller scale, differences in microclimate can be large. Loss of snow, extent of snow and ice cover, open or frozen

Table 3.5. Distinguished seasons in the Admiralty Bay area

Summer	$t_{min} \geq 0$ °C – period without freezing
Mean starting date	16 November
Autumn	$t_{max} > 0$ °C; $t_{min} < 0$ °C freezes
Winter	$t_{max} < 0$ °C; freezing time
Mean starting date	7 April
Spring	$t_{max} > 0$ °C; $t_{min} < 0$ °C

Table 3.6. Median dates and length of each season during 1978–1989

Season	Beginning	Mean length (days)	Range (days)
Summer	28 December	52	31– 96
Autumn	18 February	75	45–140
Winter	4 May	135	96–195
Spring	15 September	103	44–117

sea, exposure to north or south, different land cover by flora, and the presence of stream and ponds, all strongly influence microclimatic conditions (Kejna and Laska 1999a).

Kejna and Laska (1999b) found differences in climatic conditions, mainly in winds, between Arctowski Station and Demay Point near the entrance to Admiralty Bay. Bintanja (1992) recorded high values of ablation and air temperature on Ecology Glacier. At 100 m a.s.l., ablation is 1.5–2.0 m of water. During the summer, from February to May (117 days) ablation was 390 to 1000 mm of ice at 200 m a.s.l. (Rachlewicz 1995).

Taking into consideration mean, minimal and maximal air temperatures, Kejna (1992) distinguished the seasons by characteristics and starting dates (Tables 3.5 and 3.6).

3.4 Climatological Effects

3.4.1 Periglacial Features and Processes

The area surrounding Admiralty Bay shows many clear indications of periglacial climate. The retreat of glaciers, mainly the Ecology Glacier, has become evident for new barren soils and is under investigation from a glaciology point of view (Braun and Goßmann, Chap. 6), the development of soils (Blume et al., Chap. 7), botanical recolonisation and soil botanical feedbacks

(Olech, Chap. 12; Kappen and Schroeter, Chap. 18). Short summers and a low mean annual temperature have created permafrost (Marsz and Rakusa-Suszczewski 1987) in the long existent ice-free regions. In the area of Jardine Peak at 169 m a.s.l. during 14 to 31 of January 1998 the active layer was between 20 to 150 cm thick (B. Nagy, pers. comm.).

Near Arctowski Station during the summer, earth temperatures are above freezing in the top 50 cm (Kratke and Wielbińska 1981; Cygan 1981). By March the thaw depth increases to 75 cm, due to a homothermic warming of the soil to about 0 °C and temperature is homothermic. Between April and October the earth is below zero at all depths. The amplitude of temperature at the surface of the earth during the year is a few tens of degrees celsius; at 100 cm it is less than 2–3 °C. In 1998 the mean annual ground temperature at 1 cm depth was 0.5 °C and was higher than the air temperature of –0.7 °C. During the summer, the temperature of the surface of the ground fluctuated less in the vicinity of Admiralty Bay (9.2 °C) than in the valley near Jardin Peak (12.4 °C) where the sites without plants and with northern exposure were warmer than the sites covered with plants and close to flowing water (Zwolska 2001).

Dark rock with northern exposure can warm up to 30 °C. The temperature of shallow ponds can reach 12 °C (Rakusa-Suszczewski 1995b) and sometimes up to 17 °C (Kejna, pers. comm.). Often repeated freezing and thawing of ground and rocks causes disintegration. Precipitation higher than evaporation makes the ground permanently wet. The temperature crosses 0 °C during each month of the year, and outflow of water from the land is permanent, but especially intensive during spring. Intensity of this process has an important influence on hydrological, optical and biological conditions in the coastal water of King George Island (Rakusa-Suszczewski 1996). Outflow of water from land has an important influence on outwash and transport of organic substances and nutrients (Table 3.7) from recent penguin colonies and seal wallows, and from areas of mineralised guano of abandoned old rookeries (Tatur et al. 1997; Juchnowicz-Bierbasz 1999).

Strong winds redistribute snow and cause accumulations at particular points, with effects on local microclimatic conditions. Areas where the snow melts early have longer exposure to radiation, and remain dry and warm

Table 3.7. Range of the main chemical characteristics of interstitial fresh water around Arctowski Station during January–March 1999 according to M. Juchnowicz-Bierbasz (1999)

$P\text{-}PO_4$ (mmol dm^{-3})	$Si\text{-}SiO_4$ (mmol dm^{-3})	$N\text{-}NO_2$ (mmol dm^{-3})	$N\text{-}NH_4$ (mmol dm^{-3})	$N\text{-}NO_3$ (mmol dm^{-3})	N-urea (µg dm^{-3})	pH	Conduct. (µS cm^{-1})	TDS (mg dm^{-3})
0.02–20.14	5.14–77.76	0–4.00	0.21–282.1	0–9.8	0.01–0.361	2.6–7.3	135–1713	152–1917

longer. The season of 1997/1998 was extremely dry: the sum of precipitation in December and January was the lowest for the last 20 years.

During strong winds, sand and silt particles are blown onto the glacier and into the water of Admiralty Bay. According to Krajewski (1986), during the summer, about 22 g of mineral material per m^2 are transported at a height of 1.5 m aboveground.

Aeolian transport of mineral matter, plants and microfauna is important in the processes of colonisation of newly-exposed ground following deglaciation (Janiec 1996). Strong winds transport large amounts of seawater aerosols from Admiralty Bay to the land, affecting soil, stream and pond chemistry, and by increasing salinity, influencing thaw-freeze processes.

3.4.2 Variability in the Terrestrial Ecosystem

In the South Shetland Islands there are a few important factors which decide changes and functioning of the coastal terrestrial ecosystem. The ecosystem depends strongly on the variability of physical processes, like:

1. Increase in the minimum winter air temperature
2. Increase in numbers of the freeze-thaw cycles
3. Temperature limit
4. Decrease in summer precipitation
5. Strong deglaciation on land
6. Decrease in extent of sea-ice surface

All these are physical factors having direct influence on biological and chemical processes. The biological processes like life cycle and feeding processes of marine birds and mammals which reproduce on seashores influence the transport of organic matter from sea to land. Physical processes like air and water flow redistribute ammonia or nutrients and fertilise coastal geoecosystems (Rakusa-Suszczewski 1995b). Anthropogenic influences on marine and land ecosystems are connected with previous and recent exploitation of living resources.

The functioning of polar geoecosystems is correlated with the functioning of ecosystems on a global scale (Ackley et al. 1996). This global feedback is especially significant in Antarctic ecosystems. Polar ecosystems are, it seems, stable and elastic, but not constant (Remmert 1980), because of more or less variability in the same seasons over consecutive years. In meso- and macro-time scales, variability of the geoecosystem is manifested as:

1. Irregular interannual changes
2. Multiyear cyclic changes
3. Year-to-year directional changes
4. Anthropogenic changes

Seasonality of biological processes, namely trophic relations and species phenology in the Maritime Antarctic, is influenced by physical processes involving great energy flows. These processes are the main factors generating variability in the functioning of the geoecosystem, and in energy flow and matter cycling in a time scale of months and years.

There is a negative correlation between air temperature and extent of sea ice in the Antarctic (Ackley et al. 1996). The extent of sea ice and its duration have a crucial influence on krill (*Euphausia superba*), as the under-ice biota is one of their main food sources in winter. In winters with heavy ice, krill larvae have higher condition factors, higher lipid contents and grow, while in winters with little ice, larvae have lower lipid contents, lower condition factors and shrink (Quetin and Ross 1991). It was also observed (Rakusa-Suszczewski 1988) that cold winters are followed by high abundance of krill, while warm winters are followed by low abundance, which can be partly explained by the trophic conditions during the previous winter. The availability of krill decides the total food consumption by penguins and seals, and thus the quantity of matter transported onto the shore of Admiralty Bay. Food availability for penguins and seals depends on the water current in the northern part of the Bransfield Strait (Madejski and Rakusa-Suszczewski 1990). This current transports krill from the Bellingshausen Sea. Its velocity fluctuates and was much greater in summer than in winter. Data collected during SIBEX (1982/1983) show that the inflow of warmer water from the north into the South Shetland Islands region was accompanied by a decrease in krill biomass. Biomass changes caused by population growth during the summer (Godlewska and Rakusa-Suszczewski 1988) and by winter mortality (Rakusa-Suszczewski 1990) cannot explain irregular, year-to-year fluctuations in krill biomass exceeding two orders of magnitude (Kalinowski 1984; Kalinowski et al. 1985). It seems that the variability of krill biomass in the Bransfield Strait is, for the most part, caused by fluctuations of atmospheric circulation in that area and, consequently, by the inflow of warm northern waters which prevent the western, krill-rich waters from entering the Bransfield Strait and also affect ice conditions (Loeb et al. 1997).

Availability of krill and its distance from penguin colonies are main factors influencing their breeding success and population numbers. The numbers of Adelie penguins (*Pygoscelis adeliae*) nesting on the western shores of Admiralty Bay have decreased (Jabłoński 1984). The greatest decrease was observed in the early 1980s, when air temperature was at its highest. Numbers of Gentoo penguins (*Pygoscelis papua*) nesting in the region of Admiralty Bay show great interannual fluctuations (Myrcha 1993), mostly as a consequence of atmospheric changes during the first 2 weeks of October when this species begins its breeding season (Moczydłowski 1986) and during the incubation of eggs. In 1985, storms destroyed all the eggs in the Gentoo penguin colony situated on Thomas Point. Chinstrap penguins (*Pygoscelis antarctica*) were most numerous in 1978/1979. In 1989/1990, chinstrap penguin numbers in the whole area of Admiralty Bay were lower by a factor of 2.5 compared with the population 10 years earlier. The causes of such variability are as yet unknown (Myrcha 1993). Changes in penguin population have direct inflow on organic matter transport on shore.

Changes in numbers of pinnipeds in Admiralty Bay are also characteristic. During the years 1988–1992, the number of elephant seals and fur seals in Admiralty Bay did not differ significantly from year to year, while crabeater seal and fur seal seal numbers show only irregular changes (Rakusa-Suszczewski and Sierakowski 1993).

Changes caused by direct human interference have special significance for the Antarctic ecosystems. Fur seals (*Arctocephalus gazella*) were almost exterminated during the nineteenth century, but since their protection, their numbers have increased rapidly. Trampling of plant communities by fur seals has a well-known effect on South Georgia. If the fur seal population in South Shetland increases to the level before exploitation, plants on the shore of Admiralty Bay may also be damaged. Elephant seals (*Mirounga leonina)* on the western shore of Admiralty Bay also demolish spots of plants, but they are also a source of nutrients. Whales are now fully protected, but the whale population is increasing more slowly.

On a small scale, human impact has caused changes in land communities: the distribution of Branchinecta in small water bodies near Arctowski Station (Janiec 1991), and the introduction of the grass *Poa* spp. (Olech 1996). During a logistic operation in the summer of 1998, a few hundred pea seeds accidentally germinated on Thomas Point. Another example of anthropogenic influence are changes of the nesting places by some of the bird species (*Macronectes giganteus*, *Pygoscelis papua*) from the vicinity of Arctowski Station to more distant regions of the Ecology Glacier moraine.

Changed physical phenomena such as increase in minimum winter air temperature, more frequent crossing of the freezing/thawing temperature border, decreasing precipitation, deglaciation on land, together with anthropogenic influences are important factors of change and functioning of the coastal marine/land geoecosystem of Admiralty Bay.

Acknowledgements. I am greatly indebted to Dr B. Stonehouse for his valuable suggestions and for improvements to the manuscript.

References

Ackley S, Bentley C, Foldvik A, Clarke A, King J, Priddle J, Goodwin I (1996) A summary of global change in the Antarctic. Antarctic Global Change Research. Newsletter of the SCAR Global Change Programme 2, pp 3–15

Alberts FG (1981) Geographic names of the Antarctic. Geographic Names Data Base Division – Defence Mapping Agency, Washington, DC, 20315, pp 1–959

Battke Z (1980) Zatoka Admiralicji. Skala 1:50000. Antarktyczna Stacja im. H. Arctowskiego. Druk WZKart. 1981

Battke Z (1990) Admiralty Bay. King George Island. 1:50,000. E Romer State Cartographical Publishing House, Warsaw

Bintanja R (1992) Glaciological and meteorological investigations on Ecology Glacier, King George Island, Antarctica (summer 1990–1991). Circump J 1–2:59–71

Birkenmajer K (1980a) New place names introduced to the area of Admiralty Bay, King George Island (South Shetland Islands, Antarctica). Stud Geol Pol 64:67–88

Birkenmajer K (1980b) Geology of Admiralty Bay, King George Island (South Shetland Islands, Antarctica). An outline. Pol Polar Res 1:29–54

Birkenmajer K (1983) Polskie badania geologiczne w Antarktyce. Kosmos 32(2, 79):309–331

Birkenmajer K, Danowski W, Rolnicki K (1988) Late Holocene raised marine terrace at Arctowski station, King George Island (South Shetland Island, Antarctica). Stud Geol Pol XCV:75–80

Bongrain M (1914) Description des sites et banquises instructions nautiques. In: Charcot J (ed) Deuxieme Expedition Antarctique Francaise (1908–1910) – sciences physiques: documents scientifiques. Masson et Cie, Paris

Cygan B (1981) Characteristics of meteorological conditions at the Arctowski Station during the summer season of 1979/1980. Pol Polar Res 2:35–46

Furmańczyk K, Marsz A (1980) Szetlandy Poludniowe (South Shetland Islands) – Wyspa Króla Jerzego (King George Is.– Isla 25 de Mayo – Ostrov Vaterloo) – Zatoka Admiralicji (Admiralty Bay – Bahia Lasserre), 1:25,000. IGUG, Druk OPGK, Szczecin

Gautier C, Gang he SY, Lubin D (1994) Role of clouds and ozone on special ultraviolet-B radiation and biologically active UV dose over Antarctica. Antarct Res Ser 62:1–15

Godlewska M, Rakusa-Suszczewski S (1988) Variability of krill, *Euphausia superba* Dana 1852 (Crustacea, Euphausiacea), distribution and biomass in the western Antarctic (Bransfield Strait, Drake Passage, Elephant Island) during 1976–1987. Inv Pesq 52(4):575–586

Groeneweg WJ, Beunk FF (1992) Petrogenesis of cretaceous tertiary volcanic rocks from King George Island: implication for the magmatic tectonic evolution of the South Shetland Islands region, West Antarctica. Circump J I-2(7):10–25, 125–136

Jabłoński B (1984) Distribution and numbers of penguins in the region of King George Island (South Shetland Islands) in the breeding season 1980/1981. Pol Polar Res 5:17–30.

Janiec K (1991) A further note on *Branchinecta gaini* Daday 1910 population from the region of the Arctowski Station (King George Island, South Shetland Islands). Pol Polar Res 12:123–127

Janiec K (1996) Short distance wind transport of microfauna in maritime Antarctic (King George Island, South Shetland Islands). Pol Polar Res 17:203–211

Juchnowicz-Bierbasz M (1999) Year-round changes of nutrients in fresh water bodies near Arctowski Station (South Shetland Islands, Antarctic). Pol Polar Res 20(3):243–258

Kalinowski J (1984) The biomass of krill in the eastern part of Bransfield Strait, March 1981. Pol Polar Res 5:99–105

Kalinowski J, Godlewska M, Klusek Z (1985) Distribution and stock of krill in the Bransfield Strait and the Drake Passage during December 1983–January 1984. BIOMASS-SIBEX. Pol Polar Res 6:151–158

Kejna M (1992) Próba wydzielenia termicznych pór roku na stacji H. Arctowskiego Szetlandy Poludniowe w latach 1978–1989. Probl Klimatol Polarn 2:21–29

Kejna M (1999) Air temperature on King George Island, South Shetland Islands, Antarctica. Pol Polar Res 20:183–201

Kejna M, Laska K (1999a) Differentiation of ground temperature in the region of Arctowski Station, King George Island, Antarctica in 1996. Pol Polar Res 20:221–241

Kejna M, Laska K (1999b) Weather conditions at Arctowski Station, King George Island, South Shetland Islands, Antarctica in 1996. Pol Polar Res 20:203–220

Kowalski D (1985) Wind structure at the Arctowski Station. Pol Polar Res 6:391–403

Kowalski D, Wielbińska D (1989) Synoptic features of the severe winter 1986 at Arctowski station, King George Island, West Antarctica. Pol Polar Res 10:57–71

Krajewski K (1986) On aeolian processes near H. Arctowski Station, King George Island, South Shetlands. Biul Peryglac 31:171–181

Kratke J, Wielbińska D (1981) Co-occurrence of particular meteorological elements in the region of Arctowski station (King George Island, South Shetlands) in 1978. Pol Polar Res 2:7–21

Krężel A, Pęcherzewski K (1981) Preliminary data on total radiation in the region of Arctowski Station (King George Island, South Shetland Islands). Pol Polar Res 2:47–54

Limbert DWS (1984) West Antarctic temperatures, regional differences, and the nominal length of summer and winter seasons. Environment of West Antarctica: Potential CO_2. Induced Changes Report of a Workshop Held in Madison, Wisconsin, 5–7 July 1983. National Academy Press, Washington, DC, pp 116–139

Loeb V, Siegel V, Holm-Hansen O, Hewitt R, Fraser W, Trivelpiece W, Trivelpiece S (1997) Effects of sea-ice extent and krill or salp dominance on the Antarctic food web. Nature 387:897–900.

Madejski P, Rakusa-Suszczewski S (1990) Annual observations of icebergs as tracers of water movement in the Bransfield Strait in front of Admiralty Bay (King George, South Shetland Islands). Antarct Sci 2:259–263

Marsz A (1984) From surveys of the geomorphology of the shores and bottom of the Ezcurra Inlet. Oceanologia 15:209–220

Marsz A (1992) Struktura pogód i roczna sezonowosc klimatu Stacji Arctowskiego. Problemy klimatologii polarnej, Gdynia, pp 30–49

Marsz A, Rakusa-Suszczewski S (1987) Charakterystyka ekologiczna rejonu Zatoki Admiralicji (King George Island, South Shetland Islands). I. Klimat i obszary wolne od lodu. Kosmos 36:103–127

Martianov V, Rakusa-Suszczewski S (1990) Ten years of climate observations at the Arctowski and Bellingshausen Stations (King George Is., South Shetlands, Antarctic) In: Breymeyer A (ed) Global Change Regional Research Centres. Seminar Papers and IGBP WG 2 Report,. Inst Geogr Spatial Organ PAS:80–87

Moczydłowski E (1986) Microclimate of the nest-sites of *Pygoscelid* penguins (Admiralty Bay, South Shetland Islands). Pol Polar Res 7:377–394

Muser D (1995) Der Gletscherrückzug auf King George Island, Süd-Shetland Inseln zwischen 1956 und 1962. Eine digitale Auswertung von Karten, Luftbildern und Satellitendaten. Praca dyplomowa z Institut für Physische Geographie Albert-Ludwigs Universität we Fryburgu (Freiburg i. Br.), wykonana pod kierunkiem Prof. Dr. H. Goßmann (maszynopis), 82 pp

Myrcha A (1993) Birds. In: Rakusa-Suszczewski S (ed) The maritime Antarctic coastal ecosystem of Admiralty Bay. Department of Antarctic Biology, Polish Academy of Sciences, Warsaw, pp 129–141

Olech M (1996) Human impact on terrestrial ecosystems in West Antarctica. Proc NIPR Symp Polar Biol 9:299–306

Prosek P, Janouch M (1996) The measurement of ultraviolet radiation at the Polish Henryk Arctowski Station (South Shetlands – Antarctica) in the summer of 1994/95. Probl Klimatol Polarn 6:139–148

Prosek P, Janouch M, Kruszewski G (1996) Components of radiation balance and their regime in the Summer of 1994/1995 at H. Arctowski Station (South Shetlands). Probl Klimatol Polarn 6:107–138

Quetin LB, Ross RM (1991) Behavioral and physiological characteristics of the Antarctic krill, *Euphausia superba*. Am Zool 31:49–63

Rachlewicz G (1995) Ablacja Lodowca Ekologii, Wyspa Króla Jerzego (Szetlandy Południowe) w roku 1991. XXII Sympozjum Polarne Wroclaw 1995:61–68
Rakusa-Suszczewski S (1988) Differences in the hydrology, biomass, and species distribution of plankton, fishes, and birds in the Bransfield Strait and Drake Passage during FIBEX 1981 and SIBEX 1983/84. In: Sahrhage D (ed) Antarctic ocean and resources variability. Springer, Berlin Heidelberg New York, pp 214–218
Rakusa-Suszczewski S (1990) Seasonal changes in respiration and biomass of *Euphausia superba* Dana from Admiralty Bay (South Shetland Islands, Antarctica). Pol Arch Hydrobiol 37:305–311
Rakusa-Suszczewski S (1995a) The hydrography of Admiralty Bay and its inlets, coves and lagoons (King George Island, Antarctica). Pol Polar Res 16:61–70
Rakusa-Suszczewski S (1995b) Flow of matter in the Admiralty Bay area, King George Island, Maritime Antarctic. Proc NIPR Symp Polar Biol 8:101–113
Rakusa-Suszczewski S (1996) Spatial and seasonal variability of temperature and salinity in Bransfield Strait and Admiralty Bay, Antarctica. Pol Polar Res 17:29–42
Rakusa-Suszczewski S, Sierakowski K (1993) Pinnipeds in Admiralty Bay King George Island, South Shetlands (1988–1992). Pol Polar Res 14:441–454
Remmert H (1980) Arctic animal ecology. Springer, Berlin Heidelberg New York
Rodriquez R, Llasat CM, Rakusa-Suszczewski S (1996) Analysis of the mean and extreme temperature series of the Arctowski Antarctic base. Probl Klimatol Polarn 6:191–212
Roy CR, Gies HP, Tomlinson DW, Lugg DL (1994) Effects of ozone depletion on the ultraviolet radiation environment at the Australian stations in Antarctica. Antarct Res Ser 62:1–15
Sailing directions for Antarctica (1985) First Edition. Published by the Defence Mapping Agency Hydrographic. Topographic Centre Pub 200, 129 pp
Styszyńska A (1992) Uslonecznienie na stacji Arctowskiego (King George Island, South Shetland Islands). Probl Klimatol Polarn 2:12–20
Szeliga J, Rakusa-Suszczewski S (1994) Morphometry of Bransfield Strait, West Antarctica. Pol Polar Res 15:21–29
Tatur A, Myrcha A, Niegodzisz J (1997) Formation of abandoned penguin rookery ecosystems in the maritime Antarctic. Polar Biol 17:405–417
The Antarctic Pilot (1974) Published by the Hydrographer of the Navy, 4th edn, pp 1–336
Vosjan J, Pauptit E (1992) Penetration of photosynthetically available light (PAR), UV-A, and UV-B in Admiralty Bay, King George Island, Antarctica. Circump J 1–2:50–71
WMO (1995) The global climate system review June 1991–November 1993, WMO No 819:1–150
Zubek K (1980) Climatic condition at H. Arctowski Station (King George Island, South Shetland Islands). Pol Arch Hydrobiol 27:235–244
Zwolska I (1999) Spatial and temporal differentiation of ground surface temperature in the region of Admiralty Bay in 1998 (King George Island, South Shetland Islands.) Folia Fac Sci Nat Univ Masarykianae Brunensis, Geografia 25:61–64

4 Wilkes Land (Casey Station)

R.D. Seppelt

4.1 The Windmill Islands – Physical Setting

The Windmill Islands, at around 66°17' S, 110°32' E on the eastern side of Vincennes Bay, are a series of peninsulas and islands occupying an area of exposed or permanently ice and snow-covered ground of around 105 km^2, including an area of approximately 25 km^2 of permanently ice-covered terrain between the Løken Moraine line, as an eastern boundary, and the rock outcrops of the peninsulas. The Vanderford Glacier forms a natural southern boundary to the region (Fig. 4.1).

The islands and coastal peninsulas, outcrops and nunataks consist of low rounded hills reaching elevations up to 113 m, with intervening valleys filled with ice and snow or moraine debris and exfoliated detritus. The relief of the more southerly islands and peninsulas is more rugged than that of the north. Rock exposures vary from fresh to weathered, depending largely on the composition of the country rock.

The region was first visited by members of the Second Antarctic Development Project, Unites States Navy Task Force 39, during 1947–1948, when a landing was made on Holl Island to establish a ground control for the US Navy Operation Highjump aerial photographic mission made during 1946–1947, the photographs from which were used to produce the first maps of the area. The name, Windmill Islands, is derived from the extensive use of helicopters made by the Navy Task Force.

4.2 Geology and Geomorphology

The Windmill Islands, together with the Bunger Hills some 400 km to the west, form one of the few major areas of rock outcrop in East Antarctica that can be directly correlated with an Australian equivalent in the reconstruction of Gondwana (Oliver et al. 1983). Rocks of the Windmill Islands exposure represent one of the most easterly outcrops of a Mesoproterozoic low-pressure granulite facies terrain that extends westwards to the Archaean complexes of Princess Elizabeth Land (Tingey 1982). Since the discovery of the region there

Ecological Studies, Vol. 154
L. Beyer and M. Bölter (eds.) Geoecology of Antarctic Ice-Free Coastal Landscapes

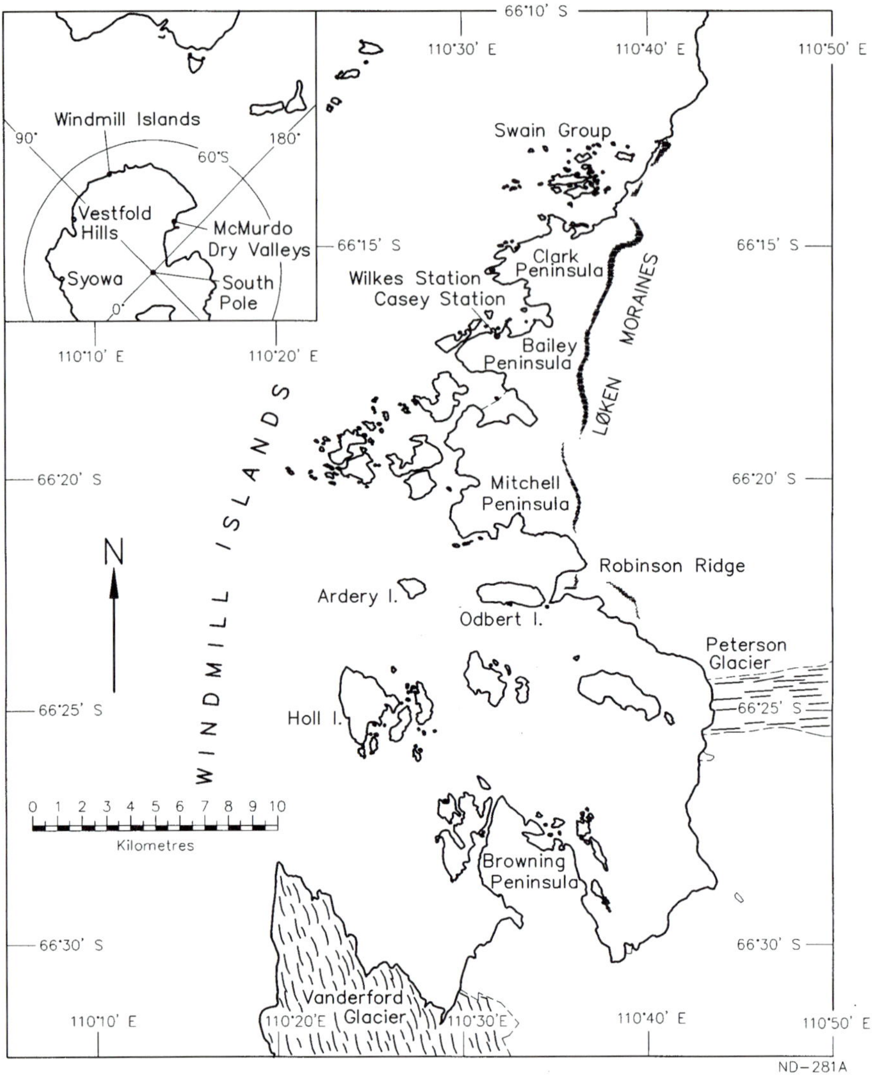

Fig. 4.1. Windmill Island region, Wilkes Land, showing locations of Casey Station, major peninsulas and islands

have been various geological studies, including those of Ravich and Voronov (1968), Starik et al. (1959, 1960), Ravich (1960), Voronov and Karasik (1963), Cameron et al. (1960), Robertson (1961), McLeod (1964), McLeod and Gregory (1966), Webb et al. (1964), Oliver (1972), Blight and Oliver (1977, 1982), Collerson and Reid (1981), Lovering et al. (1981), Oliver et al. (1983), Plimer and Lovering (1983), Williams et al. (1983), Kilpatrick and Ellis (1992), Paul et al. (1995), and Post et al. (1997).

Rocks of the region consist of a layered sequence (ranging in composition from acid to ultrabasic) of schists, gneisses and migmatites (the Windmill

Metamorphics) intruded by the Ardery Charnockite and a porphyritic granite (the Ford Granite), and cut by two swarms of easterly-trending dolerite dykes (Blight and Oliver 1977, 1982). The Windmill Metamorphics probably originally consisted of acid and basic volcanics interbedded with sediments ranging from dirty sandstones to shales and possibly intruded by basic bodies of varying type. The metamorphic grade ranges from amphibolite facies in the north (sillimanite-biotite-orthoclase), through biotite-cordierite-almandine granulite, to hornblende-orthopyroxene granulite in the south. There are no major-element chemical differences between the northern amphibolite facies and the southern granulite facies rocks (Blight and Oliver 1982). The Ardery Charnockitic Intrusives are magmatic rocks unusually rich in TiO_2, P_2O_5 and K_2O and poor in CaO and Al_2O_3 and contain the mineral assemblage orthopyroxene + alkali feldspar + biotite + clinopyroxene + quartz + plagioclase + apatite and opaques, producing a rock type prone to deep weathering which relates more to features of its origin rather than to surficial processes (J. Kilpatrick, pers. comm.). The rocks tend to crumble readily into their constituent mineral grains and are more prone to the abrasive effects of wind scouring and to freeze-thaw action. The metamorphic sequences found in the northerly parts of the region have a much more stable mineral assemblage and crystalline structure. This difference has a marked influence on the suitability of the rock types as a substrate for slow growing lichens.

The age of the rocks has been variously dated between 1600- and 3500-Ma-old (Lovering et al. 1981) and between 950- and 1400-Ma-old (Cameron et al. 1960; Webb et al. 1964; Arriens 1975). Williams et al. (1983) concluded that the protolith of at least the orthogneisses cannot be much older than 1470 Ma. Paul et al. (1995) showed evidence of repeated tectonothermal activity in the Windmill Islands region, based on structural and metamorphic relationships. Post et al. (1997), using sensitive high resolution ion microprobe zircon U-Pb geochronology, supported this hypothesis and derived metamorphic ages of 1340–1310 Ma followed by a granulite facies overprint at around 1210–1180 Ma, suggestive of metamorphism associated with the convergence of the east Antarctic Shield with the Archaean Yilgarn Block of Western Australia between 1340 and 1180 Ma.

Soils of the region are frozen for most of the year, with only the upper 30 to 60 cm thawing during the summer, when the top few centimetres refreeze almost nightly. The soils are mainly formed by cryoturbation and cryoclastic weathering. Most of the soils are influenced by biotic structure formation, humus accumulation, and chemical weathering together with acidification and brownification (Blume and Bölter 1993a,b). Polygons 0.3–10 m in diameter have been formed by cryoturbation. The intensity of recent cryoturbation is evident in the fine silt and clay content of the soils. The soils are acid, with a pH of KCl generally between 4 and 4.5. Organic matter content in the first 1–5 cm of many soils ranges from 0.5–2 %. Shallow peat accumulates sometimes in small depressions. The sandy soils are podzolised. There is a strong correlation between vegetation cover and soil type (Blume and Bölter 1993a,b), with podzols being covered by lithic and soil lichens and moss, while on the silty Leptosols and Regosols and especially on loamy Leptosols there were only scattered lithic lichens. While fine silty soils and clays may occur in areas of melt water run off or in areas subjected to cryoturbation, skeletal soils, commonly consisting of weathered rock chips of >2 mm diameter, are also widespread and have been referred to as Lithosols.

4.3 Holocene Emergence

Goodwin (1993) determined a Holocene deglaciation sequence for the Windmill Islands from ^{14}C-ages of raised marine shorelines, lake bottom sediments, and Adélie penguin remains found in abandoned colonies. Penguin remains have been dated at 4380±250 ^{14}C years B.P. (corrected age of 3289 cal years B.P.), while a dark brown compacted peat with penguin quills has been dated at 4980±80 ^{14}C years B.P. (corrected age range of 4280–4530 cal years B.P.). There is a gradient in elevation of the upper marine limit with the highest elevation being 31–32 m on Browning Peninsula and Holl Island at the southern end of the region.

The southern islands were deglaciated by about 8000 years B.P. while the northern islands were deglaciated by about 5500 years B.P. Isostatic uplift rates have been calculated at 0.5–0.6 m/100 years, with an estimated total uplift around 53 m, indicating Late Pleistocene ice sheet thickness of 200 m over the islands and 400 m over the adjacent Petersen Bank (Hollin and Cameron 1961; Goodwin 1993). The margin of the Late Pleistocene grounded ice sheet extended an estimated 8–15 km offshore, coincident with the location of the 200 m isobath (Goodwin 1993).

The extent of ice cover at the last glacial maximum and in the period since then has considerable importance in understanding both the potential origin and the abundance of the present day terrestrial biota of the Windmill Islands region. Denton and Hughes (1981) postulated that the East Antarctic ice sheet extended over the coastal zone as far as the continental shelf during the last glacial maximum, a distance of up to 100 km from the present ice margin. This implied that inner shelf regions, like the Vestfold Hills and Bunger Hills, and other coastal exposures such as the Larsemann Hills and Windmill Islands, were covered by an ice sheet of up to 1000 m in thickness. However, recent studies in these regions, such as those of Adamson and Pickard (1986), Gore and Colhoun (1995, 1997), Fabel et al. (1997), Zwartz et al. (1997) for the Vestfold Hills, Gillieson (1991) and Burgess et al. (1994, 1997) for the Larsemann Hills, Colhoun and Adamson (1989, 1992), Colhoun (1991, 1997), Adamson and Colhoun (1992), Colhoun et al. (1992), and Melles et al. (1997) for the Bunger Hills, Hollin and Cameron (1961) and Goodwin (1993) for the Windmill Islands, have indicated that either the ice sheet was much thinner or at least some areas of these present day coastal regions may have remained unglaciated during at least the Holocene period. Analysis of $\delta^{18}O$ values in ice cores from the Law Dome, east of the Windmill Islands, indicates maximal thinning of the ice sheet of less than 530 m between the last glacial maximum and the Holocene period, while similar analyses from Terre Adélie at a site 150 km inland from Dumont D'Urville Station indicate a greater thinning of around 1320 m (Grootes and Stuiver 1987).

The presence of at least some coastal ice-free areas during the last glacial maximum may be indicated by radiocarbon dates from moss found in permafrost in the Larsemann Hills at 27,950±710 years B.P. (Burgess et al. 1994). Pickard and Seppelt (1984) have reported radiocarbon dates for in situ subfossil moss from near Clear Lake in the Vestfold Hills of 7355±150 to 8400±250 years B.P. Further evidence to support a much thinner ice and at least a partially floating ice sheet (or even refugia) at some of the coastal sites is provided from a number of studies. For example, Pliocene age fossil dolphins have been found in the Vestfold Hills (Quilty 1991) adjacent to unconsolidated marine sediments that were dated within the range of 3.5–4.5 Ma from studies of molluscs, diatoms, and the racemisation of proteins from the molluscs (Adamson and Pickard 1986; Pickard et al. 1986, 1988).

4.4 Climate

The climate of the Windmill Islands is frigid-Antarctic (sensu Longton 1988). Meteorological data from Casey Station, on Bailey Peninsula, shows mean temperatures for the warmest and coldest months of 0.3 and –14.9 °C respectively, with extreme temperatures ranging from 9.2 to –41 °C. Midsummer air temperatures at Casey Station may reach 9 °C but the mean summer temperature is slightly less than 1 °C and, consequently, freezing is possible at any time. The mean annual temperature is –9.3 °C for 1957–1983 (Jacka et al. 1984). The climate is dry with a mean annual snowfall of 195 mm $year^{-1}$ rainfall equivalent. These data have changed within the last decade to –9.1 °C and mean annual precipitation to 230 mm (D. Shepherd, Australian Antarctic Division). Mean annual precipitation in the marginal 200–300 km of the Antarctic continent are generally within the range of 60–150 cm of snow (Lovering and Prescott 1979).

There are, on average, 96 days of gale-force winds per year (>17.5 m s^{-1}), predominantly from the east, east-north-east and south off the polar ice cap. Relative constancy of wind direction is a feature of the surface winds of the Antarctic plateau (Schwerdtfeger 1984). Annual mean wind speed is around 5.6 m s^{-1}. Streten (1968) concluded that Antarctic coastal stations fell into one of two categories with respect to mean annual wind speeds. Those stations with only minimal katabatic wind influence had mean wind speeds generally in the 4–6 m s^{-1} range while those with a strong katabatic influence fell within the 10–12 m s^{-1} range. Blizzards, with winds up to and >50 m s^{-1}, occur frequently, but chiefly during the winter months. There is a high incidence of daily sunshine during the summer months.

Although there is no recorded data available, the southern part of the Windmill Islands is climatically more severe than the northern part. Strong- to gale-force katabatic winds, the frequency, direction and intensity of which are determined by the local geography (Schwerdtfeger and Mahrt 1970), channel cold air from the elevated Law Dome (summit altitude 1389 m) down the Vanderford Glacier and are a pronounced feature of the local climate at Browning Peninsula and the nearby islands. Winds of approximately 10 m s^{-1} are capable of picking up and transporting loose snow. Physical abrasion of surfaces by wind-transported grit, snow and ice particles, may significantly weather rock surfaces and limit the stature of vegetation and restrict the distribution of much of the vegetation to sheltered habitats.

References

Adamson DA, Colhoun EA (1992) The Late Quaternary glaciation and deglaciation of the Bunger Hills, Antarctica. Antarct Sci 4:463–467

Adamson DA, Pickard J (1986) Cainozoic history of the Vestfold Hills. In: Pickard J (ed) Antarctic oasis: terrestrial environments and history of the Vestfold Hills. Academic Press, Sydney, pp 63–98

Arriens PA (1975) The Precambrian geochronology of Antarctica. 1st Australian Geological Convention, Abstracts, pp 97–98

Blight DL, Oliver RL (1977) The metamorphic geology of the Windmill Islands, Antarctica: a preliminary account. J Geol Soc Aust 24:239–262

Blight DL, Oliver RL (1982) Aspects of the geological history of the Windmill Islands, Antarctica. In: Craddock C (ed) Antarctic Geoscience – Symposium on Antarctic geology and geophysics, Madison, Wisconsin, 22–27August 1977. University of Wisconsin Press, Madison, pp 445–454

Blume H-P, Bölter M (1993a) Podsole, Leptosole und Regosole der Antarktis. Mitt Dtsch Bodenkundl Ges 72:843–846

Blume H-P, Bölter M (1993b) Soils of Casey Station (Wilkes Land, Antarctica). In: Gilichinski D (ed) Joint Russian-American Seminar on cryopedology and global change, Pushchino. Post Sem Proc 1st Intern Conf Cryopedology 1992, Russian Academy of Sciences Pushchino, pp 96–104

Burgess J, Carson C, Heat J, Spate A (1997) Larsemann Hills: not heavily glaciated during the last glacial maximum. In: Ricci CA (ed) The Antarctic region: geological evolution and processes. Proceedings of the VII International Symposium on Antarctic earth sciences, Siena, 1995. Terra Antarctica Publication, Siena, pp 841–843

Burgess J, Spate A, Shevlin J (1994) The onset of deglaciation in the Larsemann Hills, Eastern Antarctica. Antarct Sci 6:491–495

Cameron RI, Goldich SS, Hoffman JH (1960) Radioactive age of rocks from the Windmill Islands, Budd Coast, Antarctica. Contrib Geol (Stockholm) 6:1–6

Colhoun EA (1991) Geological evidence for changes in the East Antarctica ice sheet (60°-120°E) during the last glaciation. Polar Rec 27:345–355

Colhoun EA (1997) A review of geomorphological research in Bunger Hills and expansion of the East Antarctic ice sheet during the last glacial maximum. In: Ricci CA (ed) The Antarctic region: geological evolution and processes. Proceedings of the VII International Symposium on Antarctic earth sciences, Siena, 1995. Terra Antarctica Publication, Siena, pp 801–807

Colhoun EA, Adamson DA (1989) Former glacial lakes of the Bunger Hills, Antarctica. Austral Geogr 20:125–136

Colhoun EA, Adamson DA (1992) Raised beaches of the Bunger Hills, Antarctica. Australian National Antarctic Research Expedition Scientific Reports, No 136. Australian Antarctic Division, Kingston

Colhoun EA, Mabin MGC, Adamson DA, Kirk RM (1992) Antarctic ice volume and contribution to sea-level fall at 20,000 yr B.P. from raised beaches. Nature 358:316–319

Collerson KD, Reid E (1981) Crustal history in the Casey area, Wilkes Land, Antarctica. Aust Natl Antarct Res Exped News 8:1–2

Denton GH, Hughes PJ (1981) The last great ice sheets. Wiley-Interscience, New York

Fabel D, Stone J, Fifield LK, Cresswell RG (1997) Deglaciation of the Vestfold Hills, East Antarctica: preliminary evidence from exposure dating of three subglacial erratics. In: Ricci CA (ed) The Antarctic region: geological evolution and processes. Proceedings of the VII International Symposium on Antarctic earth sciences, Siena, 1995.Terra Antarctica Publication, Siena, pp. 829–834

Gillieson D (1991) An environmental history of two freshwater lakes in the Larsemann Hills, Antarctica. Hydrobiol 214:327–331

Goodwin ID (1993) Holocene deglaciation, sea level change and the emergence of the Windmill Islands, Budd Coast, Antarctica. Quat Res 40:70–80

Gore DB, Colhoun EA (1995) Regional contrasts in terrain and glacial sediments, Vestfold Hills, East Antarctica. In: Abstracts of the VII International Symposium on Antarctic earth sciences, Siena, 1995, p 163.

Gore DB, Colhoun EA (1997) Regional contrasts in weathering and glacial sediments suggests long term subaerial exposure of Vestfold Hills, East Antarctica. In: Ricci CA (ed) The Antarctic region: geological evolution and processes. Proceedings of the VII International Symposium on Antarctic earth sciences, Siena, 1995. Terra Antarctica Publication, Siena, pp 835–839

Grootes PM, Stuiver M (1987) Ice sheet elevation changes from isotope profiles. In: Washington ED (ed) The physical basis of ice sheet modelling. Int Assoc Hydrol Sci 170:269–281

Hollin JT, Cameron RL (1961) IGY glaciological work at Wilkes Station, Antarctica. J Glaciol 3:833–843

Jacka TH, Christou L, Cook BJ (1984) A data bank of mean monthly and annual surface temperatures for Antarctica, the southern Ocean and south Pacific Ocean. ANARE Res Note 22:1–84, Australian Antarctic Division, Hobart

Kilpatrick JA, Ellis DJ (1992) C-type magmas: igneous charnockites and their extrusive equivalents. Trans R Soc Edinb Earth Sci 83:155–164

Longton RE (1988) Biology of polar bryophytes and lichens. Cambridge University Press, Cambridge

Lovering JF, Prescott JRV (1979) Antarctica. Last of lands. Melbourne University Press, Melbourne

Lovering JF, Travis GA, Cumaford DJ, Kelly PR (1981) Evolution of the Gondwana Archaean Shield: zircon dating by ion microprobe and relationships between Australia and Wilkes Land, (Antarctica). In: Glover JE, Groves DI (eds) Archaean geology. Geol Soc Aust Spec Publ 7:193–204

McLeod IR (1964) An outline of the geology of the sector 45° to 80°E, Antarctica. In: Adie RJ (ed) Proc. 1st International SCAR-IUGS Symposium on Antarctic Geology, Capetown, 16–21 September 1963. North Holland, Amsterdam, pp 237–247

McLeod IR, Gregory CM (1966) Geological investigations along the Antarctic coast between longitude 108°E and 166°E. Australian National Antarctic Research Expedition Report, Series A (111) Geol Publ 83, pp 1–53

Melles M, Kulbe T, Verkulich SR, Pushina ZV, Hubberten HW (1997) Late Pleistocene and Holocene environmental history of Bunger Hills, East Antarctica, as revealed by fresh-water and epishelf lake sediments. In: Ricci CA (ed) The Antarctic region: geological evolution and processes. Proceedings of the VII International Symposium on Antarctic earth sciences, Siena, 1995. Terra Antarctica Publication, Siena, pp 809–820

Oliver RL (1972) Some aspects of Antarctic-Australian geological relationships. In: Adie RJ (ed) Antarctic Geology and Geophysics. Symposium on Antarctic Geology and Solid Earth Geophysics, Oslo, 6–15 August 1970. Universitetsforlaget, Oslo, pp 859–864

Oliver RL, Cooper JA, Truelove AJ (1983) Petrology and zircon geochronology of Herring Island and Commonwealth Bay and evidence for Gondwana reconstruction. In: Oliver RL, James PR, Jago JB (eds) Antarctic Earth Science. Proceedings of the Fourth International Symposium on Antarctic Earth Sciences, Adelaide, South Australia, 16–22 August 1982. Australian Academy of Sciences, Canberra/Cambridge University Press, Cambridge, pp 64–68

Paul E, Stüwe K, Teasdale J, Worley B (1995) Structural and metamorphic geology of the Windmill Islands, east Antarctica: field evidence for repeated tectonothermal activity. Aust J Earth Sci 42:453–469

Pickard J, Seppelt RD (1984) Holocene occurrence of the moss *Bryum algens* Card. in the Vestfold Hills, Antarctica. J Bryol 13:209–217

Pickard J, Adamson DA, Harwood DM, Miller GH, Quilty PG, Dell RK (1986) Early Pliocene marine sediments in the Vestfold Hills, East Antarctica. S Afr J Sci 82:520–521

Pickard J, Adamson DA, Harwood DM, Miller GH, Quilty PG, Dell RK (1988) Early Pliocene sediments: coastline, and climate of East Antarctica. Geol 16:158–161

Plimer IR, Lovering JF (1983) Manganese-rich chemical sediments from Wilkes Land, Antarctica. In: Oliver RL, James PR, Jago JB (eds) Antarctic Earth Science. Australian Academy of Science, Canberra, pp 69–72

Post NJ, Hensen BJ, Kinny PD (1997) Two metamorphic episodes during a 1340–1180 Ma convergent tectonic event in the Windmill Islands, East Antarctica. In: Riccia CA (ed.) The Antarctic region: geological evolution and processes. Proceedings of the VII international symposium on Antarctic earth sciences, Siena, 10–15 September 1995. Terra Antarctica Publication, Siena, pp 157–161

Quilty PG (1991) The geology of Marine Plain, Vestfold Hills, East Antarctica. In: Thomson MRA, Crame JA, Thomson JW (eds) Geological evolution of Antarctica. Cambridge University Press, Cambridge, pp 683–686

Ravich MG (1960) The rocks of Grearson Hills and the Windmill Islands (Grearson Oasis). Collected papers Antarctic Geology, part 2. Sci Res Inst Antarct Geol 113:53–81

Ravich MG, Voronov PS (1958) Geological structures of the coast of the East Antarctic continent (between 55°E and 110°E longitude). Sov Geol Dok Akad Nauk SSSR 2:3–26

Robertson R (1961) Preliminary report on the bedrock geology of the Windmill Islands. IGY Glaciol Rep 4:1–26

Schwerdtfeger W (1974) Weather and climate of the Antarctic. Elsevier, Amsterdam

Schwerdtfeger W, Mahrt LJ (1970) The relation between the Antarctic temperature inversion in the surface layer and its wind regime. In: Gow AJ, Keeler C, Langway CC,

Weekes WF (eds) Proceedings of the International Symposium on Antarctic glaciological exploration, Hanover, NH, September 1968. Publ No 86, International Association of Scientific Hydrology, pp 308–315

Starik IE, Ravich MG, Krylov AJ, Silin JI (1959) On absolute age of rocks of the Eastern Antarctic Platform. Sov Geol Dok Akad Nauk SSSR 126:124–146

Starik IE, Ravich MG, Krylov AJ, Silin JI, Atrashenok LJ, Lovcjus AV (1960) New data on the absolute ages of rock of the eastern Antarctic continent. Sov Geol Dok Akad Nauk SSSR 134:1441–1443

Streten NA (1968) Some features of mean annual windspeed data for coastal east Antarctica. Polar Rec 14:315–322

Tingey RJ (1982) The geological evolution of the Prince Charles Mountains: an Antarctic Archaean cratonic block. In: Craddock C (ed) Antarctic geoscience. University of Wisconsin Press, Madison, pp 455–464

Voronov PS, Karasik AM (1963) Brief geological and geophysical characteristics of Grearson Oasis in Budd Coast area of East Antarctica. Inf Bull Sov Antarct Exped 42:5–11

Webb AW, McDougall I, Cooper JA (1964) Potassium-argon dates from the Vincennes Bay region of Oates Land. In: Adie RJ (ed) Antarctic geology. Proceedings of the 1st International Symposium on Antarctic geology, Cape Town, 16–21 September 1963. North Holland, Amsterdam, pp 597–602

Williams IS, Compston W, Collerson KD, Arriens PA, Lovering JF (1983) A reassessment of the age of the Windmill Metamorphics, Casey area. In: Oliver RL, James PR, Jago JB (eds) Antarctic earth science. Proceedings of the 4th International Symposium on Antarctic earth sciences, Adelaide, South Australia, 16–22 August 1982. Australian Academy of Sciences, Canberra/Cambridge University Press, Cambridge, pp 73–75

Zwartz D, Lambeck K, Bird M, Stone J (1997) Constraints on the former Antarctic ice sheet from sea-level observations and geodynamic modelling. In: Ricci CA (ed) The Antarctic region: geological evolution and processes. Proceedings of the VII international symposium on Antarctic earth sciences, Siena, 10–15 September, 1995. Terra Antarctica Publication, Siena, pp 162–168

5 Dronning Maud Land and Its Environments

S. Ohtani and H. Kanda

5.1 Introduction

Dronning Maud Land is located in the southern ('continental') Antarctic, ranging between the Stancom-Wills Glacier at 20°00' W, and Shinnan Glacier at 44°38' E (Fig. 5.1). The climatic regime is comparable to that of the vicinities of the Australian Casey Station, Wilkes Land (Seppelt, Chap. 4). Precipitation occurs only as snow and plant growth is thus restricted to the melting process. The growing season in the Antarctic is short, lasting for 120 to 150 days in the northern ('maritime') Antarctic (El-Sayed and Fryxell 1993). In the area of the Japanese Syowa Station, at East Ongul Island (69°00' S, 39°35' E), the southern Antarctic, just 40 days allow growth in the moss habitats depending on small-scale snowdrift (Kanda 1986). In Yukidori Valley, also near Syowa Station, the growing season is considered to be about 60 days in the prominently moss vegetation. This climatic regime hampers the development of vascular plants, and vegetation is dominated by cryptogamic organisms, i.e. algae, lichens and mosses, where meltwater provides sufficient moisture. This chapter introduces geographical and climatic features in Dronning Maud Land focusing on environmental factors surrounding moss and algal habitats.

5.2 The Geographical Zone of Dronning Maud Land

Various authors have proposed demarcations of Antarctic botanical zones, e.g. Skottsberg 1904; Holdgate 1964; Wace 1964; Korotkevich 1967; Longton 1988; Kanda and Komárková 1997. These attempts can be summarised in three main areas: the northernmost treeless Subantarctic subzone, the northern ('maritime') Antarctic subzone, which is the region in which *Deschampsia antarctica* grasslands are mixed with a moss-lichen vegetation (Rakusa-Suszczewski, Chap. 3; Olech, Chap. 12), and the southern ('continental') Antarctic subzone. The subzone 'southern Antarctic' includes the Antarctic continent, the southern and south-eastern parts of the Antarctic Peninsula,

Ecological Studies, Vol. 154
L. Beyer and M. Bölter (eds.) Geoecology of Antarctic Ice-Free Coastal Landscapes

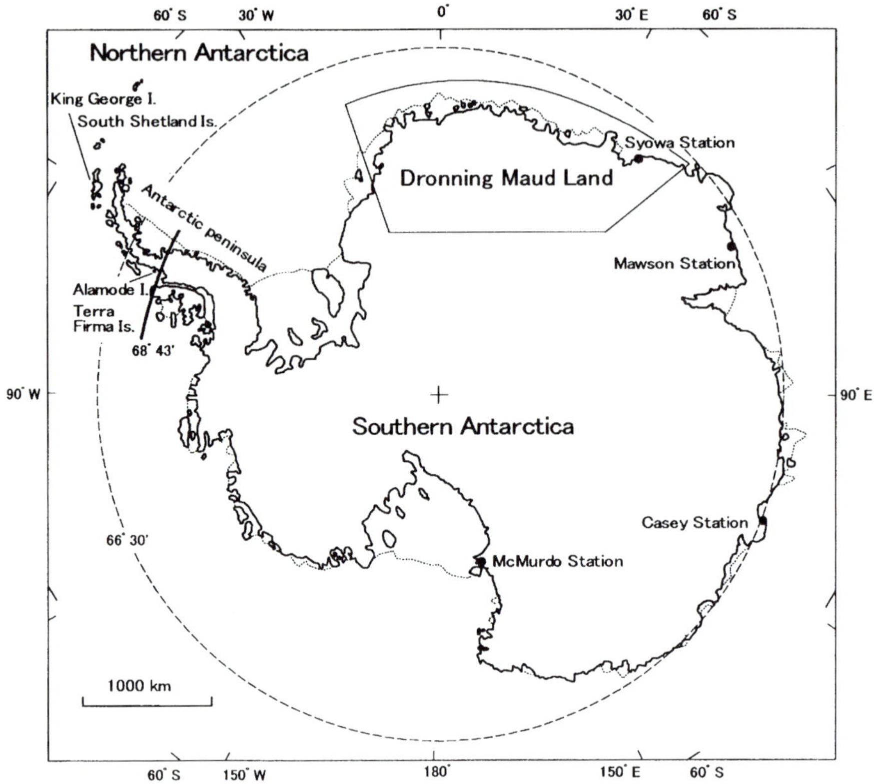

Fig. 5.1. The Antarctic and Dronning Maud Land, *dashed line* at 66°30′S shows Antarctic circle

and the associated islands bordering around 68° 43'S (Kanda and Komárková 1997; Fig. 5.1).

Dronning Maud Land is located between the terminus of the Stancom-Wills Glacier (20°00' W) and the Shinnan Glacier (44°38' E; Figs. 5.1 and 5.2). According to Weyant (1966), Dronning Maud Land includes the following areas of three geographical zones:

1. Interior Antarctic plateaus: Mizuho Station area, Dome Fuji Camp area
2. Antarctic slopes: Heimefrontfjella, Tottanfjella, Vestfjella, Robertskollen, Passat, Sverdrupfjella, Gjelsvikfjella, Mühlig-Hofmannfjella, Sør-Rondane, Belgicafjella, Yamato mountains (Dronning Fabiolafjella), Asuka Station area
3. Antarctic coasts: Novolazarevskaya Station area, Maitri Station area, Syowa Station area.

Four Japanese stations are located in East Dronning Maud Land and Syowa Station (69°00' S, 35°35' E; 29.2 m alt.) is attached to the Antarctic coastal region, Mizuho Station (70°42' S, 44°20' E; 2230 m alt.), Asuka Station (71°31' S,

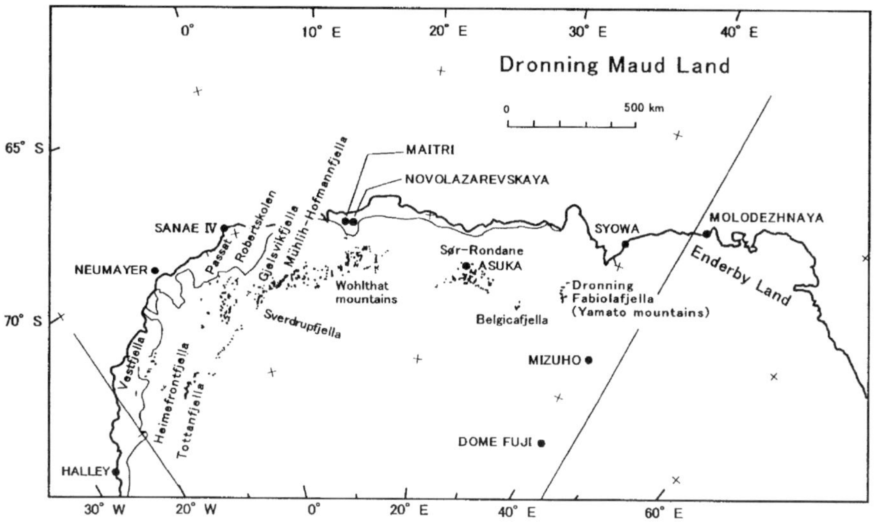

Fig. 5.2. Dronning Maud Land, ranging between the Stancom-Wills Glacier at 20°00' W, and Shinnan Glacier at 44°38' E

24°08' E; 930.4 m alt.) and Dome Fuji Camp (77°19' S, 39°42' E; 3810 m alt.) are on the interior Antarctic plateaus. Dome Fuji Camp is built on the continental ice over 1000 km away from the coast in the highlands at an altitude up to 3800 m; it is located on the interior Antarctic plateaus. The locations Sør-Rondane, Belgicafjella, and Yamato mountains (Dronning Fabiolafjella) are connected with the Antarctic slopes.

5.3 Climatic Zonations

According to Weyant (1966), the Antarctic may also be conveniently divided into three climatic zones, the interior Antarctic plateau, the Antarctic slopes, and the Antarctic coasts. The interior Antarctic plateau is less cloudy, it has light winds, little annual precipitation, comparatively little snowdrift, but extremely low temperatures. Almost all nunataks or mountains at a distance of 100–300 km away from the coast or sea ice are included in the Antarctic Slopes. Cloudiness, precipitation and temperatures are intermediate between those of the plateau and those along the coast, and increase the closer you get to the coastline. The vicinities of the stations located along the coast are included in the Antarctic coastal zones. These areas are warmer, have more clouds and precipitation than the slopes or the plateau, but they are strongly influenced by cyclonic activity (Weyant 1966).

The first international exploration which set up a base in Dronning Maud Land was a joint Swedish-British-Norwegian expedition led by Norwegian

John Gjaever in 1950. Afterwards, the Japanese Antarctic Research Expedition (JARE) constructed Syowa Station at East Ongul Island, East Dronning Maud Land, in the year of IGY in 1957. Since the establishment of this station, JARE has carried out various studies on the subject of the plant communities in the vicinity of the station.

According to Rudolph (1967), the southern Antarctic subzone has a mean January temperature of between 0 and –18 °C; only some coasts and the interior show values of –8 to –15 °C or below. Average annual precipitation is between 146 and 192 mm, and almost all of it is snow. The area of Syowa Station has precipitation only as snow, mean temperature is given as –13.4 °C, and mean maximum wind speed as 26 km h^{-1} (1989–1997). The mean January temperature between 1989 and 1997 was –0.7 °C, the mean January maximum temperature was 7.8 °C, and the mean January minimum temperature was –8.8 °C. The mean January temperature in Asuka Station was –8.6 °C over 3 years (1989–1991), and at Dome Fuji Camp it was –34.4 °C over 4 years (1995–1998). Svarthamaren, Mühlig-Hofmannfjella, Central Dronning Maud Land have been established as SSSI No. 23 in 1987 (SCAR 2000). This area stretches from 71°33'17" S, 5°09'12" E in the northwest to 71°55'58" S, 5°15'12" E in the southeast. The January air temperature has been observed to range between –5 and –15 °C.

5.4 Distribution of Mosses, Lichens and Soil Algae

Vegetation and plant communities of the whole Dronning Maud Land are discussed later (Kanda et al., Chap. 15). Here, the environments of the species occurring in the vicinities of Syowa Station are introduced, with a focus on the Yukidori Valley. Along the Sôya Coast of Lützow-Holm Bay, East Antarctica, there are some large ice-free areas, named Langhovde, Skarvsnes and Skallen (Fig. 5.3). Most snowdrifts and ice on ponds melt during the austral summer in these areas. There are no vascular plants. Moss and lichen vegetation occurs on sandy soils where there is a water supply from streams, ponds and melting snowdrifts. The most prominent vegetation develops in the Yukidori Valley, Langhovde (Matsuda 1968; Nakanishi 1977; Kanda 1981; Inoue 1989).

Almost all plant species recorded from one of the large ice-free areas, Langhovde, occur within the valley area. They include seven species of mosses, *Bryum pseudotriquetrum*, *Bryum argenteum*, *Bryum amblyodon*, *Ceratodon purpureus*, *Pottia Heimii* (= *hendeniella heimii*), *Pottia austro-georgica*, *Grimmia lawiana*, and over 40 species of lichens such as *Usnea sphacelata*, *Umbilicaria antarctica*, *Umbilicaria decussata*, *Pseudephebe minuscula*, *Xanthoria elegans*, etc. There are up to 60 species of microalgae, including species endemic to Yukidori Valley, *Cosmarium yukidoriense* and a variety of *C. clepsydra*. Although usually no visible vegetation is found at sites remote from

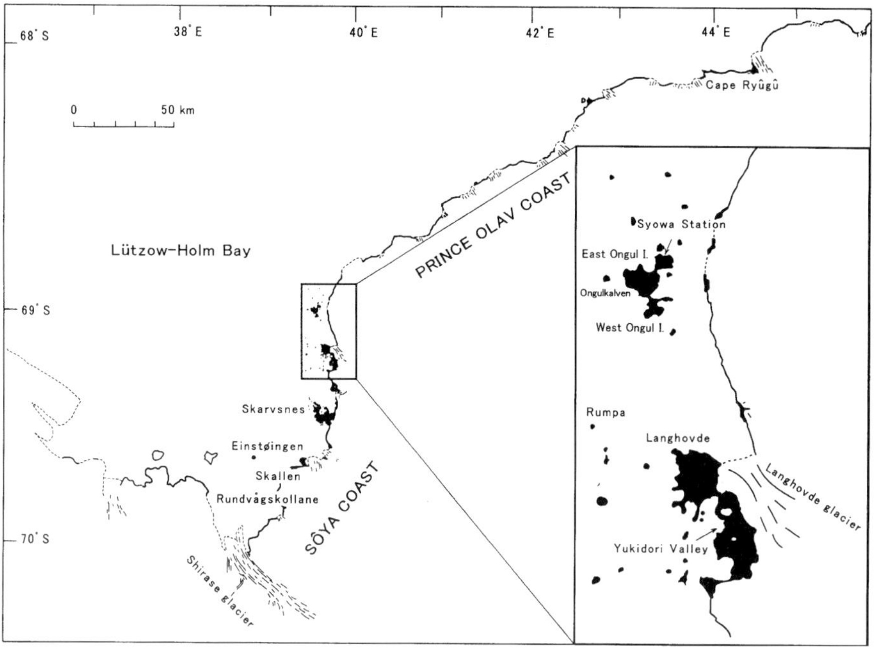

Fig. 5.3. The area of Syowa Station and Yukidori Valley, Langhovde

water supply, soil algae are often cultured on agar medium from dry sandy soils. Akiyama (1967) was the first to study the soil algal flora at the Ongul Islands in detail using cultures. The taxonomy and ecology of soil algae have since been studied in the vicinity of Syowa Station.

5.5 Soil Properties

5.5.1 Particle Size Distributions

In the vicinity of Syowa Station, most soils are without vegetation and are composed of sand and gravel (Akiyama 1967; Ayukawa et al. 1998). Akiyama (1967) reports that sand and mud (<0.02 mm) content in the soil ranges from 42.1–97.0 % and 0.4–3.4 % (with one exception of 14.3 %), respectively, in the Ongul Islands. Ayukawa et al. (1998) studied particle size distributions along a transect crossing a stream in the Yukidori Valley, Langhovde (Fig. 5.4a,b). Over 50 % of soil consists of sand in most samples. Relatively high clay contents are measured in bare soils adjacent to moss compared with soils distant from the vegetation (Fig. 5.4 c). The authors consider that clay content is an important factor for colonisation by mosses and algae

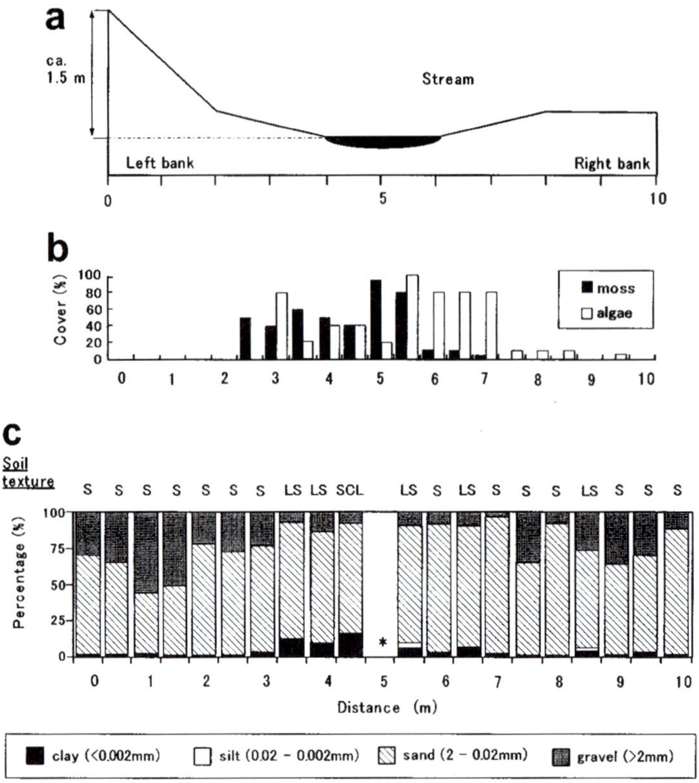

Fig. 5.4a–c. Soil texture and particle size distribution in moss vegetation along the transect in Yukidori Valley. **a** Transect profile, **b** vegetation cover along the transect in the valley and **c** soil texture and particle size distribution of soils sampled along the transect. *S* Sand, *LS* loamy sand, *SCL* sandy clay loam. * Soil was not sampled because there was not enough area of bare ground in this quadrate. (After Ayukawa et al. 1998)

because higher water-holding capacity was measured at sites with higher clay content.

5.5.2 Electrical Conductivity and pH (H_2O)

Conductivity (EC) values (soil:water=1:5) range from 0.033 to 0.73 dS m^{-1} with one exception of 2.2 dS m^{-1} in the Ongul Islands (Akiyama 1967; Ohtani et al. 2000). Akiyama (1967) attributes the differences of EC values to chloride contents of soils. Chloride is obviously derived from windblown sea-spray. Ayukawa et al. (1998) show that EC values range from 0.006 to 0.132 dS m^{-1} along a transect crossing a stream in Yukidori Valley. Relatively high EC values are measured at sites adjacent to moss vegetation, while lower EC values occur at sites distant from vegetation (Fig. 5.5).

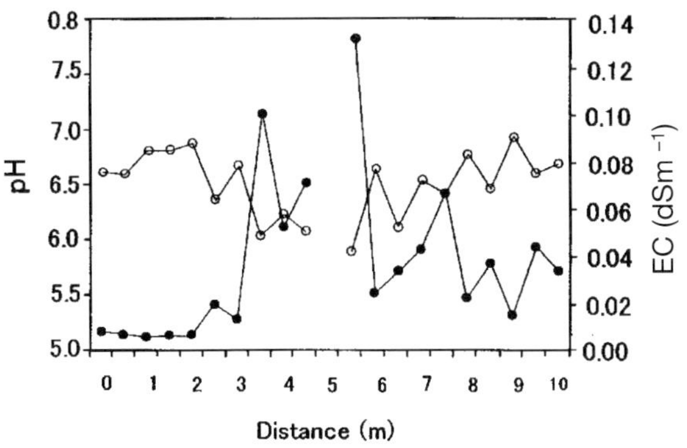

Fig. 5.5. pH and conductivity (EC) of soils sampled along the transect in Yukidori Valley (see Fig. 5.4). *Open circles* pH; *closed circles* EC. (After Ayukawa et al. 1998)

Surface soil pH (soil:water=1:2.5) ranges from 5.7 to 9.4 at the Ongul Islands (Akiyama 1967; Yamanaka and Sato 1977; Ohtani et al. 2000). Broady (1986) reports a similar range (pH 5.3–9) at the Vestfold Hills. Lower values are described for the Windmill Islands (Blume et al., Chap. 7). Relatively high values of pH 8.8–9.4 have been measured in soils collected from sites close to a urinal tank in Syowa Station (Ohtani et al. 2000) and in soils containing small shells at Skarvsnes (Yamanaka and Sato 1977). Ayukawa et al. (1998) showed relatively low pH at sites adjacent to moss vegetation and higher pH values at the sites distant from vegetation along a transect crossing a stream in Yukidori Valley, Langhovde (Fig. 5.5).

5.5.3 Water Availability

The availability of water is crucial for terrestrial organisms throughout the Antarctic (Boyd et al. 1966; Cameron 1969), especially in the extensive areas of scant precipitation depending on snow or glaciers; poorly developed soils in the southern Antarctic with low water-holding capacity and high salinity cause low water availability (Bölter and Blume, Chap. 16). During some austral summers, a relatively large amount of water is available from melting snow and ice, especially when glaciers are retreating. This meltwater is considered a reliable source of moisture only for organisms capable of withstanding its fluctuations, such as bryophytes and lichens (Kanda and Komárková 1997).

Substantial factors which define the difference of distribution are considered to be a short period for the growing season, yearly temperature regime, or an oligotrophic soil nutrition. The most important environmental factors which limit the short growing season are temperature and water supply. During the austral summer, with a plenty of radiation, the most important limiting factor is thought to be water supply, which mainly derives from meltwater of the ice cap.

Water availability has also been considered as a factor limiting distribution of moss in the vicinity of Syowa Station (Matsuda 1968; Yamanaka and Sato 1977; Nakanishi 1977; Kanda 1981, 1986; Kanda and Inoue 1994; Kanda et. al. 1990; Ayukawa et al. 1998). Kanda (1986) shows the seasonal change of water content at the moss vegetation beside the small-scale snowdrift in the isolated small island, East Ongul Island, during 1983–1984.

Compared with the seasonal changes of snowdrift, the water content of mosses shows a striking contrast as it increases after the decay of the snowdrift. Before the moss vegetation was exposed from snow cover (September to November), the water content was 0.3–2.4%. In early December, however, the water content increased to 2.4–46.0%. Judging from the air temperature and the solar radiation, the water content of mosses might begin to rise in late November. The soil near the terminal of the snowdrift became moistened on 2 December when the air temperature and solar radiation were about −5 °C and 100 h/10 days, respectively. The mean water content of mosses has a more significant correlation with solar radiation than with the air temperature. From late December to early January, the water content of mosses reaches a maximum value of 128.8% (Fig. 5.6).

In contrast, the water contents and temperatures of moss cushions have been measured throughout the year at the Yukidori Valley with prominent vegetation in 1988–1989 (Kanda et al. 1990). The maximum yearly mean water content reaches 462.5% at a site of the upper course of the valley. Figure 5.7 shows the seasonal changes of monthly mean air temperatures, and temperatures and water contents of moss cushions in the permanent monitoring sites (Kanda, unpubl.). It has also been demonstrated that bothwater content and temperature are correlated to the development of moss cushions. Compared with the small-scale moss vegetation in the East Ongul Island, it is characteristic for large-scale moss vegetation in the Yukidori Valley to be able to preserve water even in February. Thus, the yearly change of water availability causes the development of moss vegetation, and consequently it is considered to affect the ecosystem in the valley.

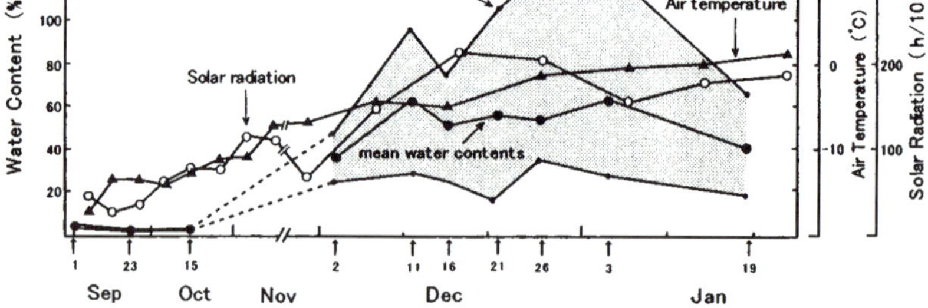

Fig. 5.6. Seasonal changes of water content in moss vegetation. Water contents were measured from early September 1983 to the middle of January 1984. (Kanda 1986)

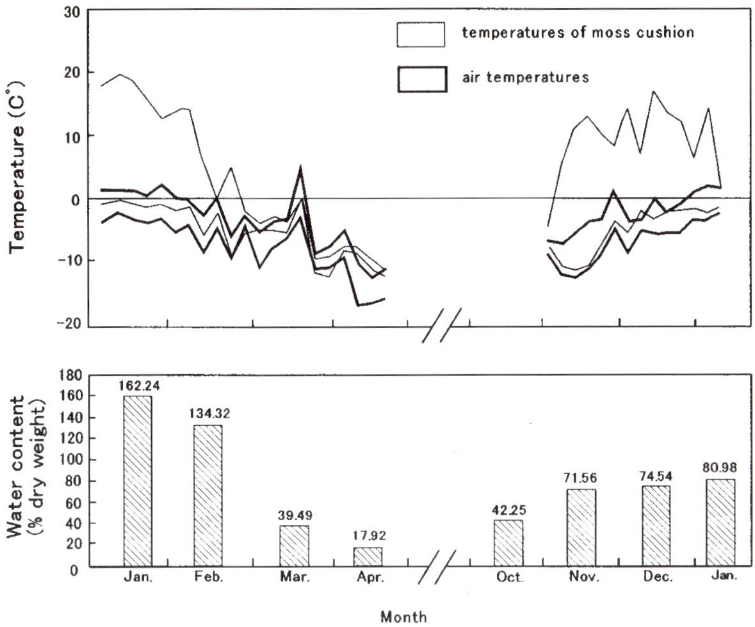

Fig. 5.7. Seasonal changes of monthly mean air temperatures, temperatures and water contents of moss cushion. Water contents and temperatures measured from early January 1988 to early January 1989. Water contents are calculated as percentages of the dry weight. (Kanda, unpubl.)

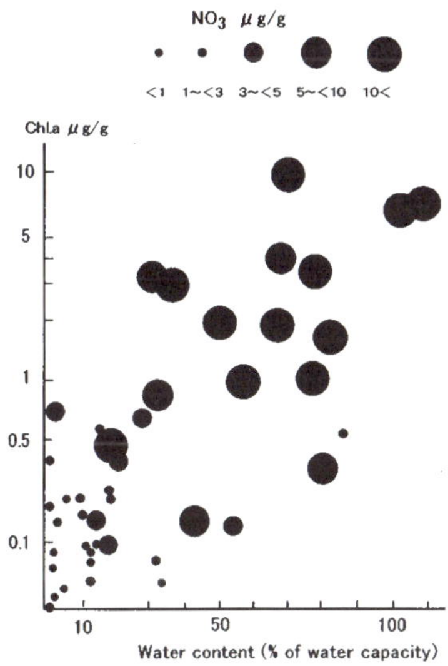

Fig. 5.8. Relationship between water content, NO_3 and chlorophyll *a* concentration in the vicinity of Syowa Station. (After Akiyama et al. 1986b)

Akiyama (1986b) studied correlations between water content, nutrients and chlorophyll *a* in soils in the vicinity of Syowa Station. He has shown a positive correlation between water content and chlorophyll *a*, and nitrate concentration is positively correlated with both water and chlorophyll *a* (Fig. 5.8). Gradual or sudden decreases in chlorophyll *a* and/or cell number of soil algae have been reported at a few metres distance from streams or lake shores (Akiyama 1986b; Ohtani et al. 1991).

5.5.4 Nutrients

Along with cold and drought, lack of nutrients is probably one of the main factors limiting plant growth in Antarctic terrestrial habitats. At Ongul Island, total carbon (TC) and total nitrogen (TN) contents of bare surface soils remote from penguin rookeries are very low. Concentrations in the area of Syowa Station, Ongul Island, range from 0.023–0.09 and 0.003–0.006 %, respectively (Yamanaka and Sato 1977; Ohtani et al. 2000). Relatively high contents of TC (0.21–0.78 %) and TN (0.012–0.08 %) are found in soils covered with moss vegetation (Yamanaka and Sato 1977). Higher TC (0.34–0.52 %) and TN contents (0.021–0.09 %) are measured from soils covered by *Prasiola crispa* at Ongulkalven (Yamanaka and Sato 1977). Maximum TC of 0.991 % and TN of 0.401 % have been measured in bare soil collected in a penguin rookery at Ongulkalven (Ohtani et al. 2000). A wide range of TC (0.17–4.43 %) and TN (0.014–0.482 %) can be obtained from soils in Langhovde (Yamanaka and Sato 1977; Ayukawa et al. 1998). High values of TC (1.36–3.78 %) and TN (0.102–0.482 %), about five times higher than those obtained from the Ongul Islands, are measured in bare soils adjacent to moss vegetation in Yukidori Valley, Langhovde (Ayukawa et al. 1998). TC data from the Windmill Islands range – with respect to individual soil types – from 0.02 % on young moraines to 30 % in soils of recent penguin rookeries (Blume et al., Chap. 7).

On the Ongul Islands, Ohtani et al. (2000) reported that total phosphorus (TP) contents of surface soils range from 0.161 to 0.233 % with one exception of a sample collected at a penguin rookery (0.809 %). Values from the Windmill Islands are found in comparable ranges (Blume et al., Chap. 7). Yamanaka and Sato (1977) report that P_2O_5 contents of surface soils without vegetation range from 7.4 to 20.3 mg/100 g in the vicinity of Syowa Station. They have measured high P_2O_5 contents of 51–295.9 mg/100 g from soils covered by *Prasiola crispa* near penguin rookeries at Ongulkalven. Campbell and Claridge (1966) reported that the P_2O_5 content of nesting areas is much higher in recently abandoned (10.5–17.3 %) and occupied soils (7.3 %) than in those longest abandoned (0.7–0.8 %) and in soils remote from the penguin rookery (0.02 %) at Inexpressible Island, Terra Nova Bay, Antarctica.

Low C/N ratios (1–4) were reported from the occupied penguin rookery and from a recently abandoned rookery at Inexpressible Island (Campbell

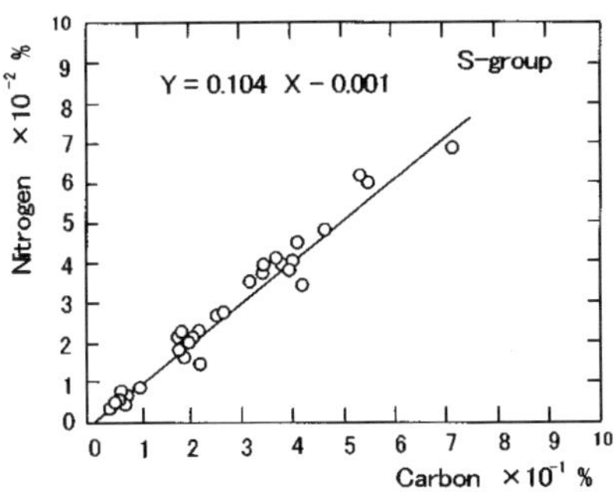

Fig. 5.9. Relationship between carbon and nitrogen content in soil samples collected from along a transect in a moss community-soil system in Rundvågskollane, Lützow-Holm Bay. (After Ino and Nakatsubo 1986)

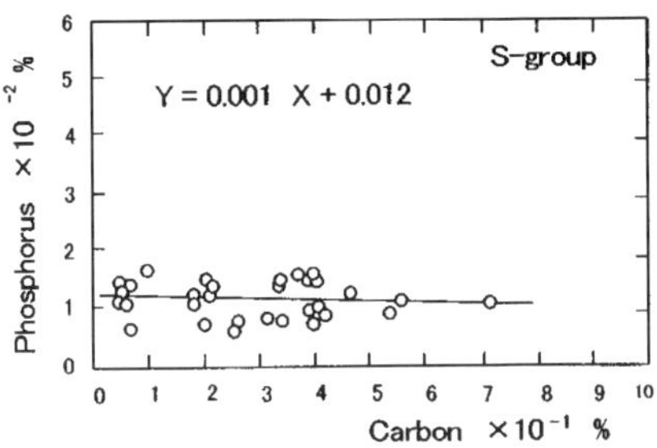

Fig. 5.10. Relationship between carbon content and phosphorus content in the same soil samples as Fig. 5.9

and Claridge 1966), and from Admiralty Bay region, King George Island (Pietr et al. 1983). Ohtani et al. (2000) reported a low C/N ratio of 2.5 from a bare soil sample collected from an occupied penguin rookery at Ongulkalven. Conversely, Ohtani et al. (2000) report that soils eutrophicated by human beings at Syowa Station had C/N ratios between 18.8 and 33.5. These higher ratios are caused by relatively low TN contents in soils near the urinal tank compared with soils of penguin rookeries.

Ino and Nakatsubo (1986) studied distributions of carbon, nitrogen and phosphorus in a moss community soil system developed at Rundovågskollane, at Lützow-Holm Bay. C/N ratios of moss and soil samples are nearly constant at ca.16 and ca. 10, respectively. A significant relationship (r=0.986) between TC and TN content in soil samples can be found in soil samples collected along a line transect crossing moss communities (Fig. 5.9). In contrast, correlations between TC and TP are very low (r=0.671, Fig. 5.10). A rise of TC content does not cause a rise of TP in either soils or moss. TP contents in soils and moss are ca. 0.012 and ca. 0.030 %, respectively.

The cyanobacterium *Nostoc commune* is the principal nitrogen-fixing terrestrial organism in cold regions. Significant nitrogen fixation has been demonstrated by cyanobacteria epiphytic on mosses with high water content in the southern Antarctic (Nakatsubo and Ino 1986, 1987; Pandey et al. 1992). In the moss samples collected from the Syowa Station area, nitrogen-fixing activities are measured by the acetylene reduction method. The cumulative amount of nitrogen fixation is 329 mg N m^{-2} during the growing periods of 40 days in East Ongul Island, the estimate for an annual nitrogen requirement is 84 mg N m^{-2} (Nakatsubo and Ino 1986). Moss communities which grow on the sand at dry habitats have a dense cover of cyanobacteria and show high nitrogen-fixing activities, while those at wet habitats near streams show weak or no activity. N-fixing cyanobacteria are common in benthic mats in pools on the McMurdo Ice Shelf (Howard-Williams et al. 1990). In contrast, at Signy Island, Fogg and Stewart (1968) and Horne (1972) have demonstrated N-fixation up to 2.1 mg N m^{-2} $year^{-1}$ by *Nostoc*, either in a free-living thaloid form or as a symbiont in lichens in base-rich wet habitats.

In the Yukidori Valley, seabirds nesting on the ice-free areas can also have considerable influence on the vegetation. At the moss community beside a lake where Antarctic skuas (*Catharacta maccormicki*) have nests and spread bones and feathers of snow petrels (*Pagodroma nivea*), there is a rich nutrient resource from bird droppings, and also favourable water conditions.

As mentioned above, in the Syowa Station area, the vegetation is rare or absent in northern and northeastern parts of the ice-free area and the western part with seaward peninsulas and small islands. One of the main environmental factors affecting lichen distribution is the wind-blown salt spray accumulated in the snow cover (Inoue 1991). Similarly, in the Vestfold Hills, the moss and lichen vegetation is distributed sparsely over the easterly inland regions, but the westerly section is strongly influenced by salt spray and virtually barren of moss and lichen (Pickard 1986; Seppelt 1986).

Hoshiai and Matsuda (1979) have studied the distribution of biomass of microalgae in soils at Ongulkalven. High values of chlorophyll *a* are found near an occupied and abandoned penguin rookery (Fig. 5.11a). High active fractions of chlorophyll *a* (the ratio of chlorophyll *a* to the sum of chlorophyll *a* and phaeophytin) are found near an occupied rookery. Meltwater is supplied from snow patches accumulated in the lee of rocks around the rookery and in a depression where meltwater is supplied from snowdrifts near abandoned rookeries (Fig. 5.11b). It is concluded that the activity and biomass of microalgae depends on the degree of water supply and the soil fertility. The latter is increased by penguin excrement.

Ohtani et al. (2000) show that the community at sites near a urinal tank in Syowa Station is different from other sites based on results of cluster analysis of species assemblages over 6 years. Correspondence analysis shows this to be due to relatively high frequencies of *Prasiola crispa* and *Phormidium autumnale* and the absence of *Crinalium* cf. *epipsammum* and *Nostoc* cf. *sphaericum* at sites near a urinal tank. As values of TC and TN are higher than at other sites, they conclude that the former two species characterise a eutrophic environment, and the latter an oligotrophic environment.

High levels of nutrients are toxic and can limit terrestrial organisms in small areas, e.g. close to seabird colonies. Sieburth (1960, 1961, 1963) and Boyd et al (1970) demonstrated the antibiosis of acrylic acid, carried from the sea to land through algae, krill, and penguins, which may inhibit vegetation development.

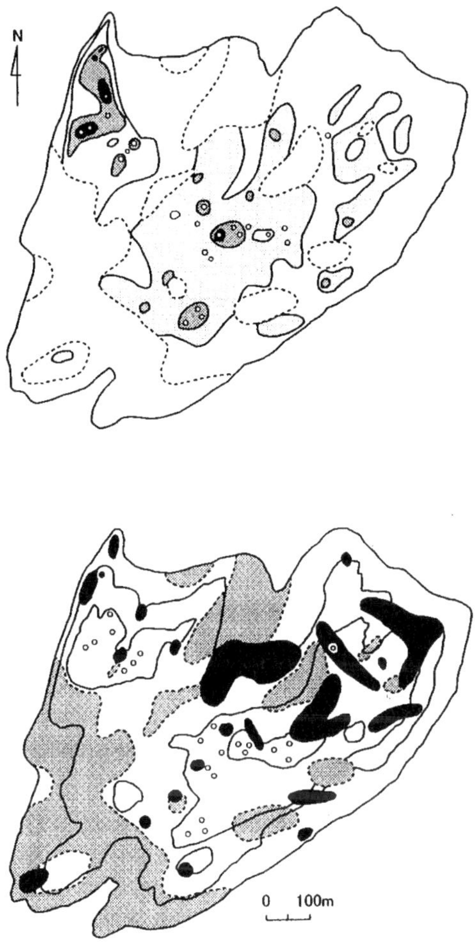

Fig. 5.11.a Distribution of the biomass of microalgae in soil. Sum of chlorophyll *a* and phaeophytin above 10 mg/ml (*black*); between 10 and 1.0 mg/ml (*thick dotted*); between 1.0–0.1 mg/ml (*thin dotted*); below 0.1 µg/ml (*white*); and snow drifts shown by *broken line*. Occupied rookery (*closed circle*, see *arrow*); abandoned rookery (*open circle*). **b** Areas of higher values in the ratio of chlorophyll *a* to the sum of chlorophyll *a* and phaeophytin. The ratio of chlorophyll *a* exceeds 30 % in the *black* parts. Occupied rookery (*closed circle*, see *arrow*); abandoned rookery (*open circle*). (After Hoshiai and Matsuda 1979)

Acrylic acid in penguin guano, derived from phytoplankton via the food chain, is known to be an antibiotic. It has an effect on the quality and occasionally on the quantity of the soil microflora (Boyd et al. 1966, Boyd 1967).

Ecological and physiological effects of penguin excrement on terrestrial algae have been studied at Ongulkalven. The occupied rookery is situated in the northwestern part of the island and an old abandoned rookery is scattered in the centre. Akiyama et al. (1986a) found acrylic acid and oxalic acid to be allelopathic substances in the excrement of Adélie penguins. These, either in combination or individually, inhibit growth of *Koliella* (Fig. 5.12). Free-living algal cells do not grow in cultures inoculated with soils from the centre of penguin rookeries and chlorophyll concentrations are low, although high concentrations of phosphate and nitrate are detected (Akiyama et al. 1986b). Matsuda (1968) reports that *Prasiola crispa* is not found in a pool at an occupied rookery but occurs at a site 50 m from the pool. This species is abundant at the

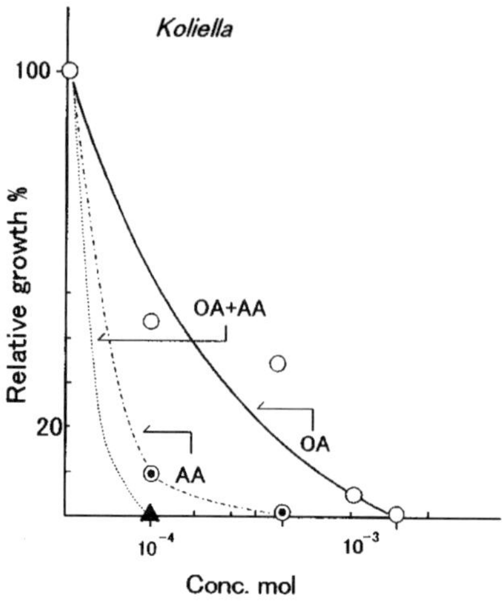

Fig. 5.12. Inhibitory effects of acrylic acid and oxalic acid on the growth of Antarctic *Koliella*. *AA* Acrylic acid, *OA* oxalic acid. (After Akiyama et al. 1986a)

recently abandoned rookery. Ohtani et al. (2000) isolated a small number of taxa in cultures of surface soils collected from an occupied penguin rookery site. *Leptolyngbya* cf. *battersii* dominated the community and associated algae are to a small extent *Botrydiopsis* spp., *Prasiola crispa* and other colony-forming algae. This species-poor community and scarcity of *P. crispa* in the occupied rookery soils might be attributed to the allelopathic substances.

5.6 Environmental Monitoring

There are two protected areas based on the Antarctic Treaty in Dronning Maud Land. One is Yukidori Valley, Langhovde, East Dronning Maud Land, and the other is Svarthamaren, Mühlig-Hofmannfjella, Central Dronning Maud Land. The Yukidori Valley (69°14'00"–69°15'00" S, 39°44'20"– 39°48'00" E) was first designated as a site of special scientific interest (SSSI No. 22) in 1987. At the present time, each site is recognised as one of ASPA (Antarctic specially protected areas) with new management plans (Figs. 5.3, 14.1). The Yukidori Valley has a breeding ground for snow petrels (*Pagodroma nivea*), which are important suppliers of nutrients. Studies on cryptogamic vegetation and microfauna have been carried out over many years by the Japanese Antarctic Research Expedition (JARE; Kanda et al. 1990; Kanda and Inoue 1994; Sugawara et al. 1995) in order to monitor ecosystem changes with respect to impacts of global and local changes in the Antarctic. In the Yukidori Valley, 24 permanent quadrats of moss communities and 23 of lichen communities were set up along the stream during 1987–1989.

Water contents and temperatures of moss cushions have been measured throughout the year (Kanda et al. 1990). Furthermore, the change of moss and lichen colonies occurring in the quadrats has been monitored during a period of about 10 years (Kanda and Inoue 1994). Comparisons are made with vegetation in the East Ongul Island depending on small-scale snowdrifts, water for vegetation beside streams and lakes will be available even in February in the Yukidori Valley. This means that currently and meaningfully temporal and spatial changes in moss vegetation offer opportunities for predicting the variation of terrestrial ecosystems affected by local and global environmental change. Other monitoring studies in the Yukidori Valley focus mainly on microclimatic and limnetic factors (Ohtani et al. 1990).

The other area is Svarthamaren, Mühlig-Hofmannfjella, Central Dronning Maud Land, which was established as SSSI No. 23 in 1987. It stretches from 71°33'17" S, 5°09'12" E in the northwest to 71°55'58" S, 5°15'12" E in the southeast. In West Dronning Maud Land, Thor (1997) established permanent plots with lichens and mosses and recorded 18 species of lichens from Heimefrontfjella and 13 species of lichens from Vestfjella. These recordings will contribute to long-term monitoring of vegetation change.

At Syowa Station, Ohtani et al. (2000) monitored the occurrence of algal species over a 6-year period at ten sites. Twenty-two species of soil algae were isolated. Macroscopic colonies of *Prasiola crispa* occur at the site closest to a urinal tank in Syowa Station. The surface soil 2.5 m from the tank has a high content of total carbon (0.301 %) and total nitrogen (0.016 %). Yamamoto et al. (1991) report that cellulose decomposition by soil microorganisms is much higher (9–25 %) near the tank than at sites 120 m from the station (1.4 %). Soils near the tank had become eutrophic during the 25 years since station construction. *Prasiola crispa* could be used as an indicator of eutrophication by human activities, because it forms green macroscopic foliose colonies on the surface of eutrophicated soils (Ohtani et al. 2000). Soil algal communities remained similar at all monitoring sites over 6 years.

5.7 Human Activity and Long-Range Dispersal

Lindsay (1973) described the introductions of lichens to the sub-Antarctic islands of South Georgia from the Northern Hemisphere by human activities. Broady and Smith (1994) reported human activity to be an effective vector for long-range dispersal of microalgae. They have found several soil algae on vegetables transported from New Zealand to Scott Base and from soils attached to boots of expedition members before departure for Antarctica. Boyd et al. (1966) found high numbers of thermophilic bacteria in soils where contamination from man and other animals has occurred in the Ross Sea region, Antarctica. These studies show both the possibility of long-range dispersal of microorganisms, and environmental changes by human activities in the Antarctic region. Broady and Smith (1994) found ten taxa at Scott Base to be genera unrecorded in the Ross Sea region. They suggest that of these genera, the spread of *Cylindrospermum* and *Eustigmatos* can be monitored relatively easily, due to their distinctive morphological and cytological features, if they become established outside the base environments. However, they found no evidence that any of the potential colonisers have been dispersed away from the base. Broady (1996) lists other distinctive and common taxa of terrestrial algae from more temperate regions which have never been recorded from Antarctica. These include *Porphyridium purpureum*, *Euglena*, the xanthophyte *Botrydium* and the chlorophytes *Protosiphon* and *Trentepohlia*. Ohtani et al. (2000) did not detect any of these algae from the monitoring sites in the vicinity of Syowa Station.

In Dronning Maud Land, there have been no reports of incidental or accidental introduction of vascular plants, so far. However, a flowering plant was found in 1995 in a rock fissure about 5 m north of a small prefab house built for penguin study 20 km south of Syowa Station. The plant was assigned as *Poa* cf. *trivialis* L, a cosmopolitan weedy plant. This report is the first record of a flowering plant growing in the southern Antarctic and the species is the first record in the Antarctic botanical zone (Kanda et al., Chap. 14).

References

Akiyama M (1967) On some Antarctic terrestrial and subterranean algae. Mem Fac Educ Shimane Univ 1:36–56

Akiyama M, Kanda H, Ohyama Y (1986a) Allelopathic effect of penguin excrements and guanos on the growth of Antarctic soil algae. Mem Natl Inst Polar Res Ser E 37:11–16

Akiyama M, Ohyama Y, Kanda H (1986b) Soil nutrient condition related to the distribution of terrestrial algae near Syowa Station (extended abstract). Mem Natl Inst Polar Res Spec Issue 44:198–201

Ayukawa E, Imura S, Kanda H (1998) Soil property and distribution of vegetation in the Yukidori Valley, Antarctica. Bryol Res 7:109–115 (in Japanese with English summary)

Boyd WL (1967) Ecology and physiology of soil microorganisms in polar region. JARE Sci Rep, Proc Symp Pacific Antarct Sci, Spec Iss 1:265–275

Boyd WL, Staley J T, Boyd J W (1966) Ecology of soil microorganisms of Antarctica. Antarct soils and soil forming processes. antarct Res 8:125–129

Boyd WL, Rothenberg I, Boyd JW (1970) Soil microorganisms at Paradise Harbour, Antarctica. Ecology 51:1040–1045

Broady PA (1986) Ecology and taxonomy of the terrestrial algae of the Vestfold Hills. In: Pickard J (ed) Antarctic oasis, terrestrial environment and history of the Vestfold Hills. Academic Press, New York, pp 165–202

Broady PA (1996) Diversity, distribution and dispersal of Antarctic terrestrial algae. Biodiv Conserv 5:1307–1335

Broady PA, Smith RA (1994) A preliminary investigation of the diversity, survivability and dispersal of algae introduced into Antarctica by human activity. Proc NIPR Symp Polar Biol 7:185–197

Campbell IB, Claridge GGC (1966) A sequence of soils from a penguin rookery, Inexpressible Island, Antarctica. N Z J Sci 9:361–72

Cameron RE (1969) Abundance of microflora in soils of desert region. JPL Tech Rep 32-1378:1-16
El-Sayed SZ, Fryxell GA (1993) Phytoplankton. In: Friedmann EI (ed) Antarctic microbiology. Wiley-Liss, New York, pp 65-122
Fogg GE, Stewart WDP (1968) In situ determinations of biological nitrogen fixation in Antarctica. Br Antarct Surv Bull 15:39-46
Holdgate MW (1964) Terrestrial ecology in the maritime Antarctic. In: Carrick R, Holdgate MW, Prevost J (eds) Biologie antarctique. Proc 1st SCAR Symposium, Paris, 1962. Hermann, Paris, pp 181-194
Horne AJ (1972) The ecology of nitrogen fixation on Signy Island, South Orkney Islands. Br Antarct Surv Bull 27:1-18
Hoshiai T, Matsuda T (1979) Adélie penguin rookeries in the Lützow-Holm Bay area and relation of rookery to algal biomass in soil. Mem Natl Inst Polar Res Spec Issue 11:140-152
Howard-Williams C, Pridmore RD, Broady PA, Vincent WF (1990). Environmental and biological variability in the McMurdo Ice Shelf ecosystem. In: Kerry KR, Hempel G (eds) Antarctic ecosystems: ecological change and conservation. Springer, Berlin Heidelberg New York, pp 23-31
Ino Y, Nakatsubo T (1986) Distribution of carbon, nitrogen and phosphorus in a moss community-soil system developed on a cold desert in Antarctica. Ecol Res 1:59-69
Inoue M (1991) Ecological notes on the differences in flora and habitat of lichens between the Syowa Station area in continental Antarctic and King George Island in maritime Antarctic. Proc NIPR Symp Polar Biol 4:91-106
Inoue M (1989) Factors influencing the existence of lichens in the ice-free areas near Syowa Station, East Antarctica. Proc NIPR Symp Polar Biol 2:167-180
Kanda H (1981) Flora and vegetation of mosses in ice-free areas of Sôya Coast and Prince Olav Coast, East Antarctica. Hikobia (Suppl 1):91-100
Kanda H (1986) Seasonal changes of some environmental factors around the moss vegetation near Syowa Station, East Antarctica. Mem Natl Inst Polar Res Ser E 37:17-26
Kanda H, Komárková V (1997) Antarctic terrestrial ecosystems. In: Wielgolaski FE (ed) Polar and alpine tundra. Ecosystems of the world, vol 3. Elsevier, Amsterdam, pp 721-761
Kanda H, Inoue M (1994) Ecological monitoring of moss and lichen vegetation in the Syowa Station region, Antarctica. Proc NIPR Symp Polar Biol 7:221-231
Kanda H, Inoue M, Mochida Y, Sugawara H, Ino Y, Ohtani S, Ohyama Y (1990) Biological studies on ecosystems in the Yukidori Valley, Langhovde, East Antarctica. Antarct Rec 34:76-93
Korotkevich ES (1967) Polar desert. Inf Bjull Sov Antarkt Eksped 65:5-29
Lindsay DC (1973) Probable introductions of lichens to South Georgia. Br Antarct Surv Bull 33-34:169-172
Longton RE (1988) Biology of polar bryophytes and lichens. Cambridge University Press, Cambridge
Matsuda M (1968) Ecological study of the moss community and microorganisms in the vicinity of Syowa Station, Antarctica. JARE Sci Rep Ser E (Biol) 29:1-58
Nakanishi M (1977) Ecological studies of the moss and lichen communities in the ice-free areas near Syowa Station, Antarctica. Antarct Rec 59:68-96
Nakatsubo T, Ino Y (1986) Nitrogen cycling in an Antarctic ecosystem. 1. Biological nitrogen fixation in the vicinity of Syowa Station. Mem Natl Inst Polar Res Ser E 37:1-10
Nakatsubo T, Ino Y (1987) Nitrogen cycling in an Antarctic ecosystem, 2. Estimation of the amount of nitrogen fixation in a moss community on East Ongul Island. Ecol Res 2:31-40

Ohtani S, Kanda H, Ino Y (1990) Microclimate data measured at the Yukidori Valley, Langhovde, Antarctica in 1988–1989. JARE Data Rep 152 (Terr Biol 1), 216p

Ohtani S, Akiyama M, Kanda H (1991) Analysis of Antarctic soil algae by the direct observation using the contact slide method. Nankyoku Shiry (Antarct Rec) 35:285–295

Ohtani S, Suyama K, Yamamoto H, Aridomi Y, Itoh R, Fukuoka Y (2000) Distribution of soil algae at the monitoring sites in the vicinity of Syowa Station between austral summers of 1992/1993 and 1997/1998. Polar Biosci 13:113–132

Pandey KD, Kashyap AK, Gupta RK (1992) Nitrogen fixation by cyanobacteria associated with moss communities in Schirmacher Oasis, Antarctica. Isr J Bot 41:187–198

Pickard J (1986) Spatial relations of the vegetation of the Vestfold Hills. In: Pickard J (ed) Antarctic oasis. Terrestrial environments and history of the Vestfold Hills. Academic Press, Sydney, pp 275–308

Pietr SJ, Tatur A, Myrcha A (1983) Mineralization of penguin excrements in the Admiralty Bay region (King George Island, South Shetland Islands, Antarctica). Pol Polar Res 4:97–112

Rudolph ED (1967) Climate. In: Bushnell VC (ed) Terrestrial life of Antarctica. Antarctic Map Folio Series, No 5, American Geographical Society, New York, pp 5–7

SCAR (2000) Measures, Decisions and Resolutions adopted at the XXIII Antarctic Treaty Consultative Meeting, Lima, Peru, 24 May –4 June 1999, SCAR Bull 136:1–5

Seppelt RD (1986) Bryophytes of the Vestfold Hills. In: Pickard J (ed) Antarctic oasis. Terrestrial environments and history of the Vestfold Hills. Academic Press, Sydney, pp 221–246

Sieburth JMcN (1960) Acrylic acid, and 'antibiotic' principle in phaeocystis blooms in Antarctic waters. Science 132:679–677

Sieburth JMcN (1961) Antibiotic properties of acrylic acid, a factor in the gastrointestinal antibiosis of polar marine animals. J Bact 82:72–79

Sieburth JMcN (1963) Bacterial habitats in the Antarctic environment. In: Oppenheimer CH (ed) Symposium on marine microbiology. Springfield, pp 533–548

Skottsberg C (1904) On the zonal distribution of the South Atlantic and Antarctic vegetation. Geogr J 24:655–663

Sugawara H, Ohyama Y, Higashi S (1995) Distribution and temperature tolerance of the Antarctic free-living mite *Antarcticola meyeri* (Acari, Cryptostigmata). Polar Biol 15:1–8

Thor G (1997) Establishment of permanent plots with lichens and mosses for monitoring local human impact on environment in Heimefrontfjella and Vestfjella, Dronning Maud Land, Antarctica. Antarct Rec 41(3):652–672

Wace NM (1964) Vascular plants. In: van Mieghem J, van Oye P, Schell J (eds) Biogeography and ecology in Antarctica. Monographiae Biologicae 15. Junk, The Hague

Weyant WS (1966) The Antarctic climate. In: Tedrow JCF (ed) Antarctic soils and soil forming processes. Antarctic Research Series, vol 8. American Geophysical Union, Washington, DC, pp 47–59

Yamamoto H, Ohtani S, Tatsuyama K, Akiyama M (1991) Preliminary report on cellulolytic activity in the Antarctic region (extended abstract). Proc NIPR Symp Polar Biol 4:179–182

Yamanaka M, Sato K (1977) Distribution of terrestrial plant communities near Syowa Station in Antarctica, with special reference to water supply and soil property. Nankyoku Shiry (Antarct Rec) 59:54–67 (in Japanese with English summary)

Part II

Soils, Soil Development and Soil Properties

Preface

A first summary of descriptions of soil formation theories was developed by Jenny (1941). The last edition of his book, published in Volume 37 of Ecological Studies (Jenny 1986), contains the 'CLORPT equation' (soil formation=f (cl, ø, r, p, t), where cl is climate, ø is biotic factor, r is topography, p is parent materials and t is time). This general figure is still valid for our current understanding of soil development, and only the human factor needs to be added (Amundsen et al. 1994). For Antarctica, it has been suggested for a long time that climate and climatic effects (cl-factor) are basic constraints for soil formation. This is most evident for the polar desert regions with high aridity (Priscu 1998) where chemical and biological factors are widely hampered by low temperature. Here, soil formation is primarily driven by physical processes, a fact to which Campbell and Claridge (1987) focus much attention when reviewing studies on Antarctic soils. The Antarctic has only a very limited extent of soils, which are at a low level of development, and they are only poorly described, hence, their taxonomy is also fairly vague. Tedrow (1977) was the first who tried to put them into a soil classification system, and this approach is still under discussion.

This chapter is mainly a monograph on the available soil information on three typical Antarctic oases. It excludes the Dry Valley region, which is not regarded as an oasis due to its aridity and therefore has a desert ecosystem character (see definitions in Chap. 1). This opinion contradicts the current literature (e.g. Tedrow 1977; Campbell and Claridge 1987; Priscu 1998). The chapter focuses on soil ecological properties within the framework of interactions between plants–soil organisms–soils–microclimate and attempts to show that the terrestrial environments in Antarctica are not fully characterised by the current knowledge on the Dry Valley ecosystems. Both the biological (ø factor) and chemical processes are much more comparable to lower latitudes as suggested earlier. The intensity of chemical weathering depends on the parent materials (p-factor), the topography (r) as well as the time (t). In this chapter we would like to show that the whole CLORPT equation in its full breadth is valid for soils of Antarctic oases.

Recently, several attempts have been made to combine traditional and newly elaborated concepts for Antarctic soils. The papers of Bockheim and

Ugolini (1990), Bockheim (1997) and Beyer et al. (1999) provide such information. The summary provides, for the first time, detailed information on soil formation for the soils in the Antarctic oases.

The locations in our presentation need special attention because of their coastal positions. During the glacier retreat after the last glaciation, they were lifted to considerable heights. Our own observations of spicules from marine sponges at our sites close to Casey Station underline this with biological evidence (Chap. 8). Wide areas of coastal regions are Pleistocene sediments, e.g. the areas of the Vestfold Hills (Pickard 1986). Uncertainty remains about the extent of the former glacial ice sheet which is not well dated, e.g. whether it just reached the present coastline. For two regions described in this part, the Windmill Islands as well as Admiralty Bay region at King George Island, terraces and moraines can be seen, but precise dates are still pending. Hence, important data for judging soil development must be obtained.

In contrast to old soils of the Dry Valley regions, which can be dated back to 10–15 million years (Claridge and Campbell 1985), the story of the sites mentioned here is much shorter and much more variable due to biological influences. Chapter 6 focuses on the most recent glacial retreat and its consequences. The actual diversity of soils, their extension and patchiness are presented by the examples of the Windmill Islands and King George Island (Chap. 7). Connections with harsh weather conditions, mainly with freeze-thaw cycles, energy balances and the impacts of chemical soil weathering are given in Chapters 8 and 9.

The tundra-like soils as well as regions with considerable peat in the maritime Antarctic have led to areas with high stocks of below-ground organic matter. Although there are areas in the continental regions also showing such patterns, they are much more restricted to small local phenomena. Nevertheless, soil-forming processes even at high levels, e.g., podzolisation, have been observed. This serves the hypothesis that the general processes of soil formation in Antarctica are similar to those of other areas of the world, but just at (smaller) local scales and need longer time scales to form visible structures. Chapter 10 focuses on the organic matter as structural matter of soils and as an indicator for biological processes. With Chapter 11, this part of the book focuses on the biological factor (ø) of the CLORPT equation, knowing full well that the effect of penguins is quite uncommon for soil systems out of Antarctica.

A prominent factor for structuring effects in coastal areas are seals and birds, i.e. their resting or breeding places. The special type of ornithogenic soils, first noted by Syreochkovsky (1969) and then generally outlined by Tedrow and Ugolini (1966), is a special feature not only of this region, and can be compared best with the South American guano fields, or even to dump fields in former agricultural regions. The unstable nature of penguin rookeries has influenced wide regions as it can be seen from typical soil layers, e.g. at King George Island (Chap. 11). Influences are mirrored by soil and vegetation patterns to different extents and these document a history of landscape development.

It is often a competition between fox and hare when discussing the point of interaction between soils and plants and which of these two preconditions the other. In the Antarctic region we can learn that both plant colonisation and soil formation act in close relationship – when the input of microbes, i.e. bacteria, cyanobacteria and algae (ø factor), is considered. Indicators for this are changing pH values with distance from glacier fronts, the coastline, or rookeries. Factors like relief (r-factor) and parent materials (p-factor), which basically means the effective heat input, water supply from snow melt or occasional paddy areas, show important constraints on the processes leading to local soil patterns rather than the general environmental condition of permafrost. Thus, the availability of water and its local action become a prominent feature of soil pattern structure and the main reason for wide ranges in parameters used to describe its condition in physical, chemical or biological terms.

References

Amundsen R, Harden J, Singer (1994): Factors of soil formation: a fiftieth anniversary retrospective. SSSA Special Publication No 33. Soil Science Society of America, Madison, Wisconsin

Beyer L, Bockheim JG, Campbell IG, Claridge GGC (1999) Genesis, properties and sensitivity of Antarctic Gelisols. Antarct Sci 11:387–398

Bockheim JG (1997) Properties and classification of cold desert soils from Antarctica. Soil Sci Soc Am J 61:224–231

Bockheim JG, Ugolini FC (1990) A review of pedogenic zonation in well-drained soils of the southern circumpolar region. Quart Res 34:47–66

Campbell IB, Claridge GGC (1987) Antarctica: soils, weathering processes and environment. Elsevier, Amsterdam

Claridge GGC, Campbell IB (1985) Physical geography – soils. In: Bonner WN, Walton DWH (eds) Key environments: Antarctica. Pergamon Press, Oxford, pp 62–72

Jenny H (1941) Factors of soil formation. McGraw-Hill, New York

Jenny H (1986) The soil resource. Ecological studies 37. Springer, Berlin Heidelberg New York

Priscu JC (1998) Ecosystem dynamics in a polar desert. The McMurdo Dry Valleys, Antarctica. Antarctic Research Series 72, American Geophysical Union, Washington, pp 1–369

Syroechkovsky EE (1959) The role of animals in the formation of primary soils under the condition of circumpolar regions of the earth (Antarctica). Zool Zhurn 38:1770–1775

Tedrow JCF (1977) Soils of the polar landscapes. Rutgers Univ Press, New Brunswick

Tedrow JCF, Ugolini FC (1966) Antarctic soils. In: Tedrow JCF (ed) Antarctic soils and soil forming processes. American Geophysical Union. Antarct Res Ser 8:161–177

Pickard J (1986) The Vestfold Hills: a window on Antarctica. In: Pickard J (ed) Antarctic oasis. Terrestrial environments and history of the Vestfold Hills. Academic Press, Sidney, pp 333–351

6 Glacial Changes in the Areas of Admiralty Bay and Potter Cove, King George Island, Maritime Antarctica

M. BRAUN and H. GOSSMANN

6.1 Climate Change Indications in the Region of the Antarctic Peninsula

The Antarctic Peninsula is a transition zone, where climatic settings vary between a temperate north and a more polar-influenced south and between a polar-maritime west coast and a continental east coast. Due to these particular climatic conditions and its location as a physical barrier to tropospheric circulation, the Antarctic Peninsula represents an area of highest sensitivity to changes in the climatic system. In recent years, a number of indications for climatic changes affecting this region have been detected.

Long-term surface air temperature records from various locations on both sides of the Antarctic Peninsula revealed significant atmospheric warming trends during recent decades (King 1994; Stark 1994; Smith et al. 1996; Harangozo et al. 1997; Skvarca et al. 1998; Rakusa-Suszczewski, Chap. 3). The highest temperature increase was reported for the Marguerite Bay record (0.067 K a^{-1}), whereas only half of this air temperature rise (0.038 K a^{-1}) could be detected in the extended meteorological readings of Bellingshausen Station on King George Island (King 1994). The long-term monthly air temperature trends are strongest in winter. However, during summer months, an increase was also determined, but was less significant. Figure 6.1 shows monthly and yearly mean values as well as the calculated linear trend of the air temperature record of Bellingshausen Station. Moreover, the examination of climatological records from early expeditions to the Peninsula showed temperatures significantly lower than those recorded today (Jones 1990). Jones suggested that the observed warming trends extend back to at least 1900. A general increase in degree-days since the 1960s was detected for the Marguerite Bay Record (Fox and Cooper 1998). They attributed the reduction in the extent of perennial snow cover to this fact. This assumption is supported by findings of Smith et al. (1998) who discovered a lowering of the ice ramp near Rothera Base, the disintegration of the ice ramp on Stonington Island (Splettstoesser 1992) and surface lowering on Moraine Corrie Glacier, Alexander Island (Morris 1999).

Ecological Studies, Vol. 154
L. Beyer and M. Bölter (eds.) Geoecology of Antarctic Ice-Free Coastal Landscapes

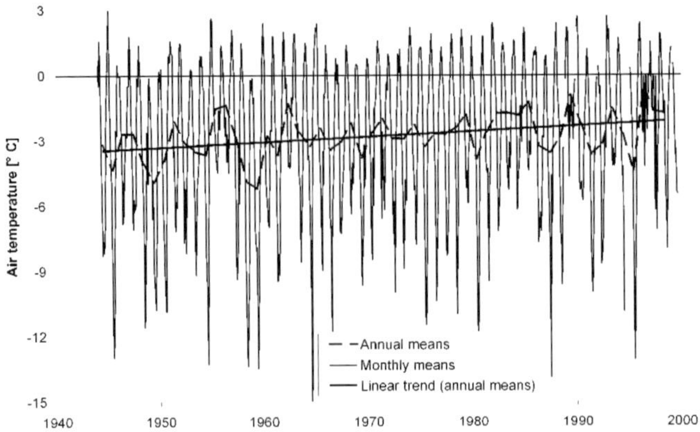

Fig. 6.1. Mean monthly and annual air temperatures (1944–1999) at Bellingshausen Station, King George Island. The linear trend is indicated by the *bold line*. Data source: BAS Jones and Limbert database

Evidence for a link between Antarctic Peninsula air temperature increase, atmospheric circulation and interaction with sea ice was presented by King (1994) and King and Harangozo (1998). Pressure indices and statistical analysis indicated that, especially during winter months, a more frequent advection of warm humid air masses from northerly directions occurred. This is in good agreement with the analysis of precipitation events on the western coastal side of the Antarctic Peninsula (Turner et al. 1997). They argued that a systematic increase in the number of precipitation events at Faraday Station since the 1950s is associated with a change in the depression tracks across the Bellingshausen Sea and with an increase in the number of depressions approaching from outside the Antarctic rather than from the west.

As glacier mass balance is triggered by the prevailing meteorological conditions, it is obvious that changes in the atmospheric circulation will also affect the glacier systems. Oerlemans and Fortuin (1992) suggest that glaciers in maritime climates are more sensitive to climatic changes than glaciers in more continental climates.

Hulbe (1997) pointed out that the enormous disintegration of ice shelves on both sides of the Antarctic Peninsula during the last decades might be a consequence of the weakening of the ice shelves caused by extreme surface melting during several consecutive extraordinary warm ablation seasons in recent years. These spectacular collapses raised public awareness and drew scientific attention to the area (Doake and Vaughan 1991; Doake et al. 1998; Lucchitta and Rosanova 1998; Rott et al. 1996, 1998).

Moreover, several studies reported obvious glacier retreat along the Antarctic Peninsula and the South Shetland Islands (e.g., Wunderle 1996; Calvet et al. 1992). For King George Island itself, substantial changes of glacial front positions have been detected since the 1950s. Wunderle (1996), Muser (1995) and Park et al. (1998) analysed aerial photography, historic maps and SPOT satellite images for the regions of Marian Cove, Potter Cove and Admiralty Bay. Simões et al. (1999) presented a study on glacier retreat for the first time for the entire island between 1956 and 1988/1995. They obtained significantly higher retreat rates for the southern than for the northern glacier margins. This seems to coincide with the predicted reaction of the ice cap to a 1-K air temperature increase (Knap et al. 1996; Oerlemanns et al. 1998). In the maritime climate of King George Island, summer ablation is a considerable mass balance parameter, as has been demonstrated by Bintanja (1995), Braun and Schneider (2000) and Braun et al. (2001a). However, during winter months, the snow cover can also almost completely disappear due to advection of warm humid air masses from northerly directions (Rachlewicz 1997).

Changes in the glacial system affect the nearby land ecosystems in many ways. Recently, ice-free areas on land have enabled plant succession and have formed new habitats for animals. Increased snow and ice melt induced by rising air temperatures results in increased freshwater input and sediment plumes in the adjacent fjords and bays. This chapter focuses on changes in ice front positions since 1956 and possible links between glacier retreat and changes in mass balance parameters due to climate changes.

6.2 Satellite Images and Historical Maps

Remote sensing enables the detection of environmental changes over large and difficult-to-reach areas as well as at times when collection of field data is impossible. The monitoring of glacier front positions over several decades is a good example of the usefulness of remote sensing data around the Antarctic Peninsula.

For the area of the South Shetland Islands, available remote sensing data span several sensor types with different nominal ground resolutions. A first extended aerial survey of the study area was undertaken by the Falkland Island Dependency Aerial Survey Expedition (FIDASE) in December 1956. In the following decades, additional aerial surveys from various nations covered selected areas on the island. However, their availability is very limited. Satellite images from optical sensors are available from LANDSAT and SPOT satellites with up to 10-m ground resolution. Although imagery from SAR satellites would be favourable for the region due to their weather independence, they are less valuable for the determination of glacier retreat. Side-looking active radar systems lead to geometric distortions. This can only be overcome by using a high-resolution digital elevation model for relief correction, which are lacking in Antarctica.

Maps of King George Island date back to the discovery of the islands in 1819, and the first sealing activities (Rakusa-Suszczewski, Chap. 3). However, their reliability on exact coastlines and glacier front positions is very questionable.

The data used for this study are summarised in Table 6.1. Since most known glacier retreat rates in the region of the South Shetland Islands are in

Table 6.1. Summary of characteristics of maps, aerial photography and satellite imagery used for the present study

Data type	Acquisition date	Scale/nominal ground resolution
FIDASE aerial photography	20/12/1956	1:28,000
Topographic map Admiralty Bay	1990, ice front positions based on aerial photography from 1978/1979	1:65,000 approximately
SPOT panchromatic satellite image	19/02/1988	10 m
SPOT panchromatic satellite image	17/11/1992	10 m
SPOT XS multispectral satellite image	26/11/1994	20 m
SPOT XS multispectral satellite image	29/03/1995	20 m
SPOT XS multispectral satellite image	23/02/2000	20 m

the order of 0.01–0.5 km^2 per decade, a ground resolution of at least 30 m is desirable; this should be better or at least in the same range as the expected retreat. The accuracy of the determination of glacier front positions is influenced by snow cover extent, seasonal oscillations of glacier termini and the process of co-registration of the imagery. Errors at the single pixel level can also result from the mixing of the reflected signal from the snow or glacier with the adjacent rock or debris cover. Retreat areas and retreat rates for the periods 1956–1979, 1979–1988 and 1988–1992 were taken from Muser (1995). Additionally, glacier retreat for the period 1992–1995 was analysed for selected areas.

6.3 Glacier Retreat in Admiralty Bay and Potter Cove

Two major glacier types can be distinguished in Admiralty Bay: glaciers terminating on land and tidewater glaciers, whose glacier fronts terminate in the sea. As previous extensive glaciological studies have shown, most parts of the King George Island ice cap can be regarded as temperate glaciers (e.g. Wen et al. 1998; Macheret and Moskalevsky 1999). According to Warren (1992), the terminus of temperate tidewater glaciers does not float.

Figure 6.2 shows the glacial basins draining into Admiralty Bay as delineated by Simões et al. (1999). The major retreat areas between 1956 and 1995 have been highlighted in white. In Table 6.2 the drainage basin characteristics, as obtained from a digital elevation model (Braun et al., submitted), and glacier retreat rates between the different acquisition dates (Muser 1995) are sum-

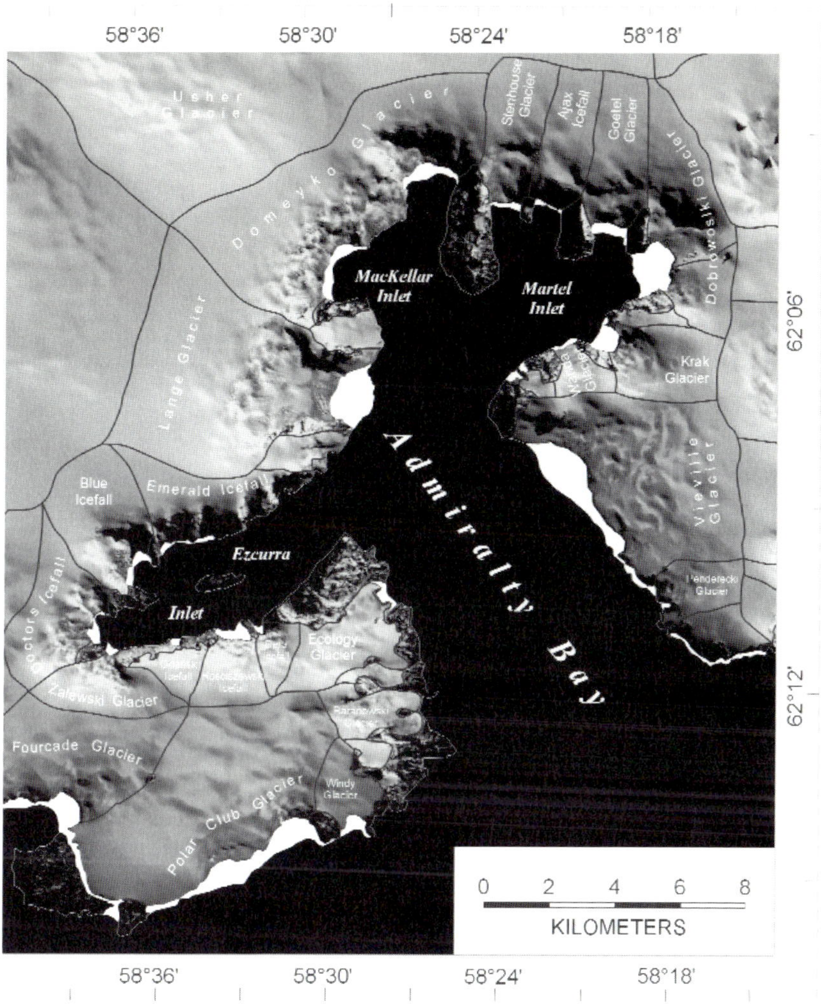

Fig. 6.2. Major areas of glacier retreat in Admiralty Bay and Potter Cove. Indicated by the *white areas* are ice front retreats between 1956 and 1995. The delineation of glacial drainage areas (*black solid lines*) was taken from Simões et al. (1999). All glaciers mentioned in the text are labelled

marised. Over the entire observation period all glaciers in Admiralty Bay prove area loss, however, large differences between the various drainage basins can be found. Highest retreat rates were measured in the first two periods, followed by variable retreat and advance in the period 1988 to 1992. Regarding the absolute retreat values in Table 6.2, the seasonal oscillation of the glacier front position has to be considered. Some of the small glacier advances during the observation period may be attributed to that fact.

Table 6.2. Characteristics of glacier drainage basins and ice front retreat rates for the major glaciers in the area of Admiralty Bay, King George Island. Basin characteristics were obtained from a digital terrain model (Braun et al., 2001b). Retreat rates for the period 1956–1992 were taken from Muser (1995). Negative values denote a glacier advance, positive values glacier retreat. *T* denotes tidewater glacier, *L* denotes glacier terminating on land, *n.d.*: not detected

	Drainage basin characteristics					Glacier retreat rates [km^2]					Ratio retreat 1956–1992/1995 to area (%)
Glacier name	Type of glacier	Area (km^2)	Mean altitude (m)	Maximum altitude (m)	Minimum altitude (m)	1956–1979	1979–1988	1988–1992	1992–1995	1956–1992/95	
Ajax Icefall	T	6.0	443	583	0	–0.09	0.07	0.04	0.06	0.02	0.3
Baranowski Glacier	L	3.7	195	366	2	–0.06	0.10	0.05	n.d.	0.09	2.4
Dera Icefall	L	2.0	287	455	51	0.02	0.04	0.00	n.d.	0.06	3.0
Dobrowolski Glacier	T	13.2	343	402	0	0.56	0.70	–0.25	0.21	1.22	9.2
Doctors Icefall	T	8.7	289	620	0	0.02	0.10	0.01	n.d.	0.13	1.5
Domeyko Glacier	T	36.1	359	627	0	–0.43	0.60	0.64	0.53	1.34	3.7
Dragon Glacier	L	0.5	212	305	69	0.04	0.17	–0.03	n.d.	0.18	36.0
Ecology Glacier	L	6.3	210	450	0	0.08	0.19	0.10	n.d.	0.37	5.9
Emerald Icefall	T	17.9	395	679	0	0.14	0.37	–0.10	n.d.	0.41	2.3
Gdansk Icefall	L	2.0	324	455	135	0.03	0.03	0.00	n.d.	0.06	3.0
Goetel Glacier	T	7.3	425	583	0	0.04	0.04	0.11	0.09	0.28	3.8
Krak Glacier	T	7.0	324	410	0	0.80	0.50	0.07	0.09	1.46	20.9
Lange Glacier	T	31.7	442	677	0	1.19	0.57	0.02	0.11	1.89	6.0
Pandereski Glacier	T	4.0	234	357	0	1.03	0.18	–0.08	n.d.	1.13	28.3
Rosciszewski Icefall	L	3.8	319	458	0	0.16	0.06	0.00	n.d.	0.22	5.8
Stenhouse Glacier	T	8.2	423	573	0	–0.10	0.09	0.05	0.09	0.13	1.6
Vieville Glacier	T	24.7	260	409	0	0.64	0.81	0.58	0.38	2.41	9.8
Wanda Glacier	L	2.0	236	361	0	0.11	0.15	0.05	n.d.	0.31	15.5
Zalewski Glacier	T	5.7	309	453	8	0.01	0.01	0.03	n.d.	0.05	0.9
Polar Club Glacier	T	28.9	243	460	0	2.29	2.29	0.61	n.d.	2.90	10.0
Fourcade Glacier	T	18.0	261	450	0	0.55	0.55	0.16	n.d.	0.71	3.9
Sum						11.73	2.06	1.56	15.35		

6.4 Tidewater Glaciers

Obviously, not all tidewater glaciers in Admiralty Bay behave similarly. Some huge tidewater glaciers, such as the Polar Club Glacier, Lange Glacier, Dobrowolski Glacier or Vieville Glacier showed remarkable retreat, whereas smaller tidewater glaciers almost did not change their glacier front positions. Notable differences in the drainage basin areas can be found. The mean altitudes range between 195 and 443 m a.s.l. Highest values of the retreat rates are encountered in the first two periods (1956–1979 and 1979–1988). This is in good agreement with the observation by Kejna et al. (1998). Over the entire

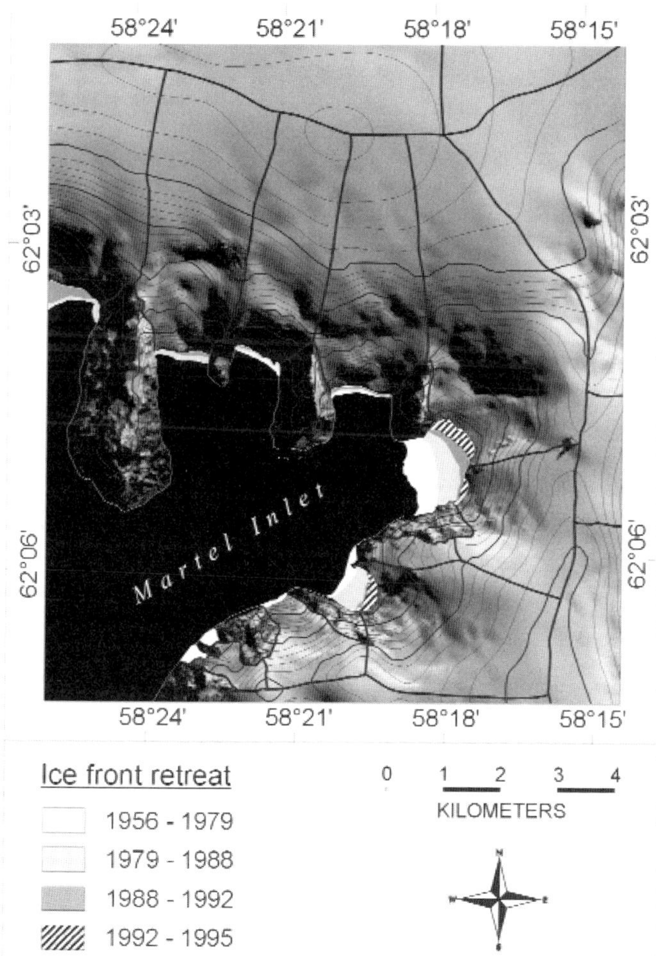

Fig. 6.3. Subset of Martel Inlet with different stages of ice front positions and drainage basins (*black solid lines*). The contour interval is 20 m

observation period, the largest glacier area was lost by Polar Club Glacier (2.9 km²), followed by Vieville (2.0 km²) and Lange Glacier (1.7 km²). Glaciers with highest mass loss also showed the largest drainage basin areas. A first idea about reasons for such glacier changes would be climate-induced changes in mass balance parameters. Normally, the accumulation area ratio (AAR) would be a proper measure for this purpose, but it is not available for the glaciers on King George Island. As a first approach, glaciers with different altitudinal characteristics should be affected differently. However, no obvious differences between strongly and less retreating glaciers can be found in the altitudinal range (Table 6.2). Lange Glacier shows highest elevations and a high mean altitude (442 m a.s.l.), whereas Polar Club Glacier reaches only a maximum altitude of 460 m and a mean altitude of 243 m a.s.l. The smaller tidewater glaciers, e.g. Ajax Icefall, Doctors Icefall or Zalewski Glacier, show a small retreat/area ratio and are located in different parts of the ice cap. In a subset of Martel Inlet, these differences are illustrated (Fig. 6.3). Krak and Dobrowolski Glacier retreated constantly since 1956 with the largest area loss in the first period. However, in the 1990s, significant ice cliff retreat could still be observed.

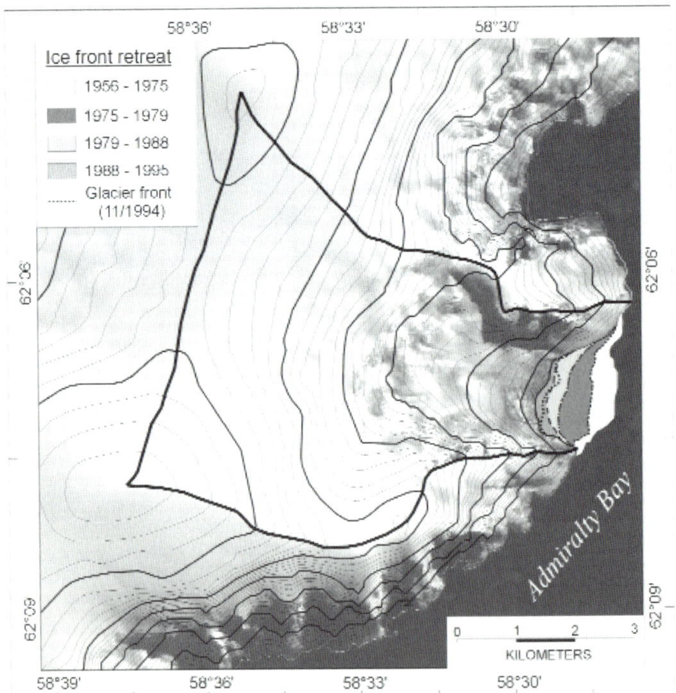

Fig. 6.4. The drainage basin of Lange Glacier – one of the biggest outlet glaciers on the southern side of the King George Island ice cap. Ice front positions are indicated by different line styles. The basin boundary is indicated by a *solid black line*. The contour interval is 20 m

The different retreat phases and difficulties in determination of glacier front position due to seasonal changes will be outlined in the example of Lange Glacier (Fig. 6.4). A first phase of retreat (440 m) was recorded for the period between 1956 and 1975. Even more striking was the retreat in the short time period between 1975 and 1979. Together, these two periods led to an ice cliff retreat of about 880 m, i.e. a loss of approximately 1.1 km^2. This process continued between 1979 and 1988, when the glacier retreated an additional 280 m, thus losing another 0.56 km^2. Between 1988 and 1995 the glacier retreated a further 180 m, i.e. it lost an area of 0.13 km^2. In total, the glacier retreated about 1380 m from 1956 to 1995, thus losing about 1.7 km^2.

It should be noted that a comparison of the satellite images of February 1988 (late summer) and November 1994 (spring) does not indicate a significant glacier retreat for this period. However, one has to be cautious when interpreting these data as a steady state or advance of the glacier terminus. When comparing ice front states of Stenhouse Glacier during 1957/1958, a strong oscillation of the glacier terminus position, depending on the season, could be observed (Noble 1965). In February/March 1957 massive calving led to ice front retreat, while during the wintertime the glacier terminus advanced 150 m. This was followed again by a retreat of the terminus in the summer months. Keeping this yearly cycle in mind, the acquisition date of aerial and satellite images becomes crucial when small changes in glacier extent are monitored.

6.5 Glaciers Terminating on Land

Regarding the retreat area, glaciers terminating on land show the smallest absolute values. However, considering the ratio between retreat area and basin area, they exhibit the largest percentages, e.g. Dragon Glacier or Wanda Glacier. These glaciers show the lowest altitudinal extents and mean altitudes. Therefore, it has to be expected that the AAR is smallest for this glacier type in the Admiralty Bay area.

Since ice front changes are very small, accuracy in determining glacier borders has to be considered. Figure 6.5 illustrates the appearance of the different imagery types, based on a subset of the Arctowski area. The FIDASE aerial photography from December 1956 shows an extended snow cover. The SPOT panchromatic scene from 19 February 1988 demonstrates optimal conditions for the determination of glacier borders and ablation areas, whereas in the scene from 18 November 1992, the almost worst-case conditions occurred, with extended snow cover over the ice-free areas as well. The inner part of Ezcurra Inlet is still covered with winter sea ice. The high reflectance of the snow cover prevents the differentiation of topographic features (e.g. ice divides) as can be distinguished in the image of 29 March 1995. Although this

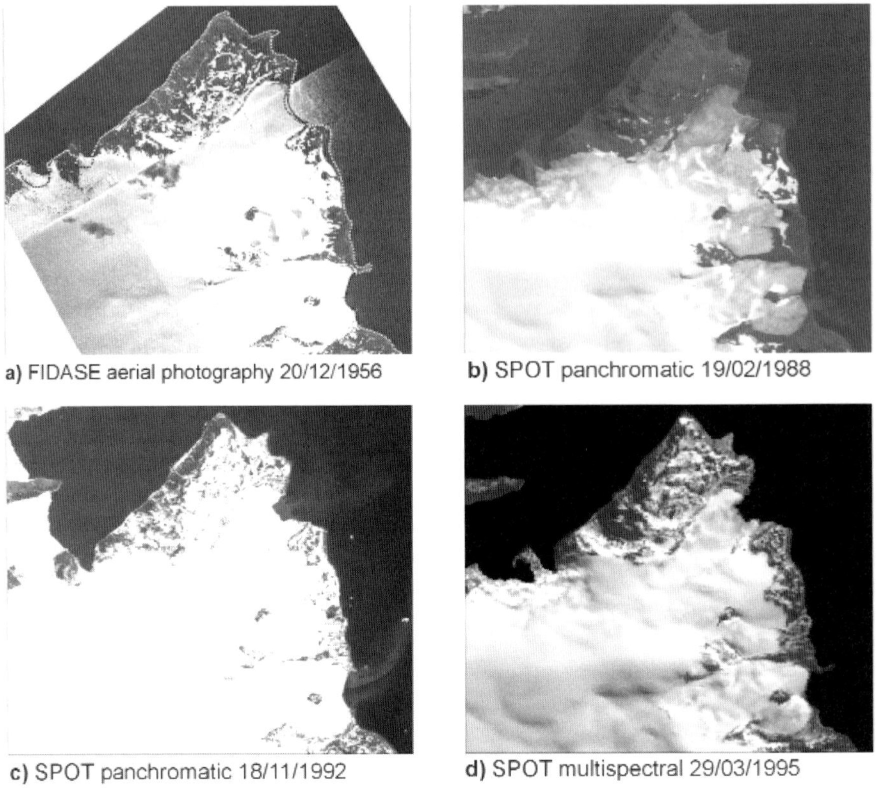

Fig. 6.5a–d. The north-eastern part of Warszawa Icefield near Arctowski Station as visible via different imagery. In **a** the *dotted line* indicates coastline and glacier front positions from 1979. The inner part of Ezcurra Inlet is still covered by winter sea ice (**c**). Glacier retreat is most notable at Ecology Glacier and Dera and Rosciszewski icefalls. Note the meltwater-induced sediment plumes in front of Ecology and Baranowski Glacier in **b** and **c** as visible by increased reflectance

last image exhibits a thin snow cover of freshly fallen snow, the glacier borders can still be discriminated. The lower ground resolution of the multispectral scene is also remarkable since single pixels appear to be visible.

The estimation of an absolute error in ice front positioning is difficult since there are various influencing factors as stated above. Muser (1995) used residual images to determine glacial changes. Using this technique, snow cover on ice-free areas can lead to errors. However, as he based his work on several panchromatic scenes from the year 1992, he could minimise these errors. Nevertheless, the absolute retreat values have to be treated with some caution. Ecology Glacier showed a retreat of about 0.37 km^2 between 1956 and 1995. This value comprises mainly the retreat in the central glacier front with a maximum recession of about 270 m between 1956 and 1992.

6.6 Possible Reasons for the Differences in Glacier Retreat of the Tidewater Glaciers in Admiralty Bay

The results presented above indicate that different mechanisms seem to drive glacier retreat in Admiralty Bay. The calving speed and calving flux of tidewater glaciers are controlled by water depth, subglacial runoff, ice velocities, ice thickness and ice cliff height, water temperature, topography, buoyancy forces and accumulated strain (Warren 1992). Calving is suppressed during the winter freeze-up and is highest in autumn. This agrees well with the observations on Lange and Stenhouse Glacier as mentioned above. However, after Warren (1992), the relationship between water depth and calving rate as well as the fjord or outlet geometry is most important for the explanation of the often catastrophic retreat of tidewater glaciers. Greater water depth seems to enforce the calving process. Often, the terminus of many stable tidewater glaciers rests on a final moraine. Once the glacier retreats from this moraine, it ends in deep water and enforced calving has to be expected until stabilisation occurs further to the interior of the drainage basin due to more favourable valley or fjord geometry.

Lange Glacier may be regarded as a good example of this situation. It showed an immense retreat at first and now has an almost stagnant glacier front position in a more protected part of the outlet. This hypothesis is supported by the fact that all glaciers showing strong ice front retreat ended in a shallow inlet or bay. The moraines, on which many of these tidewater glaciers seemed to have been grounded during the 1956 image acquisition, are still visible in recent satellite imagery (e.g. at Vieville Glacier). Re-advancement can only occur when the glacier provides sufficient sediment to lower the water depth in front of its ice cliff (Warren 1992). In front of the small glaciers, such as Ajax Icefall or Emerald Icefall, final moraines are absent. It is thought that bedrock and glacier rise directly above sea level so that the influence of water depth is minimised. This demonstrates that, for tidewater glaciers, the trigger initiating the glacier fluctuation cycle may be climatic, such as a rise in the equilibrium line altitude (ELA), but once the cycle is in progress, internal mechanisms can over-ride external forcing (Warren 1992; Paterson 1994; Benn and Evans 1998). However, the significant retreat of glaciers terminating on land indicates that changes in mass balance parameters have occurred or do still occur. Since ablation is the prevailing mass loss for these glacier types, increasing air temperatures as observed along the Antarctic Peninsula will affect the glacier mass balance considerably.

The low mean altitude of these small glaciers terminating on land, together with their small altitudinal range, favour their sensitivity towards mass balance changes. This has been demonstrated using sensitivity analysis of energy balance models by Bintanja (1995) and Braun et al. (2001a). Indications for a rise in ELA since the first measurements on King George Island are given in

Table 6.3. Summary of equilibrium line altitudes (ELA) in the region of the South Shetland Islands as reported by various authors. *DI* Deception Island, *KGI* King George Island, *LI* Livingston Island

Balance year	Location	ELA [m]	Reference
1968–1969	G1 (DI)	275	Orheim and Govorukha (1982)
1969–1970	G1 (DI)/Bellingshausen Dome (KGI)	290/140	Orheim and Govorukha (1982)
1970–1971	G1 (DI)/Bellingshausen Dome (KGI)	325/170	Orheim and Govorukha (1982)
1971–1972	G1 (DI)/Rotch Ice Dome (LI)	370/140	Orheim and Govorukha (1982)
1972–1973	Stenhouse Glacier (KGI)	<100	Curl (1980)
1972–1973	G1 (DI)/Rotch Ice Dome (LI)	330/170	Orheim and Govorukha (1982)
1973–1974	Stenhouse Glacier (KGI)	<100	Curl (1980)
1973–1974	G1 (DI)/Rotch Ice Dome (LI)	280/150	Orheim and Govorukha (1982)
1985–1986	Bellingshausen Dome (KGI)	150	Jiawen et al. (1995)
1985–1989	Nelson Island	110	Jiawen et al. (1995)
	South Shetland Islands	<100	Clapperton (1990)
1987–1988	King George Island Ice Cap	300–350[a]	Simões et al. (1999)
1991–1992	Bellingshausen Dome (KGI)	140	Wen et al. (1994)
1991–1992	Ecology Glacier (KGI)	100	Bintanja (1995), estimated
1992–1993	Hurd Dome (LI)	235	Vilaplana and Pallos (1994)
1965–1993	Hurd Dome (LI)	ca. 200	Pourchet 1996 (pers. comm.)
1996–1997	King George Island northern side	ca. 250[a]	Braun et al. (2000)

[a] Values signify an approximation of ELA by the transient snow line position revealed by satellite remote sensing.

Table 6.3. ELAs observed during the 1990s reveal considerably higher values than those measured in the 1970s. This is in good agreement with the higher mean air temperatures recorded during this decade. The data on Deception Island cannot be regarded as representative for the entire South Shetland Islands, since the data were recorded shortly after the volcanic eruptions in 1967, 1969 and 1970. Moreover, the island exhibits only a small amount of ice cover. The suggestion that glacier retreat is climatically induced coincides with the observations by Park et al. (1998) for Marian Cove. They concluded from their analysis that the sea ice conditions in Marian Cove did not seem to be responsible for the observed ice front position changes. Both Park et al. (1998) and Kejna et al. (1998) attributed glacier retreat to the observed increasing mean air temperatures in the region.

6.7 Conclusion

Not all glaciers, or even all tidewater glaciers, reacted similarly over the observed time period. Some retreated immensely, others less or even remained stable. Looking at only the most spectacular retreats of tidewater glaciers such as Lange Glacier or Vieville Glacier may over-estimate the actual state of climatically induced mass balance changes. These catastrophic retreats may be attributed to a dynamic adjustment of the glacier tongues to a new stable position. However, due to the simultaneous retreat of land-terminating glaciers, there is strong evidence that glacier retreat on King George Island was initiated by the observed climatic changes. This hypothesis is supported by the comparison of equilibrium line altitudes from various years. Data from recent mass balance years revealed sustained higher equilibrium line altitudes in the region of the South Shetland Islands. Higher ablation rates due to increasing air temperatures were already predicted. This has resulted in an increased mass loss of land-terminating glaciers as well as a destabilisation of the glacier fronts of tidewater glaciers due to meltwater.

Acknowledgements. The funding for this study was provided by the Deutsche Forschungsgemeinschaft within the project King George Island Eis (KIGEIS, contract # SA 694/1-1). The authors would like to thank the Brazilian Antarctic Program as well as the Alfred-Wegener-Institut für Polar- und Meeresforschung for their generous logistic support. Special thanks is extended to our Brazilian colleagues, J.C. Simões, F. Aquino, U.F. Bremer and F. Ferron (all Laboratório de Pesquisas Antárticas e Glaciológicas, Universidade Federal do Rio Grande do Sul), who made Royal Navy aerial photography from 1975 of Lange Glacier available to us. Helpful comments on the manuscript by Frank Rau are kindly acknowledged.

References

Antarctic Place Names Committee Foreign and Commonwealth (1986): APC Misc 64, South Shetland Islands, Sheet 1 King George Island, 1:100 000, 10th edn

Benn PI, Evans DJA (1998) Glaciers and glaciation. Arnold, London

Bintanja R (1995) The local surface energy balance of Ecology Glacier, King George Island. Antarct Sci 7:315–325

Braun M, Schneider C (2000) Characteristics of summertime energy balance on the west coast of the Antarctic Peninsula. Ann Glaciol 31:179–183

Braun M, Rau F, Saurer H, Goßmann H (2000) The development of radar glacier zones on the King George Island ice cap (Antarctica) during the austral summer 1996/97 as observed in ERS-2 SAR data. Ann Glaciol 31:357–363

Braun M, Saurer H, Vogt S, Simões JC, Goßmann H (2001a): The influence of large-scale atmospheric circulation on surface energy balance on the ice cap of King George Island. Intern J Climatol 21:21–36

Braun M, Simões JC, Vogt S, Bremer UF, Blindow N, Pfender M, Saurer H, Aquino FE, Ferron FA (2001b) An improved topographic database for King George Island. Compilation, application and outlook. Antarct Sci 13(1):41–63
British Antarctic Survey (1968): British Antarctic Territory- South Shetland Islands. Cambridge, British Antarctic Survey, D501, Sheets W62 56 and W 62 58, DOS 610, 1:200 000
Calvet J, Corbera J, Furada G (1992) Variacion del frente glaciar en Bahia Sur y Punta Siddons entre 1956 y 1991, Isla Livingston, Islas Shetland del Sur. In: López-Martinez J (ed) Geología de la Antártida Occidental. III Congreso Geológico de Espana y VIII Congreso Latinoamericano de Geología, Salamanca, Espana, pp 283–292
Clapperton CM (1990) Quaternary glaciations in the southern ocean and Antarctic Peninsula area. Quat Sci Rev 9:229–252
Curl JE (1980) A glacial history of the South Shetland Islands, Antarctica. Inst of Polar Studies, Ohio, Report no 63
Doake CSM, Vaughan DG (1991) Rapid disintegration of the Wordie Ice Shelf in response to atmospheric warming. Nature 350:328–330
Doake CSM, Corr HFJ, Rott H, Skvarka P, Young NW (1998) Break-up and conditions for stability of the northern Larsen Ice Shelf, Antarctica. Nature 391:778–780
Fox AJ, Cooper APR (1998) Climate-change indicators from archival aerial photography of the Antarctic Peninsula. Ann Glaciol 27:636–642
Harangozo SA, Colwell SR, King JC (1997) An analysis of a 34-year air temperature record from Fossil Bluff (71°S, 68°W), Antarctica. Antarct Sci 9:355–363
Hulbe CL (1997) Recent changes to Antarctic Peninsula ice shelves: what lessons have been learned? Nat Sci 1(6)
Jiawen R, Dahe Q, Petit JR, Jouzel J, Wenti W, Chen L, Xiaojun W, Songlin Q, Xiaoxiang W (1995) Glaciological studies on Nelson Island, South Shetland Islands, Antarctica. J Glaciol 41:408–412
Jones PD (1990) Antarctic temperatures over the present century – a study of the early expedition record. J Climate 3:1193–1203
Kejna M, Láska K, Caputa Z (1998) Recession of Ecology Glacier (King George Island) in the period 1961–1996. Polish Polar Studies 25th International Symposium, Warszawa, 1998. Institute of Geophysics of the Polish Academy of Sciences, Warszawa, pp 121–128
King JC (1994) Recent climate variability in the vicinity of the Antarctic Peninsula. Intern J Climatol 14:357–369
King JC, Harangozo SA (1998) Climate change in the western Antarctic Peninsula since 1945: observations and possible causes. Ann Glaciol 27:571–575
Knap WH, Oerlemans J, Cadée M (1996) Climate sensitivity of the ice cap of King George Island, South Shetland Islands, Antarctica. Ann Glaciol 23:154–159
Lucchitta BK, Rosanova CE (1998) Retreat of northern margins of George VI and Wilkins Ice Shelves, Antarctic Peninsula. Ann Glaciol 27:41–46
Macheret YuYu, Moskalevsky MYu (1999) Study of Lange Glacier on King George Island, Antarctica. Ann Glaciol 29
Morris EM (1999) Surface ablation rates on Moraine Corrie Glacier, Antarctica. Glob Planet Change 22:221–231
Muser D (1995) Der Gletscherrückzug auf King George Island, South Shetland Islands zwischen 1956 und 1992. Eine digitale Auswertung von Karten, Luftbildern und Satellitendaten. Unpublished thesis, Albert-Ludwigs-Universität Freiburg, 82 pp
Nakladem Instytutu Ekologii (1990): Admiralty Bay, King George Island. Warszawa
Noble HM (1965) Glaciological observations at Admiralty Bay, King George Island, in 1957–1958. Br Antarct Surv Bull 5:1–11
Oerlemans J, Fortuin JPF (1992) Sensitivity of glaciers and small ice caps to greenhouse warming. Science 258:115–117

Oerlemans J, Anderson B, Hubbard A, Huybrechts Ph, Johannesson T, Knap WH, Schmeits M, Stroeven AP, van de Wal RSW, Wallinga J, Zuo Z (1998) Modelling the response of glaciers to climate warming. Climate Dynam 14:267–274

Orheim O, Govorukha LS (1982) Present-day glaciation in the South Shetland Islands. Ann Glaciol 3:233–238

Park B-K, Chang S-K, Yoon HI, Chung H (1998) Recent retreat of ice cliffs, King George Island, South Shetland Islands, Antarctic Peninsula. Ann Glaciol 27:633–635

Patterson WSB (1994) The physics of glaciers, 3rd edn. Elsevier, Amsterdam

Rachlewicz G (1997) Mid-winter thawing in the vicinity of Arctowski Station, King George Island. Pol Polar Res 18:15–24

Rott H, Skvarka P, Nagler T (1996) Rapid collapse of northern Larsen Ice Shelf, Antarct Sci 271:788–792

Rott H, Rack W, Nagler T, Skvarka P (1998) Climatically induced retreat and collapse of northern Larsen Ice Shelf, Antarctic Peninsula. Ann Glaciol 27:86–92

Skvarca P, Rack W, Rott H, Ibarzábal y Donángelo T (1998) Evidence of recent climatic warming on the eastern Antarctic Peninsula. Ann Glaciol 27:628–932

Simões JC, Bremer UF, Aquino FE, Ferron FE (1999) Morphology and variations of glacial drainage basins in the King George Island ice field, Antarctica. Ann Glaciol 29

Smith AM, Vaughan DG, Doake CSM, Johnson AC (1998) Surface lowering of the ice ramp at Rothera Point, Antarctic Peninsula, in response to regional climate change. Ann Glaciol 27:113–118

Smith RC, Stammerjohn SE, Baker KS (1996) Surface air temperature variations in the western Antarctic Peninsula region. Antarct Res Ser 70:105–121

Splettstoesser J (1992) Antarctic global warming? Nature 355:503

Stark P (1994) Climatic warming in the central Antarctic Peninsula area. Weather 49:215–220

Turner J, Colwel SR, Harangozo S (1997) Variability of precipitation over the coastal western Antarctic Peninsula from synoptic observations. J Geophys Res 102 (D12):13999–14007

Vilaplana JM, Pallàs R (1994): Características y evolución del manto nivoso en Isla Livingston. In: Cacho J, Serrat D (ed) Actas del V Simposio Español de Estudios Antárticos. Madrid, Comisión Interministerial de Ciencia y Tecnología, pp 279–290

Warren CR (1992) Iceberg calving and the glaciomarine record. Progr Physic Geogr 16:253–282

Wen J, Kang J, Xie Z, Han J, Lluberas A (1994) Climate, mass balance and glacial changes on small dome of Collins ice cap, King George Island, Antarctica. Antarct Res 5:52–61

Wen J, Kang J, Han J, Xie Z, Liu L, D Wang D (1998) Glaciological studies on King George Island ice cap, South Shetland Islands, Antarctica. Ann Glaciol 27:105–109

Wunderle S (1996) Die Schneedeckendynamik der Antarkische Halbinsel und ihre Erfassung mit aktiven und passiven Fernerkundungsverfahren. Freiburger Geographische Hefte 48, 172 pp

7 Soils and Soilscapes

H.-P. Blume, D. Kuhn and M. Bölter

7.1 Introduction

Despite basic research on soil science in the Antarctic, overall views on this continent and its adjacent islands are rare. Detailed soil descriptions refer mostly to the areas of McMurdo and the Dry Valleys, a few places around research stations or individual, occasional landings (e.g. O'Brien et al. 1979; Leonardi et al. 1987; Blume and Bölter 1993; Campbell et al. 1998; Bölter et al. 1994, 1999, 2000; Ohtani and Kanda, Chap. 5). These data have been used for syntheses and some regionalisations of the Antarctic ice-free regions (Campbell and Claridge 1987; Bockheim and Ugolini 1990; Blume et al. 1997, 1998). Nevertheless, soil maps of representative sites of the Antarctic are rare, and could show intrinsic variabilities of landscapes and soils. Such data become important when attempts are made to explain recent vegetation patterns, or when discussions arise about soil development, or their use as data archives (e.g. Tatur and Myrcha 1993; Fabiszewski and Wojtún 1993). The data for this chapter were mainly obtained during two expeditions to the Windmill Islands (Casey Station, 1991/1992) and King George Island (Arctowski Station, 1994/1995), but data from earlier expeditions to these places are also used. This chapter describes, at a local landscape scale, the distribution of a great variety of soils with the aim of showing soil associations, and presents their summarised properties obtained by standard laboratory methods. Soil classifications based on different schemes are used. This approach also refers to a merging of the terms soils and landscape to soilscape.

7.2 Sites and Analyses

In brief, the parent material of the Windmill Island site (66°17'S, 110°32'E) consists of a layered sequence of gneiss with schist and migmatites, the Windmill Metamorphics (Blight and Oliver 1977). They are partly covered by

Ecological Studies, Vol. 154
L. Beyer and M. Bölter (eds.) Geoecology of Antarctic Ice-Free Coastal Landscapes

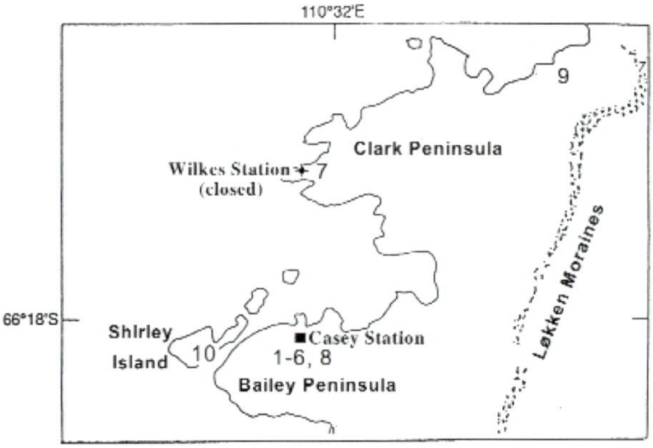

Fig. 7.1. Locations of the investigation sites, Windmill Islands, Continental East Antarctic

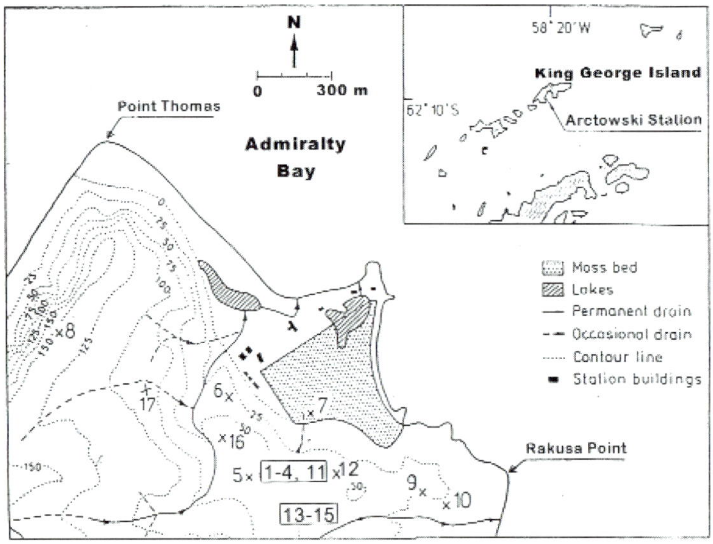

Fig. 7.2. Locations of the investigation sites, King George Island, maritime Antarctic

neoglacial solifluction deposits and moraines. The region lost its ice cover 5500 years B.P. and was lifted during that time. The surveyed area is a 6.4-ha large hill with meltwater lakes in the north and the south, and altitudes between 30 and 45 m a.s.l. (Fig. 7.1). The hill is covered by vegetation above 40 m in the SE direction to 30 m in the NE direction. The hill is free of snow for 2 to 3 months, because the highest part of the hill has its snow blown off into the valleys behind.

The parent material of the region at King George Island site (62°10'S, 58°28'W; Fig. 7.2) is mainly tertiary tholeiite basalt (Birkenmajer 1980) covered by neoglacial (decades to centuries old). There are newly developed moraines and older moraines near the Ecology Glacier which show solifluetion deposits. Volcanic ash from neighbouring islands (mainly Deception Island) was deposited during the Quaternary. The surveyed area of 130 ha is covered with vegetation to an altitude between 70 and 90 m a.s.l., but very young moraines (10 to 50 years) of the Ecology Glacier are also free of vegetation.

Soil descriptions are available from about 200 soil pits from both Windmill Islands and King George Island, by texture, structure and moisture in the field according to Schlichting et al. (1995), and soil colour according to Munsell charts. All pits are located along toposequences at Casey and Arctowski. From each layer small samples are analysed for contents of stones, gravel and water, pH ($CaCl_2$ or KCl,), electrical conductivity ($\mu S\ cm^{-1}$ at 25 °C, 10 g soil in 25 ml water, actual data are transformed to saturation extract). Redox potentials (Eh) were measured in different layers at about 60 locations 8–12 h after the installation of platinum electrodes. Results are transformed to those of normal hydrogen electrode, into rH values, and relate to pH.

Laboratory analyses are available for representative soils according to Schlichting et al. (1995) for:

1. Bulk density (d_b: drying and weighing of 100-ml soil samples);
2. Particle size (sieve and pipette method after H_2O_2 (and partly dithionite/citrate) extraction, and $(NaPO_3)_6$ dispersion);
3. Organic carbon (TOC, dry combustion at 1200 °C and coulometric CO_2 measurements);
4. Total nitrogen (N_t, Kjeldahl digestion, colorimetric measurement of NH_4);
5. Carbonates (H_3PO_4 treatment at 80 °C and CO_2 measurement like TOC, transformation into $CaCO_3$);
6. Pedogenic oxides (dithionite-citrate-bicarbonate extraction of Fe_d; sodium pyrophosphate extract of $Fe_{p)}$;
7. CEC and exchangeable cations (exchange of Ca_a, Mg_a, K_a and Na_a with $BaCl_2$ (after extracting the water-soluble cations, measurement by AAS);
8. Exchangeable H_a and Al_a (pH measurement in Ca-acetate pH 7.2);
9. CEC=H_a (+Al_a)+Ca_a+Mg_a+K_a+Na_a, and calculation of the base saturation intensity (=b.s.);
10. Available or weatherable PO_4 (P_a: extract with NH_4-lactate acetic acid, or P_v:extract with hot HCl after organic matter oxidation at 500 °C, and colorimetric measurement).

7.3 Soil Conditions

7.3.1 Windmill Islands (Casey Area)

All the soils of the ice-free areas of the Windmill Islands have permafrost at depths between 30 and 80 cm. Most soils show desert pavement as well as cryoturbation phenomena like stone circles and tonguing on soil surfaces, together with particle sorting inside (=gelundic phase after FAO 1988;

Fig. 7.3). There are greater skeletal or sand contents in the surface layers and greater clay contents in the subsurface horizons above permafrost (Table 7.1, sites 1–3). However, there are no ice- or sand-filled wedges. These findings are in agreement with Bockheim (1995). Vesicular structure can be observed within the first 1–10 cm of the soils, similar (but deeper) to those of hot deserts, like the Sahara (Blume et al. 1984).

There is no active cryoturbation in sandy soils with vegetation cover (Table 7.1, sites 3–5, 9). As loamy soils beside them (Table 7.1, sites 2, 8) show only covers with short-living algae but no lichens, it can be concluded that cryoturbation prevents growth of slow-growing cryptogams (Blume and Bölter 1996).

Sites with too short periods without snow cover (Table 7.1, site 1), and soils of very young moraines (Table 7.1, site 7) also do not have plant cover. Large boulders and rocky outcrops are covered by various lichens. In co-operation with cryptoendolithic organisms which penetrated rock fissures, they act as moderators for chemical weathering (Chen et al. 2000). Most soils are clearly acidified; pH values are around 4. Only soils of extremely young moraines (<100 years; Table 7.1, site 7), and those of active penguin rookeries (Table 7.1, site 10) show an almost neutral reaction. These data are significantly lower than described for the Ongul Island (Ohtani and Kanda, Chap. 5). Most soils can be classified as mesotrophic according to their base saturation (b.s.: 30–60 %). Soils of young moraines and of active bird rookeries are eutrophic (b.s.: >60), and some of the identified Podzols (Table 7.1, site 9) are dystrophic (b.s: <30).

Organic matter is accumulated in all soils. Those of very young moraines contain only 0.2–0.4 g kg^{-1} total organic matter (TOC, Table 7.1, site 7), soils which are covered with snow for more than 11 months per year have 1–4 g kg^{-1} TOC (Table 7.1, site 1), loamy soils with a strong active cryoturbation and no vegetation cover are therefore 2–9 g kg^{-1} TOC (Table 7. 1, sites 2, 8), such data are in comparison to those from Ongul Island (Ohtani and Kanda, Chap. 5). Podzolised soils with vegetation cover contain 5–40 g kg^{-1} TOC (Table 7.1, sites 3, 4, 9), and peaty soils as well as those of recent rookeries even 160–300 g kg^{-1} TOC (sites 5, 6, 10). There are no pronounced differences between surface and subsurface soils, due to strong cryoturbation of loamy soils and a strong podzolisation of sandy to gravely soils (Blume et al. 1997). Some soils have brown-coloured Bw horizons (Table 7.1, site 8; see also sites D2 and B11, in Beyer and Bölter 1999). Chemical weathering together with formation of iron oxides (see Fe_{d-p} contents, Table 7.1) and clay minerals (Blume et al. 1997) have taken place concomitantly with strong physical

Fig. 7.3. Periglacial phenomena of the Windmill Islands, Antarctic. **a** Stones circle on a moraine; **b** tonguing of soil creep on moraine slope; **c** soil within a stone circle: **d** moss cushion on lee side of boulder

Table 7.1. Soil properties at Casey, Windmill Islands (*s.d.* sampling depth; *skelet* stones+gravels; *sa* sand, *si* silt, *cl* clay; pH in $CaCl_2$, soils 2 and 3 in KCl, *h* humification intensity after von Post)

Horizon	depth (cm)	Skel. (%)	Color moist	pH	d_b (kg l^{-1})	sa (%)	si (%)	cl (%)	TOC (g kg^{-1})	C/N	Fe_p (g kg^{-1})	e.c. (mS)	b.s. (%)	$Fe_{d\text{-}p}$ (g kg^{-1})	P_a (g kg^{-1})
1 (202) Hapli-turbic Cryosol with gelundic phase on stony moraine with soil algae															
Az	0–1	54	2.5Y4/2	4.6	1.6	66	33	1.4	1.8	7.8	0.54	14	72	2.4	0.13
AC1	−5	17	2.5Y4/2	4.5	1.4	54	42	4.0	1.4	9.8	0.58	2.3	55	3.0	0.14
AC2	−20	41	2.5Y4/2	4.4	1.7	67	28	4.7	1.8	5.5	0.76	0.4	48	2.5	0.21
Cw1	−50	66	2.5Y5/2	4.4		70	25	5.4	3.9	7.9	0.82		59	1.5	0.15
Cw2	−70	62	2.5Y5/2	4.0		83	11	6.3	3.0	5.8	0.92		63	1.9	0.17
2 (2) Stagni-leptic Cryosol, on basic gneiss to schist, with gelundic phase, with soil algae; 44 m a.s.l.															
Ah	0–0.5	33	2.5Y5/2	4.4	1.3	63	22	15	7.2	12	1.4	2.1	55	2.1	0.06
AB	−5	20	2.5Y6/4	4.4	1.3	36	31	34	4.9	10	6.2	1.0	55	2.1	0.10
Bg1	−20	13	7.5YR5/8	4.3	1.3	33	29	38	6.6	9.7	5.5	0.3	53	1.3	0.26
Bg2	−30	10	GY6/1	4.4		31	24	45	2.6	9.0	7.0		55	7.0	0.17
3 (25) Lepti-gelic Podzol with gelundic phase on leuco gneiss with lichens; 44 m a.s.l.															
AE	0–1	44	10YR5/2	5.0	1.6	95	3	2	29	7.6	0.5	2.1	62	0.7	0.02
Bh	−5	38	5YR3/1.5	4.2	1.6	90	7	3	21	5.0	1.5	1.o	35	1.4	0.13
Bhs	−26	28	9YR3.5/3	4.4	1.4	66	29	5	29	5.6	2.2		43	1.9	0.35
4 (210) Skeleti-gelic Podzol of a former penguin rookery on leuco gneiss, under lichens; 38 m a.s.l.															
AE	6–8	99	10YR6/2	4.2	1.8	88	7	5	39	24	1.5	3.8	65	0.9	0.33
Bh	−11	85	10YR3/2	4.3	1.8	83	12	5	41	11	5.9	1.5	60	1.0	0.47
Bsh1	−18	85	10YR3/4	4.4	1.6	67	19	13	24	5.0	9.6	1.1	67	1.4	0.86
Bsh2	−30	88	8YR4.5/5	4.6		"	"	"	29	8.0	6.5	1.1	67	0.3	
Bsh3	−40	85	10YR4/5	4.8		78	17	5	30	7.3	4.5	2.0	76	1.9	1.36
Cgw	−50	70	2.5Y5.5/4	5.3		"	"	"	3	10	3.1	1.4	93	1.6	n.d.
5 (137) Foli-gelic Histosol on gneiss beneath lichens upon died mosses, 38 m a.s.l.															
LH	0–3	0		4.7	0.3				290	19		3.5	88	1	0.18
H1	−8	6	5YR3/2	3.9	0.4				200	9.1		2.8	73	3–5	0.17
H2	−16	9	5YR2/1.5	3.8	0.6				190	9.1		0.8	60	8–9	0.25
HC	−28	21	5YR2/1.5	3.8					160	12		0.4	46	9	0.23

Table 7.1 (*Continued*)

Horizon	depth (cm)	Skel. (%)	Color moist	pH	d_b (kg l^{-1})	sa (%)	si (%)	cl (%)	TOC (g kg^{-1})	C/N	Fe_p (g kg^{-1})	e.c. (mS)	b.s. (%)	Fe_{d-p} (g kg^{-1})	P_a (g kg^{-1})
6 (148) Fibri-gelic Histosol on gneiss of a meltwater lake; water 20 cm above surface; 33 m a.s.l.															
H1	0–1	0	5Y6/1	6.0	0.1				270	20		1.8	85	1–2	2.1
H2	–8	0	5GY5/1	5.3	0.1				260	21		4.9	68	1–2	0.63
H3	–18	0	5GY5/1	4.9	0.2				160	18		0.6	83	2–3	0.001
7 (216) Skeleti-turbic Cryosol with gelundic phase on blocky moraine on glacier ice with soil algae															
Au	0–3	46	2.5Y5/1	5.9	1.8	81	27	1.7	0.2	15		1.0	78	1.4	0.01
Cw1	–15	42	2.5Y5/2	6.2	1.8	65	21	13	0.4	13		0.6	85	1.5	0.01
Cw2	–25	64	2.5Y5/2	6.3	1.8	81	26	2.4	0.2	7.5		0.6	85	1.4	0.01
8 (155) Stagni-turbic Cryosol on basic gneiss with gelundic phase and soil algae															
Ah	0–1	32	2.5Y4/2	4.2	1.4	76	15	6.6	8.0	6.3	1.5	2.9	63	1.7	0.20
Bw	–20	36	2.5Y3/4	4.1	1.7	68	25	7.2	9.4	7.4	1.7	0.3	52	2.9	0.27
Bg	–40	38	1.5Y4/4+	4.2	1.8	86	9.2	5.2	5.3	6.2	0.9	0.4	58	1.3	0.15
R/Bg	–60	40	+7YR5/6	4.0	1.8	76	18	6.2	3.7	8.6	3.6	0.5	51	4.5	0.28
9 (217) Skeleti-gelic Podzol with gelundic phase on fluvio-glacial gravels with lichens															
AE	0–3	55	10YR7/3	4.1	1.9	99	0.5	0.9	13	24	0.3	2.7	24	0.5	0.01
Bh	–8	48	8YR5/4	4.3	1.8	97	2.0	1.2	9.1	10	1.1	0.9	10	0.1	0.04
Bhs	–15	53	8YR6/5	4.4	1.9	98	1.5	0.4	5.3	11	1.0	0.4	11	0.2	0.04
Bsw	–25	37	10YR4/3	4.5	1.8	72	26	2.4	2.4	9.2	0.4	0.5	9.2	0.4	0.06
10 (211) Skeleti-gelic Regosol of a recent penguin rookery without vegetation															
Ah1	0–0.8	>70	9YR6/4												
Ah2	–20	>70	5YR3/2	6.5					158	1.4			96	0.32	4.3

weathering. In some soils, Bw horizons are transformed into mottled Bg horizons (Table 7.1, sites 2, 8) due to oxygen deficiency by water stagnation above the permafrost or rock.

Sandy to gravely soils with vegetation cover are podzolised. Antarctic Podzols are characterised by a very small bleached E-horizon, which is not more than 1–3 cm beside larger B horizons (Table 7.1, sites 3, 9). Only Podzols with extremely high gravel contents of more than 900 g kg^{-1} show larger E-horizons (Table 7.1, site 4). Many soils were influenced by birds, mainly penguins, in the past before the landscape was lifted and closer to the sea. They show high contents of available P (>0.1–0.12 g kg^{-1}, Table 7.1) which corresponds with 1.5 g kg^{-1} citrate extract P_2O_5. Originally, this is typical for an anthropogenic epipedon (after Soil Survey Staff 1998). Corresponding data from Ongul Island show similar values (Ohtani and Kanda, Chap. 5). Only the studied soils of Clark Peninsula (Fig. 7.1, sampling sites 7 and 9) seem to be free of the influence of birds, as they do not show elevated phosphate levels. However, the area of Clark Peninsula, especially the middle ridge, is covered by relic nesting sites (Seppelt, pers. comm.). Most studied soils are enriched with airborne salts from the sea in the first 1–3 cm (e.c. values, Table 7.1). The salts are accumulated in the surface soil by capillary rise during a longer dry period.

7.3.2 King George Island (Arctowski Area)

All the studied soils of Arctowski area (Fig. 7.2) are frozen for at least 7–9 months and occasionally longer, like those above 100 m a.s.l. (site 8). Frozen ground is mainly found at depths between 0.5 and 1.5 m at the end of the vegetation period. However, water in soils of the young moraines near the Ecology Glacier (Fig. 7.2) with heights of 3 m seem to melt totally in some cases. Some soils show visible cryoturbation phenomena and periglacial features like stone rings (Table 2, site 10), garlands on slopes (sites 6, 16), involutions (after Catt 1986) of soil layers (site 13), together with stone sorting (site 2) and irregular depth distributions of organic matter (sites 2, 8). Most soils with a dense vegetation cover show no sorting phenomena and less TOC contents with depth (sites 1, 7). In total, recent cryoturbation and solifluction processes seem to be less important when compared with the Windmill Islands (Fig. 7.4).

Most soils have high total phosphate contents (>2 g kg^{-1} P; Kuhn 1997) and weatherable phosphate (>0.5 g kg^{-1}; Table 7.2). These values cannot be explained by the parent material (tholeiite basalt has 0.7 g kg^{-1} P). Hence, they reflect incorporation of guano, mainly from penguins (Tatur 1989). Some soils are abandoned rookeries, like site 12, where bones of penguins can be found in combination with extremely high contents of phosphate. Most soils with a dense vegetation cover are very strongly enriched with organic matter (sites 1, 4, 7, 12, 13) and are moderate to strongly (sites 12, 13) acidified. Only profile 10 had some carbonates and therefore reacted neutral.

Some soils, like site 1, have a mollic epipedon (for definition see Soil Survey Staff 1998). They are characterised by a coprogenic fabric together with a high activity of soil animals, especially nematodes and collembolans (Bölter et al.

Fig. 7.4.a Region of the Arctowski Station seen from Panorama Ridge (site 8, Fig. 7.2) to Shag Point with partly flooded lowlands; **b** frost patterns on barren soils (site 6, Fig. 7.2), *bar* = 25 cm; **c** soil edge along a meadow (sites 1–4, Fig 7.2), *bar* = 25 cm; **d** soil edge (Skeleti-gelic Regosol: Tab. 7.3) in an Adélie penguin colony

Table 7.2. Soils at Arctowski, King George Island, Maritime Antarctica (pH $CaCl_2$; other symbols see Table 7.1)

Horizon	Depth (cm)	Color moist	Skel. (%)	sa (%)	Silt (%)	cl (%)	pH	TOC ($g\,kg^{-1}$)	carb ($g\,kg^{-1}$)	d_b ($kg\,l^{-1}$)	C/N	e.c. (mS)	b.s. (%)	Fe_p ($g\,kg^{-1}$)	Fe_{d-p} ($g\,kg^{-1}$)	P_v ($g\,kg^{-1}$)
1 Molli-gelic Cambisol (or hapli-gelic phaeozem) on basalt with grasses and mosses 38 m a.s.l.																
Ah1	0–8	10YR3/3	24	73	18	9	5.5	53	0	1.0	9.0	3.0	87	1.2	13.2	0.87
Ah2	–19	10YR3/2.5	39	74	16	10	5.3	11	0	1.2	8.8	1.2	82	1.7	12.4	0.85
Bw	–49	8YR3/2.5	43	75	19	6	5.7	2.5	0	1.4	4.6	0.7	89	1.1	16.0	0.87
Cw	–70	10YR3/3	44	74	19	7	6.2	2.0	0	1.4		0.8	90	0.6	17.8	0.90
2 Vitri-regic Cryosol on basaltic solifluction deposits with 10 % grass, 40 m a.s.l.																
AC	0–1	10YR3/2	48	75	13	12	5.2	2.5	0	1.5	4.1	0.4	84	0.7	12.9	0.69
Cw1	–8	10YR3/3	48	75	15	10	5.3	1.8	0	1.3	6.0	0.3	85	0.6	12.9	0.72
Cw2	–30	10YR3/3	40	77	13	11	5.5	2.1	0	1.2	13	0.3	88	0.6	13.8	0.79
C	–54	2.5Y3/3	41	73	16	12	5.6	2.0	0	1.4		0.3	89	0.7	13.5	0.81
4 Skeleti-gelic Podzol on basalt with grasses, >40 cm rock, 40 m a.s.l.																
Ah	0–2	9YR3/3	9	48	18	34	4.3	31	0	1.1	7.9	1.2	56	3.0	16.6	1.7
AE	–13	6YR3/3	12	64	15	21	4.2	27	0	0.9	5.0	2.5	47	3.8	16.6	1.5
Bh	–15	9YR5/4	30	64	15	21	4.1	32	0	1.0	12	0.8	50	7.1	16.4	4.4
Bhs	–18	8YR3/3	37	76	11	13	3.9	44	0	1.3	7.4	0.5	39	7.0	16.6	5.1
Bms	–25	5YR4/3	60	76	11	13	4.0	9.6	0	1.7	12	0.6	47	3.0	17.0	1.6
Bsw	–34	8YR3/4	70	83	6	11	3.9	2.1	0	1.4		0.8	50	2.0	11.4	1.4
5 Orthithioni-gelic Leptosol on basalt with 50 % grass+lichens, >49 cm rock, 55 m a.s.l.																
AB	0–2	10YR3/4	37	73	17	10	5.8	2.0	0.08	1.3		0.4	92	0.2	8.1	0.6
Bw1	–8	10YR3/4	53	64	26	10	6.0	2.4	0	1.5		0.4	94	0.2	8.5	0.6
Bw2	–20	6YR3/3	90	76	16	8	6.2	2.0	0	1.6		0.6	95	0.2	8.3	0.8
Cw	–49	10YR5/4	90	79	15	6	**3.7**	1.9	0	1.6		0.6	**20**	2.7	9.2	1.3
7 Dystri-gelic Fluvisol on sea sands and gravel, with mosses, high ground water table 3 m a.s.l.																
H	5–0	peat														
Ah1	0–7	10YR3/2	26	89	6	5	4.4	71	0	0.9	10	0.7	16	2.8	6.5	
Ah2	–12	2.5Y4/2	14	90	6	4	4.1	17	0	1.1	2.4	1.0	30	5.0	6.4	
C	–48	5Y2.5/2	46	98	1	1	4.4	2.2	0	1.1	1.7	1.1	21	1.9	4.4	

Table 7.2 (*Continued*)

Horizon	Depth (cm)	Color moist	Skel. (%)	sa (%)	Silt (%)	cl (%)	pH	TOC ($g\,kg^{-1}$)	carb ($g\,kg^{-1}$)	d_b ($kg\,l^{-1}$)	C/N	e.c. (mS)	b.s. (%)	Fe_p ($g\,kg^{-1}$)	Fe_{d-p} ($g\,kg^{-1}$)	P_v ($g\,kg^{-1}$)
8 Vitri-regic Cryosol on tholeiite basalt ± vulcan. glasses, without vegetation, 122 m a.s.l.																
Ah	0–4	10YR3/2	29	82	10	8	6.4	o.7	0.07	1.3	3.5	1.3	91	0.1	12.5	0.4
Cw	–25	10YR3/3	31	72	19	9	6.4	0.7	0.07	1.4	2.5	1.2	93	0.1	12.6	0.4
C	–50	10YR3/3	63	81	14	5	6.5	0.7	0.03	1.4	7.0	o.7	96	0.1	7.2	0.3
10 Calcari-regic Cryosol on young moraine without vegetation, 18 m a.s.l.																
Ah	0–4	2Y3/2	33	57	30	13	6.4	1.0	0	1.4		0.6	81	0.0	12.6	0.9
BwC	–30	2Y3/2	56	50	32	18	7.1	0.5	0.2	1.4		0.7	99	0.1	12.3	0.7
Cw	–80	5Y3/2	39	71	19	10	7.1	0.6	0.2	1.4		1.0	95	0.1	13.6	0.8
12 Humi-gelic Umbrisol of former penguin rookery on fluvio glacial sediment, with moss a. grass, 45 m a.s.l.																
Ah1	0–19	7YR3/3	30	85	12	3	4.2	80	0	1.3	9.3	0.3	9	4.6	13.6	7.5
Ah2	–36	10YR5/3	31	75	16	9	4.2	19	0	1.5	6.1	0.3	7	7.6	13.2	26.5
Ah3	–64	10YR5.5/4	25	60	29	11	4.1	24	0	1.2	5.4	0.2	7	10.0	16.3	37.0
Ah4	–86	8YR5/6	62	91	5	4	4.1	26	0	1.4	4.0	0.4	7	6.6	7.3	35.4
13 Gleyi-turbic Cryosol on fluvio glacial sediments with mosses and grass; permafrost at 60 cm, 48 m a.s.l.																
H	0–5	6YR5/7	3	48	42	10	3.5	128	0		20	1.2	7	2.0	6.1	
Ah/B	–19	7.5YR3/2	3	96	3	1	3.7	15	0		8.8	0.3	6	1.8	6.0	1.1
Agh	–29	5YR3/3	0	74	20	6	3.6	74	0		7.0	1.9	5	3.9	2.5	2.3
Bg	–39	10YR4/2	1	71	28	1	3.6	9.2	0		3.7	0.2	8	2.0	6.2	1.4
Br	–64	2.5Y3/3	0	94	6	1	3.7	12	0		6.2	0.3	6	1.9	1.3	2.3
Bir	–80	N2	2	sil	ty	mud	5.7	12	0		6.0	0.1	23	2.9	11.8	17
FAir	–102	N2	0	sil	ty	mud	6.1	75	0.2		8.0		61	3.5	12.7	18
16 Skeleti-stagnic Cryosol from basaltic solifluction deposits without vegetation, 58 m a.s.l., Fe_o																
Aw	0–3	10YR3/4	42	52	34	14	5.0	1.3	0	1.2		o.7	85	3.7	2.3	0.8
Bg1	–32	4Y5/4+	39	48	34	18	5.5	0.4	0	1.2		0.3	86	2.7	1.6	1.1
Bg2	–58	+5YR4/6	63	48	34	18	5.6	0.5	0	1.2		0.3	83	2.1	1.6	0.7
Cg	–80	5Y5/2.5	54	53	31	16	5.8	0.6	0	1.2		0.4	91	1.9	1.1	0.8

1997). Profile 4 shows patterns of podzolisation, e.g. there is no albic E horizon, but the B horizon is enriched with organically bound iron (see Fe_p in Table 7.2) in combination with mobile humus (high ODOE values, Kuhn 1997). Particles are partly cemented by coatings similar to Ortstein. Volcanic glass has been identified by microscope analysis and allophane by higher contents of oxalate extractable Al, Fe and Si (Kuhn 1997).

7.4 Soil Classifications

The soils of Casey are classified according to the revised legend of the World Soil Map (FAO 1988) as gelic subunits of Histosols, Leptosols, Podzols and Regosols in the field (Blume and Bölter 1993). In addition, boulders and rock outcrops with an ecto- and endolithic flora are classified as Lithosols. Similarly, the soils at Arctowski are classified after the draft of the World Base for Soil Resources (WRB; ISSS/ISRIC/FAO 1994). Most units are gelic subunits of Fluvisols, Histosols, Leptosols, Cambisols, Regosols and Umbrisols. Soils with visible cryoturbation phenomena inside the solum indicated by inversions (=swirl-like pattern) of layers (e.g. site 13, Table 7.2) are classified as thixotropic Cryosols (Kuhn 1997). For this publication, we transfer the classifications into the new WRB system (ISSS/ISRIC/FAO 1998), and in addition into the new US concept of Gelisols (=soils with permafrost; Soil Survey Staff 1998; see Table 7.3).

At Casey, most soils are classified as Cryosols (=permafrost within 1 m + cryoturbation phenomena) with the following subunits: lithic (hard rock within 10 cm), leptic (hard rock within 1 m), skeletic (coarse soil 40–90 % within 1 m), turbic (stone circles or garlands at surface, or texture sorting or involution inside the profile), and stagnic (mottles within 50 cm). Histosols and Podzols are not classified as Cryosols, because the first does not show any cryoturbation patterns, and the latter are not influenced by cryoturbation for at least hundreds of years (Podzols are completely covered with up to 5-cm-high bush lichens, which grow very slowly). Skeleti-gelic Podzols of abandoned penguin rookeries (e.g. site 4, Table 7.1) have no visible cryoturbation phenomena. Other Skeleti-gelic Podzols like site 9 (Table 7.1) are part of a gravelly pattern ground with 50–100 m polygons, and lepti-gelic Podzols (site 3, Table 7.1) are located in the centre of 1–3 m wide stone circles. No Podzol shows any active cryoturbation. The latter is the case with active rookeries of Adélie penguins (site 10, Table 7.1) where the birds heaped up coarse gravels and filled the interspace with their excrement. Most of the soils are acidified, but have (except site 9, Table 7.1) b.s.-values >50 %, and thus should be classified as eutric subunits (in contrast to Beyer and Bölter 1999).

We further recommend studying boulders and rock outcrops which are covered by epilithic as well as endolithic floras of lichens, algae and bacteria. These

Table 7.3. Classification of the soils at Casey (*C*), Windmill Islands, and Arctowski (*A*), King George Island

WRB system ISSS/ISRIC/FAO 1998	US-Soil Taxonomy Soil Survey Staff 1998	Area
Lithic Cryosol	Lithic Haplorthel	C
Stagni-leptic Cryosol	Lithic Aquiturbel	C
Vitri-leptic Cryosol	Lithic Haplorthel	A
Skeleti-turbic Cryosol	Typic Haploturbel	C
Hapli-turbic Cryosol	Typic Haploturbel	A C
Umbri-turbic Cryosol	Typic Haploturbel	A
Gleyi-turbic Cryosol	Cumulic Umbriturbel	A
Stagni-turbic Cryosol	Lithic Aquiturbel	C
Skeleti-stagnic Cryosol	Typic Aquiturbel Lithic Haplorthel	A
Hapli-regic Cryosol[a]	Typic Haploturbel	A
Calcari-regic Cryosol[a]	Typic Haploturbel Lithic Haplorthel	A
Vitri-regic Cryosol[a]	Typic Haplorthel	A
Fibri-gelic Histosol	Lithic Fibristel	C
Foli-gelic Histosol	Lithic Folistel	C
Dystri-gelic Histosol	Fluvaquentic Sapristel	A
Molli-gelic Cambisol	Typic Haplorthel	A
Eutri-gelic Cambisol	Typic Haplorthel	A
Dystri-gelic Cambisol	Typic Umbrorthel	A
Skeleti-gelic Phaeozem[b]	Typic Mollorthel	A
Dystri-gelic Fluvisol	Typic Psammorthel	A
Sali-gelic Fluvisol	Typic Psammorthel	A
Eutri-gelic Leptosol	Lithic Haplorthel	A
Skeleti-gelic Leptosol	Lithic Haplorthel	A
Vitri-gelic Leptosol	Lithic Haplorthel	A
Orthithioni-gelic Leptosol	Lithic Haplorthel	A
Lepti-gelic Podzol	Lithic Haplorthel	C
Skeleti-gelic Podzol	Typic Haplorthel	C
Ornithogeni-gelic Podzol[c]		C
Skeleti-gelic Regosol or Ornithogeni-gelic Regosol[c]		C
Humi-gelic Umbrisol or	Cumulic Umbrorthel	A
Ornithogeni-gelic Umbrisol[c]		A
Hapli-gelic Lithosol[d]		C
Proti-gelic Lithosol[d]		C

[a] Regic Cryosols are recommended by Tarnocai et al.(2000) as Cryosols without further diagnostic horizons.

[b] Gelic phaeozems are recommended by Blume and Kuhn, because Phaeozems do exist under permafrost conditions.

[c] Soils of current and abandoned penguin rookeries were earlier named as ornithogenic soils (e.g. Heine and Speir 1989; we recommend a subunit of other units, which is characterised by bird bones and high P content.

[d] Lithosols are recommended by Blume and Bölter (1993) for boulders and rock surfaces with a ectolithic and endolithic flora. Proto-gelic L. with small biotic and chemical change of the top cm, and hapli-gelic L. with strong change.

environments show physical and chemical changes inside the first millimetres to centimetres of the rock surface (Friedmann 1982; Chen et al. 2000) due to the different weathering processes. Thus, they may be regarded as separate soil bodies. Blume and Bölter (1993; Table 7.3) described them as Lithosols.

At Arctowski we have to separate further subunits of the Cryosols. Soils lacking development (=soils with A- and C-like profiles 2, 8 and 10 in Table 7.2) are classified as Regic Cryosols following a recommendation of Tarnocai et al. (in press). Further, there are soils with vitric features (=>10 % volcanic glass + other primary minerals in the fine soil, and >30 cm thick, and d_b>0.9 kg l^{-1}, <10 % clay, or >0.4 % Al_o+1/2 Fe_o (o means oxalate extract) or >25 % P retention, after ISSS/ISRIC/FAO 1998). The vitri-gelic subunits (profiles 2 and 8) have lower contents of volcanic glass + weatherable minerals like olivine, augite and andesite, but 0.7 % Al_o+1/2 Fe_o, and a high CEC of the clay fraction (240 $cmol_c$ kg^{-1}; Kuhn 1997). Besides the Cryosols, we can classify gelic subunits of Cambisols (site 1, Table 7.2), Fluvisols (site 7), Leptosols (5), Podzols (4) and Umbrisols (12). An Orthithioni-gelic Leptosol (5 in Table 7.2) is characterised by a sulphuric horizon (ISSS/ISRIC/FAO 1998) with an extremely low pH. We also found soils with a mollic epipedon of more than 25 cm, which have to be classified as Gelic Phaeozems.

Three of the studied soils are of current (site 10 in Table 7.1) and abandoned (site 4 in Table 7.1, site 12 in Table 7.2) penguin rookeries. They are characterised by elevated P contents (comparable to an anthropogenic epipedon with values >1.5 g kg^{-1} citrate extractable P_2O_5, according to Soil Survey Staff 1998), bird bones, feathers, or sorted coarse gravel. We recommend, therefore, classifying such soils as ornithogeni-gelic subunits (Table 7.3).

The main problem of the classification of the soils at Arctowski is the unknown depth of permafrost. Cryosols should have an upper boundary of permafrost in the first metre (whereas the upper boundary for gelic subunits should be in the second metre after WRB). In (continental) Canada, the boundary between continuous and discontinuous permafrost is at around –5 °C of mean annual temperature (MAT, after Brown 1978). Under the maritime conditions of King George Island, this boundary should be higher. However, a dramatic rise of MAT from –3.5 °C during the 1940s to –2.0 °C in the 1990s (Braun and Goßmann, Chap. 6) makes it possible that many former Cryosols have to now be classified as gelic subunits, at least the ones exposed to the sun. Some of the gelic subunit soils have to be cancelled due to a boundary deeper than 2 m or to a total disappearance of permafrost, especially at young, stony moraines in the neighbourhood of the sea. This, however, needs to be proved in future when considering both ecological and soil genetic problems. Hence, it is recommended that the related suborders of the new Gelisol order of the US Soil Taxonomy (Soil Survey Staff 1998) should be worked out in more detail before they are properly applied for Antarctic soils (Table 7.3).

7.5 Soilscapes

7.5.1 Windmill Islands

Figures 7.5 and 7.6 show a soil association map at Casey. Distinct units can be established from metre to metre, and the very rough scale of 1:1250 did not allow an indication of areas of separate pedotopes. They are shown as separate soil associations or soilscapes.

The hill is formed by leuco gneiss with vertically standing interruptions by basic gneiss and schist in small long-lined depressions due to their stronger weatherability (soilscape 9). The hard rock material is covered by crashed moraine material in the upper parts of the hill, whereas the lower eastern part in particular is covered by a blocky moraine material with more than 1-m thickness near the valley. The main part of the hill is free of snow for 2–3 months and is therefore densely covered by vegetation. Soilscapes 8 in the west and 5 in the east are covered by snow for 11–12 months. Consequently, no vegetation exists, only some lichens grow on big boulders.

Soilscape 6 is characterised by stone circles and stripes at slopes with Proto-gelic Lithosols at boulders and Skeleti-gelic Cryosols with very low TOC contents and without brownification inside the polygons (Fig. 7.6a). The soilscape 5 is built up similarly, but with Hapli-gelic Lithosols and with slightly higher TOC contents of the Cryosols (Figs. 7.6a, b, c). Mainly at the top of the hill (soilscape 1) Hapli-gelic Lithosols dominate with 40–60 % at rock outcrops beside very thin Lithi-gelic Leptosols and slightly deeper developed Podzols on hard rock (Fig. 7.6 c). All units are covered by lichens. The loamy Leptic and Turbic Cryosols of soilscape 2 have mainly stagnic properties (rH 18–25=slightly reduced conditions) in addition, and occupy a small area of schist and basic gneiss, beside outcrops of leuco gneiss with Hapli-gelic Lithosols (Fig. 7.6d). At soilscape 3, 1–5-m-wide stone circles are formed by a thin moraine cover upon hard rock. Inside the circles are sandy Gelic Podzols with a dense vegetation of lichens, or Leptic Cryosols with sandy loam and a scarce vegetation in combination with Lepti-gelic Histosols in small depressions and a dense cover of moss (Fig. 7.6 c, d). Folic Histosols are aerated (rH 28), but Terric Histosols are not (rH 15). Soilscape 7 is dominated by gravely Gelic Podzols of abandoned penguin rookeries (site 4 in Table 7.1). The Podzols of soilscape 4 (Fig. 7.6d) seem to have been strongly influenced by birds in the past due to their high P content (data not presented). The beaches of the meltwater lakes (soilscape 10) are characterised by Gelic Lithosols on big boulders and thin Fibri-gelic Histosols with reduced conditions (rH 15–19), beside the first centimetre (rH 33), which are covered with water 10–40 cm deep (Fig. 7.6 c; see also Beyer 2000).

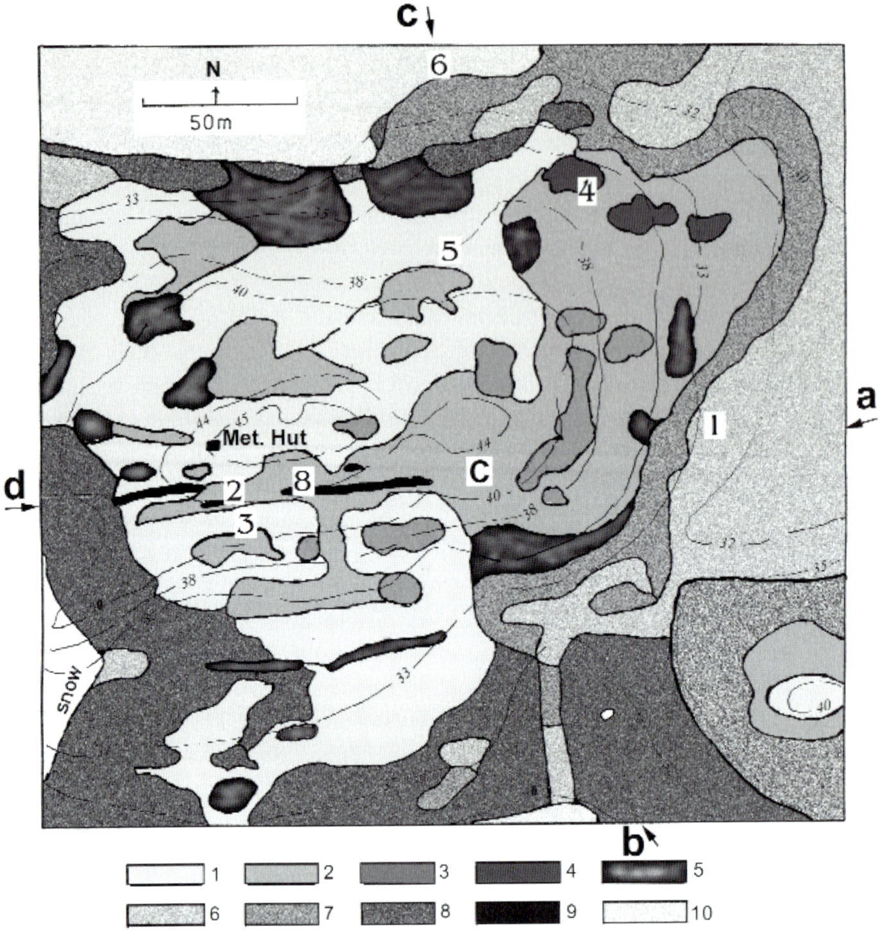

Fig. 7.5. Soilscapes on Mable Hill near Casey Station, Windmill Islands, (after Blume and Bölter 1993; classification of soils after ISSS/ISRIC/FAO 1998). *Numbers 1–6* and *8* indicate sampling sites (see Fig. 7.1), the location *C* indicates the centre of the lines to the four directions (see Fig. 7.6a-d). *Arrows* and *letters* along the sides indicate the transects (see Fig. 7.5a–d) directed to the centre *C*. The different soilscapes are indicated by individual tones: *1* Hapli-gelic Lithosols+Lithic Cryosols (s)+Skeleti-gelic Podzols (s) on leuco gneiss. *2* Hapli-gelic Lithosols+Leptic Cryosols (sl)+Lepti-gelic Podzol (s)+Gelic Histosols on rocks with few moraine. *3* Skeleti-gelic Podzols (s)+Hapli-gelic Lithosols on moraine covered rocks. *4* Skeleti-gelic Podzols (s)+Hapli-gelic Lithosols of former penguin rookeries. *5* Hapli-gelic Lithosols+Leptic Cryosols (sl)+Turbic Cryosols of rocks with few moraine schist. *6* Skeleti-turbic Cryosols (ls-sl)+Proto-gelic Lithosols of blocky moraines. *7* Hapli-+Skeleti-turbic Cryosols (sl)+Hapli-gelic Lithosols of blocky moraines. *8* Hapli-gelic Lithosols+Leptic Cryosols (sl) of rock outcrops in valleys. *9* Hapli-gelic Lithosols+Stagni-turbic Cryosols (l-cl)+Leptic Cryosols on basic gneiss+schist. *10* Fibri-gelic Histosols+Gelic Lithosols of lakes and blocky lake beaches. *s* Sand, *sl* sandy loam, *cl* clay loam, *ls* loamy sand

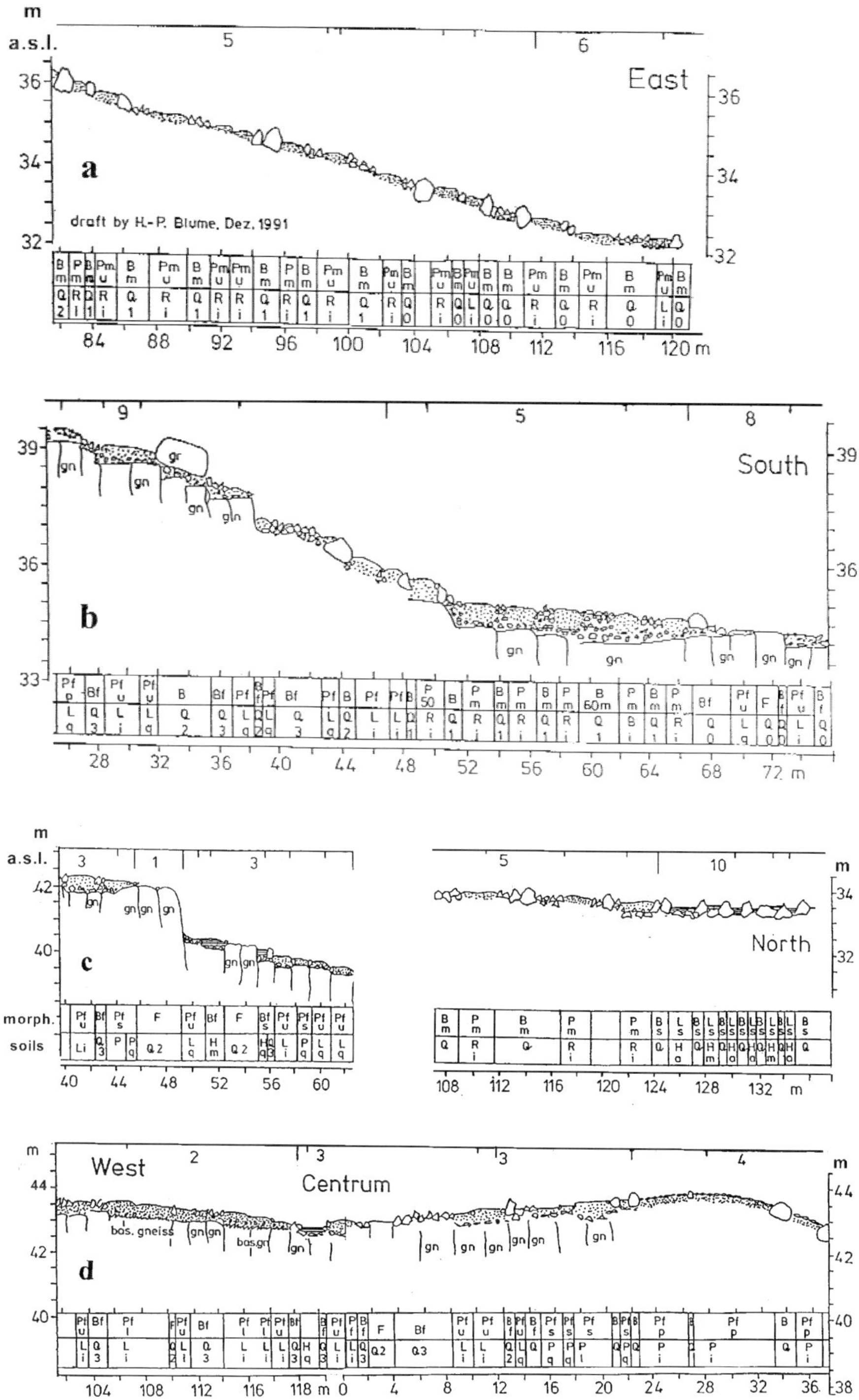

Fig. 7.6. (*Legend see page 108*)

Fig. 7.6. (*page 107*) Transects **a-d** at Mable Hill and descriptions of soil types and morphologies (after Blume and Bölter 1993). Transects are displayed from the central point C (see Fig. 7.5) to the east (**a**), to the south (**b**), to the north (**d**) and west (**e**). Morphological types: *F* rock outcrop; *B* big boulder above rocks; *Bf* accumulations of stones above rock; *Pf* polygon from weathered rock+moraine schist; *Bm* moraine boulder and stone; *Pm* silt in polygon centre to gravel at borders; *Bs* boulder in lake; *Ls* phytogenic lake sediment; *s* sandy to gravelly material; *u* loamy sand; *l* loamy material; *p* coarse gravel without polygon. Soil types: *L* Cryosols (<30 cm on rock); *Lq* Lithic Cryosol (<10 cm on rock); *Li* Stagni-/Turbi-leptic Cryosol (30 cm solum); *Ri* Turbic Cryosol on moraine; *P* Podzol; *Pl* Lepti-gelic Podzol (<30 cm on rock); *Pq* Lithi-gelic Podzol (<10 cm on rock); *Pi* other Gelic Podzol; *H* Gelic Histosol; *Hm* Histosol with peat from moss; *Ha* Histosol with peat from algae; *Ls* Lithi-gelic Histosol (<10 cm peat on rock); *Q* Gelic Lithosols, *1* <5 % plant cover, *2* 5–70 % plant cover, *3* >70 % plant cover

7.5.2 King George Island

Figure 7.7 shows the soil map at Arctowski. The northwest and the central part of the area is situated on rocks, mostly of volcanic origin, basaltic and andesitic with sedimentary intrusions (Birkenmajer 1980), partly with solifluction deposits. The southeast is occupied by moraines of the Ecology Glacier which are mainly formed from basaltic rocks and volcanic tephra. The shore of the Admiralty Bay and the lagoon close to the station are covered by coarse sandy to gravelly sea sediments.

At soilscape 1 (Fig. 7.7) eutrophic (pH 5–6) Vitri- and Lithi-gelic Leptosols occur on basalt at steep slopes (10–50°) together with Hapli-gelic Lithosols covered by lichens. Vitri-leptic Cryosols on solifluction deposits with cryoturbation features are located at moderate slopes (5–10°). No plant cover exists apart from some grass. At soilscape 2 on gentle to moderately steep slopes (2–15°) with eutric (pH 5–7) Lithi-gelic Leptosols and Vitri-endoleptic Cryosols (hard rock at a depth of25–50 cm). In these areas, garlands and other cryoturbation phenomena also occur beside Proto-gelic Lithosols at rock outcrops and boulders. Some grass and lichens (about 10 % of the area) occur below 90 m a.s.l., some grass between 90 and 100 m, and no vegetation above 100 m. TOC content of these soils is very low, <0.1 %. Soilscape 3 is situated between 50 to 100 m a.s.l. and characterised by Vitri-gelic Leptosols on basalt. Further, Leptic as well as Regic Cryosols of solifluction deposits with cryoturbation phenomena are present. Most soils are eutric, but some Orthithioni- gelic Regosols (site 5 in Table 7.2) have a strongly acidified subsoil, possibly due to oxidised sulphides. The soils are moderately to strongly covered by lichen and grass (20–80 %), Cryosols to lesser extent. Soilscape 4 is situated on northern exposed rock terraces (3–7° slope) with solifluction deposits. Most soils are strongly influenced by cryoturbation, hence Cryosols dominate. The loamy soils are eutric (pH 5.5–6.5), have low TOC contents (0.2–0.5 %), and some of them have stagnic properties. Plant cover is rare (<5 % lichens and grass). Soilscape 5 represents

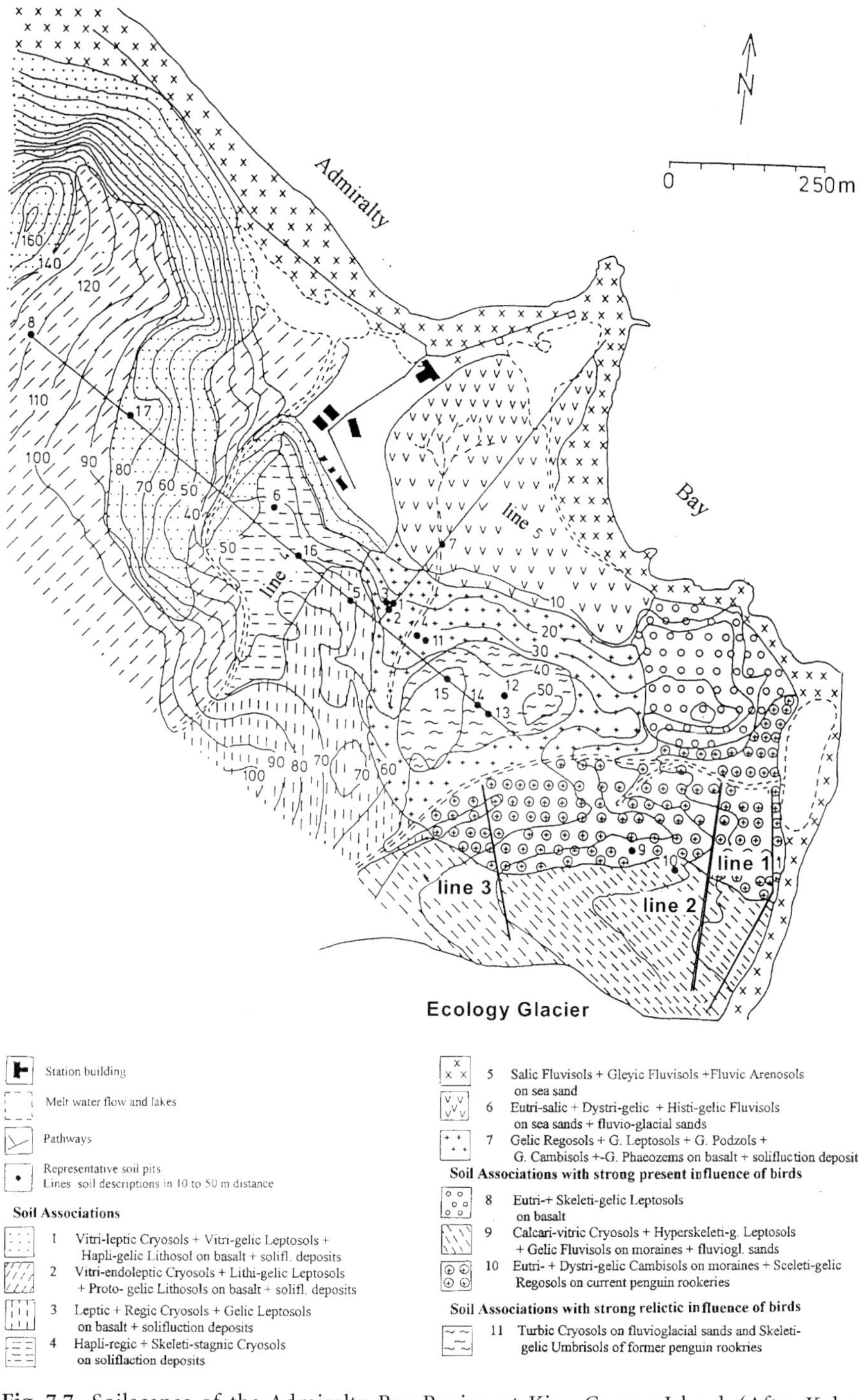

Fig. 7.7. Soilscapes of the Admiralty Bay Region at King George Island. (After Kuhn 1997)

Salic Fluvisols of coarse beach sands and Fluvic Arenosols on gravely strand walls (scarcely covered by grass) along the coastline of Admiralty Bay. It is strongly influenced by the tidal sea water, normally 1.5 m. Some Salic Fluvisols are characterised by a black coloured reduced subsoil, probably influenced by the excrement of birds and sea elephants. Soilscape 6 is situated behind soilscape 5 at a lagoon near Arctowski Station. The Eutri-salic Fluvisols near the beach as well as the Dystri- and Histi-gelic Fluvisols behind have developed from gravely sea sands and coarse-textured fluvi-glacial sediments. Unfortunately, we have no data about present permafrost, but due to the influence of the sea these soils seem to be free of permafrost. The Histic and Dystric Fluvisols have a dense cover of moss and grass. They are enriched with organic matter, whereas the Eutri-salic Fluvisols are less covered by grass and mosses (cover about 30 %) and thus have lower TOC content. Soilscape 7 is situated on moderate to steep slopes (5–25°) of basalt and solifluction deposits. These soils have vitric properties due to higher contents of tephra and allophane (Kuhn 1997). Loamy Eutri- and Molli-gelic Cambisols on basalt with a dense grass cover do occur. The mollic epipedon is enriched with humus (10–50 g kg^{-1} TOC) and is sometimes thicker than 20 cm, so that Gelic Phaeozems do also exist, due to high activity of collembolans, nematodes and microbes (Bölter et al. 1997). Vitri-regic Cryosols from solifluction deposits with low plant cover also exist next to Dystri-gelic Podzols – without a bleached E horizon, but with Ortstein in the subsoil.

Soilscape 9 is situated on extremely young neoglacial moraines which are probably not older than 50 years, according to Braun and Goßmann (Chap. 6). Fluvi-glacial sands occur between them and between the first and the Ecology Glacier. The Calcari-vitric Cryosols of the moraines show texture sorting and are slightly calcareous (1–3 g kg^{-1}). Some of the soils are strongly enriched with coarse gravel and stones (>500 g kg^{-1}). They do not show any cryoturbation phenomena, hence they are classified as Endosceleti-gelic Leptosols. Dark coloured, sulphidic and Saline Fluvisols are located close to the first moraines, which are influenced by meltwater of the glacier and by seawater spray as well. The humus content of all soils is low, due to a just developing plant cover, actually <50 g kg^{-1} at the first and up to 300 g kg^{-1} on the older moraines. Soilscape 10 is located on older moraines (probably 50 to 200 years old). The Eutri- and Dystri-gelic Cambisols have a denser plant cover of 30–100 % (Olech 1993, Olech, Chap. 12). They are decalcified and acidified (pH 3.5–4.5!), brownified and enriched with humus (10–20 g kg^{-1} TOC). Soilscape 8 is located on a current penguin rookery. Eutri- and Skeleti-gelic Leptosols with high contents of nutrients, organic matter, and an alkaline reaction occur. The northeast of soilscape 10 is also occupied by penguins (Gentoo Penguins), but their colony is widespread which means that this area is less influenced by their nesting sites (Tatur 1989). Soilscape 11 is an extremely acid-, humus- and phosphate-enriched Skeleti-gelic Umbrisol, mixed with Gleyi-turbic Cryosols and situated at a great abandoned penguin rookery. Beside soilscapes 8 and 11, which are extremely influenced by birds, other soilscapes (except 9 and the higher situated parts of soilscapes 1 and 2), are probably also influenced by birds due to their elevated P contents (see also Tatur 1989).

7.6 Conclusions

The soils of the ice-free Continental East Antarctic as well as most of those of the Antarctic islands are influenced by permafrost. The area is characterised by a large variety of very different soils. Influences of different parent rocks, very different intensities of weathering, cryoturbation and solifluction as well as distinct differences in the degree of vegetation cover from zero to complete are the main reasons for this observation. Soil conditions may change over a short distance, in many cases from metre to metre. This happens due to strong influences of cryoturbation and solifluction which result in stone ring polygons on plateaux, and garlands in slope positions. Many loamy to clay soils show the distinct influence of active cryoturbation in the first metre, thus they have to be classified as Cryosols according to WRB. However, cryoturbation phenomena of many sandy to stony soils seem to be relics, especially those in areas with dense vegetation cover. They have to be classified as gelic subunits of Histosols, Podzols, Cambisols and Umbrisols.

Two soil units were found which were difficult to classify. The first unit can be described by widespread boulders and rock outcrops, which are covered by an epilithic as well as by an endolithic flora of lichens, algae and bacteria (to a depth from 1 mm to about 1 cm). Actually, they should be classified as Gelilithic Leptosols, but the term *Gelic Lithosol* has been chosen because no fine soil material exists. The second unit comprise soils which are strongly influenced by penguins. These birds heap up coarse gravel, together with excrement from their seafood diet. The soils are strongly enriched with phosphates, similar to Anthrosols. There are no possibilities of classifying such soils: the best way would be, in our opinion, to classify them as ornithogenigelic subunits of Cambisols, Umbrisols or Podzols.

Classifications of Antarctic soils according to the US Soil Taxonomy can be performed by using the subunits of the new order of Gelisols. However, this classification needs to be worked out in more detail before it can be applied properly for Antarctic soils.

Acknowledgements. We thank the Australian National Antarctic Research Expedition, the Casey summer crew 1990/1991, the Polish Academy of Science, the Arctowski summer crew 1994 and the German Research Council (DFG), for help and financial support. Mrs. Kneesch conducted excellent laboratory analyses.

References

Beyer L (2000) Properties, formation, and geo-ecological significance of organic soils in the coastal region of East Antarctica (Wilkes Land). Catena 39:79–93

Beyer L, Bölter M (1999) Formation, ecology, and geography of Cryosols of an ice-free oasis in coastal East Antarctica near Casey Station (Wilkes Land). Aust J Soil Res 37:209–244

Beyer L, Bölter M (2000) Chemical and biological properties, formation, occurrence and classification of spodic Cryosols in a terrestrial ecosystem of East Antarctica. Catena 39:95–120

Birkenmajer K (1980) Geology of Admiralty Bay, King George Island. An outline. Pol Polar Res 1:29–54

Blight DL, Oliver RL (1977) The metamorphic geology of the Windmill Islands, Antarctica, a preliminary account. J Geol Soc Aust 24:239–262

Blume H-P, Alaily F, Smettau U, Zielinski G (1984). Soil types and associations of SW Egypt. Berl Geowiss Abh (A) 50:293–302

Blume H-P, Bölter M (1993) Soils of Casey Station (Wilkes Land, Antarctica). In: Gilichinski D (ed) Joint Russian-American seminar on cryopedology and global change. Pushchino. Post-Sem Proc 1st Int Conf Cryopedology 1992, Russian Academy of Sci, Pushchino, pp 96–104

Blume H-P, Bölter M (1996) Wechselwirkungen zwischen Boden- und Vegetationsentwicklung in der Kontinentalen Antarktis. Verh Ges Ökol 25:25–34

Blume H-P, Beyer L, Bölter M, Erlenkeuser H, Kalk E, Kneesch S, Pfisterer U, Schneider D (1997) Pedogenic zonation in soils of the Southern Circumpolar region. Adv GeoEcol 30:69–90

Blume H-P, Beyer L, Schneider D (1998) Soils of the southern circumpolar region and their classification. Euras Soil Sci 31:477–485

Bockheim JG (1995) Permafrost distribution in the southern circumpolar region and its relation to the environment. Permafrost Periglacial Process 6:27–45

Bockheim JG, Ugolini FC (1990) A review of pedogenic zonation in well-drained soils of the southern circumpolar region. Quart Res 34:47–66

Bölter M, Blume H-P, Erlenkeuser H (1994) Pedologic, isotopic and microbiological properties of Antarctic soils. Polarforschung 64:1–7

Bölter M, Blume H-P, Schneider D, Beyer L (1997) Soil properties and distributions of invertebrates and bacteria from King George Island (Arctowski Sation), aritime Antarctic. Polar Biol 18:295–304

Bölter M, Blume H-P, Kuhn D (1999) Soils and their microbiological properties from a transect from Cape Horn to the Antarctic Peninsula. Polar Biosci 12:54–67

Bölter M, Beyer L, Seppelt RD, Pingpank (2000) Studies on floristic diversity, soil organic matter, and soil microbes from the Windmill Islands, East Antarctica. Bibl Lichenol 75:421–432

Brown R (1978) Permafrost. In: Hydrologic atlas of Canada. Fisheries and Environment Canada, Ottawa

Campbell IB, Claridge GGC (1987) Antarctica – soils, weathering processes and climate. Elsevier, Amsterdam

Campbell IB, Claridge GGC, Campbell DI, Balks MR (1998) The soil environment of the McMurdo Dry Valleys, Antarctica. In: Priscu JC (ed) Ecosystem dynamics in a polar desert. The McMurdo Dry Valleys, Antarctica. Abtarct Res Ser 72, Am Geophys Union, Washington DC, pp 297-322

Catt IA (1986) Soils and Quarternary geology. Clarendon Press, Oxford

Chen J, Blume H-P, Beyer L (2000) Weathering of rocks induced by lichen colonization – a review. Catena 39:121–146

Fabiszewski J, Wojtún B (1993) Peat-forming vegetation. In: Rakusa-Suszczewski S (ed) The maritime Antarctic coastal ecosystem of Admiralty Bay. Department of Antarctic Biology, Polish Academy of Sciences, Warsaw, pp 189–195

FAO-UNESCO (1988) Soil map of the world, revised legend with corrections. FAO, Rome

Friedmann EI (1982) Microorganisms in Antarctic cold dessert. Science 225:1045–1053

ISSS/ISRIC/FAO (1994) World reference base for soil resources. Draft. FAO, Rome

ISSS/-WRB [International Soil Science Society – World Reference Base for Soil Resources] (1998) World Reference base for soil resources. World Soil Resources Reports 84, ISSSS-ISRIC-FAO, Rome

Kuhn D (1997) Genese, Ökologie und Soziologie einer Bodengesellschaft in einem Periglazialgebiet der King George Insel West-Antarktis. Schriftenr Inst Bodenkd Univ Kiel 40:1–173

Leonardi JM, Marchetti C, Monticelli L, Osterrieth M (1987) Caracterizacion preliminar de un histosol Antartico bajo gramineas. Contr Cient Inst Antart Argentino 340:1–17

O'Brien RMG, Romans JCC, Robertson L (1979) Three soil profiles from Elephant Island, South Shetland Islands. Br Antarct Surv Bull 47:1–12

Olech M (1993) Lower plants. In Rakusa-Suszczewski S (ed) The marine Antarctic coastal ecosystem of Admiralty Bay. Department of Antarctic Biology, Polish Academy of Sciences, Warsaw, pp 173–179

Schlichting E, Blume H-P, Stahr K (1995) Bodenkundliches Praktikum. 2. Aufl. Blackwell, Berlin

Soil Survey Staff (1998): Keys to soil taxonomy, 8th edn, US Dept Agric – Nat Res Counc, Washington, DC

Tarnocai C, Broll G, Blume H-P (2002) World reference base for soil resources: classification of permafrost-affected soils (in press)

Tatur A (1989) Ornithogenic soils of the maritime Antarctic. Pol Polar Res 10:481–532

Tatur A, Myrcha A (1993) Ornithogenic soils. In: Rakusa-Suszczewski S (ed) The maritime Antarctic coastal ecosystem of Admiralty Bay. Department of Antarctic Biology, Polish Academy Science, Warsaw, pp 161–165

8 Weathering and Soil Formation

H.-P. Blume, L. Beyer, E. Kalk, and D. Kuhn

8.1 Introduction

In the coastal regions of Antarctica, soil formation occurs to a greater extent than previously thought (Blume et al. 1997; Beyer et al. 1999). Soil formation can be found in accordance with humid climate conditions, as physical and chemical weathering which results in acidification, brownification, clay formation, organic matter accumulation, as well as partial podzolisation and redoximorphism. Such phenomena have been described during studies of soils at Wilkes Land near Casey Station (Blume and Bölter 1993; Beyer et al. 1998; Blume et al., Chap. 7), and at King George Island near Arctowski Station (Kuhn 1997; Blume et al., Chap. 7). In contrast, soils of the cold deserts of the Ross Sea section are characterised only by physical weathering in combination with an accumulation of carbonates and salts, but there is nearly no humus accumulation (Campbell and Claridge l987; Bockheim and Ugolini 1990; Beyer et al. 1999). In recent reviews some more chemical and mineralogical data of representative soils have been given (Blume et al., Chap 7). This contribution discusses the main reasons for soil formation in Antarctic areas.

8.2 Sites and Methodology

8.2.1 Soil Analyses

Three soils (Hapli- and Stagni-turbic Cryosols, Skeleti-gelic Podzol) near Casey, Wilkes Land in Continental East Antarctica, and five soils (calcari-regic and Skeleti-stagnic Cryosols, Molli-gelic Cambisol, Skeleti-gelic Podzol and Humi-gelic Umbrisol) near Arctowski, King George Island in maritime Antarctica were studied. The site and special soil properties are described by Blume et al., Chap. 7. Methods are described in detail by Blume et al., Chap. 7. Oxalate extractable Fe_o, Al_o, Mn_o and Si_o, pyrophosphate-extractable Al_p, Fe_p, partly exchangeable Al_p, dithionite-citrate extractable Al_d and Fe_d, hot-acid (HCl) extractable Fe_v and Mn_v fractions. Optical density of oxalate extract

Ecological Studies, Vol. 154
L. Beyer and M. Bölter (eds.) Geoecology of Antarctic Ice-Free Coastal Landscapes

(ODOE) as well as the optical density of the pyrophosphate extract (ODPE) are required to unravel soil weathering and formation. Fe_v–Fe_d ($Fe_{v\text{-}d}$) reflects the iron of the clay minerals. Fe_d–Fe_o ($Fe_{d\text{-}o}$) stands for well-crystallised iron oxides like goethite and lepidocrocite. Fe_o–Fe_p ($Fe_{o\text{-}p}$) is poorly-crystallised iron oxides like ferrihydrite, but also iron carbonates. Fe_p reflects organically bound (and exchangeable) iron as well as iron phosphates. The same pattern is valid for Al and Mn compounds, but not as good as for Fe. Occasionally, with high content of iron phosphates, data for Fe_p may be higher than for Fe_o. The reason for this is that Fe- (and Al-) phosphates are not soluble at pH 3.25. Weathering and clay formation is normally associated with an enrichment of Fe-, Al- and Mn-oxides. However, their contents are normally highly correlated with the clay content. Therefore, the impact of podzolisation or gleying on soils is better reflected by using clay-related (oxides/clay ratio) depth functions than by pure oxide-related depth functions. Si_o represents weathering products of silicates in combination with $Al_{o\text{-}a}$ allophanes (Schlichting et al. 1995). The mineralogy of some soils is studied by texture fractionation after H_2O_2 and $(NaPO_3)_6$ treatment; X-ray diffraction of clay and fine silt (2–20 µm); microscope analysis of coarse silt (20–63), fine (63–200) and medium (200–630 µm) sand fractions. Thin sections are prepared from some soil horizons (Schlichting et al. 1995).

General physical and chemical property characterisations of the studied soils are given in Blume et al. (Chap. 7). Further, mineralogical information about these soils will be given in the following tables.

8.2.2 Lithology of the Sites and Mineralogy of Soils

The bedrock of the first site at Wilkes Land consists of a layered sequence of gneiss with schist and migmatite, the Windmill Metamorphics (Blight and Oliver 1977), they are partly covered by neoglacial solifluction deposits and moraines. This region was deglaciated 5500 years B.P. and has been lifted since this time (Goodwin 1993). The three soils studied at Wilkes Land are frozen most of the time during the year, and permafrost has been found at 50 to 80 cm depth. The Hapli-turbic Cryosol in a downslope position is free of snow only in warm years, and even then not for more than 1 month. This means that no establishment is possible and a vegetation cover is missing. The two other profiles are in an upper slope position and therefore free of snow every year for 1–2 months.

The rocks of the second site on King George Island are mainly Tertiary tholeiite basalt (Birkenmajer 1980), covered by neoglacial (decades to centuries old) and older moraines as well as by solifluction deposits. Volcanic ash from neighbouring Deception Island was occasionally deposited during the Quaternary.

The Windmill Islands Region. The Hapli-turbic Cryosol is located in the centre of a stone garland. This location, combined with the discontinuous depth functions of texture fractions and the organic matter (Table 7.1, soil 1), suggests a strong impact of cryoturbation. The mineralogy of the soil (Table 8.1) is characterised by high amounts of quartz, feldspars and mica geologically derived from leuco gneiss. Such primary minerals also dominate the clay fraction, suggesting a strong cryoclastic impact of weathering. However, a low pH level, occurrences of some mineral grains with weathered surfaces, pedogenic iron and very little smectite already reflect an initial chemical weathering stage as well as formation of new minerals.

The parent material of the Stagni-turbic Cryosol consists of schist and basic gneiss. Therefore, soil mineralogy composition results in high clay (Table 7.1, soil 2) and low quartz contents (Table 8.2), especially in the subsoil, whereas the topsoil is somewhat influenced by the neighbouring leuco

Table 8.1. Mineralogy of a loamy Hapli-turbic Cryosol; Casey. (After Blume et al. 1997)

a) Mean of 20–630 µm ∅ in % (*unw.* unweathered, *w.* medium weathered)

Horizon	Depth (cm)	Chlorite unw.	Chlorite w.	Pyroxenes	Amphiboles	Mica unw.	Mica w.	Weath. min.	Opaque min.[a]
Az	0–1	4.2	0	0.9	0.9	16.6	0	0	2.0
AC	5–20	4.2	0	0	3.3	14.9	0	0.2	1.7
Cw2	50–70	5.3	0	0.7	2.0	13.8	0	0.2	1.6

Horizon	Depth (cm)	Plagiocl.	Alkali-feldsp.	Weathered feldsp.	Epidote	Quartz	Garnet	Rock fragm.	Others
Az	0–1	22.6	16.9	0.9	0.7	28.6	0.2	3.8	1.6
AC	5–20	22.0	16.3	1.3	0.4	29.5	0.7	4.4	1.1
Cw2	50–70	21.1	16.9	1.8	1.8	30.0	1.3	2.4	1.1

b) Clay and fine silt fraction

Horizon	Depth (cm)	Quartz	Feldspars	Pyroxene	Olivine serpent.	Chlorite	Illite/mica	Smectite
Clay fraction (<2 µm ∅) in %, semiquantitative								
Az	0–1	20	43	0	7	<1	30	<1
Cw2	50–70	14	56	0	0	<1	30	<1
Silt fraction (2–20 µm ∅) in %, semiquantitative								
Az	0–1	27	43	11	0	9	10	<1
Cw2	50–70	29	20	0	10	3	36	<1

[a] Anatase and rutile also present.

Table 8.2. Mineralogy of a loamy Stagni-leptic Cryosol, on basic gneiss to schist; Casey (Blume et al. 1997, changed and added)

a) Mean of 20–630 μm ∅ in % (*unw.* unweathered, *w.* medium weathered)

Horizon	Depth (cm)	Chlorite unw.	Chlorite w.	Pyroxenes	Amphiboles	Mica unw.	Mica w.	Weath.	Opaque min.[a]	Garnet
Ah	0–1	3.1	0	0.2	2.5	15.8	7.1	0	1.3	1.3
AB	1–5	6.2	0	1.5	2.8	25.6	6.6	0	1.1	1.5
Bg2	20–30	5.6	0	1.8	2.0	20.2	1.1	0	1.1	1.6

Horizon	Depth (cm)	Plagiocl.	Alkali-feldsp.	Weath. feldsp.	Epidote	Quartz	Rock fragm.	Others
Ah	0–1	17.2	8.9	4.7	0.7	29.5	4.4	3.3
AC	1–5	16.2	11.0	0.9	0.4	21.6	2.8	1.8
Bg2	20–30	18.6	14.0	3.3	0.5	26.6	2.2	0.9

b) Clay and fine silt fraction

Horizon	Depth (cm)	Magnetite	Quartz	Feldspars	Chlorite	Illite/mica	Smectite	Mixed layers	Kaolinites
Clay fraction (<2 μm ∅) in %, semiquantitative									
Ah	0–1	<1	20	58	<1	22	<1	0	0
AB	1–5	<1	17	16	4	43	20	<1	0
Bg2	20–30	<1	11	58	0	31	0	0	<1
Silt fraction (2–20 μm ∅) in %, semiquantitative									
Ah	0–1	6	22	34	5	33	0	0	0
AB	1–5	14	20	30	6	30	<1	0	0
Bg2	20–30	24	17	13	11	35	<1	0	0

[a] Anatase and rutile also present.

gneiss weathering. Stones and sands abound in the topsoil and the depth function of organic carbon is discontinuous due to a strong cryoturbation impact. Clay is formed by strong cryoclastic weathering. The soil data suggest that this also promotes chemical weathering: much pedogenic iron (Fe) was found (Table 8.3) as well as large amounts of partly weathered minerals, e.g. feldspars (Table 8.2); smectite is also formed. The acidic conditions in combination with variable redox conditions (strong mottling of subsoil: see Munsell colours, Chap. 7, Table 7.1, soil 2) may have influenced this process. Fe_p contents are higher than Fe_o contents (Table 7.1, soil 2), which suggests that large amounts of iron phosphates (together with Al phosphates) are present in the soil. Beyer and Bölter (1999) found 1.3–1.8 g kg^{-1} citrate-extractable P, which is regarded as a high amount of phosphate in a similar

Table 8.3. Chemical conditions of soils near Casey, Wilkes Land, Continental Antarctica

Horizon	cm	Fe_p	Fe_{o-p} (mg g^{-1})	Fe_{d-p}	o:d	Mn_o	Al_d (mg g^{-1})	Al_p	ODPE	Fe_p	Al_d in mg g^{-1} of clay	Mn_o	Eh (mV)
1 Hapli-turbic Cryosol on stony moraine													
Az	0–1	0.5	n.d.	3.0	n.d.	2.6	0.5	0.5	0.018	36	36	0.18	650
AC	–30	0.7	n.d.	3.0	n.d.	3.6	0.6	0.6	0.027	15	13	0.08	640
Cw	–70	0.9	n.d.	2.6	n.d.	2.7	0.5	0.6	0.040	14	8	0.04	600
2 Stagni-leptic Cryosol on basic gneiss to schist													
Ah	0–0.5	1.4	–0.8	2.1	0.2	22	3.3	n.d.	0.016	9.4	22	0.15	n.d.
AB	–5	6.6	–5.0	2.2	0.2	26	17.6	n.d.	0.026	19	52	0.08	490
Bg1	–20	5.7	–3.7	1.3	0.3	30	16.1	n.d.	0.040	15	42	0.08	670
Bg2	–30	7.3	–5.5	7.3	0.1	23	18.6	n.d.	0.033	16	41	0.05	420
4 Skeleti-gelic Podzol of a former penguin rookery on leuco gneiss													
AE	6–8	1.5	–0.9	0.9	0.25	5	0.4	0.6	0.37	30	8.0	0.1	n.d.
Bh	–11	5.9	–3.4	1.0	0.36	11	2.3	2.5	1.48	118	46	0.2	n.d.
Bsh1	–18	9.6	–4.9	1.4	0.43	20	3.9	4.7	1.12	74	30	0.2	n.d.
Bsh2	–30	6.5	–2.8	0.3	0.55	20	3.2	3.7	0.52	50	25	0.2	n.d.
Bsh3	–40	4.5	–1.1	1.9	0.53	28	3.5	3.4	0.41	90	70	0.6	n.d.
Cgw	–50	3.1	–0.9	1.6	0.47	16	1.1	2.2	0.06	62	22	0.3	n.d.

Table 8.4. Mineralogy of a Skeleti-gelic Podzol (former penguin rookery) on gneiss; Casey

a) Mean of 20–630 μm ∅ in % (*unw.* unweathered, *w.* medium weathered)

Horizon	Depth (cm)	Vulc. glas	Apatite	Chlorite		pyr.+ amp.		Micas		Strongly w. min.
				unw.	w	unw.	w.	unw.	w.	
AE	6–8	1.1	0.7	9.8	0	1.5	0.4	12.6	0	1.3
Bh	8–11	0	0	9.4	4.2	3.3	0	12.5	0.4	1.1
Bsh	11–30	0	0	16.7	29.8	0.4	0	17.5	0.4	3.8
Cgw	40–50	0	0	6.9	0	3.1	0	16.9	0.9	1.6

Horizon	Depth (cm)	Plagiocl.	Alkali-feldsp.	Weath. feldsp.	Quartz	Chalce-done	Opaque min[a]	Rock fragm.	Others[b]	Biota[c]
AE	6–8	18.6	14.2	1.8	30.8	0.9	2.1	1.6	1.9	many
Bh	8–11	19.3	17.7	2.2	25.1	0	0.6	1.8	1.9	0
Bsh	11–30	8.2	6.0	3.8	7.5	0.2	2.7	1.1	1.9	0
Cgw	40–50	21.5	13.3	2.7	18.9	1.3	1.3	7.3	4.3	0

b) Clay and fine silt fraction

Hor	Depth (cm)	Quartz	Feld-spars	Pyro xene	Mica illite	Mixed layers	Apa-tite	Stren-gite	Vivia-nite	Kaoli-nite	Chlo-rite
Clay fraction (<2 μm ∅) in %, semiquantitative											
AE	0–8	41	33	8	9	0	4	5	0	<1	0
Bh	8–11	26	36	17	6	0	15	0	0	0	0
Bsh	11–30	0	16	0	<1	0	80	4	0	0	0
Cgw	40–50	24	32	0	31	<1	6	0	7	0	0
Silt fraction (2–20 μm ∅) in %, semiquantitative											
Bh	8–11	40	13	5	15	0	27	0	0	<1	0
Bsh	11–30	10	10	0	23	0	45	10	0	0	2

[a] And anatase, ilmenite, rutile, zircon.
[b] Garnet, epidote, sillimanite.
[c] Sponge spicules.

soil. A high phosphate content can be related to impacts of seabirds on soil chemistry. The Fe_{d-p} contents are higher in the Bg2 horizon (Table 8.3), which is correlated with the occurrence of strong orange-coloured mottles (see Munsell colour, Table 7.1, soil 2). Both observations suggest that well-crystallised lepidocrocite is formed under stagnic, but not gleyic, moisture regime conditions.

The third profile, a Podzol, was formed at an abandoned penguin rookery. The birds collect coarse gravels and stones (Table 7.1, soil 4). Further, excrement, bones and feathers from the rookery sites (R.I. Lewis Smith, BAS, pers. comm.) can be observed. Their soil chemistry results in extremely high amounts of different phosphate minerals, mainly apatite but also strengite and vivianite in the clay fraction. The occurrence of the latter suggests that reduced conditions exist in the deeper subsoil during part of the year (Table 8.4). We believe that active cryoturbation was absent in the soil during the last centuries, because otherwise the clear horizontation and a strong enrichment of seaborne sponge spicules only in the topsoil would not exist. Their origin must be seen in relation to the upward movement of the sea level during the Holocene. A strong chemical weathering of special mineral species (e.g. chlorites, feldspars, micas) occurred, probably due to the formation of nitric and sulphuric acids from urea and the proteinaceous bird excrement (Beyer et al. 1997).

King George Island. The five studied soils of King George Island were formed under warmer conditions. These soils are covered by a dense vegetation. Only a Skeleti-stagnic Cryosol on a steep slope with strong solifluction and cryoturbation and a Calcari-regic Cryosol on a very young moraine (perhaps not more than 100 years) were sparsely covered by plants (Table 7.2, soils 10 and 16).

The Calcari-regic Cryosol was formed on a gravelly to blocky moraine of mainly tholeiite basalt origin: many plagioclases, especially andesine, zeolites and pyroxenes (Table 8.5) as well as volcanic glasses and ashes, and some carbonates. An accumulation of stones at the soil surface together with a weak stone circle formation is noted by Kuhn (1997) at this site. For this reason, the soil can be classified as a Cryosol, although the soil is not permanently frozen in most parts of the profile above 2 m during warm summers. Weathered minerals (feldspars, pyroxenes), substantial amounts of smectites (Table 8.5) and relatively high contents of pedogenic iron (Table 8.6a, soil 10) can be found. Some chalcedone and secondary quartz are also found (Table 8.5). This suggests that chemical weathering and formation of new minerals have taken place under former warmer climate conditions before glacier regression. Andic properties are reflected by volcanic glasses (Table 8.5) as well as by the definition of $Al_o+1/2Fe_o$ being 0.7 (Table 8.6b, soil 10). However, an Andic Cryosol should show a ratio>2 (ISSS/ISRIC/FAO 1998).

Table 8.5. Mineralogy of a Calcari-regic Cryosol on stony moraine without vegetation, Arctowski

a) Mean of 63–630 μm ∅ in % (*unw.* unweathered, *w.* medium weathered)

Horizon	Depth (cm)	Volc. glass	Olivine	Serpen-tine	Chlorite	Mica unw.	Mica w.	Pyroxene unw.	Pyroxene w.	Strongly w. min.
Aw	0–4	0.3	0	4.7	0	0	0	5.4	0	15.6
Cw2	10–30	0	0.3	2.0	0	1.0	1.7	5.2	0	25.1
Cw4	50–80	0.7	0.7	1.3	0.7	0.6	0	5.3	0	29.0

Horizon	Ande-sine	Plagiocl.	Alkali-feldsp.	Weath. feldsp.	Zeo-lites	Chalc. sec. qua.	Opaque min.	Rock fragm.	Apatite	Others
Aw	22.4	10.8	2.	4.3	4.4	0.3	1.7	25.6	0.3	0.8
Cw2	16.7	6.5	1.7	10.0	2.4	0.7	0.3	21.4	0	0
Cw4	12.0	12.4	1.0	12.0	2.7	0	2.3	17.1	0	2.1

b) Clay fraction

Horizon	Depth (cm)	Serpen-tine	Feldspars	Chlorite	Illite/mica	Olivine	Smectite	Pyroxene
Fraction (<2 μm ∅) in %, clay semiquantitative								
Aw	0–4	15	26	13	9	0	24	13
Cw1	–10	5	12	4	11		54	14
Cw2	–30	10	27	10	9	12	15	17
Cw4	50–80	20	19	31	12	9	0	9

Table 8.6. Pedogenic oxides in soils on King George Island. (After Kuhn 1997)

a) Iron, silica and organic matter

Horizon	Depth (cm)	Fe_p (mg g^{-1})	Fe_{o-p} (mg g^{-1})	Fe_{d-o} (mg g^{-1})	Fe_{v-d} (mg g^{-1})	Fe_o:Fe_d	Fe_o (mg g^{-1})	Si_o (mg g^{-1})	ODOE	mS:fS[a]
Site 10: Calcari-regic Cryosol on young moraine without vegetation										
Ah	0–4	0	7.1	5.5	29.6	0.56	3.9	1.3	0.02	0.8
BwC1	4–10	0	7.4	4.7	34.6	0.61	3.7	1.3	0.03	1.1
BwC2	10–30	0.1	4.5	8.0	24.2	0.36	2.5	1.8	0.04	1.0
IICw2	50–80	0.1	1.6	4.2	23.6	0.13	1.9	1.7	0.03	2.0
Site 16: Skeleti-stagnic Cryosol from basaltic solifluction deposits without vegetation										
Ah	0–3	n.d.	3.7	2.3	18.9	0.61	2.6	4.6	n.d.	1.7
Bg1	–32	n.d.	2.7	1.6	14.8	0.62	1.3	3.1	n.d.	1.7
Bg2	–58	n.d.	2.1	1.6	20.7	0.55	1.2	3.4	n.d.	n.d.
Cg	–80	n.d.	1.9	1.1	15.0	0.64	1.2	3.5	n.d.	2.0
Site 1: Molli-gelic Cambisol on basalt (>7 dm) with grasses and mosses										
Ah	0–8	1.2	3.4	9.8	20.2	0.32	5.1	n.d.	0.15	2.4
Ah2	–19	1.7	3.6	9.1	24.3	0.37	6.6	n.d.	0.14	2.7
II[b]Bw	–49	1.1	4.4	11.6	31.1	0.34	7.9	n.d.	0.10	1.9
Cw	–70	0.6	0.8	17.0	31.2	0.08	2.0	n.d.	0.07	1.9
Site 4: Skeleti-gelic Podzol on basalt (>4 dm) with grasses										
Ah	0–2	3.0	4.0	12.6	21.0	0.36	2.1	6.6	0.19	2.0
AE	–13	3.8	3.1	13.5	26.2	0.34	3.3	n.d.	0.22	n.d.
Bh	–15	7.1	1.3	15.1	24.5	0.36	4.0	4.8	0.37	2.6
Bhs	–18	7.0	2.2	14.5	28.6	0.39	7.1	4.3	0.47	n.d.
Bms	–25	3.0	6.3	10.1	20.7	0.46	7.1	n.d.	0.11	3.2
Bsw	–34	2.1	4.6	6.7	36.5	0.50	6.1	n.d.	0.08	n.d.
IIBw	–40	3.3	4.4	7.7	29.0	0.49	7.0	2.8	0.11	2.1

Table 8.6. (*Continued*)

Horizon	Depth (cm)	Fe_p (mg g^{-1})	Fe_{o-p} (mg g^{-1})	Fe_{d-o} (mg g^{-1})	Fe_{v-d} (mg g^{-1})	Fe_o:Fe_d	Fe_o (mg g^{-1})	Si_o (mg g^{-1})	ODOE	mS:fS[a]
Site 12: Humi-gelic Umbrisol of former penguin rookery with vegetation										
Ah	0–19	4.6	3.0	7.6	33.3	0.42	25.0	6.6	0.20	5.0
IifAh	–36	7.6	1.6	11.6	19.4	0.44	10.0	4.8	0.42	2.1
IIIfAh	–64	10.0	5.7	10.7	15.2	0.60	14.0	4.3	0.11	1.4
IVfAh	–86	6.6	3.8	3.5	20.7	0.75	26.0	2.8	0.08	2.4

b) Aluminium and manganese

Horizon	Al_a (mg g^{-1})	Al_{p-a} (mg g^{-1})	Al_{o-p} (mg g^{-1})	Al_d (mg g^{-1})	Al_o+1/2 Fe_o	Mn_o (mg g^{-1})	Mn_{d-o} (mg g^{-1})	Mn_{v-d} (mg g^{-1})	Al_o (in mg g^{-1} clay)	Mn_o
Site: 10 Calcari-regic Cryosol on young moraine without vegetation										
Ah	n.d.	n.d.	3.5[c]	1.4	7.0	0.23	0.04	0.65	19	1.3
BwC1	n.d.	n.d.	3.1	1.1	6.8	0.23	0.02	0.81	16	1.2
BwC2	n.d.	n.d.	5.3	1.4	7.6	0.12	0.12	0.68	24	0.5
IICw2	n.d.	n.d.	5.1	1.5	5.9	0.09	0.14	0.61	57	1.0
Site: 16 Skeleti-stagnic Cryosol from basaltic solifluction deposits without vegetation										
Ah	n.d.	n.d.	4.1[c]	0.66	6.0	0.59	–0.23	0.36	29	4.2
Bg1	n.d.	n.d.	2.0	<0.4	3.4	0.43	–0.20	0.39	11	2.4
Bg2	n.d.	n.d.	2.2	<0.4	3.3	0.46	–0.16	0.44	n.d.	n.d.
Cg	n.d.	n.d.	1.7	<0.4	2.7	0.42	–018	0.36	11	2.1

Table 8.6. (*Continued*)

Horizon	Al_a (mg g^{-1})	Al_{p-a} (mg g^{-1})	Al_{o-p} (mg g^{-1})	Al_d (mg g^{-1})	Al_o+1/2 Fe_o	Mn_o (mg g^{-1})	Mn_{d-o} (mg g^{-1})	Mn_{v-d} (mg g^{-1})	Al_o (in mg g^{-1} clay)	Mn_o (in mg g^{-1} clay)
Site: 1 Molli-gelic Cambisol on basalt (>7 dm) with grasses and mosses										
Ahi	0.6	0.3	4.5	2.5	7.7	0.19	0.40	0.04	54	1.9
Ah2	1.4	0	3.8	3.0	7.8	0.23	0.39	0.24	65	2.9
II[b]Bw	0.4	0.5	3.0	2.7	6.6	0.27	0.28	0.23	56	3.9
Cw	0.3	0.7	3.2	3.0	4.9	0.36	0.29	0.20	60	5.1
Site: 4 Skeleti-gelic Podzol on basalt (>4 dm) with grasses										
Ah	0.9	0.5	4.3	5.3	9.2	0.17	0.24	0.19	17	0.5
AE	1.7	2.6	2.0	6.2	8.8	0.13	0.32	0.23	22	0.6
Bh	1.5	2.0	3.3	7.5	11.0	0.10	0.09	0.59	32	0.5
Bhs	1.8	2.8	2.3	8.1	12.5	0.08	0.19	0.38	53	2.1
Bms	2.0	–0.5	2.6	6.3	9.7	0.23	0.36	0.17	31	1.8
Bsw	2.5	–1.2	1.8	3.5	6.5	0.20	0.46	0.16	28	1.8
IIBw	2.7	–1.2	2.0	3.8	7.3	0.39	0.38	0.17	32	3.5
Site: 12 Humi-gelic Umbrisol of former penguin rookery with vegetation										
Ah	1.3	0.6	2.7	5.8	8.4	0.03	0.07	0.50	154	1.0
IifAh	1.5	1.5	3.7	7.5	11.3	0.02	0.13	0.44	75	0.2
IIIfAh	2.3	0	4.7	5.9	14.7	0.04	0.26	0.19	62	0.4
IVfAh	2.7	0	3.6	3.8	11.5	0.04	0.06	0.42	158	1.0

[a] Medium sand:fine sand.
[b] New lithogenic layer.
[c] Al_o.

A Skeleti-stagnic Cryosol is formed from solifluction deposits without vegetation cover. Strong influences of solifluction and cryoturbation are documented by heterogeneous depth functions of stones, organic matter (Table 7.2, soil 16) and medium to fine sand quotients (Table 8.6a, soil 16). The silt fraction is dominated by pyroxine, feldspar, chlorite and mica, whereas the clay fraction also contains smectites (Table 8.7, soil 16).

The molli-gelic Cambisol is formed on basalt under a dense vegetation cover. The medium to fine sand quotients show that there is a texture discontinuation between top and subsoil (Table. 8.6a, soil 1). However, the humus contents decrease and the pH values increase regularly with depth (Table 7.2, soil 1), indicating stable conditions and no cryoturbation during recent centuries. A formation of smectite (Table. 8.7, soil 1) together with iron oxides (Table 8.6a, soil 1) is found in the Bw horizon.

The Skeleti-gelic Podzol is formed on tholeiite basalt. The mineralogy of this soil reflects this (Table 8.8). Permafrost exists at least in the second metre and the effects of cryoturbation can be observed in the field. Higher contents of stones and sand at the top, combined with decreasing contents of silt and clay with soil depth support this suggestion (Table 7.2, soil 4). The soil is strongly acidified and enriched with pedogenic iron: crystallised oxides as well as organically bound compounds (Table 8.6a, soil 4). To a great extent strongly weathered grains with a maximum in the topsoil were found: feldspars, mica and pyroxenes. However, a large amount could not be identified due to strong weathering (Table 8.8). In addition, obvious smectites were formed. Easily weatherable volcanic glasses and olivine without any weathering can be found only in the topsoil. This fact probably reflects the impact of recent volcanism. The last eruption of a volcano at Deception Island (around

Table 8.7. Mineralogy of a Cryosol and a Cambisol on King George Island

Horizon	Depth cm	Quartz	Olivine	Serpentine	Feldspar	Chlorite	Illite/mica	Pyroxene	Smectite
Site: 16 Skeleti-stagnic Cryosol from basaltic solifluction deposits without vegetation									
a) Silt fraction (2–20 μm) in %									
Aw	0–3	0	0	0	5	50	15	30	0
Bg2	32–58	0	0	0	30	10	30	30	0
b) Clay fraction in %									
Aw	0–3	0	0	0	30	40	30	0	0
Bg2	32–58	0	0	0	5	25	15	30	25
Site: 1 Molli-gelic Cambisol from basalt with dense vegetation									
Clay fraction in %									
Bw	19–49	0	0	10	20	0	10	20	40

Table 8.8. Mineralogy of a loamy Skeleti-gelic Podzol with andic phase on moraine of basalt, Arctowski

a) Mean of 20–630 μm ∅ in % (*unw.* unweathered, *w.* medium weathered)

Horizon	Depth (cm)	Volc. glass	Olivine	Serpen-tine	Chlorite	Mica unw.	Mica w.	Pyroxene unw.	Pyroxene w.	Strongly w. min.
Ah	0–2	1.6	6.8	2.0	1.2	1.6	2.8	6.0	0	36.8
Bhs	15–18	1.7	1.4	2.4	0.3	1.7	0	7.9	0	9.6
CBw	25–34	0	0	2.3	0	0	0	5.3	2.3	5.7

Horizon	Ande-sine	Plagiocl.	Alkali-feldsp.	Weathered feldsp.	Zeo lites	Chalc. sec. qua.	Opaque min.	Others	Rock fragm.
Ah	2.0	6.8	5.6	9.2	0.4	3.2	2.0	1.6	10.8
Bhs	11.6	1.7	2.7	8.9	0.2	3.8	4.1	0.3	41.6
CBw	19.7	1.0	1.7	5.3	3.7	0.7	1.3	0.3	50.2

b) Clay fraction (<2 μm ∅) in %, clay semiquantitative

Horizon	Depth (cm)	Quartz	Feldspars	Chlorite	Illite/mica	Olivine	Smectite	Pyroxene
Ah	0–2	<10	30	<5	10	<5	35	5
Bhs	15–18	< 5	30	25	<5	<5	25	10
CBw	25–34	< 5	10	30	<5	<5	40	<5

Table 8.9. Mineralogy of a Humi-gelic Umbrisol (former penguin rookery) on basalt, Arctowski

Depth (cm)	Smectite	Illite	Chlorite	Variscite	Vivianite	Apatite	Other phosphates
Fraction (<2 µm ∅) in %, clay semiquantitative							
43–48	<10	<10	<10	<10	10–20	<10	>20
64–86	<10	<10	<10	<10	10–20	<10	>20

160 km away) was recorded in the 1960s. The essential influence of birds is suggested from the high content of total phosphorus (Table 7.2, soil 4).

The Humi-gelic Umbrisol developed at an abandoned penguin rookery; the occurrence of penguin bones at a depth of between 20 and 80 cm (Kuhn 1997), the extremely high content of up to 12 % total P, a high gravel content, as well as a layering confirm this observation (Table 7.2, soil 12). Variscite, vivianite and apatite are found, as well as other phosphate minerals, which could not be identified by X-ray measurements (Table 8.9). Additionally, Tatur (1985) and Myrcha and Tatur (1991) describe Arctowskite $(Al_{8.4}Fe_{0.6})$ $(OH)_3$ $(PO_{4)8})$, Taranakite $[(K, NH_4)_3Al_5H_6(PO_{4)8}]$ and some amorphous Al-phosphates at similar sites (see also Chap. 10).

8.3 Soil-Forming Processes and Their Products

8.3.1 Weathering, Acidification and Mineral Formation

Strong physical weathering is normally combined with strong cryoturbation processes in permafrost regions (Bockheim 1995). Cryoturbation is primarily based on the freeze-thaw cycling of soil water. Rock fragments and minerals of the topsoil are crushed by freeze-thaw intervals. Cryoturbation mainly occurs during the summer months with no snow at the soil surface (Campbell and Claridge 1987). Such conditions exist for only 1 or 2 months on hill backs in Wilkes Land, at down-slopes for not more than 1 month. Depressions may remain snow covered during the whole year. At King George Island 3–4 months are favourable for soil formation. Freeze-thaw cycles are less pronounced due to higher average ambient temperatures and lower temperature amplitudes (Beyer and Bölter, Chap. 2). In addition, frost impact is reduced by a thick plant cover (below 60 m a.s.l.). Only soils of very young moraines with Calcari-regic Cryosols and on steep slopes above 50 m a.s.l. with the Skeleti-stagnic Cryosols are less insulated by vegetation carpets.

Frost weathering is mainly restricted to the topsoil. In the loamy soils, like the stagnic Cryosols, frost impact is valid for only some centimetres, whereas it extends to several decimetres in the gravel-rich soils, like the Podzol at Wilkes Land, due to their low water content and the low thermal capacity of the soil surfaces. Nevertheless, weathering in loamy soils is more intensive because of intensive cryoturbation during summer. Fresh soil particles are permanently translocated to the top from active cryoturbic processes which form new targets. Elevated clay contents of the Stagni-leptic Cryosols in the subsoils are probably not a result of stronger weathering because coarse soil particles are translocated to the soil surface. This promotes their disintegration.

Sandy to gravelly soils could have a well-developed lichen cover, which may insulate the mineral soil against freeze/thaw cycle impact (Melick et al. 1994). The extremely high gravel contents of the Podzols prevent frost weathering and only temperature weathering occurs. However, from an ecological point of view, the high gravel content reduces the water supply in soil for plants and microorganisms to a certain extent.

Soils from loose sediments are mostly Haplic Cryosols at Wilkes Land and King George Island. At the latter site soils contain volcanic glass and porous lime which disintegrate particularly easily. In the Stagni-leptic Cryosols gneiss and schist layers stretch vertically and favour penetration and freezing of water in fissures. High content of primary minerals in the clay fraction demonstrate that physical weathering also occurs in the very fine clay fraction. This is true for easily fissionable minerals, like feldspar and mica, but is also true for stable quartz. These findings are in full agreement with Bockheim and Ugolini (1990) as well as Beyer et al. (1999).

Nearly all investigated soils of the humid maritime and East Continental Antarctic are strongly acidified (Blume et al.1997; Beyer et al. 1999). Only soils on very young moraines with carbonates are neutral to weakly alkaline. This is commonly found in the soils of the dry Antarctic Cold Desert (Campbell and Claridge 1987; Beyer et al. 1999). Organic acids are present, but primarily nitric acid and sulphuric acid promote the proton production. Snow and soil algae are supposed to be the reason for acidification at the vegetation-free sites with Haplic Cryosols. Biomass mineralisation and subsequent processes can produce nitric and sulphuric acid (Blume and Bölter 1996), e.g. during nitrification and probably sulphurication. In addition, cyanobacteria promote these processes by nitrogen fixation. The recently more expressed acidification of soils from leuco gneiss is based on its low neutralisation capacity. Normally, soils from basalt and schist show higher neutralisation capacity. However, current data suggest that low temperatures retard neutralisation under cold conditions.

Bird excrement is alkaline, but finally promotes an acidification of the soil (Tatur 1989). Guano is base-rich and abandoned rookeries often show luxuriant vegetation. Phosphate analyses of many soils around Arctowski have shown that almost all sites below 60 to 70 m a.s.l. are affected. Huge areas were lifted in the Holocene due to melting of ice up to 100 m (Myrcha and Tatur 1991). Only soils of very young moraines are less or not influenced (Blume et al., Chap. 7).

Chemical weathering occurs in many soils. Selected mineral species like feldspar, pyroxene and mica show corrosive surfaces. The occurrence of pedogenic iron oxides (see also Beyer and Bölter 1999) as well as illites and smectites in the clay fraction reflect a soil solution-derived chemical weathering. As pointed out in the introduction, it is questionable whether chemical weathering occurs at low temperatures. Results from physical weathering show new surfaces for acid attacks (Kopp and Kowalkowski 1990). Water is available for soil-solution derived weathering in the warm summer, as under humid climate conditions. In addition, significant amounts of plant-derived organic complexes occur in soils despite small biomass production. Further, biogenic crusts, algal carpets as well as the lichen and/or moss cushions release organic compounds to soils. These organic moieties are often chemically aggressive (Beyer and Bölter 1999) and are able to bind weathering products and promote solution. Additionally, they are involved in the podzolisation processes (Blume et al. 1997; Beyer et al. 1997; Beyer and Bölter 2000). Myrcha and Tatur (1991) postulate for the Antarctic Peninsula region that iron is entirely leached from the soil by the strong action of guano solutions, which results in the formation of bleached soil horizons. In summary, these findings on acidification, brownification and sometimes podzolisation under cold but humid periglacial conditions are in agreement with observations made by Kopp (1972), Kowalkowski and Borzyskowiski (1973), and others who suggest similar conditions in central Europe during the late Pleistocene.

However, the question still remains, why do other researchers think that chemical weathering does not play a remarkable role in polar desert and subantarctic tundra soils? Sometimes their observations are similar to ours, but their interpretation is different. For instance, O'Brien et al. (1979) analysed soil on an old moraine from metamorphic rocks at Elephant Island. This soil is extremely acid and most of the minerals of the sand fraction are intensively weathered with no possibility of identification. O'Brien et al. (1979) believe in a low intensity of weathering, because they did not find differences between topsoil and subsoil. We think that the difference was not detectable because horizons were very strongly mixed by cryoturbation.

Mineral formation is also suggested to be meaningful in some Antarctic soils. In Stagni-leptic Cryosol from schist and the Gelic Podzol from basalt, illite and smectite are identified in the clay fraction, but no kaolinite. This can be attributed to a mineral reorganisation by potassium impoverishment from micas, while no formation of new kaolinite from weathering products of feldspars occurs. However, new formations of smectites from solution products of pyroxene in soils from basalt has to be taken into account because mica contents are only low. In addition, iron oxides are formed: Goethite is identified and brownification of the soils with vegetation cover takes place. The reddish brown colour of the B horizons (e.g. 7.5 YR 5/8, Table 7.1b) suggests the occurrence of ferrihydrite. However, the content of Fe_o is generally low, whereas the content of Fe_p is often high, combined with a high ODPE value. This suggests the formation of organically bound iron forms beside iron phosphates. The process of 'rubification' mentioned by Bockheim and Ugolini (1990) is therefore suggested to be mainly a formation of organic iron complexes which we found in our soils (Beyer et al. 1997). In addition, iron oxides are also formed on rock as well as on stones under the influence of lichens within the first few millimetres (Chen and Gong 1995; Chen et al. 2000).

8.3.2 Humus Accumulation and Biotic Aggregation

The humus content of soils without vegetation, covered by snow for more than 11 months and sometimes all year-round, is low in soils of Wilkes Land. The carbon level of 0.2 – 0.4 % total organic carbon (TOC, Table 7.1) is derived from snow and soil algae and carbon is probably translocated and preserved by cryoturbation into the subsurface soil (Blume and Bölter 1996). Bockheim (1993) describes 0.1 % TOC in soils of the Antarctic cold desert. However, the observed contents of 0.2–0.4 % are high in comparison to most hot desert soils (Blume et al. 1984). Two to 3 % TOC within the entire the profile occurs in sandy soils which have no snow cover for at least 2 or 3 months and which are covered by lichens (Blume et al., Chap. 7, Table 7.1). In loamy soils with many soil algae but no lichens, approximately 1 % TOC occurs. This contradicts the findings of Bockheim and Ugolini (1990) who suggest that only the occurrence of well-drained 'ahumic' soils for coastal regions as at Wilkes Land.

On slope terraces, where there are steps of bedrock with meltwater stagnation, peat is formed from mosses at Wilkes Land (Blume et al., Chap. 7, Table 7.1, soil 5). The peat is strongly humidified (h8 to h9, after v. Post) in the subsurface soil horizons and it contains substantial amounts of aromatic, phenolic and carboxylic compounds (Beyer et al. 1995). The mean radiocarbon-dated age of the surface layer is 850 (±40) years and the age of the subsurface layer 1420 (±50) years (Blume et al. 1997).

At the bottom of shallow meltwater lakes a 10- to 20-cm-thick layer of dead algae sometimes accumulates (Blume et al., Chap. 7, Table 7.1, soil 6) which shows no visible humification (h1–2). The mean radiocarbon dated age of the top centimetre is about 290 (±35) years, of deeper layers 850 (±40) years and 2110 (±50) years (Blume et al. 1997). This refers to a long period of organic matter accumulation. Due to the high heat capacity of water, the soil temperature at the bottom of the lake does not rise above 2–4 °C, and the microbial activity is supposed to be extremely low. However, as soon as such peat soils (Fibric Histosols) become dry, aerated and warm, microbes decompose organic matter (OM) and humification occurs (Blume and Bölter 1996). An accumulation of 5.2 kg TOC m^{-2} during a suggested age of such lakes of 300 years reflects a mean annual deposition rate of 1.8 g TOC m^{-2} a^{-1}. This is 2–4 % of the quantity which is accumulated under temperate climate conditions.

Soils at King George Island show a similar accumulation pattern of OM. Poorly drained Histosols and Gleysols have 20 to more than 80 kg TOC m^{-2} in the Subantarctic Tundra and 5 to more than 20 kg in the Polar Desert (Blume et al. 1997). Well-drained Cambisols and Podzols can show even higher amounts than these soils. Only soils without any vegetation cover and a short time without snow cover show low amounts of between 1 to 2 kg TOC m^{-2} at both sites.

We suggest three reasons for these observations of OM accumulation:

1. Biomass production is low in cold regions, but higher than micro-biological activity in soil;
2. The OM is translocated from the topsoil into subsoil by cryoturbation in loamy soils and by podzolisation in sandy soils, i.e. a preservation against decomposition;
3. Penguins and other seabirds transfer biogenic organic matter from sea to soil with their droppings. However, there is not only an external OM accumulation at the recent rookeries. A fair amount of abandoned rookeries exist at higher landscape positions, due to a lifting during the Holocene by the melting of the ice cover (Myrcha and Tatur 1991).

Additionally, some soil animals like acari, collembolans and nematodes may produce small excremental grains with loose structure in the Gelic Cambisols and Phaeozems and Podzols (Fig. 8.2, No. 5) of King George Island (Bölter et al. 1997). In contrast to these soils with biological activity, many Cryosols have a dense structure without biogenic aggregates (e.g. Fig. 8.2, No. 4).

8.3.3 Podzolisation

One of the most striking features at the observed landscapes is the occurrence of soils with podzolisation at King George Island (Kuhn 1997; Blume et al. 1997) as well as at Wilkes Land (Blume 1994; Beyer and Bölter 2000), which is in contrast to the currently reviewed soil materials (e.g. Campbell and Claridge, 1987; Bockheim and Ugolini, 1990). These soils show a slightly leached surface soil layer (AE horizon) or an already well-developed leached (albic!) E horizon and an organic matter accumulated (spodic!) Bh horizon in the subsurface soil (ISSS/ISRIC/FAO 1998). At King George Island mobile organic iron complexes are translocated in the landscape (Fig. 8.1, No. 4).

A Skeleti-gelic Podzol with some andic properties at King George Island is characterised by high ODOE values and high contents of pedogenic iron, the required properties to be classified as a Podzol (ISSS/ISRIC/FAO 1998). In

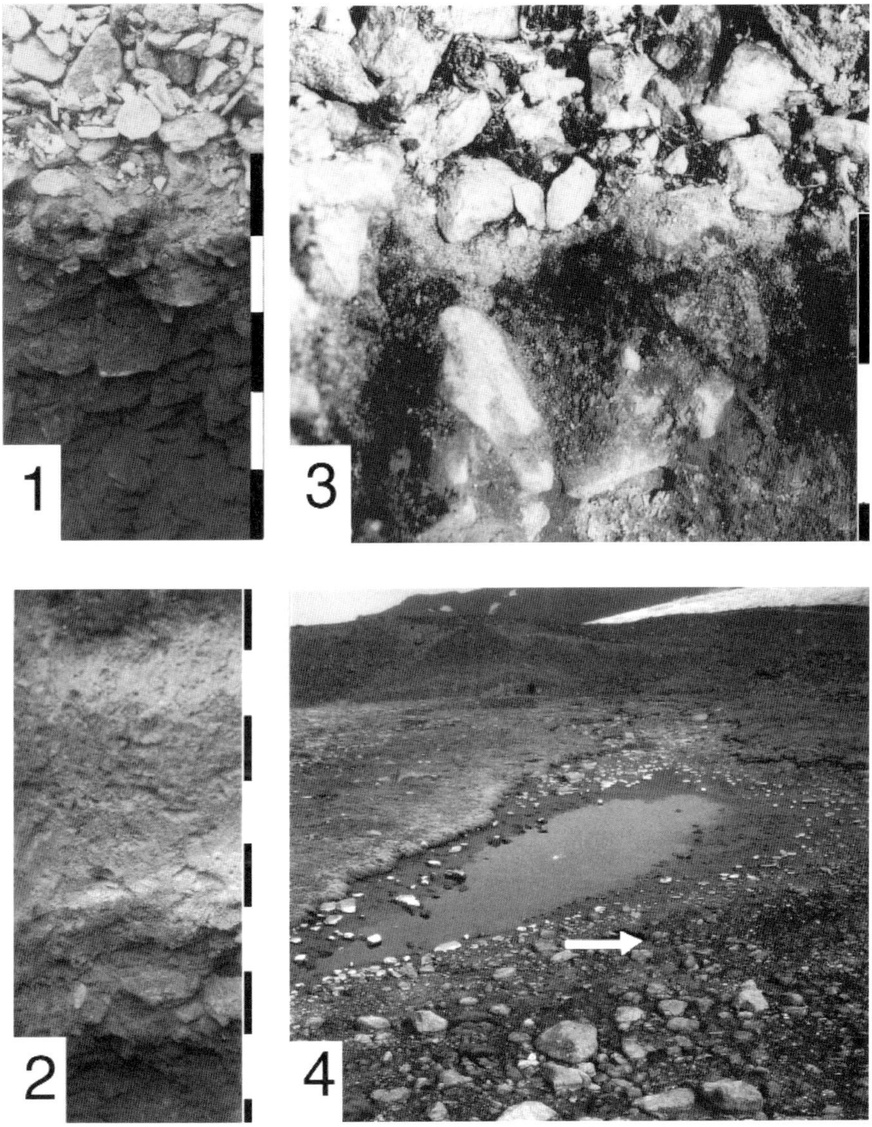

Fig. 8.1. Antarctic soils and sites. **1** Skeleti-gelic Podzol of a former penguin rookery, Wilkes Land (Blume et al., Chap. 7., Table 7.1, soil 4); **2** Humi-gelic Umbrisol of a former penguin rookery, King George Island (Table 7.2, soil 12); **3** Lepti-gelic Podzol, Wilkes Land (Table 7.1, soil 3); **4** meltwater lake and shore with enrichment of iron oxides on soil surface (*arrow*) at King George Island (scales in dm)

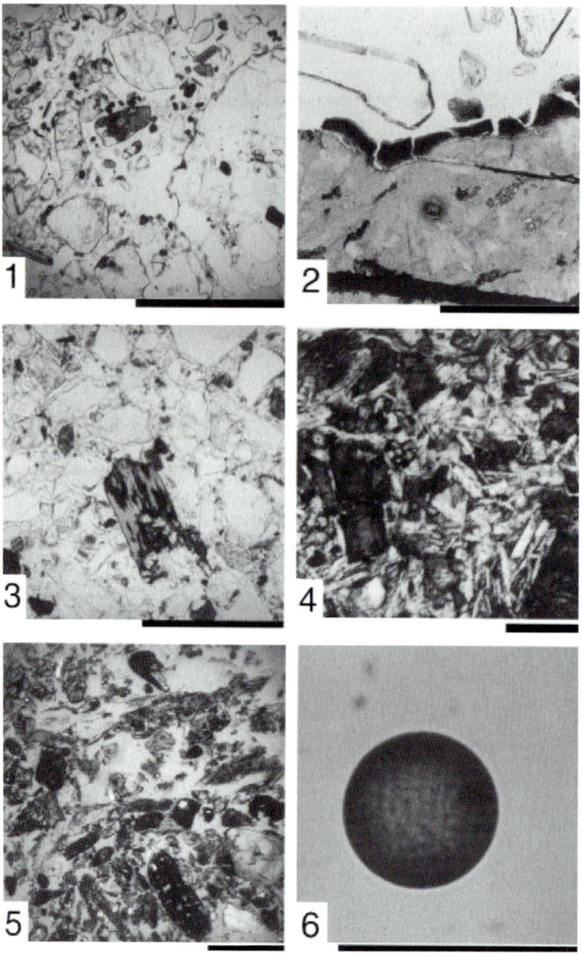

Fig. 8.2. Micromorphology of Antarctic soils: **1** and **2** Bh horizon, **3** Bhs horizon of a Lepti-gelic Podzol (Table 7.1, soil 3); **4** Ah horizon of a Hapli-turbic Cryosol (Table 7.2, soil 4); **5** Ah horizon and **6** volcanic glass of a Skeleti-gelic Podzol of King George Island (Table 7.2, soil 4). **1, 3–5,** scale 1 mm; **2,** scale 0.1 mm; **6,** scale 0.025 mm

addition, an Ortstein layer, an extremely expressed spodic layer, and coatings on grain are observed (Table 8.2, soil 4; Table 8.6; soil 4). At Wilkes Land the thickness of the albic E horizon is normally very shallow, less than 1–3 cm in most cases (e.g. Table 8.1. soil 3; Fig. 8.1 c). The spodic Bh horizon of such soils have typical coatings of precipitated metallo-organic compounds on the grains (Fig. 8.2a–c). Podzols with thicker albic E are only formed at abandoned penguin rookeries (Fig. 8.1a; Table 8.3, soil 4). These Podzols are less acid and are characterised by narrow C/N ratios than their counterparts at non-ornithogenic sites (Beyer and Bölter 2000). This is due an enrichment of proteins, which is not found in temperate Podzols (Blume et al. 1997). Moreover, humic acids and their salts (humates) with high aromaticity dominate the SOM, whereas fulvic acids and their salts (fulvates) occur to a greater extent in temperate Podzols (Beyer and Bölter 2000). Many of the investigated Podzols are found at abandoned penguin rookeries, which means that more or less humified excremental products affect the podzolisation processes. Nevertheless, non-ornithogenic Podzols occur, e.g. the Skeleti-gelic Podzol on fluvioglacial gravels at Wilkes Land (Blume et al., Chap. 7, Table 7.1, soil 9).

8.3.4 Carbonisation, Salinisation and Alkalisation

No accumulation of secondary carbonates or any alkalinity is observed within the survey of 200 soils at Wilkes Land and King George Island. Some carbonates were found in soils on very young moraines near glaciers at King George Island from carbonate-containing till (e.g. Blume et al., Chap. 7, Table 7.2, soil 10). However, associated and similar soils at older moraines are already decarbonated and more or less acidified. Only the shore soils along the coastal line at Wilkes Land are enriched with salts from the ocean in the first centimetres (Blume et al., Chap. 7, Table 7.1) after longer periods of dryness. Salt accumulation in the first centimetres of the soil is caused by an upward movement due to transpiration of soil moisture within the 4 to 6 weeks of thawing and aridity. In some years salts in the snow cover might be enriched at the soil surface due to evaporation of snow without melting. Occasionally, elevated concentrations of NH_4 and NO_3 can be monitored. Some nitrogen compounds may derive from the mineralisation of necrotic cyanobacteria in snow and nitrogen-fixing lichens on soil (Bölter and Blume 1996). Nevertheless, in summary, we suggest that humid conditions and soil water leaching dominate most sites of Wilkes Land and King George Island, and carbonisation, salinisation and alkalisation are of minor importance or do not occur. This is in contrast to the reviewed data of the arid Antarctic Cold Desert of the Dry Valleys given by Campbell and Claridge (1987).

8.3.5 Redoximorphism

Subsurface horizons in loamy soils frequently have Fe and Mn mottles which reflect redoximorphic features at King George Island (Blume et al., Chap. 7, Table 7.2, soil 16) as well as at Wilkes Land (Blume et al., Chap. 7, Table 7.1, soils 2, 4 and 8). Obviously, aerobic and anaerobic soil conditions alternate with each other and occasionally water saturation occurs. Vivianite is only persistent under anaerobic conditions, but is found in the subsoil of a Skeleti-gelic Podzol from a former penguin site (Table 8.4). This means that microbial activity is high enough in the compacted subsoil with stagnic water to express oxygen deficiency and reduction of Fe^{3+}. In addition, peat is accumulated in landscape positions at King George Island and Wilkes Land (Blume et al. 1997; Blume et al. Chap. 7), where well-drained soils are found in other parts of the world. A low evapotranspiration of the precipitation, combined with the occurrence of stagnic water are suggested to be the predominant reasons for such observations. This confirms that mostly humid and not arid conditions dominate in the soils during summer in maritime Antarctica as well as the coastal regions of East Antarctica.

8.4 Conclusions

The mineralogy of the sand and silt fractions and, in part, of the clay fraction too, reflects the geology of the sites: leuco gneiss, basic gneiss and schist near Casey, tholeiite basalt and volcanic ashes near Arctowski. Therefore, cryoclastic weathering, in combination with cryoturbation are the main soil forming processes. Most of the soils studied are medium to highly acidified, down to pH 4. Further, chemical weathering takes place, especially in soils with dense vegetation cover. Surface corrosion leads to some mineral species like feldspar, pyroxene and mica. The appearance of pedogenic iron oxides, illites and smectites in the clay fraction indicates relatively strong chemical weathering processes. This chemical weathering under cold conditions is promoted by strong physical weathering and cryoturbation. A speciality of many Antarctic soils is the occurrence of phosphate minerals like apatite, Arctowskite, strengite, varicite and vivianite due to a large influence of birds, especially penguins.

Sites in the Antarctic are characterised by an extremely low biomass production, on the one hand, and relatively high organic matter content, on the other, which may reach several percent by weight and which may appear down to several decimetres below the soil surface. This strong accumulation of humus probably results from a translocation of water-soluble organic matter from the topsoil to the subsoil by water percolation (=podzolisation) in sandy soils, and by strong cryoturbation in loamy soils. Further, biological activity in the soil is much lower in relation to biomass production on the soil due to a lower mean temperature. A large transfer of biogenic organic matter from the sea to soils by penguins and other seabirds can be assumed.

There is nearly no accumulation of carbonates and water-soluble salts in soils of the Antarctic islands and of continental East Antarctica in contrast to soils of continental Antarctic deserts. This obviously results from humid climate conditions with down ward movement of soil water during the Antarctic summer.

Acknowledgements. This work was financially supported by the DFG in Germany and logistically by the Australian National Research Expeditions (with the 1991 Casey personnel), and by the Polish Academy of Science (with the 1995 Arctowski personnel).

References

Beyer L, Sorge C, Blume H-P, Schulten H (1995) Soil organic matter composition and transformation in gelic Histosols of coastal continental Antarctica. Soil Biol Biochem 27:1279–1288

Beyer L, Blume H-P, Knicker H, Bölter M (1997) Soil organic matter of suggested spodic horizons in relic ornithogenic soils of coastal Continental Antarctica (Casey Station, Wilkes Land) and spodic horizons in soils of Germany. Soil Sci 162:518–527

Beyer L, Pingpank K, Bölter M, Seppelt RD (1998) Small-distance variation of carbon and nitrogen in mineral Antarctic Cryosols near Casey Station (Wilkes Land). Z Pflanzenernähr Bodenk 161:211–220

Beyer L, Bockheim JG, Campbell IB, Claridge, GGC (1999) Genesis, properties and sensitivity of Antarctic Gelisols. Antarct Sci 11:387–398

Beyer L, Bölter M (1999) Formation, ecology, and geography of Cryosols of an ice-free oasis in coastal East Antarctica near Casey Station, (Wilkes Land). Aust J Soil Res 37:209–244

Beyer L, Bölter M (2000) Chemical and biological properties, formation, occurrence and classification of Spodic Cryosols in a terrestrial ecosystem of East Antarctica (Wilkes Land). Catena 39:95–119

Birkenmajer K (1980) Geology of Admiralty Bay, King George Island (South Shetland Islands, Antarctica). An outline. Pol Polar Res 1:29–54

Blight DL, Oliver RL (1977) The metamorphic geology of the Windmill Islands, Antarctica. J Geol Soc Aust 24:239–262

Blume H-P (1994) Soils of Casey, Antarctica. 15th World Congr Soil Sci AcapulcoTransact 6b, pp 382–383

Blume H-P, Alaily F, Smettan U, Zielinski G (1984) Soil types and associations of SW Egypt. Berl Geowiss Abh (A) 50:293–302

Blume H-P, Beyer L, Bölter M, Erlenkeuser H, Kalk E, Kneesch S, Pfisterer U, Schneider D (1997) Pedogenic zonation in soils of the southern circumpolar region. Adv GeoEcol 30:69–90

Blume H-P, Bölter M (1993) Soils of Casey Station (Wilkes Land, Antarctica). In: Gilichinski D (ed) Joint Russion-American seminar on cryopedology and global change. Post-Sem Proc 1st Intern Conf. on Cryopedology 1992. Russian Academy of Sci., Pushchino, pp 96–103

Blume H-P, Bölter M (1996) Wechselwirkungen zwischen Boden- und Vegetationsentwicklung in der kontinentalen Antarktis. Verh Ges Ökol 25:25–34

Bockheim J (1993) Global change and soil formation in the Antarctic region. In: Gilichinski D (ed) Post-Sem Proc 1st Int Conf Cryopedol, Russian Academy of Science; Pushchino, pp 132–140

Bockheim J (1995) Permafrost distribution in the Southern circumpolar region and its relation to the environment. Permafrost Periglacial Process 6:27–45

Bockheim J, Ugolini F (1990) A review of pedogenic zonation in well-drained soils of the Southern circumpolar region. Quat Res 34:47–66

Bölter M (1992) Environmental conditions and microbiological properties from soils and lichens from Antarctica (Casey Station, Wilkes Land). Polar Biol 11:591–599

Bölter M, Blume H-P, Schneider D, Beyer L (1997) Soil properties and distributions of invertebrates and bacteria from King George Island (Arctowski Station), maritime Antarctic. Polar Biol 18:295–304
Campbell IB, Claridge CGG (1987) Antarctica – soils, weathering processes and climate. Elsevier, Amsterdam
Chen J, Blume H-P, Beyer L (2000) Weathering of rocks induced by lichen colonization – a review. Catena 39:121–146
Chen J, Gong ZT (1995) Role of lichens in weathering and soil-forming processes in Fildes Peninsula, Antarct Pedosph 5:305–314
Goodwin J (1993) Holocene deglaciation sea level change and the emergence of the Windmill Islands, Budd Coast, Antarctica. Quatern Res 37:10–18
ISSS-WRB [International Soil Science Society – World Reference Base for Soil Resources] (1998) World Reference base for soil resources. World Soil Resources Reports 84, ISSSS-ISRIC-FAO, Rome
Kopp D (1972) Periglaziale Umlagerungs- (Perstruktions-)zonen im nordmitteleuropäischen Tiefland und ihre bodengenetische Bedeutung. Tag-Ber deutsche Landwirtsch-Wiss Berlin 102:55–81
Kopp D, Kowalkowski A (1990) Cryogenic and pedogenic perstraction in Tertiary and Quaternary deposits, as exemplified in the outcrop of Sternebeck. Quat Stud Poland 9:1–71
Kowalkowski A, Borzyskowiski J (1973) The role of periglacial and extraperiglacial perstructions in the formation of soil profile in central Europe. Folia Quat 49:37–45
Kuhn D (1997) Genese, Ökologie und Soziologie einer Bodengesellschaft in einem Periglazialgebiet der König-Georg-Insel (West-Antarktis). Schriftenr Inst Pflanzenern Bodenkd, Univ Kiel, No 40
Melick D, Hovenden M, Seppelt RD (1994) Phytogeography of bryophyte and lichen vegetation in the Windmill Islands. Continental Antarctica. Vegetation 111:71–87
Myrcha A, Tatur A (1991) Ecological role of the current and abandoned penguin rookeries in the land environment of the maritime Antarctic. Pol Polar Res 12:3–24
O'Brien R, Romans G, Robertson L (1979) Three soil profiles from Elephant Island, South Shetland Islands. Br Antarct Surv Bull 47:1–12
Schlichting E, Blume H-P, Stahr K (1995) Bodenkundliches Praktikum, 2. Aufl. Blackwell, Berlin
Tatur A (1989) Ornithogenic soils of maritime Antarctic. Pol Polar Res 10:481–532

9 Refractory Soil Organic Matter – Formation, Accumulation, Translocation and Transformation

L. Beyer, D.M. White, H.-P. Blume, M. Bölter, D. Kuhn, K. Pingpank, and B. Vogt

9.1 Introduction

Soil formation in the Antarctic is a recognised process. The occurrence of accumulations of organic matter despite a hostile climate, and formation of organic soils by debris of mosses, lichens and algae is widespread (Campbell and Claridge 1987; Beyer et al. 1999). This leads to mineral soils which may contain large amounts of soil organic matter in the maritime Antarctic (Blume et al. 1996). In contrast, patches of organic soils are associated locally with shallow mineral soils in coastal regions of East Antarctica (Beyer et al. 1999).

In general, very little information is available on soil organic matter (SOM) of Antarctic terrestrial ecosystems in its broader sense. Reports about this soil constituent consider either the bulk matter, estimated by combustion, or individual dissolved carbohydrates (e.g. Tearle 1987; Roser et al. 1992). Most terrestrial carbon and nitrogen stocks in Antarctic ecosystems should be understood as fragile factors in the environment (Chen and Blume 1998; Campbell et al. 1998). For soil ecological and biological studies, specific information on SOM is required to quantify the geo-ecological impact of SOM on plant adaptation (e.g. Broady and Weinstein 1998; Seppelt and Broady 1988) and microbial colonisation (Wynn-Williams 1990, 1993; Bölter 1997; Bölter et al. 1997). However, the knowledge of SOM composition and formation as well as transformation processes under the extreme Antarctic climate conditions are still incomplete with respect to bulk soil samples and entire profiles (Wilson et al. 1986; Calace et al. 1995).

To provide a better understanding of SOM accumulation in Antarctica, carbon and nitrogen distributions of selected mineral soils and organic soils in the coastal region of the Windmill Islands near Casey Station (East Antarctic) and on King George Island near Arctowski Station (maritime Antarctic) were investigated. In addition, an integrated approach using wet-chemical analyses, CPMAS ^{13}C-NMR spectroscopy and modern pyrolysis tools (Py-FIMS and

Ecological Studies, Vol. 154
L. Beyer and M. Bölter (eds.) Geoecology of Antarctic Ice-Free Coastal Landscapes

Py-GC/MS) was used to characterise SOM composition, transformation and its translocation from the surface into the subsurface horizons. The results are compared with well-described soil formation processes under temperate climate conditions.

9.2 Sites and Analyses

Sites. The study site near the Australian Casey Station in the Windmill Islands (WMI) is described in detail by Seppelt (Chap. 4). Parent materials are weathered gneiss and schist, moraine deposits and outwash gravel (Paul et al. 1995). The study site in Maritime Antarctica close to the Polish Arctowski Station at King George Island (KGI) is presented by Rakusa-Suszczewski (Chap. 3). The geology is characterised by tertiary basalts, andesites and volcanic eruption materials (Birkenmajer 1989; Tingey 1991), however, the soil parent material is mainly moraine rubble, eolian deposits and volcanic ashes (Campbell and Claridge 1987). Both regions are formed by permanent land lifting after the retreat of the ice masses and the former coastline is found nowadays some 100 m further inland. Relic penguin rookeries are scattered within the whole ice-free area. Most soils are permafrost-affected and the active layer maximum is approximately 50–100 cm (Kuhn 1997; Beyer et al. 1999).

Vegetation. Vegetation of the Windmill Island (WMI) region consists of mosses (e.g. *Bryum pseudotriquetrum, Ceratodon purpureus, Grimmia antarctici*) and lichens (e.g. *Usnea sphacelata, Pseudephebe minuscula, Umbilicaria decussata*). Algae (e.g. *Prasiococcus calcareus, Desmococcus olivaceus*) and cyanobacteria (e.g. *Nostoc* sp., *Phormidium* sp.) cover barren ground (Seppelt, Chap. 13; Olech, Chap. 12). Vegetation establishment and growth are slow because of low temperatures (Seppelt and Broady 1988; Smith 1990) and a limited availability of free water (Kennedy 1993). The current vegetation in the ice-free areas of King George Island (KGI) is characterised by the occurrence of the higher plants *Deschampsia antarctica* and *Colobanthus quitensis*. According to Komárková (1985), the vascular plants have their ecological optimum in the Admiralty Bay region. Mosses (bryophyta), some liver mosses, and lichens form wide stands on moraines and/or in wet areas (Olech, Chap. 12). Nevertheless, huge parts of the landscape are non-vegetated and the occurrence of vegetation shows a high variability and an extreme patchiness.

Sampling and Methods. Sampling in the KGI region was carried out in January/February 1992 and in the WMI region in January/February 1996. Soils are classified according to the Soil Taxonomy using the Gelisol order (Soil Survey Staff 1998). All chemical and physical analyses and determinations were carried out according to Schlichting et al. (1995). The 'cross-polarisation magic angle spinning' (CPMAS) ^{13}C-magnetic resonance spectra were recorded from milled bulk soil samples (<200 μm). The spectra were obtained at 2.3 T (25.2 MHz) in a Bruker MSL 100 spectrometer equipped with a commercial 7-mm CPMAS probe at a rotation frequency of 4 kHz. It has been shown that under the experimental conditions used the relative signal area can be quantitatively related to the C concentration (Fründ and Lüdemann 1989). Details are described in Beyer et al. (1997). The pyrolysis-field ionisation mass (Py-FIMS) spectra were recorded using a double-focusing Finnigan MAT 731 mass spectrometer, details are described in Beyer et al. (1992). For the pyrolysis gas chromatography mass spectrometry (Py-GC/MS) subsamples with a mass of 25 mg were placed in quartz tubes (2 cm × 2 mm ∅) and quantified using a Mettler microbalance. The pyrolysis reactor was mounted on a GC, HP 5890 Series II. Details of the method used are described in Beyer et al. (2001) and White and Beyer (1999).

9.3 Soil Organic Matter – Distribution and Accumulation

WMI Region. The mean carbon storage in mineral soil is close to 3 kg m^{-2}, whereas for nitrogen, the storage was approximately 300 g m^{-2} (Table 9.1). Local nitrogen concentrations can be found at abandoned penguin nest sites, leading to a mean C/N ratio around 8. However, there is great variation within numerous mineral soils and almost no differentiation is possible at soil order/suborder or level. Only spodic soils and Haploturbels with cryoturbic features due to the effects of thawing and freezing show significantly higher carbon pools. However, the occurrence of the Haploturbels is very rare, whereas all mineral soils with a cover of crusts are influenced by podzolisation (Beyer et al. 1999, 2000a; Beyer and Bölter 2000). No significant distinction of ornithogenic soils from other mineral soils is possible. More detailed SOM investigations for spodic soils should provide a better data base for pedogenic and soil ecological purposes in the WMI region. Additionally, organic soils can be found to a small extent (Beyer 2000), but they are characterised by an extremely high SOM accumulation, maximum >45 kg m^{-2} (Table 9.1). These soils are probably of major concern with respect to SOM as a carbon and nitrogen sink.

KGI Region. The mean carbon storage in soil is twice as high (factor 2.1) as in the WMI region, whereas the factor for nitrogen is 3.4 (cf. data in Tables 9.1 and 9.2), mean C/N ratios are around 7 (Table 9.2). This suggests a great and widespread ornithogenic impact on most soils (Blume et al. 1997, Beyer et al. 2000a). In contrast to the WMI region, no organic soils are found. This is in agreement with the observations made by Blume et al. (1996). However, min-

Table 9.1. Carbon and nitrogen storage in soils of the coastal Windmill Islands (WMI) region near Casey Station in East Antarctic

		TOC			N_t			C/N		
	n	Min.	Max.	∅	Min.	Max.	∅	~~Min.~~	Max.	∅
		(g m^{-2})								
Mineral soils										
Over all	31	108	8297	2906[a]	14	1396	264[a]	4	31	12[a]
Haplorthel	25	108	8297	2736[a]	14	1396	270[a]	4	31	12[a]
Lithic	20	174	8297	2822[a]	19	1396	269[a]	4	31	12[a]
Typic	4	1809	4621	2962[a]	205	389	305[a]	6	13	10[a]
Spodic	7	1345	8160	3557[a,b]	78	389	271[a]	6	26	14[a,b]
Ornithogenic	6	864	7684	2499[a]	78	1396	402[a]	4	17	8[a]
Aquiturbel	3	767	3445	2315[a]	115	307	193[a]	5	15	11[a]
Haploturbel	3	3416	6126	4329[b]	285	347	307[a]	10	22	15[a,b]
Organic soils										
Fibristel/ Hemistel	6	5200	45600	25497[c]	244	2583	1536[b]	12	22	18[b]

[a,b] Same exponent means no significant difference

Table 9.2. Carbon and nitrogen storage in soils of the King George Island (KGI) region near Arctowski Station in the Maritime Antarctic. *n.d.* not detected

		TOC			N_t			C/N		
	n	Min.	Max.	∅	Min.	Max.	∅	Min.	Max.	∅
		(g m^{-2})								
Over all	17	88	38872	6192[a]	20	5121	888[a]	3	11	7[a]
Typic Haplorthel[c]	7	252	5490	2080[a]	49	908	336[a]	3	11	7[a]
All Turbels	6	214	38872	10538[a]	40	5121	1333[a]	4	10	7[a]
Aquiturbel	1	n.d.	n.d.	280	n.d.	n.d.	75	n.d.	n.d.	4
Spodic soils	2	3655	5490	4573[a]	497	655	576[a]	6	11	9[a]
Lithic soils	4	88	6821	2212[a]	20	743	267[a]	4	9	6[a]
Ornitho-genic soils	3	18176	38872	25266[b]	2154	5121	3560[b]	5	9	7[a]
Cumulic soils	2	18176	38872	28524[b]	3404	5121	4263[b]	5	6	7[a]

[a,b] Same exponent means no significant difference
[c] Including Psammorthel

Table 9.3. Selected properties of humic and spodic soils in the Maritime Antarctica near Arctowski Station (King-George Island) and in East Antarctica near Casey Station (Windmill Islands region). CEC_p Potential cation exchange capacity in $mmol_c\ kg^{-1}$; *BS* base saturation in % of CEC_p; P_c citrate-soluble phosphorus; *TOC* total organic carbon; N_t total nitrogen; *ODOE* optical density of the oxalate extract (Soil Survey Staff 1998); Fe_d dithionate-extractable iron; Fe_{ox} oxalate-extractable iron; Fe_{org} NaOH-extractable iron; *n.d.* not determined

Horizon	Depth (cm)	Colour (moist)	pH ($CaCl_2$)	CEC_p	BS (%)	P_c	TOC (mg g^{-1})	N_t	C/N	ODOE	Fe_d	Fe_{ox}	Fe_{org}
Loamy-skeletal lithic Haplorturbel *(C5)* from solid rock under mosses (Casey)													
A1	0–5	7.5YR2/1	3.5	340	14	0.4	116.3	6.70	17.3	663	3.5	2.1	1.3
A2	–10	5YR2.5/1	3.7	328	14	0.4	123.6	5.17	23.9	584	3.5	2.0	1.7
AC1	–15	7.5YR3/1	3.7	263	28	0.5	55.8	2.12	26.3	541	3.7	1.9	1.6
AC2	22.5	7.5YR3/2	3.7	132	10	0.6	33.0	1.40	23.6	432	3.5	1.6	1.0
R	>22.5	Bare rock											
Sandy-skeletal typic Spodorthel[a] *(WP5)* from glacial outwash sediments under mosses (Casey)													
AE	0–4	10YR4/1	3.8	163	19	0.2	39.5	2.96	13.4	164	1.5	0.8	0.7
Bh	-7	10YR2/1	3.8	222	12	0.4	64.7	4.99	13.0	414	2.2	1.4	1.3
C	–20	10YR4/3	4.1	84	10	n.d.	19.0	1.95	10.9	165	1.3	0.7	0.5
Cf	>20	Permafrost											

Table 9.3. (*Continued*)

Horizon	Depth (cm)	Colour (moist)	pH ($CaCl_2$)	CEC_p	BS (%)	P_c	TOC (mg g^{-1})	N_t	C/N	ODOE	Fe_d	Fe_{ox}	Fe_{org}
Loamy-skeletal Spodorthel[a] *(P4)* from basaltes under *Deschampsia* and *Colobanthus* (Arctowski)													
Ah	0–2	10YR3/3	4.3	450	56	n.d.	31.4	3.97	7.9	190	2.0	7.0	3.0
AE	−13	6YR3/3	4.2	426	47	n.d.	27.2	5.40	5.0	220	20.4	6.9	3.8
Bh	−15	10YR5/4	4.1	436	50	n.d.	31.8	2.75	11.6	370	23.5	8.4	7.1
Bhs	−18	5YR3/2.5	3.9	401	39	n.d.	44.1	5.99	7.4	470	23.7	9.2	7.0
Bs	−25	5YR4/3	4.0	419	47	n.d.	9.6	0.81	11.9	110	20.0	9.3	3.0
CBw	−34	5YR3/4	3.9	417	50	n.d.	2.1	<0.5	n.d.	80	13.4	6.7	2.0
BwC	−40	10YR5/5	3.8	400	49	n.d.	2.7	<0.5	n.d.	110	15.6	7.7	3.3
R	>40	Bare rock											
Dystric loamy/sandy cumulic Umbriturbel *(P13)* from outwash sediments under *Deschampsia* and mosses (Arctowski)													
Oi	1–0	Undecomposed moss layer											
Ah	0–5	6YR5/7	3.5	846	7	n.d.	128.3	6.34	20.2	280	8.1	3.9	2.0
Bw	−19	7.5YR3/2	3.7	240	6	1.1	15.1	1.72	8.8	150	7.8	3.7	1.8
2A	−29	5YR3/3	3.6	534	5	2.3	74.3	10.7	7.0	480	6.4	5.7	3.9
2Bw	−39	10YR4/2	3.6	228	8	1.4	9.2	2.47	3.7	150	3.0	3.0	2.0
A/Bf	−64	2.5Y3.5/3	3.7	291	6	2.3	12.2	1.97	6.2	110	3.2	2.0	1.9
B/Af	−80	N2	5.7	245	23	16.9	12.0	2.01	6.0	210	14.7	10.9	2.9
3Af	102	N2	6.1	415	61	18.2	75.3	9.47	8.0	220	16.2	14.9	n.d.

[a] Taxa for spodic Gelisol proposed from Beyer and Bölter (2000).

erals soils often contain large amounts of carbon and nitrogen close to the level of the organic soils in the WMI region. An ornithogenic impact is directly reflected in the carbon and nitrogen pools in soil.

Table 9.3 shows that SOM can also be found at great depth in selected representative soils. Highest carbon and nitrogen contents are not found only in the surface horizons (e.g. soils WP5, P4, P13). The data of carbon and nitrogen distribution within individual profiles suggest that:

1. Not all SOM derives from the recent vegetation, but from other sources, e.g. penguin droppings (Table 9.3, soil P13: ornithogenic soil),
2. SOM is buried due to an active cryoturbation (Table 9.3, soil C5: Haploturbel), or
3. SOM is translocated from the surface horizons into some subsurface horizons due to leaching processes (Table 9.3, soil WP 5 and P4: Podzols).

In summary, all three processes have great impact on SOM accumulation and preservation.

9.4 Soil Formation and SOM Translocation

In mineral soils fine material is sand to sandy loam. For example, (1) a sandy-skeletal Lithic Haplorthel (sample C11) is located in a small depression between single bare rock fragments under a moss and lichen carpet with 100 % surface covering, the parent material was weathered gneiss bare rock; (2) a sandy-skeletal Typic Haplorthel (WP5) is located at the edge of a relic and dried-up meltwater under a thick cushion of mosses, the parent materials are glacial outwash sands and gravel; (3) a loamy-skeletal Typic Haplorthel (D3) is located at an upper hill slope underneath a thick gravel pavement, typical for relic penguin rookery sites, no vegetation is present here. All three soils are typical shallow Podzols (spodic soils), they represent 20–30 % of the mineral soils in the Casey area (Beyer and Bölter 2000).

In addition to the SOM accumulation in a variety of Antarctic soils, the spodic soils show a specific sequence of horizons: AE-Bh (Bhs, Bs)-C (Table 9.3). Maximum values of iron fractions, of optical density of the oxalate extract (ODOE), total contents of organic carbon (TOC) and total nitrogen (N_t) are present in the spodic Bh/Bhs horizons. This is a characteristic feature of podzolic soils (Soil Survey Staff 1998). Such characteristic depth functions of these parameters suggest a translocation of SOM into the subsoil (McKeague et al. 1983).

Table 9.4. Soil organic matter composition according to CPMAS ^{13}C-NMR spectroscopy of spodic soils in East Antarctica (signal range in ppm shift to the reference TMS). *n.d.* Not determined

Horizon	Alkylic carbon			*O*-alkylic carbon				Aromatic carbon			Carboxylic carbon		
	0–25	–45	Σ	45–60	–90	110	Σ	110–140	–160	Σ	160–190	–220	Σ
C11: Soil from weathered gneiss rock under mosses and lichens in a small depression													
AE	6	10	16	5	15	12	33	19	11	30	14	7	21
Bh	11	13	24	6	15	8	29	11	11	22	15	10	25
WP5: Soil from glacial outwash sediments under mosses at a dried-up meltwater lake side													
AE	10	15	25	8	25	12	45	10	6	16	11	3	14
Bh	11	17	28	8	25	11	14	9	5	14	10	4	14
C	10	15	25	7	20	10	37	14	9	23	12	3	15
D3: Soil under gravel pavement with nil vegetation at an abandoned penguin rookery													
AE	8	11	19	5	19	12	36	14	10	24	15	6	21
Bh	11	14	25	5	16	9	32	13	10	23	15	5	20
Bhs	9	14	23	5	12	10	27	14	11	25	18	7	25
2Bs	15	18	34	6	15	7	28	11	7	18	15	5	20
3Cw	7	19	12	5	11	6	22	22	6	36	18	5	23
Organic parent materials													
Mosses	n.d.	n.d.	23	8	24	20	52	9	5	13	n.d.	n.d.	12
Algae	n.d.	n.d.	20	9	32	23	64	6	2	8	n.d.	n.d.	8
Guano	n.d.	n.d.	29	n.d.	n.d.	n.d.	37	n.d.	n.d.	20	n.d.	n.d.	14

9.4.1 ^{13}C-NMR Spectroscopy

Table 9.4 shows the SOM composition of the different horizons of some podzolic soils in the WMI region resulting from ^{13}C-NMR spectroscopy. The increase in the summarised alkyl carbon and simultaneous decrease in *O*-alkyls is comparable to patterns observed in Podzols in temperate climate regimes (Post et al. 1988). However, the high level of carboxylic carbon in all soil horizons is very uncommon. The abrupt change of the SOM composition between the Bhs, the 2Bs and the 3Cw horizons of the Podzol at the penguin rookery (D3) is probably due to the growing of layers, droppings and deposits from different materials by the penguins. Beyer et al. (1997) suggest for the spodic horizons that in contrast to those in Germany, the SOM of this soil is characterised by a high percentage of amino derivatives from proteins, polysaccharides, urates and chitin. This results in a mean C/N ratio of 10 and a high content of carboxylic carbon units, which probably derive from amino and other organic acids.

The podzolic soil from solid rock under mosses and lichens (C11) shows an aromatic carbon pattern more similar to Podzols in temperate climate regions. However, the aromatic content in the AE-horizon is very high and cannot be explained with residual preservation from the mosses. We know nothing about other sources of SOM in the area as relic organic matter or eolian inputs. The greatest similarity of SOM composition to Podzols from temperate climates is found in the podzolic soil from outwash sediments under mosses (WP5). Comparing the SOM composition of the soil horizons and selected organic parent materials, correlation coefficients (Table 9.5) suggest that in soil WP5 algal and moss carbon sources have influenced SOM to a great extent, whereas this effect was much less in soils D3 and C11.

Table 9.5. Linear correlation coefficients (r) between NMR subunits of the SOM in spodic soils of East Antarctica and organic matter of mosses and algae (for data see Table 9.4, n=9)

Soil	Horizon	r Value$_{algae}$	r Value$_{mosses}$
C11	AE	0.488	0.589
	Bh	0.480	0.610
WP5	AE	0.931***	0.966***
	Bh	0.911***	0.950***
	C	0.800**	0.856**
D3	AE	0.789**	0.807**
	Bh	0.613	0.730*
	Bhs	0.356	0.500
	2Bs	0.498	0.624
	3Cw	0.041	0.149

*p<0.05; **p<0.01; ***p<0.001

9.4.2 PyGC/MS Analysis

Table 9.6 shows the SOM composition obtained by Py-GC/MS analyses. The SOM of the albic AE and spodic Bh horizons of sample C11 is composed of a large percentage of unresolved compounds. This soil is from weathered gneiss rock under mosses and lichens. Most compounds composing the unresolved fraction appear to be heterocyclic or cyclic molecules. In addition, NMR analysis suggests 20–30 % aromatic carbon nature in soil C11 (Table 9.4), which are probably derived from larger humic molecules. The more deeply located horizons are dominated by unresolved compounds (BsC) and lipids (2C). The decrease in phenol precursors and AHP is in agreement with the decrease in aromatic carbon units within the NMR analyses (Table 9.4).

Lipids are abundant in all samples of soil WP5, a soil from glacial outwash sediments under mosses (Table 9.6). As shown for soil C11 from solid rock, the SOM patterns of soil WP5 of the upper AE and Bh horizons are similar. They consist of protein, phenol precursors, carbohydrates, trace amounts of amino sugars, but low amounts of unresolved compounds. The SOM in the parental C horizon has higher amounts of unresolved carbon units, lipids and amino sugars, in contrast to the C horizon of the Podzol C11. Proteins are only found in the two topsoils, probably as a consequence of microbial colonisa-

Table 9.6. Soil organic matter compound classes according to the pyrolysis GC/MS experiments in spodic soils in East Antarctica. *n.d.* None detected

Horizon	Protein	Phenol pre-cursors	Carbo-hydrates	Aromatic hydro-precursors	Lipids	Amino-carbo hydrates	Unresolved
		Percent of assigned signals					
C11: Soil from weathered gneiss rock under mosses and lichens in a small depression							
AE	9	9	14	3	20	n.d.	46
Bh	8	n.d.	30	n.d.	15	n.d.	48
BsC	11	n.d.	n.d.	n.d.	11	n.d.	78
2C	n.d.	n.d.	n.d.	n.d.	79	n.d.	21
WP5: Soil from glacial outwash sediments under mosses at a dried-up melt water lake side							
AE	5	14	23	n.d.	22	4	32
Bh	8	10	33	n.d.	25	3	21
C	n.d.	n.d.	n.d.	n.d.	39	20	42
Soil D3 under gravel pavement with nil vegetation at an abandoned penguin rookery							
AE	n.d.	n.d.	19	n.d.	16	33	32
Bh	9	13	24	4	10	10	29
Bhs	7	n.d.	30	n.d.	n.d.	22	41
2Bs	n.d.	n.d.	43	n.d.	34	n.d.	23
3Cw	n.d.	n.d.	27	n.d.	n.d.	17	56

tion. Elevated protein contents correlate with the greater occurrence of algal colonisation, which produce chlorophylls in soil (Bölter 1995, 1997).

A high concentration of amino carbohydrates can be seen in the non-vegetated spodic soil D3 located in an abandoned penguin rookery, except the 2Bs layer. None of these samples contain high concentrations of amino sugars. This may be due to some degradation products of uric acids, or uric acid itself may be hidden in the unresolved fraction (White and Beyer 1999). This probably reflects the influence of an organic input from guano. Proteins are found in the upper soil layers as seen in the other soils. NMR and Py-GC/MS-analyses suggest the occurrence of a very specific SOM pattern in the IIBs horizon. The data combinations also indicate that large amounts of aromatic carbon are hidden in the unresolved Py-GC/MS-fraction. Further, NMR data show a modification of SOM from the parent organic material, the guano, to deeper soil horizons by a decrease of alkylic carbon and an increase of carboxylic carbon units.

Significant differences in SOM composition and SOM depth function occur in mineral soils. The parent material and different carbon sources obviously influence their mechanisms of humification and translocation in the soil. This is probably not due to a special vegetation, as it is known from temperate climate regimes with heather, but more to microclimate and soil microbial activity. For example, Beyer et al. (2000b) found that total bacterial counts (TBC) of vegetated soils are correlated with selected organic compounds. As such, TBC and terminal methyl groups (0–25 ppm) show weak relationships, but the alkyl-C/O-alkyl-C ratio is well correlated with TBC. On the other hand, negative relations between TBC and signals of olefinic and alkyl aromatics (110–140 ppm) also indicate interactions between soil microbes and aromatic SOM compounds. These relationships suggest that bacterial cell carbon affects the SOM composition or, the other way around, as known from the temperate climate regions, that microbial activity and SOM composition are well correlated and interactive (e.g. Capriel et al. 1995).

Soil formation processes in the spodic soils from gneiss rock under moss (C11) indicate a strong decrease in *O*-alkylic carbon, but increases in aromatic and carboxylic carbon units from the recent moss vegetation and probably some soil algae during the humification process. NMR data and some data from Py-GC/MS-analyses confirm different SOM composition of the bleached AE horizon and dark-coloured spodic Bh horizons. This suggests their classification as Podzols (ISSS-WRB 1998). Proteins are mainly located in the topsoils, probably due to a favourite colonisation with microbes (Beyer et al. 2000b). Alkylic carbon units are translocated from the AE to the Bh and probably into deeper horizons as indicated by high abundance of lipids in the subsoil. Aromatic compounds are preserved in the uppermost topsoil layer.

The NMR data of soil WP5 in the spodic soil from glacial sediments under mosses also indicate a decrease in *O*-alkylic C but only a weak increase in aromatic and carboxylic carbon units from recent soil alga colonisation. Surprisingly, neither NMR nor Py-GC/MS data indicate different SOM compositions between the bleached AE horizon and the dark-coloured spodic Bh horizon, which suggests the Podzol classification (Soil Survey Staff 1998; ISSS-WRB 1998). SOM chemistry does not reflect soil morphology but a translocation of alkylic carbon during the podzolisation process.

A high abundance of amino carbohydrates occurs in the non-vegetated spodic soil at an abandoned penguin rookery (D3), except in the two Bs horizons. None of the other investigated soils contains such high concentrations of amino carbohydrates. NMR data refer to a SOM modification from the parent organic matter guano to deeper soil horizons by decrease of alkylic carbon and increase of carboxylic carbon units. However, an interpretation of the SOM data with respect to soil formation processes is complicated due to impacts from non specified sources (Beyer et al. 2000b). Py-GC/MS shows in most of the five soil horizons a completely different SOM pattern. In contrast to soil WP5, soil D3 SOM chemistry reflects the soil morphology with the assignment of a bleached AE and the dark-coloured spodic Bh horizon.

9.5 Soil Formation and SOM Transformation

The most common organic soil in the Casey area is the Lithic Hemistel from moss debris (Beyer et al. 2000b). The top layer of a sampled soil example, located at the end of a slope in a small, moist depression, (Oi) consists of dead moss (*Ceratodon purpures*) colonised by the lichen *Usnea antarctica*. In deeper layers, the moss thalli are more humified and coloured more brownish to black (Table 9.7). However, plant fibres are still visible among sapric organic material. Another investigated organic soil derives from a Lithic Fibristel from algal debris, located at the edge of a meltwater lake. This soil is found only locally, forming small spots near surface water (Blume et al. 1997). The complete soil consists of dead algal material. Peat is not or only weakly decomposed and has a greenish to grey colour. For most of the time during the year, this soil is covered by water or ice. During the summer of our visit, the first 10–20 cm was above the water level, creating a lower redox potential than the drier lithic Hemistel from moss debris (Table 9.7).

Table 9.8 summarises some SOM properties of the organic soils. Total organic carbon (TOC) concentrations and losses on ignition – as criteria for SOM content – decrease with profile depth. TOC is enriched in the soil organic matter (SOM) of deeper horizons as indicated by an increase in the TOC/SOM ratio. The OCH_3-concentration, indicator for lignins (Beyer et al. 1993), is very low in the entire soil profile compared with peaty soils under temperate climate conditions (Beyer et al. 1992). In the whole profile, soluble, inorganic NO_3^--N and NH_4^+-N fractions are of minor importance. The Lithic Fibristel from algal debris shows decreasing TOC concentrations with

Table 9.7. Selected properties of peaty horizons in organic soils in East Antarctica. $EC_{2.5}$ Electrical conductivity in 1:2.5 water extract. E_{h7} Redox potential measured in the field and converted on pH 7 level. (Schlichting et al. 1995)

Horizon	Depth (cm)	Colour (Munsell)	Post (degree)[a]	Moisture (content)[b]	$EC_{2.5}$ (mS)	E_{h7} (mV)	pH ($CaCl_2$)
Dystric Lithic Hemistel from moss debris							
Oi	0–3	n.d.	1–2	Dry	0.72	n.d.	4.8
Oe	–8	5YR2.5/1.5	3–5	Dry	0.61	530	3.9
Oa1	–16	5YR2.5/1.5	8–9	Moist	0.20	640	3.8
Oa2	–28	5YR2.5/1.5	8–9	Moist	0.13	n.d.	3.8
Euic Lithic Fibristel from algal debris							
Oi	0–1	5GY5/1	1–2	Very wet	1.8	620	4.8
Oe1	–8	5GY5/1	1–2	Very wet	4.9	250	4.8
Oe2	–18	5GY5/1	1–3	Very wet	0.65	260	4.8

[a] Degree of humification (Schlichting et al. 1995): 1 means not humified (fibric), 10 means completely humified (sapric).
[b] Moisture status estimated in the field.

increasing profile depth, already shown for the Hemistel (Table 9.8). Further, OCH_3-C concentrations are very low in the entire profile. Nitrogen is mainly determined as α-NH_2-N. Other N compounds are of minor importance. Elevated NH_4^+ concentrations compared with the Lithic Hemistel correspond to the low redox potentials which restrict nitrification (Beyer 2000; Beyer et al. 2000b).

Table 9.9 shows integrated areas below the ^{13}C-NMR spectra. Methoxyl, aromatic and phenolic structures are of less importance in all soil horizons of the Hemistel compared with acetal, hydroxyl and alkylic structures. Carboxylic carbon compounds increase, especially in the Oe and Oa1 horizons. Likewise, alkylic carbon increases with soil depth, whereas *O*-alkylic carbon (glucose units) decreases. This pattern indicates a decomposition of carbohydrates and an enrichment of carboxylic and alkylic carbon during soil formation processes in the Antarctic Hemistel from moss debris. The detection of aromatic carbon is of special interest as mosses do not contain lignins (Hurst and Burges 1967). For the Fibristel, the aromatic and alkylic peaks in the ^{13}C-NMR spectra are less intensive (not shown), this profile pattern shows no depth functions of characteristic ^{13}C-NMR subunits.

More detailed information on the molecular-chemical composition of SOM can be obtained for the lithic Hemistel by the Py-FIMS-results (data not shown). Dead moss material is the initial organic matter of soil formation in the Lithic Hemistel. According to the NMR data, aliphatic carbon units (acetal, hydroxyl, alkyl) dominate in the SOM. Tentative assignments by Py-FIMS indicate that the alkyl compounds consist of lipids, fatty acids and sterols.

Table 9.8. Selected SOM properties of peaty horizons in organic soils in East Antarctica N_{org} N_t-(α-NH_2-N+NO_3^- N+NH_4^+-N); *tr.* traces

Horizon	TOC	N_t	TOC	LOI	C/N (LOI)	OCH_3 (mg)[a]	α-NH_2-N (mg)[a]	α-NH_2-N (%)[b]	NO_3^--N (ng g^{-1})	NH_4^+-N (ng g^{-1})	N_{org} (mg)[a]	N_{org} (%)[b]
Lithic Hemistel from moss debris												
Oi	287	15.4	19	553	0.52	1.48	8.7	57	4.5	79	6.7	43
Oe	195	21.8	9	378	0.52	0.63	8.0	47	105.3	67	13.9	53
Oa1	188	20.7	9	318	0.59	0.86	5.9	29	31.2	28	14.7	71
Oa2	159	13.1	12	259	0.61	1.09	5.0	38	7.3	32	8.0	62
Lithic Fibristel from algal debris												
Oi	266	13.2	20	729	0.37	1.00	11.1	85	3.4	167	2.0	15
Oe1	264	10.2	26	550	0.48	1.36	11.0	100	1.2	201	tr.	<<1
Oe2	164	9.0	18	329	0.50	0.65	7.7	86	0.9	132	1.3	14

[a] mg g^{-1} soil.
[b] in % of N_t.

Table 9.9. ^{13}C-NMR experiment subunits of peaty horizons in organic soils in East Antarctica

Horizon	Alkylic 0–46	Methoxyl 46–60	Carbon (in % of TOC) Hydroxyl 60–80	Acetal 80–110	Aromatic 110–140	Phenolic 140–160	Carboxylic 160–210
Lithic Hemistel from moss debris							
Oi	23.1	8.1	23.9	20.0	8.6	4.7	10.6
Oe	29.6	8.6	19.2	17.5	8.0	4.4	12.6
Oa1	34.9	8.1	16.3	14.3	7.4	5.0	14.1
Oa2	36.4	8.5	16.9	14.8	7.2	5.1	11.0
Lithic Fibristel from algal debris							
Oi	19.6	8.7	32.3	22.8	6.0	2.3	8.4
Oe1	20.2	8.8	34.3	21.9	5.1	2.1	7.6
Oe2	22.6	9.0	32.3	20.3	5.2	2.2	8.4

O-alkylic compounds are partly identified as carbohydrates. Aromatic compounds are not lignin-originated, because typical signals in Py-FIMS are missing. Py-FIMS data suggest that alkylaromatics are present in the necrotic mosses. Under temperate conditions these compounds are typical structural building blocks of humic substances (Schnitzer and Schulten 1992).

Furthermore, only scarce humification is visible in the Oi-horizon (Table 9.7). Therefore, the high amount of carboxyl groups are probably derived from fatty acids and proteins (Wilson 1987). With regard to other peaks in the NMR spectra, aromatic structures possibly derive from aromatic amino acids like histidine ($C_{carboxyl}$: 174 ppm, C_{β}: 28 ppm, $C_{aromatic}$: 117–136 ppm) or tyrosine ($C_{carboxyl}$ 172–175 ppm, C_{β}: 27–28 ppm, $C_{aromatic}$ 114–138 ppm) (Breitmaier and Voelter 1989). Carbohydrates are decomposed during the soil formation in the Hemistel from mosses. The degradation of the nitrogen-free polysaccharides is confirmed by the drastic change in the C/N ratio from Oi to Oe (Table 9.7). The most striking feature of the ^{13}C-NMR data is the drastic increase of alkyl carbon. Py-FIMS enables a deeper insight in the behaviour of alkyl structures in the profile of the Hemistel from moss debris: alkyl-diesters, fatty acids, alkenes, alkanes, alkyl-monoesters and sterols in the mass range from m/z 202–732 (Beyer et al. 1999).

The decomposition of organic matter is not as intensive as in soils under a temperate climate (Kögel-Knabner 1993). The governing influence is the extremely cold conditions which strongly retard the transformation of fresh organic residues (Wilson 1990). Therefore, carbohydrates dominate the SOM in Antarctic organic soil as well as the deeper horizons, together with recalcitrant alkylic compounds. Schnitzer et al. (1990) observed similar effects in the organic layer of Canadian subarctic soils. Although this environment is dominated by mosses, aromatic structures identified by ^{13}C-NMR and by Py-FIMS show that lignin input is not necessary in order to form aromatic humic structures. The source of these compounds can derive from aromatic amino acids (Breitmaier and Voelter 1989). On the other hand, the synthesis of these compounds is also possible from ketonic and carboxylic structures (Ziechmann 1980). The occurrence of aromatic constituents in SOM of an Antarctic Gelisol is in line with results of Wilson et al. (1986), who determined aromatics in humic acid fractions from Antarctic soils.

The greenish-grey soil colour in all horizons of the lithic Fibristel corresponds to the wet-chemical and spectroscopic data and indicates only very weak transformation processes in this soil. The organic nitrogen is mostly bound in amino compounds. These results suggest that proteins are of major importance in soil formation. Obviously, carboxyl carbon derives from fatty acids and proteins (Breitmaier and Voelter 1989; Wilson 1990), because significant humification is not visible. Alkyl units probably derive from ether-linked *n*-alkyl chains (Gelin et al. 1993). These highly non-hydrolysable biomacromolecules are located in the outer cell wall of several species of algae (Largeau et al. 1990) and have a high preservation potential in soils and sediments (Gelin et al. 1993). However, the very slight increase of alkylic carbon units in the Fibristel does not reflect their selective preservation as described in literature (Gelin et al. 1993; Largeau et al. 1990; Nip et al. 1986).

Due to the proximity of a meltwater lake, the entire soil profile is very moist. Consequently, the thermal capacity of this organic soil is high. However, solar energy input in Antarctica is not sufficient to increase soil temperature in such wet soils significantly. Therefore, transformation processes of organic matter are not detectable. In contrast, in water-saturated soils under temperate climate conditions the initial SOM can be slowly transformed, because soil temperatures in summer are high enough to stimulate microbial decomposition processes (Blume et al. 1991). In contrast, the Hemistel is characterised by a lower moisture level, and soil temperatures during the Antarctic summer are obviously high enough to enable transformation processes of organic matter. Bölter et al. (1994) measured soil temperatures up to 12 °C in neighbouring soils during the Antarctic summer. Much higher values, up to 40 °C, were recorded in a moss bed by Lewis Smith (1986). Microbial activity is much higher under such conditions and transformation processes of the organic matter lead to the brownish to black colour of the deeper horizons (Lewis Smith 1990; Blume et al. 1997).

9.6 Conclusions

In the WMI region mean carbon storage in mineral soils is about 3 kg m^{-2} and storage of nitrogen is around 300 g m^{-2}, whereas in the organic soils carbon can exceed 45 kg m^{-2} and nitrogen 2.5 kg m^{-2}. In the KGI region there are no organic soils. But in mineral soils, carbon and nitrogen storage can reach very high levels (means: 6.2 kg m^{-2} for carbon and 900 g m^{-2} for nitrogen). These large C and N amounts suggest that soils in Antarctica act as sinks for carbon and nitrogen, comparable to well-known processes in the Arctic (e.g. Bockheim et al. 1999).

However, great variation exists within the numerous profiles, between the horizons and within samples of one profile. The detection of high concentrations in subsurface horizons indicates that not all SOM is derived from the recent vegetation, but from former times and from eolian transport. Additionally, cryoturbation and leaching processes have a great impact on SOM accumulation and preservation in Antarctic soils (Tearle 1987; Roser et al. 1992).

SOM composition is characterised by a high level of carboxylic carbon and lipidic moieties in all soil horizons in spodic soils of coastal East Antarctica. This is very uncommon compared with temperate climate regions. Correlation of selected SOM compounds and bacteria on the vegetated soils suggests that algal and moss carbon sources influence SOM to a great extent. The great variety in SOM composition and SOM depth distribution in the profiles show that both parent material and carbon sources influence the mechanisms of humification and translocation in the soil. Thus, not only vegetation but also microclimate and soil microbial transformations are important process of the geochemistry of these soils.

The patterns of long-chain alcohols, acids and ketones observed by the Py-GC/MS technique are not typical for soils in general. Most of the long chains in the Antarctic soil samples appear to contain multiple oxygen- and nitrogen-containing functional groups, cyclic ionised and heterocyclic structures and alkylations, which is confirmed by the occurrence of 20–40 % aromatic NMR carbon subunits. However, the complex and easily ionised compounds and the low carbon content made Py-GC/MS mass spectral identification very difficult and resulted in a high number of unresolved fractions.

With respect to soil formation in some spodic soils, alkyl carbon units are translocated from the surface AE into the subsurface Bh and Bs horizons, whereas aromatic compounds are preserved in the uppermost topsoil layer. Occasionally, SOM composition based on wet chemistry and ^{13}C-NMR does not reflect soil morphology, whereas pyrolysis data indicate a translocation of alkylic carbon moieties during the podzolisation process.

We suggest that podzolisation in Antarctic soils is evident, but chemically it is an ill-defined soil formation process. Formation and translocation of SOM shows a great variety, which is also affected by parent materials, physical and mechanical properties, the microclimate and local moisture regime as well as microbial activity.

The integrated approach, including morphological observations, wet-chemical analyses, CPMAS ^{13}C-NMR, and pyrolysis allows new insights into soil organic matter composition and transformation of the investigated Antarctic permafrost-affected soils. The SOM of the Hemistels consists mainly of carbohydrates and lipid-derived compounds. In water-saturated organic soils, transformation processes are not evident, but weak decomposition and transformation processes of SOM can be observed under less moist conditions. This is in line with observations made by Calace et al. (1995). In particular, carbohydrates are absent while alkylic compounds are enriched. The detection of aromatics in the SOM of Antarctic organic soils is of special interest. Thus, our data give evidence that aromatic lignin constituents are not necessary for the formation of aromatic humic substances.

Finally, we think that it is important to determine the origin of the Antarctic SOM. This should be one of the main future research aspects in Antarctic soil science. Additionally, our results suggest that SOM transfer plays an important role in nutrient cycles in Antarctic soil systems. However, for a final conclusion, it is necessary to unravel water-soluble SOM chemistry (Chapman et al. 1994) and to verify its movement in the landscape. This is more of a technical than a methodological problem due to the extreme stoniness and/or shallowness of most soils in these terrestrial Antarctic ecosystems. Moreover, investigations of soil formation processes in Antarctica are necessary for understanding the complex pedogenic processes under temperate climate conditions, where ecosystems are more structurally complex than the Antarctic terrestrial systems.

References

Beyer L (2000) Properties, formation and geo-ecological significance of organic soils in the coastal region of East Antarctica (Wilkes Land). Catena 39:79–93

Beyer L, Bölter M (2000) Chemical and biological properties, formation, occurrence and classification of spodic Cryosols in a terrestrial ecosystem of East Antarctica (Wilkes Land). Catena 39:95–119

Beyer L, Schulten H-R, Fründ R (1992) Properties and composition of soil organic matter in forest and arable soils in Schleswig-Holstein. 1.Comparison of morphology and results of wet chemistry, CPMAS ^{13}C-NMR spectroscopy and pyrolysis mass spectrometry. Z Pflanzenernähr Bodenkd 155:345–354

Beyer L, Wachendorf C, Koebbemann C (1993) A simple wet chemical extraction procedure to characterize soil organic matter. 1. Application and recovery rate. Comm Soil Sci Plant Anal 24:1645–1663

Beyer L, Blume H-P, Knicker H, Bölter M (1997) Soil organic matter of suggested spodic horizons in relic ornithogenic soils of coastal continental Antarctica (Casey Station, Wilkes Land) and spodic horizons in soils of Germany. Soil Sci 162:518–527

Beyer L, Bockheim JG, Campbell IB, Claridge GGC (1999) Genesis, properties and sensitivity of Antarctic Gelisols. Antarct Sci 11:387–398

Beyer L, White DW, Bölter M (2001) Organic matter composition, and transformation, and microbial colonisation of Gelic Podzols in the coastal region of East Antarctica. Austr J Soil Res 39:543–563

Beyer L, Pingpank K, Wriedt G, Bölter M (2000a) Soil formation in coastal continental Antarctica. Geoderma 95:283–304

Beyer L, Bölter M, Seppelt RD (2000b) Nutrient and thermal regime, microbial biomass, and vegetation of Antarctic soils in the Windmill Island region of East Antarctica (Wilkes Land). Arct Antarct Alp Res 32:30–39

Birkenmajer K (1989) A guide to tertiary geochronology of King George Island, West Antarctica Pol Polar Res 10:555–579

Blume H-P, Beyer L, Friedrich F (1991) Correlations between microbial activity and water, air temperature and nutrient status of different soils under different land use. In: Esser G, Overdieck D (eds) Modern ecology: basic and applied aspects. Elsevier, Amsterdam, pp 321–346

Blume H-P, Schneider D, Bölter M (1996) Organic matter accumulation in and podzolization of Antarctic soils. Z Pflanzenernähr Bodenkd 159:411–412

Blume H-P, Beyer L, Bölter M, Erlenkeuser H, Kalk E, Kneesch S, Pfisterer U, Schneider D (1997) Pedogenic zonation in soils of Southern circumpolar region. Adv GeoEcol 30:69–90

Bockheim JG, Everett IR, Hinkel KM, Nelson FE, Brown J (1999) Soil organic carbon storage and distribution in Arctic Tundra, Barrow, Alaska. Soil Sci Soc Am J 63:934–940

Bölter, M (1995) Distribution of bacterial numbers and biomass in soils and on plants from King George Island (Arctowski Station, maritime Antarctica). Polar Biol 15:115–124

Bölter M (1997) Microbial communities in soils and on plants from King George Island (Arctowski Station, Maritime Antarctica). In: Battaglia B, Valencia J, Walton DWH (eds) Antarctic communities: species, structure and survival. University Press, Cambridge, pp 162–169

Bölter M, Blume H-P, Erlenkeuser H (1994) Pedogenic, isotopic and microbiological properties of Antarctic soils. Polarforschung 64:1–7

Bölter M, Blume H-P, Schneider D, Beyer L (1997) Soil properties and distributions of invertebrates and bacteria from King George Island (Arctowski Station), maritime Antarct Polar Biol 18:295–304

Breitmaier E, Voelter W (1990) Carbon-13 NMR spectroscopy Verlag Chemie, Weinheim

Broady PA, Weinstein RN (1998) Algae, lichens and fungi in La Gorce Mountains, Antarctica. Antarct Sci 10:376–385

Calace N, Campanella L, de Paolis F, Petronto BM (1995) Characterization of humic acids isolated from Antarctic soils. Int J Environ Anal Chem 60:71–78

Campbell IB, Claridge GGC (1987) Antarctica: soils, weathering processes and environment. Elsevier, Amsterdam

Campbell IB, Claridge GGC, Campbell DI, Balks MR (1998) Short- and long-term impacts of human disturbances on snow-free surfaces in Antarctica. Polar Rec 34:15–24

Capriel P, Beck T, Borchert H, Grönholz A, Zachmann G (1995) Hydrophobicity of the organic matter in arable soils. Soil Biol Biochem 27:1453–1458

Chapman BE, Roser DJ, Seppelt RD (1994) ^{13}C-NMR analysis of Antarctic cryptogam extracts. Antarct Sci 6:295–305

Chen J, Blume H-P (1998) Impact of human activities on the terrestrial ecosystems of Antarctica – a review. Polarforschung 65:83–92

Fründ R, Lüdemann H-D (1989) The quantitative analysis of the solution- and CPMAS ^{13}C-NMR spectra of humic material. Sci Total Environ 81/82:157–168

Gelin F, Gatellier J-PLA, Sinninghe Damsté JS, Metzger P, Derenne S, Largeau C, de Leeuw JW (1993) Mechanisms of flash pyrolysis of ether lipids isolated from the green microalga *Botryococcus braunii* race A. J Anal Appl Pyrol 27:155–168

Goodwin ID (1993) Holocene deglaciation, sea-level change, and the emergence of the Windmill Islands, Budd Coast, Antarctica. Quat Res 40:70–80

Hurst HM, Burges NA (1967) Lignin and humic acids. In: McLaren AD, Peterson GH (eds) Soil biochemistry, vol 1. Marcel Dekker, New York, pp 260–286

ISSS-WRB [International Soil Science Society - World Reference Base for Soil Resources] (1998) world Reference base for soil resources. World Soil resources Reports 84, ISSS-ISRIC-FAO, Rome

Kögel-Knabner I (1993) Biodegradation and humification processes in forest soils. In: Bollag, JM, Stotzky G (eds) Soil biochemistry, vol 8. Marcel Dekker, New York, pp 101–135

Komárková, V (1985) Two native Antarctic vascular plants, *Deschampsia antarctica* and *Colobanthus quitensis*: a new southernmost locality and other localities in the Antarctic Peninsula area. Arct Alp Res 17:401–417

Kuhn D (1997) Genese, Ökologie und Soziologie einer Bodengesellschaft in einem Periglazialgebiet der King George Insel, West-Antarktis. Schriftenr Inst Bodenkd Univ Kiel 40:1–173

Largeau C, Derenne S, Casadevall E, Berkaloff C, Corolleur M, Lugardon B, Paynaud JF, Connan J (1990) Occurrence and origin of 'ultra laminar' structures in 'amorphous' cerogens of various source rocks and oil shales. Org Geochem 16:889–895

Lewis Smith RI (1990) Plant community dynamics in Wilkes Land, Antarctica. Proc NIPR Symp Polar Biol 3:229–244

McKeague JA, DeConnick F, Franzmeier DP (1983) Spodosols. In: Wildung LP, Smeck NE, Hall GF (eds) Pedogenesis and soil taxonomy. II. The soil orders. Elsevier, Amsterdam, pp 217–252

Nip M, Tegelaar EW, de Leeuw JW, Schenck EW (1986) A new non-saponifiable highly aliphatic and resistant biopolymer in plant cuticulas. Naturwissenschaften 73:579–585

Paul E, Stüwe K, Teasdale J, Worley B. (1995) Structural and metamorphic geology of the Windmill Islands, east Antarctica: field evidence for repeated tectonothermal activity. Aust J Earth Sci 42:453–469

Post B, Hempfling H, Klamberg H, Schulten H-R (1988) Zur Charakterisierung von Boden-Huminstoffen. Fresenius Z Anal Chem 331:273–281

Rakusa-Suszczewski S, Mietus S, Piasecki J (1993) Weather and Climate. In: Rakusa-Suszczewski S (ed) The maritime Antarctic ecosystems of Admiralty Bay. Department of Antarctic Biology, Polish Academy of Sciences, Warsawa, pp 19–25

Roser DJ, Melick DR, Ling HU, Seppelt RD (1992) Polyol and sugar content of terrestrial plants from continental Antarctica. Antarct Sci 4:413–420

Schlichting E, Blume H-P, Stahr K (1995) Bodenkundliches Praktikum. 2. Aufl. Blackwell, Berlin

Schnitzer M, Schulten H-R (1992) The analysis of soil organic matter by pyrolysis-field ionization mass spectrometry. Soil Sci Soc Am J 54:98–105

Schnitzer M, Tarnocai C, Schuppli P, Schulten H-R (1990) Nature of the organic matter in Tertiary Paleosols in the Canadian Arctic. Soil Sci 149:257–267

Seppelt RD, Broady, PA (1988) Antarctic terrestrial ecosystems: the Vestfold Hills in context. In: Ferris JM, Burton HR, Bayly IAE (eds) Biology of the Vestfold Hills, Antarctica. Kluwer, Dordrecht, pp 177–184

Soil Survey Staff (1998): Keys to soil taxonomy, 8th edn, US Dept Agric, Nat Res Counc, Washington, DC

Tearle PV (1987) Cryptogamic carbohydrate release and microbial response during spring freeze-thaw cycles in Antarctic fellfield fines. Soil Biol Biochem19:381–390

White D M, Beyer L (1999) Pyrolysis-GC/MS and GC/FID of three Antarctic soils. J Anal Appl Pyrol 50:63–76

Wilson MA (1987) NMR techniques and applications in geochemistry and soil chemistry. Pergamon Press, Oxford

Wilson MA (1990) Application of nuclear magnetic resonance spectroscopy to whole soil. In: MacCarthy P, Clapp CE, Malcolm RL, Blomm PR (eds) Humic substances in soil and crop sciences. American Soil Science Society, Madison, Wisconsin, pp 221–260

Wilson MA, Goh KM, Collin PJ, Greenfield LG (1986) Origins of humus variations. Org Geochem 9:225–231

Wynn-Williams DD (1990) Ecological aspects of Antarctic microbiology. Adv Microb Ecol 11:71–146

Wynn-Williams DD (1993) Microbial processes and initial stabilization of fellfield soil. In: Miles J, Walton DWH (eds) Primary succession on land. Blackwell, Oxford, pp 17–32

Ziechmann W (1980) Huminstoffe. Verlag Chemie, Weinheim

10 Ornithogenic Ecosystems in the Maritime Antarctic – Formation, Development and Disintegration

In memory of Andrzej Myrcha

A. Tatur

10.1 Colonisation of Antarctic Coast by Sea Vertebrates

Antarctic sea vertebrates – seals, penguins, flying birds – periodically use the Antarctic coast for resting, moulting, breeding and, in the case of flying birds, hunting. Consequently, the visited land is manured, creating new conditions for development of terrestrial biota. The formation, duration and disintegration of manured terrestrial ecosystems have been strongly related to geological events. Palaeoclimate deterioration on the Eocene/Oligocene boundary created a new polar environment with the structure of marine palaeoecosystems comparable to the present one. Quaternary climate fluctuations determined the size, shape, position and lasting time of the ice-free patches of land, and this regulated access of sea animals to the land, and affected the intensity of land manuring.

10.1.1 Marine Vertebrates in the Maritime Antarctic Palaeoecosystem

The structure of the maritime ecosystem in a shape comparable to the present one was formed simultaneously with the global Cenozoic cool mode. It started at the beginning of middle Eocene (E2) and was clearly marked by the sharp decline in marine water temperatures near the Eocene/Oligocene boundary. This last cooling event has been evidenced by micro-palaeontological and dropstone records in sediment cores of the open ocean around the Antarctic continent (data from Deep Sea Drilling Project and Ocean Drilling Program). Climatic deterioration is also well documented in fossil-rich, shallow marine clastics of the La Meseta Formation on Seymour Island (Antarctic Peninsula). The change of molluscan bottom fauna to a colder assemblage (Zinsmeister and Camacho 1982), stable carbon isotope record in molluscan shells (Gazd-

Ecological Studies, Vol. 154
L. Beyer and M. Bölter (eds.) Geoecology of Antarctic Ice-Free Coastal Landscapes

zicki et al. 1992), sedimentological and geochemical data (Dingle et al. 1998), suggest a climatic deterioration and drop in seawater temperature in the middle part of this formation (about 42 Ma). This event may be correlated, on the basis of mammal stratigraphy (Marenssi et al.1994), with the 'Middle Eocene' beginning of global cooling of Prothero (1992). From the co-occurrence of vertebrate fossils in the La Meseta Formation, we can conclude that warm-blooded toothed whales replaced sharks because they avoid cold water (Gazdzicki and Tatur, unpubl.). Thus, the newly created trophic pyramid in the sea after the cooling event consisted of phytoplankton >possible krill >fish >penguins >and toothed whales at the top around the Antarctic (Myrcha et al. 1990; Noriega et al. 1995; Tatur 1996), and did not differ (except of lacking seals) from the modern one.

Since the Eocene, the only known other evidence of Antarctic vertebrates comes from marine plain sediments of the Sirius Formation (Pliocene warming event in Antarctica) and has yielded a dolphin (Quilty 1990). Hitherto, exclusively Holocene penguin guano deposits were known from coasts of the Antarctic continent and only a late Holocene one from the maritime Antarctic (Baroni and Orombelli 1994; Emslie 1995; Tatur et al. 1997; Emslie et al. 1998). Therefore, our assumptions about the fate of manured terrestrial ecosystems in the Quaternary glacial/interglacial climatic cycles is based on the contemporary behaviour of marine vertebrates.

10.1.2 Colonisation of Land by Sea Vertebrates During Glacial/Interglacial Cycles

The rhythm of Quaternary climate changes in Antarctica (Anderson 1990) proceeds (as everywhere) according to Milankovitch astronomic cycles (Dawson 1992). Freshwater bound up in ice fields during glacial intervals resulted in global lowering of the sea level and depressed lands in the Polar regions. During the last glaciation, sea level dropped to 120–160 m (Denton and Hughes 1981). The ice sheet expanded to the margin of the shelf (Fig. 10.1) with the ice cliffs almost exclusively forming shorelines (Clapperton 1990), and the pack-ice zone enlarged (Sugden 1982).

Restricted availability of ice-free shorelines and short breeding seasons during the last glaciation (18 ka B.P.; Fig. 10.2), reduced the number of marine vertebrates breeding on land: penguins, fur seals and to a lesser degree sea elephants reproducing also on the ice. During glaciations, seals might have migrated for reproduction northward to the distant Subantarctic Islands and Tierra del Fuego. However, penguins (especially *Pygoscelis adeliae* and *P. papua*) were not able to do so, because of their strong physiological adaptation to low temperature (limited by the Antarctic Convergence). Therefore, most probably they had to reduce their numbers significantly.

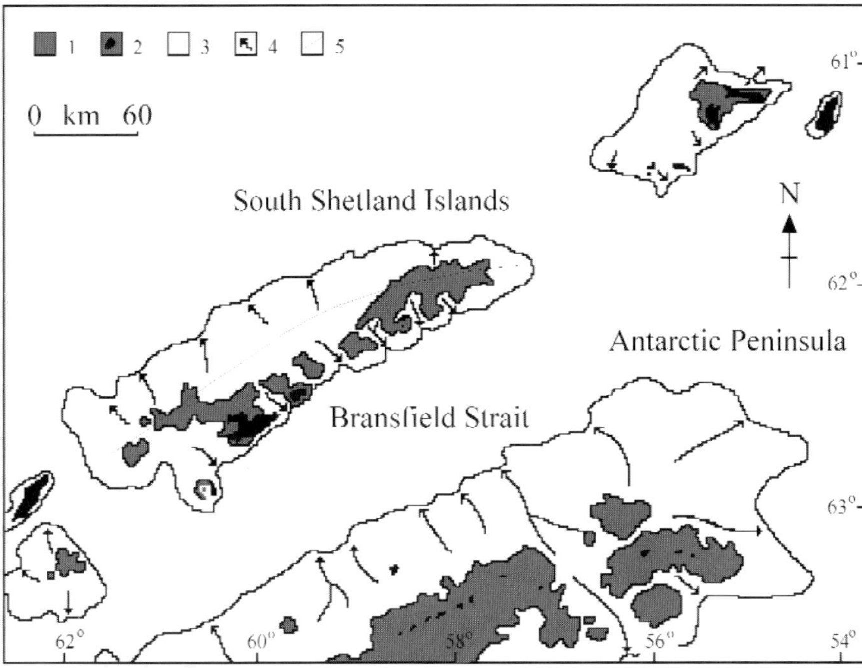

Fig. 10.1. Extent of glaciers during the last glaciation (after Clapperton 1990, simplified). *1* Today's land under ice-cap during last glaciation, *2* nunataks, *3* glaciers and shelf glaciers, *4* direction of ice movement *5* ice cap axis

Restricted ice-free patches of coast appeared during the global Holocene deglaciation, 10 ka B.P. The Antarctic coast was flooded over due to the eustatic rise of the seawater level and Early Holocene nesting places were located at least 50 m above the current sea level. Progress of deglaciation and the following glacio-isostatic uplift of land during the Holocene enlarged the area accessible for marine vertebrates. This process accelerated during the last 6 ka, when wide rings of raised beaches at the foot of early Holocene cliff were commonly formed. Thus, the Late Holocene breeding areas of vertebrates were often shifted to these near-sea raised beaches, and the former higher places, situated often on the steep Early Holocene cliffs, were abandoned (Martinez-Macchiavello et al. 1996; Tatur et al. 1997).

Small-scale climate fluctuations during the Holocene changed the extent of pack and fast ice, and determined the access of sea vertebrates to land during summer (Baroni and Orombelli 1994). However, based on field observation and keeping in mind today's global warming, one may suppose that the availability of nesting places has recently become a less important factor controlling the numbers of vertebrates (particularly penguins) populations. Thus,

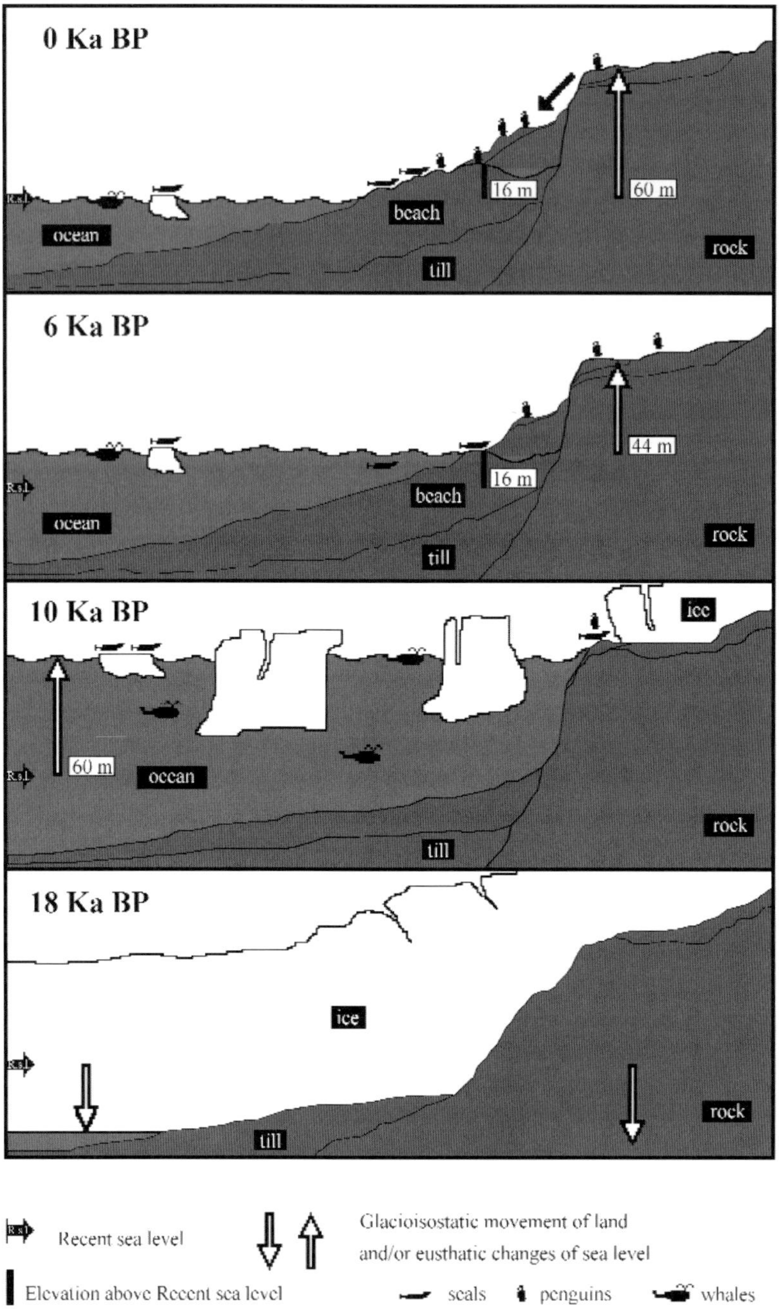

Fig. 10.2. Development of Holocene landscape in the maritime Antarctic against availability of land to sea vertebrates

the impact of these sea animals on terrestrial ecosystem formation may be currently higher than ever before during the Holocene.

10.1.3 Impact of Sea Animals on Terrestrial Ecosystem Formation

Behaviour of particular vertebrates determines their impact on terrestrial ecosystem formation. The activity of seals is usually restricted to the flat beach areas. Less mobile southern sea elephants gather in large (dozens of animals) groups in shallow ponds located close to the sea, whereas Antarctic fur seals and occasionally other seal species are more homogeneously dispersed along the coast. Penguin rookeries used to be situated in about the same dry, elevated places year after year, usually not more than 350 m inland and no higher than 60 m a.s.l. Nests of flying birds are dispersed predominantly along the coast (giant petrel, and Wilson's storm petrel, black-bellied storm petrel, cape petrel, south polar skua, brown skua, southern black-

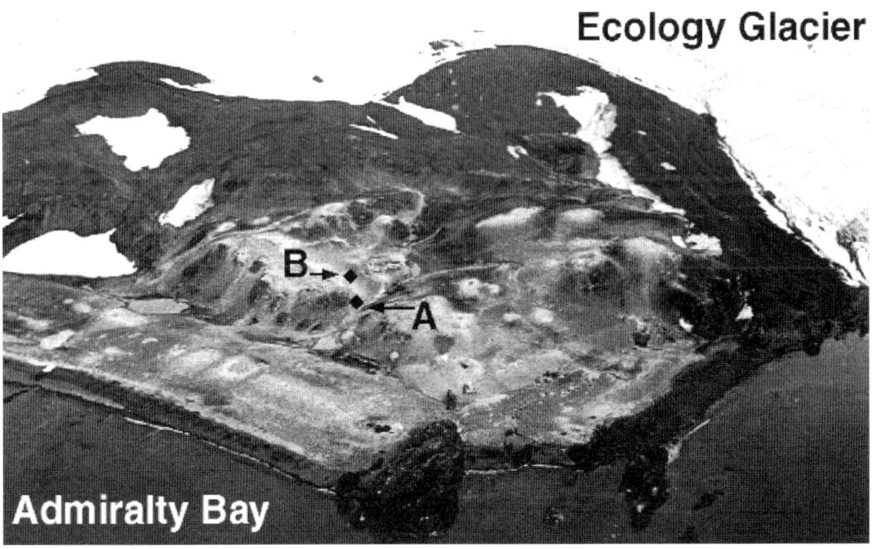

Fig. 10.3. Penguin rookery at Llano Point, King George Island – horizontal differentiation of ornithogenic soils. (photo: Furmańczyk, alteredusing a computer; data of soil differentiation based on Tatur and Myrcha 1984). The following ornithogenic soil units may be distinguished on the basis of colour differentiation: *1* soils of penguin rookeries, nests form dense dots on the top of hills. *2* Soils of deep phosphatisation zone, light coloured bands surrounded nesting places. *3* Soils in the marginal zone of direct penguin influence, dark patches away from rookeries but situated inside the watershed occupied by penguins. *4* Soils derived from accumulation of clastics eroded from higher situated ornithogenic deposits, foot of hummocks occupied by penguins: flat beaches including pool and lake sediments. Points *A* and *B* mark localisation of profiles described in Fig. 10.6

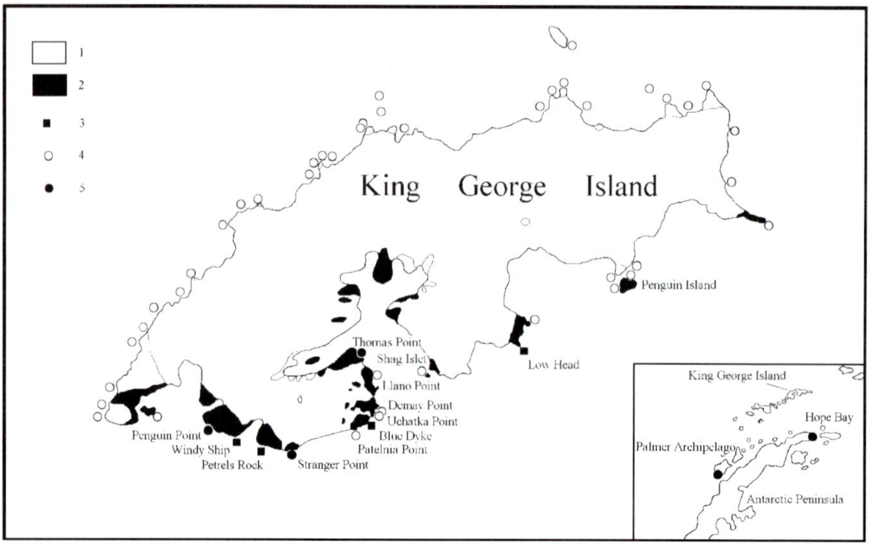

Fig. 10.4. Penguin rookeries on King George Island (after Jabłoński 1984, modified). *1* Ice-free areas, *2* area searched by Polish team, *3* abandoned rookeries, *4* rookeries shifted down, *5* other active rookeries. Two places searched on the Antarctic Peninsula are shown in the *inset*

backed gull, blue-eyed shag), some of them around penguin rookeries (parts of skua and gull populations feeding on penguins and American sheathbill). However, other birds (mainly Antarctic terns, sometimes skuas or giant petrels) may nest in the central parts of ice-free landscapes hundreds of metres from the shoreline.

Thus, birds manure not only the coastal zone, but also the inland areas. Dense populations of seals, penguins, shags, giant petrels and gulls entering new areas usually devastate former biota (if it existed before) by trampling and toxic manuring. However, vegetation around nests of smaller flying birds (small petrels, Antarctic tern), or dispersed around single nests of skuas are not devastated, but rather fertilised.

Among the vertebrates considered, the breeding activity of penguins yields the most prominent and spectacular ecological effects along the coast in the maritime Antarctic, including Admiralty Bay on King George Island (Fig. 10.3). The number of breeding pairs of penguins, located in five rookeries functioning from mid-November to the end of January (Fig. 10.4), ranged between 30,000 and 50,000. According to Rakusa-Suszczewski (1980), such a breeding population deposits 6.35 ton of dry excreta on the land daily (for comparison, all flying birds only deposit 0.14 tons). The intensity of manuring on the rookery area reached 10 kg of dry excreta per square metre during the breeding season (Tatur and Myrcha 1984).

10.2 Ornithogenic (Penguin) Soils in the Maritime Antarctic

Penguins feed in the open sea, but reproduce, moult and rest on the land. Thus, primary production of the marine phytoplankton (mainly diatoms) being transferred in the food chain to chitin, protein and lipids in bodies of krill, is then carried onto the land by penguins and deposited as excreta, which fertilises terrestrial biota. This excreta accumulates every breeding season on the coast forming 'ornithogenic soils' (Syroechkovsky 1959), called lately in the FAO-UNESCO taxonomic mode as 'Ornithosols' (Blume et al. 1997, Blume et al., Chap. 7). Boundaries of the ornithogenic soils constrained the extent of ornithogenic ecosystems. Thus, descriptions of ornithogenic soils set up the base for ecological studies.

The term 'ornithogenic soils' has been commonly accepted in the Antarctic pedological literature (Tedrow and Ugolini 1966; McCraw 1967; Campbell and Claridge 1969; Allen and Heal 1970; Ugolini 1972; Tedrow 1977; Everett 1976; Campbell and Claridge 1987; Bockheim and Ugolini 1990; Blume et al. 1997). Originally, it was applied to organic (guano) soils occurring in the Antarctic continent. Due to the abundance of liquid water in the wet climatic zone of the maritime Antarctic (in contrast to the Antarctic continent) intense microbiological decomposition of guano, leaching, and rock phosphatisation processes proceed around nesting places (Bölter et al., Chap. 11). Therefore, ornithogenic soils of the maritime Antarctic of penguin rookeries are different from the continental taxonomic equivalents (Tatur and Myrcha 1984; Myrcha et al. 1985; Tatur 1987; Tatur and Myrcha 1988; Tatur 1989; Myrcha et al. 1991; Myrcha and Tatur 1991; Tatur and Myrcha 1993; Tatur et al. 1997).

Almost 50 ha of land in Admiralty Bay on King George Island (Rakusa-Suszczewski, Chap. 3) is exposed to direct manuring by penguins, leading to current ornithogenic soil formation. Taking into account surrounding surfaces affected by diluted guano solutions (ca. 30 ha), and the area of relic ornithogenic soils in the abandoned breeding places (ca. 20 ha), the overall area affected directly by penguins increases to ca. 1 km^2 (Tatur and Myrcha 1991). That is about 5 % of the ice-free terrain (Marsz and Rakusa-Suszczewski 1987).

10.2.1 Mineralisation of Guano

A stage of rapid microbiological mineralisation of penguin excrement occurs in 1–3 weeks. About 50 % of carbon to the atmosphere volatilises during this time, whereas mineralised excreta are transformed into a uniform, semi-liquid mass, composed mainly of finely grained, weathered chitin detritus mixed with a fluorapatite suspension, with C:N:P ratios comparable to the organic matter (fresh guano layer) of the soil surface (Pietr et al. 1993). Relatively rapid decomposition results from the protein-rich substrate, surprisingly high soil surface temperatures during summer (often above 10 °C), and usually favourable soil moisture.

Particulate products of mineralisation may be easy washed down into rock debris and/or washed away from the nesting sites. A thin film of mineralised guano often covers the path from rookery to the sea during the breeding season. Anyway, due to rapid mineralisation and intense washing, the stony surface of the soil is cleaned after a few rainy days.

Ornithogenic soils are formed mainly due to microbiological (bacterial) activity, which drives the main soil-forming processes. The role of the other organisms (fungi, soil fauna, vegetation) is restricted mainly to the marginal zone of breeding area. Calcium phosphate released during chitin mineralisation precipitates as earthy masses of fluorapatite, which is the most important component of guano. Phosphatisation of underlying rock debris is also explained by microbiological activity responsible for the change of alkaline guano solutions to acid ones during slow percolation throughout the soil column. Details related to microbiological activity are described in other chapters (Bölter et al., Chap. 11; Bölter and Blume, Chap. 16).

10.2.2 Ornithogenic Waters

There are two extremely different weather conditions constraining the chemical composition of waters which wash penguin rookeries (Tatur and Myrcha 1983; Fig. 10.5). During heavy precipitation, surface runoff of penguin rookeries dominates. The running water carries high loads of guano suspension and dissolved nutrients. Mineral nitrogen occurs exclusively as ammonium ions, and the pH of the water is neutral or alkaline (Table 10.1, columns C, D, E).

During dry periods, water from previous precipitation percolates through the soil and drips with delay from the weathered springs commonly dispersed at the foot of hummocks occupied by penguins (Table 10.1, columns F, G). Soil solutions are free of suspension, the concentration of nutrients is several

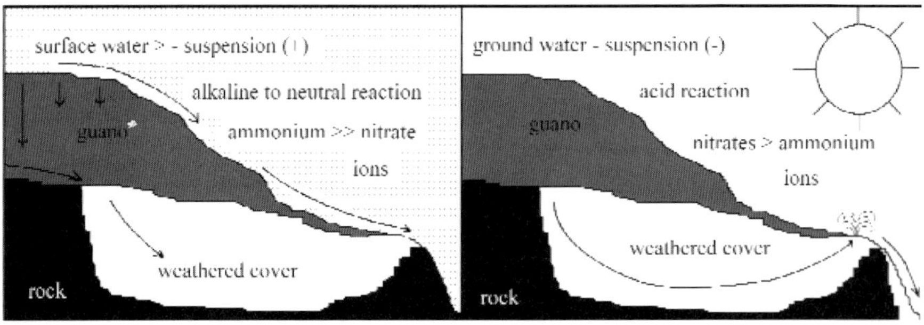

Fig. 10.5. Chemical composition of ornithogenic waters during heavy precipitation (*left*) and during a dry period (*right*)

Table 10.1. Chemical composition of ornithogenic water, Llano Point, after Tatur and Myrcha (1983) and Tatur and Barczuk (1984). Data in mg l^{-1}, except for conductivity (μS) and pH (units). *L* Low content, *H* high content, – no data

Parameters	Location[a] A	B	C	D	E	F	G
C/suspension	L	H	4800	1120	L	0	0
Conductivity	–	–	–	2800	2140	320	808
pH	–	3.8–>7.4	>7.4	>7.4	>7.4	4.0	4.0
$N-NH_4$	0.2	2–170	7000	365	360	4	450
$N-NO_3$	0.8	0–43	0.05	0.04	48	43	157
$N-NO_2$	0.00	<0.02	0.06	0.05	2.8	0.02	1.8
$P-PO_4$	0.03	2–38	300	100	44	6	220
DOC	–	–	–	–	16	<3	80
Cl	–	–	–	–	568	4	161
Ca	1.2	5–35	–	–	24	–	22
Mg	0.8	3–20	–	–	16	–	65
K	1.1	4–125	–	–	–	–	–
Na	13.0	10–110	–	–	–	–	–
Al	<0.1	<0.1–1.5	–	–	–	–	–
Fe	0.1	0.1–1.1	–	–	–	–	–

[a] A, Control stream; B, biologically enriched stream, range from 18 measurements covering entire summer 1979/1980; C, surface water runoff from penguin rookery during heavy rain, slope amongst penguins; D, surface water runoff from penguin rookery during heavy rain, foot of hummock occupied by penguins; E, semi-surface water at the point of struvite crystallisation; F, weathered spring water at the foot of hummock occupied by penguins during dry period; G, groundwater at the base of ornithogenic soil profile (Fig. 10.6, profile *A*, 90–95 cm)

times lower than in surface water, the nitrate content is usually higher than ammonium ions, the pH equals about 4 and represents a stabilisation of equilibrium constants by the buffering mineral phase changes (Cole and Jackson 1950). Such conditions also occurred in groundwater at the base of the ornithogenic soil profile (Table 10.1, columns F and G; Fig. 10.6, profile A, 90–95).

Penguins carry seawater with food onto the land therefore ornithogenic waters usually contain an enhanced concentration of seawater salts. Additionally, some cations may come from the by-products of rock phosphatisation processes (Table 10.1, columns A and B).

Described chemical variability of ornithogenic waters usually occurs around Adélie penguin rookeries, which are located inland on hummocks surrounded by vast weathered covers. Antarctic penguins often nest on steep rocky coastlines and poorly decomposed guano is easily washed away from such places.

Analyses of nutrients in water extracts from relic ornithogenic soils indicate slowly continuous dissolving of soil phosphates recorded many years after the rookery was abandoned by penguins (Myrcha and Tatur 1991).

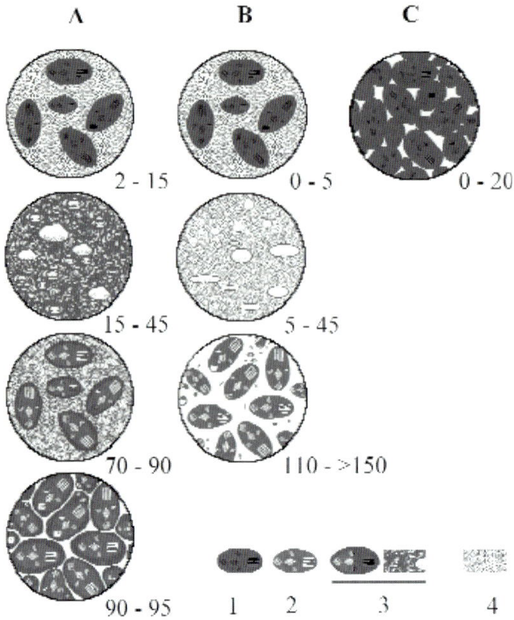

Fig. 10.6. Modes of phosphatisation process. *1* Unweathered clasts of volcanic rock. *2* Phosphatised clasts of volcanic rock (soft pseudomorphs of phosphates, silicate clay and silica after volcanic clasts). *3* Indurated framework and coats of amorphous aluminium-iron phosphate. *4* Soft phosphatic-silicate clay. Soft soil matrix: homogeneous phosphates (*lighter*) or phosphates in the mixture with silicates (*darker*) or guano (the *darkest one*). Description: **A** Profile 5 in Tatur and Myrcha (1984), localisation marked in Fig. 3. 0–2 cm, guano suspension (neutral or alkaline reaction). 2–15 cm, unweathered pebbles in leucophosphite clay. 15–45 cm, porous, indurated skeleton of amorphous aluminium phosphate caring minyulite microcrystalline mass, forming druses in the empty spaces left by entirely removed clastics (changeable pH). 45–70 cm, gradual transition. 70–90 cm, porous, indurated minyulite with phosphatised grains of clastic material coated by amorphous aluminium phosphate crust (permanently acid reaction). 90–95 cm, loose gravel of phosphatised clasts coated by amorphous aluminium phosphate crust (minyulite leached by poorly mineralised groundwater connected with stream). **B** Profile 6 in Tatur and Myrcha (1984), localisation marked in Fig. 3. 0–5 cm, guano amongst unweathered pebbles. 5–45 cm, earthy mass amongst deluvial debris is composed of leucophosphite plus silicate clay mixture and bears hard grains build of amorphous aluminium phosphates coats empty inside, or filled with remains of phosphatised clasts. 45–110 cm, gradual transition. 110 to at least 150 cm, below standing groundwater. White clay of taranakite bearing soft phosphatised clasts. **C** Profile P-24 in Tatur (1989). 0–20 cm, raised beach (12 m a.s.l.). Relic ornithogenic soil. Unweathered pebbles coated and glued by indurated amorphous aluminium-iron phosphate

10.2.3 Phosphatisation – the Main Ornithogenic Soil-Forming Process in the Maritime Antarctic

Phosphatisation is usually defined (for example Simpson 1972) as the process by which phosphorus is emplaced in a host (silicate or carbonate) material. Phosphorus occurs in a phosphatised rock zone as Ca and Al-Fe phosphates formed due to the action of a chemically aggressive solution released from weathering phosphate deposits (including guano) on underlying rock. This term is commonly used in petrologic descriptions of ornithogenic deposits in the humid Temperate Zone (Hutchinson 1950; Altschuler 1973; Flicoteaux and Lucas 1984; White and Warin 1964), but is unknown in pedology. However, if we accept 'ornithogenic soils of maritime Antarctic' as a formal unit included in Antarctic soil taxonomy, we should consequently consider phosphatisation as a main process responsible for the formation of these soils in the maritime Antarctic. The microbiological nature of processes controlling the course of phosphatisation in ornithogenic soils should be underlined. Decomposition of guano creates phosphatising solutions, in which the chemical power (resulting from changeable pH) is regulated by the intensity of ammonification against nitrification processes (Tatur and Myrcha 1983), as well as organic matter decomposition (finally producing oxalic acid) and bacterial oxidation of sulphides (producing sulphuric acid).

The phosphatisation, unlike the other soil-forming processes, has a 'double nature'. It comprises chemical degradation and dissolution of rock and simul-

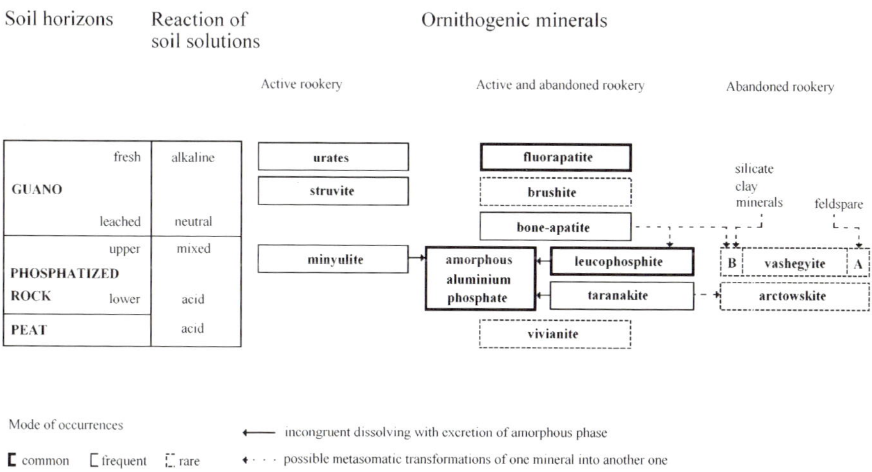

Fig. 10.7. Vertical distribution of phosphate minerals in ornithogenic soil profiles, descriptions of minerals are given in Fig. 10.8

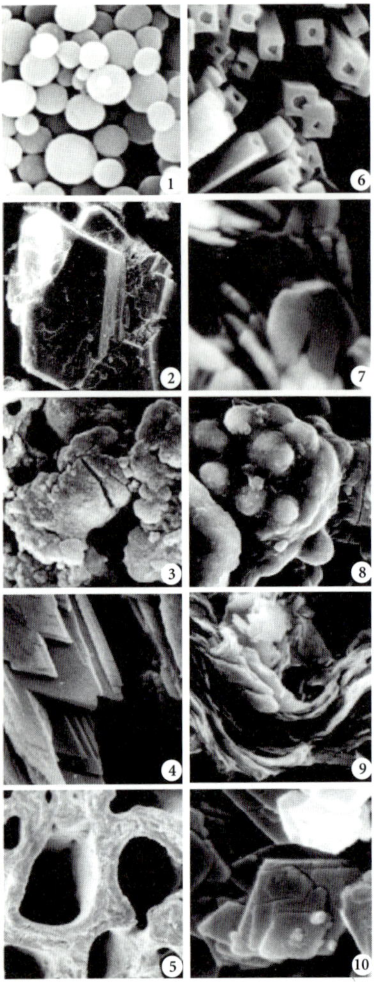

Fig. 10.8. Minerals forming ornithogenic soils. Guano layer minerals: Urates $(NH_4 >> K, Na)_{0.8}$ $(Mg, Ca, Fe)_{0.1}$ $C_5 H_3 O_3 N_4$. Occur in nodules (2–20 μm in diameter) with spherulitic structure (**1**); form during the evaporation of liquid excretion; occur only on the surface of current rookery. Struvite: $Mg (NH_4) PO_4 \cdot 6 H_2O$. Forms thin crystalline film on guano surface, crust on stones with fluorapatite, loose agglomeration of crystals (**2**), occurs only in fresh guano layer of current rookery. For formation, needs excess ammonia and magnesium supplied from seawater. Fluorapatite: $Ca_5 (PO_4)_3 F$. Main guano phosphate (current and relic soils) occurring in earthy mass glued in aggregates (**3**), local crust on stones with struvite. Formed due to precipitation of elements released from mineralised penguin excrement rich in krill remains. Brushite: $Ca HPO_4 \cdot 2 H_2O$. Occurs as well-defined crystals (**4**), in guano layer of fish-eating (F poverty) shag colonies. Minerals of phosphatised zone: Leucophosphite: $(NH_4, K)_2 (Fe, Al)_4 (PO_4)_4 (OH, F)_2 \cdot 4 H_2O$. Important component of phosphatic clay in the upper horizon of phosphatised layer in current and relic soils. Pseudomorphs of pure leucophosphite after penguin bone apatite in buried ornithogenic soil horizon were recognised in one place (**5**). Occurs in crystals (X-ray) of submicroscopic dimension. Needs excess alkali, and variable redox conditions for mobilisation and immobilisation of iron ions. Minyulite: $K, Al_2 (F>OH) (PO_4)_2 \cdot 4 H_2O$. Occurs in current soils as very fine crystals forming micro-druses in the upper part of phosphatised layer (**6**) close to largest rookeries. Needs excess alkali, abundance of fluorine and reducing conditions supporting iron leaching. Taranakite: $(K, NH_4)_3 Al_5 H_6 (PO_4)_8 \cdot 18 H_2O$. Occurs in current and relic soils as very fine loose crystals forming phosphatic clay at the bottom of phosphatised zone in permanently moist, reducing and acid conditions (**7**). Needs excess alkali for formation; highly hydrated, basic, amorphous aluminium phosphate. $Al_{10} F_9 (PO_4)_7 \cdot n H_2O$ – in co-occurrence with fluorapatite as crust on stone laying in the way of surface runoff from nesting places (**8**). For formation needs variable mineralisation of running water, high load of ornithogenic elements (P, F) $Al_{10} (F>OH)_3 (PO_4)_9 \cdot 43 H_20$ – common in all ornithogenic soils. Indurate crusts and clay matrix. Effect of incongruently dissolving of alkali-bearing composite phosphates or direct precipitate from alkali-poor solutions. Minerals found exclusively in phosphatised zone of relic soils: Vashegyite A: $(Al_{0.9}, Fe_{0.1}) PO_4 \cdot 3.15 H_2O$ and Vashegyite B $Al_{11} (OH, F)_3 (PO_4)_{10} \cdot 35 H_2O$. Forms during slow phosphatisation of clay minerals (**9**) in alkali-poor environment, may substitutes bone-apatite. "Arctowskite": $(Al_{8.4}, Fe_{0.6})_9 (OH)_3 (PO_4)_8 \cdot 27 H_2O$. Forms well-defined crystals (**10**), while ageing of taranakite ? in alkali-poor environment. Name is not formalised. Phosphates from the surrounding areas: Vivianite: $Fe^{II}_3 (PO_4)_2 \cdot 8 H_2O$. Occurs as well-defined crystals in clay intercalating peat bank. Possi-

bly, diluted ornithogenic solutions react with silicate clay. Identification of minerals is based on X-ray diffraction patterns, supported by observation under electron microscope, and chemical analyses resulting in calculation of several given formulas. For detailed information, see Tatur and Barczuk (1884, 1985), Tatur (1987), Tatur and Keck (1990)

---

taneously the formation of secondary phosphates in its place (Tatur and Barczuk 1984 1985; Tatur 1989; Tatur and Keck 1990). Thus, guano is both a subject and a source of weathering (Altschuler 1973). Phosphatisation may proceed only during heavy manuring in the presence of water, and is clearly marked if the host material is tractable for alteration (for example basalts or limestones), but is insignificant if granite or quartz sands occur in the bedrock.

The alteration of basalt-andesite bedrock to phosphates plus clay minerals and amorphous silica proceeds under an acid reaction, while further dissolving of clay and silica takes place under an alkaline reaction. However, the intensity of both processes differs down the soil profile (examples in Fig. 10.6). Therefore, soil structure and vertical stratification of diversified aluminium-iron phosphates in genetic horizons of the soil profile strongly depend on the reaction of solutions percolating throughout the soil (Fig. 10.7). Additionally, the degree of water mineralisation, alkali, iron, fluorine contents, and Eh value should also be considered as factors constraining the formation of particular minerals (Fig. 10.8).

Formation and transformation of soil phosphates proceed in several modes of the phosphatisation process:

1. Simple precipitation from supersaturated soil solutions due to: evaporation accelerated by strong wind, lyophilisation in daily freezing/melting cycle, or just Eh or pH change during percolation through the soil,
2. Excretion of thin film of phosphates (for example amorphous aluminium phosphates) on the contact surface between aggressive guano solutions and silicate minerals susceptible to dissolving,
3. Metasomatic processes (with minimal participation of the liquid phase) lead to the formation of phosphate pseudomorphs substituting silicate minerals and biological tissues,
4. Incongruently dissolving composite aluminium-iron phosphates bearing potassium and ammonia ions to amorphous aluminium phosphate is the most common transformation while ageing. This process proceeds due to periodical leaching of soils with slightly mineralised water (Cole and Jackson 1950). Amorphous aluminium phosphate-coated composite phosphates in this process protect them against further rapid dissolution.

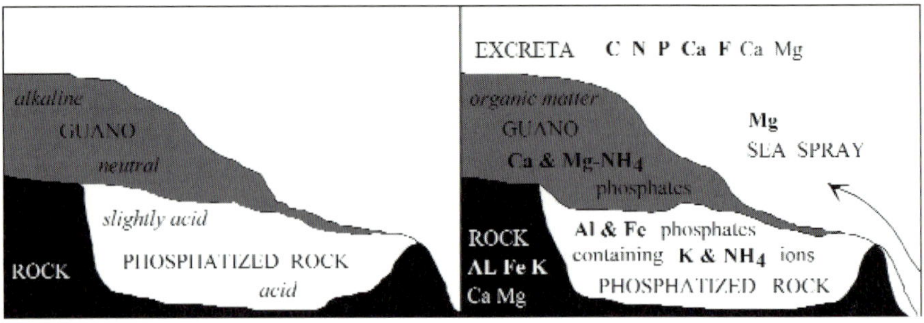

Fig. 10.9. Ornithogenic soils in the maritime Antarctic, vertical zonation of soil profile (*left*) and sources of elements forming soil phosphates (*right*)

10.2.4 Ornithogenic Soils of Active Rookeries

Mineralised penguin excretion forms the guano layer, whereas guano leachates reacting with underlying rocks create a phosphatised rock layer (Fig. 10.9). Both these processes responsible for soil formation in the maritime Antarctic (guano mineralisation and rock phosphatisation) are provided by microbiological activity and proceed in the liquid water environment.

The guano layer occurs amongst pebbles gathered by penguins during nest construction. The thickness of guano usually equals about 0.5 m in areas of nesting sites and is much thinner and free of pebbles along the slopes which surround a rookery. Phosphatised rock layers (sometimes more than 2 m thick) develop in stony weathering covers derived from basalt-andesite rocks which usually form the bedrock on King George Island. The main essence of ornithogenic soils inheres in the finest fraction forming a plastic or indurated matrix amongst clastics. It is composed of composite and/or simple aluminium phosphates mixed in varying proportions with silicates, only in exceptional cases does the entire replacement of silicates by phosphates proceed (profile A in Fig. 10.6; Table 10.2).

Detailed vertical zonation of guano and phosphatised rock layers into horizons results from colour, structure and mineral differentiation. The guano layer is usually subdivided to fresh and leached guano horizons. The fresh guano horizon (black when fresh) consists mainly of mineralised excrement and urates mixed with fluorapatite and sometimes struvite. Leached guano (lighter) contains mainly fluorapatite mixed with weathered chitin detritus.

Particular horizons of phosphatised layers are also characterised by different modes of phosphatisation (Fig. 10.6) and contain varying mineral proportions in their fine fraction. Particular ornithogenic phosphates occur at the depth defined by conditions listed in Figs. 10.7 and 10.8. Usually one or two, sometimes three, phosphates mixed with silicate clay make up the fine fraction of one horizon (Tatur and Barczuk 1984).

Table 10.2. Chemical composition of chosen ornithogenic soil samples including phosphorus and cations in aqua regia extract from grain size fraction <1 mm in diameter (data in percent). Localisation of profiles A and B in Fig. 10.3, description of profile and images of structure in Fig. 10.6. Data taken from Tatur and Myrcha (1984), Tatur (1989)

Parameters	Profile and depth								
	A					B			C
	0–2	2–15	15–45	70–90	90–95	0–5	5–45	110 –150	0–20
Fine matrix among stones (%)	10	20	80	90	70	10	50	50	10
<1 mm fraction in matrix (%)	70	33	90	87	68	73	17	71	50
H_2O	11.74	1.67	4.25	17.99	21.44	15.23	3.10	15.48	-
pH in H_2O	6.9	4.5	5.0	4.3	4.4	6.5	4.1	4.4	-
C	13.85	3.91	3.60	5.75	6.36	15.36	5.68	3.54	1.39
N	2.33	0.94	0.84	0.93	0.92	3.08	1.36	1.42	0.22
C/N	5.9	4.2	4.3	6.0	6.9	5.0	4.2	2.5	6.3
Aqua regia insoluble	25.88	33.27	8.89	5.96	9.10	23.14	47.95	32.45	–
P	6.80	8.83	13.32	12.17	10.85	6.85	7.20	10.36	9.57
Ca	6.10	0.36	0.34	1.20	1.42	5.90	0.20	0.06	0.38
Mg	1.61	0,20	0,17	0.08	0.09	2.20	0.14	0.08	0.38
Al	2.90	6.95	11.20	10.90	10.45	2.10	5.40	6.20	9.00
K	0.75	3.90	6.60	2.60	0.80	0.65	5.10	2.15	0.22
Na	0.78	0.34	0.35	0.40	0.36	0.52	0.29	0.18	0.16
Fe	1.70	3.45	0.95	0.50	0.85	1.25	5.55	5.10	7.04
Mn	0.02	0.02	0.01	0.01	0.01	0.02	0.01	0.01	0.00
Sr	0.13	0.00	0.00	0.15	0.15	0.11	0.01	0.01	0.01

Horizontal boundaries of ornithogenic soil units are defined by the intensity of manuring and by the watershed pattern in areas occupied by penguins (Fig. 10.3). Ornithogenic soils should be distinguished from the 'normal soils covered by typical zonal vegetation, although fertilised by guano solutions. By the separation line, one may propose the presence of still unspoiled primary vegetation.

10.2.5 Ornithogenic Soils of Abandoned Rookeries

Areas of abandoned rookery (Figs. 10.10a, b) are quickly colonised by vegetation (algae, lichens, mosses, liverworts, and fungi), which form a thin surface horizon of plant humus distinguished by a black colour and C/N, C/P ratios much higher than in ornithogenic materials (Tatur et al. 1997). Such newly formed soil derived from the weathered ornithogenic substratum and covered by vegetation has been named 'relic ornithogenic soil' (Tatur 1989).

Fig. 10.10. The sequence of abandoned nesting sites at Stranger Point: **a** on raised beaches, **b** on early Holocene cliff, raised beaches are seen behind. Inhibited nests of penguins form dark "dotted" circles surrounded by white bands of dry guano and phosphatic clay, but abandoned ones form entirely white circles. *Numbers* mark the sequence of abandoned nesting sites from currently inhabited colony *0* to the oldest abandoned ones *4*, environmental changes presented detailed in Fig. 10.11

The guano layer is the first to be removed by erosion and leaching in the abandoned nesting sites. Remains of guano found in the area of the former nests are usually deeply hidden under a pavement of loose pebbles (Fig. 10.11). Long-lasting (several hundreds or thousands of years) weathering of guano leads to decreasing chitin concentrations, resulting in changes of the C/P ratio about twice, due to the constant behaviour of fluorapatite. This also leads to increasing C/N ratio values above 5 in relic guano (Tatur et al. 1997). Easy dissolved struvite is absent in relic soils.

After the guano has gone, light-coloured clay containing composite aluminium-iron phosphates usually appears on the surface in the area of an abandoned rookery. Amongst these phosphates, minyulite seems to be the most tractable for alteration, therefore it has never been found in relic soils. Leuco-phosphite is less common in relic than in recent soils. In contrast, taranakite and aluminium phosphates (both amorphous and crystalline) are relatively more abundant. Crystalline aluminium phosphates (vashegyite and arctowskite) are recognised exclusively in relic soils.

Phosphatic loams forming a phosphatised rock layer, are usually mechanically segregated during gravity flows, solifluction, macro- and micro-sorting in frost processes, and may be included in the lateral moraines of expanding glaciers. Buried ornithogenic horizons were also recognised. Selective dissolving of phosphates and recrystallisation of new ones in another place after short transport in solution was observed. Newly precipitated phosphates occur as a fill in rock fissures, as a surface film of efflorescence or as incrustations on stones in weathering springs (Tatur 1989).

10.3 Development of Terrestrial Ornithogenic Ecosystems

Terrestrial ornithogenic ecosystems develop thanks to biomass by birds (mainly penguins) from the sea (Rakusa-Suszczewski 1985; 1999). The process of intense microbiological decomposition of penguin excreta, supported by disintegration of organic matter due to soil invertebrate faunal activity, dominate in ornithogenic ecosystems of active rookeries. Nematoda, Collembola and Acarina, permanently occur at the edges of rookeries at the places where the guano suspension washed down from the rookery accumulates, and are periodically abundant between breeding periods in the area of nesting sites (Myrcha et al. 1991). Besides trampling by penguins, a high concentration of nutrients in mineral forms released during organic matter mineralisation inhibits colonisation of the rookery by vegetation. Mainly the leafy alga *Prasiola crispa* develops in such a fertile environment, forming extensive mats in or around penguin rookeries. The other algae species also occurring in fresh guano (Prof. Komarek, pers. comm.) are less important. The grass *Deschampsia antarctica* grows only in the marginal zone of manuring.

However, the ornithogenic impact on the terrestrial environment does not strictly keep to the bounds of soil units. Wind and penguin predators disperse ornithogenic nutrients all over the island. Ammonia ions volatilised to the atmosphere during guano mineralisation and carried by wind, reach rocks located downwind, initiating development of nitrophilous lichens (see Olech, Chap. 12). Atmospheric ammonia also supports the growth of unicellular algae that form crypto-plankton agglomerations in the melting snow. Surface absorption of ammonia by remote soil covers leads to an increase in their fertility (Elster, Chap. 17). Penguin predators (skua, gull) feed on the rookery but nest on the surrounding hills, dispersing nutrients and fertilising local inland points. Hydrobiological action creates manuring of freshwater bodies situated in the area of penguin activity (Jones et al. 1993; Hansson and Håkansson 1992; Izaguirre et al. 1993; Zale 1994; Van de Vijer and Beyens 1997; Kawecka and Olech 1993; Kawecka et al. 1998).

Abandonment of the nesting places by penguins is a turning point for ornithogenic ecosystem development. Penguin predators disappear with their

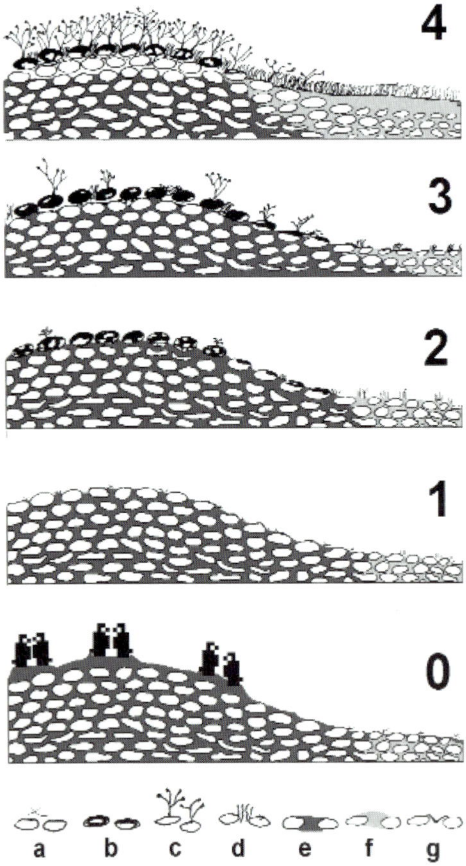

Fig. 10.11. Development of abandoned rookery ecosystems. Relic soil formation (relocation of phosphates matrix), succession of vegetation and entering of lichens and mosses are marked along the sequence of abandoned nesting sites (from **0** to **4**, see Fig. 10 for localisation). **a** *Prasiola crispa*; **b** crustous and foliose lichens; **c** *Usnea* sp.; **d** mosses and peat; **e** guano among pebbles; **f** clay bearing composite aluminium-iron phosphates; **g** plant humus

prey, and the export of ammonia via the atmosphere declines. The autochthonous microflora may rapidly replace the specific one for penguin rookeries when supplies of fresh excreta have vanished (Sieburth 1963; Pietr 1986, 1993). From that point on, the spatial ornithogenic impact is restricted to the environments directly fertilised by waters which wash abandoned rookeries (Tatur 1989; Myrcha and Tatur 1991). The vegetation entering abandoned rookeries uses ornithogenic nutrients released from previously accumulated phosphates.

The succession of vegetation in the sequence of breeding sites abandoned by penguins during the last millennium was presented by Tatur et al. (1997). It began with an invasion of *Prasiola crispa* (Fig. 10.11, site 1). Loose pebbles on the abandoned nesting places were initially colonised by crustaceous, epilithic lichens with nitrophilous species (site 2) dominating, and then by more diversified associations with the subordinate participation of ornithogenic lichens. Gradually, lichens entirely covered the surface of all pebbles in the abandoned nesting area (site 3). However, the depressions around nests were persistently colonised by extended patches of mosses, which also gradually entered the nesting sites, covering soils among pebbles. Fruticose lichens (*Usnea* sp.) ultimately covered pebbles on the nesting places to form a complete coat as the final stage of the observed succession (sites 4 and 5). In many other environmental conditions a dense carpet of *Deschampsia antarctica* covering relic ornithogenic soils was documented and their chemical composition reflects the soil nutrient status (Tatur et al. 1997).

10.4 Biogeochemical Cycle of Ornithogenic Matter on the Land/Sea Boundary

Cycles of ornithogenic matter on the land/sea boundary belong to the classic problems considered by pioneers of biogeochemistry (Hutchinson 1950). Massive faecal concentration occurs only when seabirds (including penguins) feed over fertile oceanic waters (usually in the zone of upwelling) and return to limited coastal sites for nesting. Hutchinson (1950) has estimated that all seabirds in the world carry 10^4–10^5 tons of phosphorus from sea to land annually. In this calculation, penguins living in Antarctica were either not taken into account or not in an adequate proportion, since knowledge about their numbers and food requirements was very scanty at that time.

Pygoscelis penguins feed on krill captured dozens of kilometres from the breeding place. The entire population of *Pygoscelis adeliae* and *P. antarctica* living in Antarctica is estimated as 30 millions pairs and 30 million non-breeding individuals of each species (Wilson 1983; Laws 1985). Just the population of pygoscelid penguins living in the maritime Antarctic deposits annually about $1 \cdot 10^6$ tons of dry excreta on land. This amount corresponds to about 1.5–2.0 · 10^4 tons of phosphorus (Myrcha and Tatur 1991) and represents an important part of the global participation of birds in the transfer of phosphorus from sea to land. However, contrary to the dry areas of the world, where the guano deposits have accumulated, most of the nutrients carried to land with penguin food rapidly return back to the sea as partially mineralised guano suspension and an important fraction of the nitrogen volatilises to the atmosphere. A less important part of ornithogenic nutrients is incorporated into ornithogenic soils. Under favourable conditions (e.g. the penguin rookery at Llano Point, Fig. 10.3) more than 11 % of cycling phosphorus may be left on land (Myrcha et al. 1991).

From a biogeochemical point of view, penguin activity leads to a temporary concentration of sea nutrients bound in organic matter and phosphates in the ornithogenic soils, followed by delayed return of most of them in an organic suspension or a mineralised form to the sea along the shoreline (Tatur and Myrcha 1991). This transfer, concentration and increasing turnover may be important for the productivity, structure and zonation of marine shoreline ecosystems poor in nutrients, for example in the Arctic Sea (Golovkin and Gurevic 1973; Golovkin and Garkovaja 1975). However, it does not seem to be of any importance for marine ecosystems in the maritime Antarctic where mixed oceanic water is abundant in nutrients. Only in the intertidal zone has a eutrophic influence yet been observed (Nikolajev 1980) and may probably reach high ecological significance (Bosman and Hockey 1986).

However, penguin activity is important for creation and maintaining biodiversity of Antarctic terrestrial biota. The duration of terrestrial ornithogenic ecosystems may be considered in hundreds and thousands years, and seems to be shorter in the wet maritime Antarctic than on the continent, because ornithogenic deposits there may be more easily washed away and/or dissolved (Tatur et al. 1997). Nevertheless, according to Milankovitch forecasting, during the coming glaciation, soil phosphates will have to be eroded by expanding glaciers or dissolved by seawater during an inundation event at the beginning of the next interglacial period. Only global warming produced by human activity may break this natural destination.

Acknowledgements. I am very grateful to Prof. Anna Hilbricht-Ilkowska, Prof. Manfred Bölter, Prof. Stanislaw Pietr, Prof. Eliza Prot, Dr. Izabella Wilpiszewska and the late Prof. Arturo Amos for constructive comments.

References

Allen SE, Heal OW (1970) Soils of the maritime Antarctic zone. In: Holdgate MW (ed) Antarctic ecology, vol 2. Academic Press, New York, pp 693–701
Altschuler ZS (1973) The weathering of phosphate deposits – geochemical and environmental aspects. In: Griffith EJ, Beeton A, Spencer JM, Michell DT (eds) Environmental phosphorus handbook. Wiley, London, pp 33–96
Anderson JB (1990) Marine record of Late Quaternary glacial–interglacial fluctuations in the Ross Sea and evidences for rapid episodic sea level changes due to marine ice collapse. In: Bindschadler RA (ed) West Antarctic Ice Sheet Initiative. NASA Conf Publ 3115(2):87–110
Baroni C, Orombelli G (1994) Abandoned penguin rookeries as Holocene paleoclimatic indicator in Antarctica. Geology 22:23–26
Blume H-P, Beyer L, Bölter M, Erlenkeuser H, Kalk E, Kneesch S, Pfisterer U, Schneider D (1997) Pedogenic zonation in soils of the Southern circumpolar region. Adv GeoEcol 30:69–90
Bockheim JG, Ugolini FC (1990) A review of pedogenic zonation in well-drained soils of the Southern circumpolar region. Quat Res 34:47–66
Bosman AL, Hockey PAR (1986) Seabird guano as a determinant of rocky intertidal community structure. Mar Ecol Prog Ser 32:247–257
Campbell IB, Claridge GGC (1969) A classification of frigic soils – the zonal soils of the Antarctic continent. Soil Sci 107:75–85
Campbell IB, Claridge GGC (1987) Antarctica: soils, weathering processes and environment. Elsevier, Amsterdam
Clapperton CM (1990) Quaternary glaciations in the Southern Ocean and Antarctic Peninsula area. Quat Sci Rev 9:229–252
Cole CV, Jackson ML (1950) Solubility equilibrium constants of dihydroxy aluminium dihydrogen phosphate relating to a mechanism of phosphate fixation in soils. Soil Sci Soc Am Proc 15:84–98
Dawson AG (1992) Ice age earth. Routledge, London New York
Denton H, Hughes TJ (1981) The last great ice-sheet. Wiley, New York
Dingle RV, Marenssi SA, Lavelle M (1998) High latitude Eocene climate deterioration: evidence from the northern Antarctic Peninsula. J South Am Earth Sci 11:571–579
Emslie SD (1995) Age and taphonomy of abandoned penguin rookeries in the Antarctic Peninsula. Polar Rec 31: 09–418
Emslie SD, Fraser W, Smith RC, Walker W (1998) Abandoned penguin colonies and environmental changes in the Palmer Station area, Anvers Island, Antarctic Peninsula. Antarct Sci 10:257–268
Everett KR (1976) A survey of the soils in the region of the South Shetland Islands and adjected parts of Antarctic Peninsula. Ohio State Univ, Inst Polar Stud Rep 58:1–44
Flicoteaux R, Lucas J (1984) Weathering of phosphate minerals. In: Niagru JO, Moore PB (eds) Phosphate minerals. Springer, Berlin Heidelberg New York, pp 292–317
Gazdzicki A, Gruszczynski M, Hoffman A, Malkowski K, Marenssi S, Halas S, Tatur A (1992) Stable carbon and oxygen isotope record in the Paleogene of La Meseta Formation Seymour Island Antarctica. Antarct Sci 4:461–468
Golovkin AN, Garkovaja GP (1975) Udobrenie pribrelnych vod murmanskogo poberelija ekskrementami ptic v blizi kolonij razlilnych tipov. Biol Morja 5:49–57
Golovkin AN, Greviz VI (1973) Vydelenije i ocenka gidrohimileskih anomalij v more metodom linejnych diskriminantnych funkcji (na primere barencevomorskih ptilnych bazarov). Okeanol 13:804–808

Hansson L-A, Håkansson H (1992) Diatom community response along a productivity gradient of shallow Antarctic lakes. Polar Biol 12:463–468

Hutchinson GE (1950) Biogeochemistry of vertebrate excretion. Bull Am Mus Nat Hist 96:1–554

Izaguirre I, Mataloni G, Vinocur A, Tell G (1993) Temporal and spatial variations of phytoplankton from Boeckella Lake (Hope Bay, Antarctica Peninsula). Antarct Sci 5:137–141

Jabłoński B (1984) Distribution and numbers of penguins in the region of King George Island (South Shetland Islands) in the breeding season 1980/1981. Pol Polar Res 5:17–30

Jones VJ, Juggins S, Ellis-Evans JC (1993) The relationship between water chemistry and surface sediment diatom assemblages in maritime Antarctic lakes. Antarct Sci 5:339–348

Kawecka B, Olech M (1993) Diatom communities in the Vanishing and Ornithologist Creek, King George Island, South Shetlands, Antarctica. Hydrobiologia 269/270:327–333

Kawecka B, Olech M, Nowogrodzka-Zagórska M, WojtuĖ B (1998) Diatom communities in small water bodies at H. Arctowski Polish Antarctic Station (King George Island, South Shetlands, Antarctica). Polar Biol 19:183-192

Laws RM (1985) The ecology of the Southern Ocean. Am Sci 73:26–40

Marenssi SA, Reguero MA, Santillana SN, Vizcaino SF (1994) Eocene mammals from Seymour Island, Antarctica: palaeobiogeographical implications. Antarct Sci 6:3–15

Marsz A, Rakusa-Suszczewski S (1987) Charakterystyka ekologiczna rejonu Zatoki Admiralicji (King George Island, South Shetland Islands). I. Klimat i obszary wolne od lodu. Kosmos 36:103–127

Martinez-Macchiavello JC, Tatur A, Servant-Vildary S, del Valle R (1996) Holocene environmental changes in a marine–estuarine–lacustrine sediment sequence, King George Island, Antarctica. Antarct Sci 8:312–322.

McCraw T (1967) Soils of Taylor Dry Valley, Victoria Land, Antarctica, with notes on soils from other localities in Victoria Land. NZ J Geol Geophys 10:498–539

Myrcha A, Ochyra R, Tatur A (1991) Site of Special Scientific Interest no. 8, Western shore of Admiralty Bay, King George Island, South Shetland Islands. In: Klekowski RZ, OpaliĖski KW (eds) First Polish-Soviet Antarctic Symp. 'Arctowski 85'. Institute of Ecology Publ Office, Warsaw, pp 157-168

Myrcha A, Pietr SJ, Tatur A (1985) The role of pygoscelid penguin rookeries in nutrient cycles at Admiralty Bay King George Island. In: Siegfried WR, Condy PR, Laws RM (eds) Antarctic nutrient cycle and food webs. Springer, Berlin Heidelberg New York, pp 156–162

Myrcha A, Tatur A (1991) Ecological role of the current and abandoned penguin rookeries in the land environment of the maritime Antarctic. Pol Polar Res 12:3–24

Myrcha A, Tatur A, del Valle R (1990) A new species of fossil penguin from Seymour Island (West Antarctica). Alcheringa 14:195–205

Nikolayev VA (1980) Littoral diatoms from Fildes Peninsula, King George Island. Bot Zh 65:107–112

Noriega JI, Tambussi CP, Jadwiszczak P, Myrcha A, Tatur A, Gazdzicki A (1995) Los pinguinos (Aves: Sphenisciformes) del Terciario de la isla Seymour, Antartida: revision y nuevos aportes. XI Jornadas Argentinas de Paleontologia de Vertebrados, Tucuman, Abstr

Pietr SJ (1986) The physiological groups of microorganisms in different soils at Admiralty Bay region (King George Island, South Shetland Islands, Antarctica). Pol Polar Res 7:395–406

Pietr SJ (1993) Soil microorganisms In: Rakusa-Suszczewski S (ed) The maritime Antarctic coastal ecosystem of Admiralty Bay. Dept of Antarctic Biology, PAS, Warsaw, pp 167–172

Pietr SJ, Tatur A, Myrcha A (1983) Mineralization of penguin excrements in the Admiralty Bay region (King George Island, South Shetlands, Antarctica). Pol Polar Res 4:97–112

Prothero DR (1994) The Eocene-Oligocene transition: paradise lost. Columbia Univ Press, New York

Quilty PG (1990) Significance of evidence for changes in the Antarctic marine environment over the last 5 million years. In: Kerry KR, Hempel G (eds) Antarctic ecosystems: ecological changes and conservation. Springer, Berlin Heidelberg New York, pp 3–8

Rakusa-Suszczewski S (1980) The role of near-shore research in gaining and understanding of the functioning of Antarctic ecosystem. Pol Arch Hydrobiol 27:229–233

Rakusa-Suszczewski S (1995) Flow of matter in the Admiralty Bay area, King George Island, maritime Antarctic. Proc NIPR Symp Polar Biol 8:101–113

Rakusa-Suszczewski S (1999) Ekosystem morskiej Antarktyki, (zmiany i zmiennoŌĺ). PWN, Warszawa

Sieburth JMcN (1963) Bacterial habitats in the Antarctic environment. In: Oppenheimer CH (ed) Symposium on marine microbiology. CC Thomas Publ House, Springfield, IL, pp 533–548

Simpson DR (1972) Phosphatization. In: Fairbridge RW (ed) The encyclopaedia of geochemistry and environmental sciences. Van Nostrand Reinhold, New York, pp 940–942

Sugden D (1982) Arctic and Antarctic. A modern geographical synthesis. Blackwell, Oxford

Syroechkovsky EE (1959) The role of animals in the formation of primary soils under the condition of circumpolar regions of earth (Antarctica). Zool Zhur 38:1770–1775

Tatur A (1987) Fluorine in ornithogenic soils and minerals on King George Island, West Antarctica. Pol Polar Res 8:65–74

Tatur A (1989) Ornithogenic soils of the maritime Antarctic. Pol Polar Res 10:481–532

Tatur A (1996) Reakcja ekosystemu Antarktyki na zmiany klimatu (Response of Antarctic ecosystem to climate changes) Prob Klimatol Pol 5:149–157

Tatur A (1997) Ornithogenic soils of the maritime Antarctic: taxonomic position. Pol. Polar Stud, 24 Polar Symp 1997, Warszawa, pp 297–299

Tatur A, Barczuk A (1984) Phosphates of ornithogenic soil on the volcanic King George Island (Maritime Antarctic). Pol Polar Res 5:61–97

Tatur A, Barczuk A (1985) Ornithogenic phosphates on King George Island in the maritime Antarctic Zone. In: Siegfried RP, Condy WR, Laws RM (eds) Antarctic nutrient cycle and food webs. Springer, Berlin Heidelberg New York, pp 163–168

Tatur A, Keck A (1990) Phosphates in ornithogenic soils of the maritime Antarctic. Proc NIPR Symp Polar Biol 3:133–150

Tatur A, Myrcha A (1983) Changes in chemical composition of waters running off from the penguin rookeries in Admiralty Bay region (King George Island, South Shetland Islands, Antarctica). Pol Polar Res 4:113 –125

Tatur A, Myrcha A (1984) Ornithogenic soils on King George Island, South Shetlands Islands (Maritime Antarctic Zone). Pol Polar Res 5:31–60

Tatur A, Myrcha A (1988) Soils and vegetation in abandoned penguin rookeries (Maritime Antarctic Zone). Proc NIPR Symp Polar Biol 2:181–189

Tatur A, Myrcha A (1993) Ornithogenic soils. In: Rakusa-Suszczewski S (ed) The maritime Antarctic coastal ecosystem of Admiralty Bay. Dept. of Antarctic Biology, Pol Acad Sci, Warsaw, pp 161–165

Tatur A, Myrcha A (1991) Biogeochemical cycle of ornithogenic matter on land-sea boundary in maritime Antarctic zone. In: Klekowski RZ, OpaliĖski KW (eds) First Polish-Soviet Antarctic Symposium 'Arctowski 85'. Institute of Ecology Publ Office, Warsaw, pp 169–176

Tatur A, Myrcha A, Niegodzisz J (1997) Formation of abandoned penguin rookery ecosystems in maritime Antarctic. Polar Biol 17:405–417

Tedrow JCF (1977) Soils of the polar landscape. Rutgers Univ Press, New Brunswick

Tedrow JCF, Ugolini FC (1966) Antarctic soils. In: Tedrow JCF (ed) Antarctic soils and soil forming processes. Am Geophys Union Antarct Res Ser 8:161–177

Ugolini FC (1972) Ornithogenic soils of Antarctica. In: Llano GA (ed) Antarctic terrestrial biology. Am Geophys Union Antarct Res Ser 20:181–193

Van de Vijver B, Beyens L (1997) Freshwater diatoms from some islands in the maritime Antarctic region. Antarct Sci 9:418–425

White WC, Warin ON (1964) A survey of phosphate deposits in the Southwest Pacific and Australian waters. Dept of National Dev, Bureau of Mineral Res, Geology and Geophysics, Bull 69, pp 1–173

Wilson GJ (1983) Distribution and abundance of Antarctic and sub-Antarctic penguins: a synthesis of current knowledge. BIOMASS Sci Ser 4:1–46

Zale R (1994) Changes in size of the Hope Bay Adelie penguin rookery as inferred from Lake Boeckella sediment. Ecography 17:297–304

Zinsmeister WB, Camacho HH (1982) Late Eocene (to possibly earliest Oligocene) molluscan fauna of the La Meseta Formation of Seymour Island, Antarctic Peninsula.. In: Craddock C (ed) Antarctic geoscience. Univ Wisconsin Press, Madison, pp 299–304

Part III

Soil Microbes, Plants, and Soil Fauna

Preface

Antarctic soils are well and alive – in adapting the slogan of the ACME research group: (Antarctic Cryptoendolithic Microbial Ecosystem Research Group, led by Prof. E.I. Friedmann), although they are working at a different (mostly lower) level of life. At first glance, many features of the microbial world are hampered and governed by a cold, desert-like, nutrient-poor environment. This impression is proved by low biodiversity in higher plants and soil dwelling animals. On the other hand, spots with elevated temperatures and concentrations of (available) organic matter, stimulated by (fluid) water show biomass levels and turnover rates comparable to soil elsewhere in the world. But it has to be taken into consideration that such conditions do not last for a long time and the strategy of getting familiar with, or even adapting to, low temperatures and an impoverished nutrient pool is widely used (Vincent 1988). Despite this fact, the community structure within the consortia of bacteria, yeasts and fungi is also similar to that of temperate soils (Vishniac 1993). A combined analysis of microbial communities and their activities is presented in Chapter 12.

Wide areas of the continent, but also of islands of the maritime Antarctic, can be defined as polar deserts in the sense of Aleksandrova (1980), i.e. plant associations dominated by lichens, mosses, and liverworts as well as a few herbaceous plants; the latter are limited to the maritime region. Recent reviews of the coastal ecosystems of the continent as well as of the sub-Antarctic islands by Lewis Smith (1993a,b) and Kanda and Komárková (1997) emphasise the local patchiness and the lack of interconnections between individual communities and populations.. Except those places in the vicinity of large stations, like Syowa, Casey, Mawson or Davis, comprehensive biological information about this vast area is still limited and reports relate only to occasional expeditions. Nevertheless, collections of data from various individual studies have been presented in several monographs on special tasks and has increased the knowledge on organism groups by taxonomic or physiological studies. Chapters 13 to 15 present results of plant community studies of the sites described in this volume; Chapter 16 focuses on soil-dwelling animals and reviews their role in the Antarctic terrestrial systems.

References

Aleksandrova VD (1980) The Arctic and Antarctic: their division into geobotanical areas. Cambridge University Press, Cambridge.

Kanda H, Komárková V (1997) Antarctic terrestrial ecosystems. In: Wielgolaski FE (ed) Ecosystems of the world, vol 3. Polar and alpine tundra. Elsevier, Amsterdam, pp 721–761

Lewis Smith RI (1993a) Dry coastal ecosystems of Antarctica. In: Maarel E vd (ed) Ecosystems of the world, vol 2 A. Dry coastal ecosystems. Elsevier, Amsterdam, pp 51–71

Lewis Smith RI (1993b) Dry coastal ecosystems on sub-Antarctic islands. In: Maarel E vd (ed) Ecosystems of the world, vol 2 A. Dry coastal ecosystems. Elsevier, Amsterdam, pp 73–93

Vincent WF (1988) Microbial ecosystems of Antarctica. Cambridge University Press, Cambridge

Vishniac HS (1993) The microbiology of Antarctic soils. In: Friedmann EI (ed) Antarctic microbiology. Wiley-Liss, New York, pp 297–341

11 Heterotrophic Microbes, Microbial and Enzymatic Activity in Antarctic Soils

M. Bölter, E. Kandeler, S.J. Pietr, and R.D. Seppelt

11.1 Introduction

Defining the soil microbial community or its activity is often like a fight against Servante's windmills. Despite the many methods which are available for describing individual figures of this 'black box', many uncertainties remain in this important ecological compartment. Today's methodologies still do not allow us to obtain the full picture of the microbial world. The approach to understanding the microorganism's role has to consider several restrictions, e.g. the number of organisms that are actually involved in metabolic processes, or the amount of available substrates for metabolic processes which cannot be determined as overall parameters. The figures obtained by individual methods thus cannot describe 'the microbial activity' but have to be considered as a certain underestimation or overestimation – depending on the methods applied. An underestimation depends on the reliability of data on total fluxes of matter as well as on the relationships between groups of organisms (Jackson and Volk 1970). Nevertheless, several methods have been applied so far to meet the ecological purpose and to describe the system at an acceptable level – within the limits of the test conditions (Smith and Swift 1983; Raven 1990).

Although, or just because, Antarctic soils have been considered simple ecosystems with low numbers of acting bodies, good descriptions of microbes and their activities are needed for a good understanding of the system. General conditions describing characteristics for Arctic ecosystems as noted by Heal et al. (1981) find application in the Antarctic regarding appropriate scales and the isolated distributions on islands or areas along the coasts of the coldest continent. During this chapter we will describe the microbial community mainly on the basis of epifluorescence microscopy of bacteria, fungi and algae, cultural methods, microbial activity in terms of enzymatic processes and the uptake/use of ^{14}C-labelled glucose in relation to estimations of naturally available substrates, as well as total respiration.

Ecological Studies, Vol. 154
L. Beyer and M. Bölter (eds.) Geoecology of Antarctic Ice-Free Coastal Landscapes

11.2 The Microbial Community

Studies on microbial communities in Antarctic soils deal mainly with bacteria. Their occurrence is proved in nearly all environmental niches. Their presence, however, often strikes at the borders of methodological detection levels (e.g. Line 1988; Vestal 1988a,b). Only a few comprehensive data are available about contributions of fungi and algae or even protozoa which might allow comparisons on the level of biomass or numbers of individuals (Tearle 1987; Bölter 1997a,b). Thus, it is not only difficult to find directly comparable studies on these organisms, it is often impossible to compare data of bacterial counts because of methodological inconsistencies (Bölter et al. 1993; Bölter 1994). As bacteria are frequently growing on particles, i.e. in close relation to their basic substrate,detergents or mechanical forces are used to have them as free cells in a solution (Fry 1988; Paul and Clark 1989). Further, for statistical reasons, an even distribution of the organisms is a necessity for obtaining a reliable number.

11.2.1 Bacterial Counts

The counts obtained from various soil samples from the Windmill Islands (Casey Station area) and King George Island (Arctowski Station area) are found in a range from 10^7–10^9 cells g^{-1} dry weight of soil (Bölter 1989, 1990a, 1995). The most evident exceptions from this pattern can be found in direct relation to plants, i.e. to root systems, to their living aboveground organs or to freshly produced litter. Similarly, effects of animals, e.g. bird colonies and seal wallows, can lead to extraordinary figures with counts higher than 10^{10} cells g^{-1} dry weight (Bailey and Wynn-Williams 1982; Ramsay 1983; Ramsay and Stannard 1986; Roser 1993). Heatwole et al. (1989) found colony-forming bacteria at sites of the Windmill Islands between $1.2 \cdot 10^{-4}$ and $3.1 \cdot 10^8$ g^{-1} due to the load of organic matter. A compilation of some of direct counts is depicted in Table 11.1.

An important descriptor is the distribution of the bacterial community into individual size classes. It was found during the studies of soil from Arctowski and Casey that most bacteria cluster into small cocci (<0.5 μm) and small rods (0.5–1.5 μm). The high contribution of small bacteria (a small coccus of a diameter 0.2 μm equals a volume of $4/3\ \pi\ 0.1^3=0.0042\ \mu m^3$) strongly influences the total bacterial biovolume and further figures of bacterial biomass. Cells longer than 3 μm were seldom in the soils itself and found only in direct relation to litter or roots (Bölter 1989, 1990a, 1995). Hence, the distribution of bacterial size classes may be used as fingerprints of the available organic matter along with its sustainability.

The matter of changing environmental conditions can be traced by changes in the community structure. Barren soils, and similarly deep layer

Table 11.1. Data on total bacterial counts of various habitats in Antarctica

Reference	Location	Soil description	Bacterial count ($n \cdot 10^9\ g^{-1}$ dry wt.)
Ramsay (1983)	Ross Island	Barren soil	1.1
Ramsay and Stannard (1986)	Ross Island	Abandoned penguin rookery	16
Bailey and Wynn-Williams (1982)	Signy Island	Barren soils on fjell	0.4–6.3
		Meadow	13–23
Wynn-Williams (1985a)	Signy Island	Barren soils	0.45–2.21
Roser et al. (1993)	Windmill Islands	Ornithogenic soils	0.84–10.66
French and Smith (1986)	Marion Island (sub-Antarctic)	High altitude fjaeldmark,	11.3
		Ice plateau	0.5
Bölter (1990a)	Windmill Islands	Barren soils	0.01–0.05
		Moss/lichen cushions	0.03–0.1
Bölter (1992)	Windmill Islands	Barren soils	0.001–0.008
		Moss/lichen cushions	0.01–0.03
Bölter (1995)	King George Island	Soil surfaces	0.003–0.62
Bölter et al. (2000)	Windmill Islands	Soil surfaces	0.003–4.66
Beyer and Bölter (2000)	Windmill Islands	Various soils/ soil horizons	0.02–1.06
Beyer et al. (2000)	Windmill Islands	Soil surfaces	0.2–1.23

soils, are dominated by small cocci with diameters below 0.5 μm. Such size class distributions have been monitored in soils from the maritime Antarctic as well as from continental sites (Bölter 1990a, 1995a, 1996, 1997b). Similarly, nutrient-depleted soils from Arctic desert and tundra soils show comparable community structure (Bölter 1998). Ramsay and Stannard (1986) report that 48 % of their populations were in the size class of such small cocci. They can even make up more than 75 % of the total community (Bölter et al. 1997). In combination with the smallest size class of rod-shaped bacteria, such a figure is the rule at fjell sites.

Shifts in size classes can be attributed to changing environmental conditions, mainly to variations in soil organic matter due to enhanced growth of soil algae, plant cover or enhanced accumulation of litter, detritus and other organic material (Bölter 1998).The large numbers of small cocci, however, can be interpreted in two contrasting ways:

1. They can be regarded as dwarf cells and cell size shrinks according to low nutrient status (Morita and Moyer 1989), or

2. They reflect an adaptation to low nutrients by increasing the relationship between cell surface and cell volume (Jenkinson et al. 1976).

During the microscope inspections of soils from various biotopes from the maritime and continental Antarctic, it was a general pattern to monitor small colonies of cocci and rods. However, arrangements of typically 'dividing cells' as used for determinations of biomass production according to Hagström et al. (1979) were only found very seldom and not found in significant amounts. However, direct effects of soluble organic substances were monitored during a study of bacterial growth efficiency in soils from Bailey Peninsula (Bölter 1993a). Here, cells of freshly grown colonies were significantly larger than single cells.

11.2.2 Bacterial Biomass

The description of bacterial biomass occurs predominantly in carbon units (µg C) as derived from figures of biovolumes. The critical steps during this process occur during the evaluation of the mean biovolume or during the counting of cells and putting them into individual size classes with defined biovolumes. Further, low numbers – typical for barren, nutrient-poor soils – often present statistical problems while scaling up procedures from the counting grid to the subsample and further to the sample unit. Individual laboratories have their special features, thus, it is often difficult to draw direct and unbiased lines from sampling sites of different research programs (Bölter 1994).

Data on bacterial biomass were analysed during several research projects conducted in the landscapes around Arctowski and Casey Station. They mark biomass values in the range between 0.5 and 50 µg C g^{-1} soil; higher values, up to an order of magnitude, can be obtained when analysing plants or root environments (Bölter 1992, 1995), in some cases, however, biomass figures of barren soils are often only in ng ranges (Bölter 1990). Similar values can be calculated from data of Ramsay and Stannard (1986) who analysed soils from Ross Island or from French and Smith (1986) from Marion Island. Roser et al. (1993) determine bacterial biomass from direct counts between 7.5 and 53 µg g^{-1} soil for sites which are not occupied by penguins; for ornithogenic soils he finds values up to 1400 µg g^{-1}.

Bacterial biomass, however, contributes only to a small part of total microbial biomass. Data obtained by measures of ATP and subsequent extrapolations to microbial carbon show that only 0.014% of the determined carbon can be related to bacteria (Bölter 1989). Roser et al. (1993) describe higher ratios when comparing biomass determinations by converting data from direct counts, measurements of ATP or substrate induced respiration (SIR), i.e. in about that range to which bacteria contributed to total microbial biomass which was mainly represented by algae and fungi.

11.2.3 Fungi and Yeasts

The microscopic identification and differentiation of organisms larger than typical bacterial cells often become difficult. Many of those cells mostly identified as large cocci (diameter >1 μm) can be attributed to small cyanobacteria or yeasts. There are several reports dealing with large shares of coccoid cyanobacteria (e.g. Broady 1979) and yeasts (Fletcher et al. 1985; Line 1988; Heatwole et al. 1989; Ray et al. 1989) in the microbial community. An important share in biomass can also be formed by mycelia-forming fungi (Vincent 1988).

The general picture, however, shows that such influences are only of minor importance for overall descriptions (Roser et al. 1993). They assume that yeasts contribute only 0.01 to 0.1 % of microbial biomass in soils of the Windmill Islands, even in soils which showed a rich fungal contribution. Microscope inspections of soils, however, show only a few hyphae. Roser et al. (1993) state that fungal hyphae are very rare during their microscope inspections and account for <1 % of the total microbial biomass. This can be related to the habitats of fungi which are normally organic-rich aggregates or other high quality litter. There are, however, no estimates available which might give any idea of their actual activity or their share of the degradation of organic matter in this region.

Data of Heatwole et al. (1989), which are based on colony counts, show much higher ratios between bacteria and yeasts, up to nearly 50 % at Nelly Island. However, there is evidence that yeasts are possibly closely related to human influences as described by an index derived from total number of bacteria and yeasts at Syowa Station by Toyoda et al. (1986). The observation of Smith and Walton (1985) shows a relation between tundra plant species with both bacterial and fungal populations. Although higher plants in the maritime Antarctic *(Deschampsia antarctica* and *Colobanthus quitensis*) have been excluded from mycorrhizas (Christie and Nicolson 1983), recent investigations on hepatics in continental Antarctica (Windmill Islands) show their abundance in *Cephaloziella exilifora* (Williams et al. 1994).

Microfungi have been isolated from a wide range of habitats throughout Antarctica (e.g. Tubaki 1961; Tubaki and Asano 1965; Wicklow 1968; Sun et al. 1978; Vishniac and Hempfling 1979; Pugh and Allsopp 1982; Baublis et al. 1991; Montemartini Corte and Gestro 1994; Onofri et al. 1994; Azmi and Seppelt 1998; McRae et al. 1999). There are major taxonomic difficulties with many of the fungal groups. Any sensible comparison of the fungal flora from different localities, however, is not possible due to differing taxonomic interpretations, the isolated nature of collection sites, and differences in thoroughness of collections and isolations.

The genera *Acremonium, Cladosporium, Chrysosporium, Verticillium, Penicillium, Phoma, Thelebolus, Geomyces, Epiococcum, Hormoconis, Mortierella* and *Mucor* have been identified in soils and on plant material from east

Antarctica (Greenfield and Wilson 1981; Fletcher et al. 1985; Line 1988; Heatwole et al. 1989; Del Frate and Caretta 1990; Kerry 1990a, b). *Oidiodendron, Chrysosporium, Mortierella* and *Verticillium* have been described from Signy Island by Bailey and Wynn-Williams (1982) as well as by Pugh and Allsopp (1982). A more extensive list of soil microfungi, including species of Ascomycetes, Zygomycetes and Deuteromycetes has been presented by Zabawski and Czarnecki (1981) as well as by Czarnecki and Bialasiewicz (1987) by sampling the aerosphere in the area of Arctowski Station. Reviews by Baublis et al. (1991) and Vishniac (1993) show a much larger diversity even in soils of cold deserts. Their role in soil stabilisation and thus the formation of basic substrates for further successors in the colonisation process has been stressed by Wynn-Williams (1993).

Isolations made in the Windmill Islands have indicated that around 120 taxa may be present (Azmi, McRae and Seppelt, unpubl.; McRae et al. 1999). Around half the taxa are sterile mycelia, a feature common to other studies of the fungal flora of Antarctic regions (Pugh and Allsopp 1982). Zabawski and Piasecki (1981) reported up to 92 % sterile mycelia from mineral soils at Admiralty Bay (King George Island) as well as Fletcher et al. (1985) who reported 47 and 60 % of isolates being sterile from MacRobertson Land and Enderby Land respectively.

A suite of taxa, all common commensals of sites of human occupation, has been isolated in Admiralty Bay at King George Island and in the Windmill Islands region only from the vicinity of occupied and abandoned stations (Zabawski and Piasecki 1981; Azmi and Seppelt, unpubl. data). It is not known if these have become established in the local environment or if diaspores are re-inoculated annually with the station resupply. The introduction of microbial organisms to Antarctica is a major environmental threat. For example, *Aspergillus fumigatus*, the causal agent of avian aspergillosis, has been isolated from ornithogenic soils at Cape Hallett, northern Victoria Land, where there are 75,000 breeding pairs of Adélie penguins. Human occupation at the (now abandoned) station was implicated in the introduction of this pathogen (Wicklow 1968).

A number of fungal taxa appear to be confined to the immediate vicinity of Casey Station (Azmi and Seppelt, unpubl. data). *Phialophora fastigiata* is thought to have been introduced to the area on softwood packing crates. The species is also common around Davis Station in the Vestfold Hills (Kerry 1990b). Species of *Mortierella*, *Cladosporium*, *Mucor*, and *Penicillium* are also apparently confined to Station limits. Line (1988) reported the *Hormoconis resinae* state of *Amorphotheca resinae* from fuel-contaminated soils at Mawson Station. Hydrocarbon decomposing bacteria have also been isolated from such soils.

11.3 Microbial Activity

There is a wide array of physiologically based soil processes which have been investigated in Antarctic soils and can be determined in different ways. Measurements on respiration provide information about the response of active microbial biomass to easily decomposable material. Bacterial activity can be followed after the addition of specific ^{3}H- or ^{14}C-labelled substrates. In addition, several methods for the determination of soil enzymes have been applied to Antarctic soils, since the life of soil microorganisms is composed of a series of enzymatic reactions which are responsible for nutrient cycling. Besides the characterisation of specific steps in nutrient cycling, several methods were used to characterise physiologically based soil processes, e.g. decomposition of organic matter, ammonification, nitrification, N_2 fixation and phosphorus mineralisation. Although many of these methods have been used in Antarctic soils, reliable comparisons of such data are still difficult because of individual methodological considerations.

11.3.1 Bacterial Activity

^{14}C-labelled glucose can be used as a model substrate of easy available dissolved organic matter. While accepting this model, it can be extended to describe an actual metabolic process by including its ratio to ^{12}C-glucose. Further, respiration and incorporation of the material can be determined separately and data can be obtained for substrate turnover, bacterial biomass production and remineralisation rates (cf. Meyer-Reil 1978; Bölter 1981). Other substances, e.g. various monosaccharides or amino acids, can be used. Advantages of this method are short incubation times and only small changes in the natural substrate concentration. Nevertheless, the disturbance of the environment during sampling and incubation under standard conditions in the laboratory should be born in mind. Therefore, these methods give evidence for potential rather than actual activity.

11.3.2 Turnover Times of Glucose

Results obtained by this method are available from the soil of Casey and Arctowski regions (Bölter 1990a, 1991, 1992a). Short turnover times are evident for surface samples independent of incubation temperatures. The values found lie between 3 and 10 h for soils from Casey and between 24 and 87 h at Arctowski (Bölter 1990a, 1992a). The latter values relate to those from Grobler et al. (1987) who applied the kinetic approach for the use of glucose in soils from Marion Island. For samples from the Casey area, a slight optimum can be established for an incubation method of 15 °C, whereas samples from Arctowski show an increasing trend up to 25 °C (the maximal incubation temperature). Hence, it can be argued that the bacterial community at Casey shows a trend to a psychrophilic or psychrotrophic behaviour whereas the community at Arctowski is more related to a mesophilic metabolism.

Such trend, however, would need verification by more detailed studies. Roser et al. (1993) show a typical mesophilic metabolism during substrate-induced respiration (SIR) experiments for soils from the Windmill Islands, rather than psychrotrophic or even psychrophilic behaviour. A fact which can be supported by soil temperatures. They can rise in top horizons to more than 30 °C, although only for short times. Isolated bacteria from McMurdo showed optima between 10 and 21 °C and growth maxima up to 37 °C (Siebert and Hirsch 1988; Gallikowski and Hirsch 1988). The highest numbers of colony forming units (c.f.u.) of bacteria enumerated on pepton-yeast medium at Admiralty Bay are insignificantly different in temperature range from 4 to 20 °C. At a temperature of 30 °C, similar enumeration gives around an order of magnitude lower numbers (Pietr 1993). Strains identified as *Arthrobacter* sp. from Schirmacher Oasis show an optimum at 22 °C and a maximum at 37 °C (Shivaji et al. 1989).

11.3.3 Respiration of Glucose

The amount of respired glucose generally exceeds 50 % of the gross uptake, this holds true for soils from Casey as well as for samples from Arctowski. A slight relationship with temperature can be seen only for samples from Casey (Bölter 1990a, 1991, 1992a). Such high respiratory activity points to a high share of a basal metabolism, which leaves only a few parts left for biomass production. Although such values can be found in data from other environments (e.g. marine: 50–80 %; Billen et al. 1980), there are data showing the opposite ratio, 5–10 % respiration of glucose in beach sediments from temperate regions (Meyer-Reil 1978).

Considering the actual glucose concentration, it becomes possible to assess a figure for the remineralisation rate. Using this figure, which describes a more realistic data base, the respired material goes down to 12–17 % of its total amount. Highest values of the remineralisation rate go along with high amounts of organic matter, as found in surface horizons. Maximal values in soils can be found at Casey soils with 38 μg C g^{-1} h^{-1} (Arctowski soils: 32 μg C g^{-1} h^{-1}). Data of nearly one order of magnitude higher were observed when incubating thalli of lichens from Arctowski area: 289 μg C g^{-1} h^{-1} (Bölter 1992a).

11.3.4 Biomass Production Based on Glucose Metabolism

By subtracting the respired material from the gross uptake it becomes possible to evaluate a figure of the net uptake and from this the next step is a measure of the biomass production – under the assumption that this material is converted into cell biomass (Meyer-Reil 1978; Bölter 1981). Highest values for this descriptor were monitored for topsoils enriched with organic material from plants.

At Casey, the biomass production was found up to 8 $\mu g\ C\ g^{-1}\ h^{-1}$ at 15 °C, such a figure at Arctowski is 2 $\mu g\ C\ g^{-1}\ h^{-1}$; maximal values were established at values of 36 $\mu g\ C\ g^{-1}\ h^{-1}$ for lichen thalli from Arctowski, i.e. their Aufwuchs (Bölter 1992a). More realistic values, however, can be obtained from the median values, which vary between 0.02 and 0.72 $\mu g\ C\ g^{-1}\ h^{-1}$ in relation to the organic matter contents of the soils. The strong heterogeneity of the soils and their strong patchiness, however, does not allow a realistic calculation into area related values. Further, these data are obtained under optimal conditions with regard to temperature and moisture.

These data give a picture of a potentially healthy and active bacterial community. Taking into account the data of bacterial biomass, a turnover time can be calculated. This ranges, with respect to optimal conditions, between 0.25 and 0.4 h in surface samples and between 0.72 and 2.1 h in deeper samples (Bölter 1990a). Such figures obviously cannot be maintained for a longer period due to a restricted supply of nutrients and limitations of temperature.

11.4 Enzymatic Activity

Free enzymes, particle- or organism-bound enzymes act as mediators between particulate and dissolved organic matter, e.g. structural and monomeric carbohydrates. Thus, enzymes serve as important mediators between complex substrates and osmotrophic microorganisms (Corpe and Winters 1972). Many of these enzymes are inactive because they suffer from co-products or because they are trapped and immobilised by organic or inorganic colloidal material. Nevertheless, significant changes in the structure of organic matter can be attributed to cell-free enzymes, mostly performed by hydrolytic enzymes (Ladd 1985; Paul and Clark 1989). The determination of enzymatic activity needs – as above, the determination of bacterial activity – a drastic change in the soil constitution and in the substrate supply. Hence, the data have to be considered as potential rather than actual process descriptions.

The use of fluorescein diacetate (FDA), methylumbelliferyl (MUF) coupled substrates, as well as other approaches have gained data about enzymatic

Table 11.2. Comparison of the range of enzyme activities of Antarctic and temperate soils differing in vegetation type and soil type. Data of Antarctic soils are given in *bold letters*

Soil enzyme activity	Range of activities	Soil type	Reference
Xylanase activity (mg glucose g^{-1} 24 h^{-1})	**0.001–0.500**	Typic Haplorthel (USDA)	Tscherko et al. (2002)
	0.90–2.93	Typic Umbrorthel (USDA) ornithog.	Tscherko et al. (2002)
	1.8–3.0	Orthic Luvisol (FAO)	Kandeler and Eder (1993)
	0.75–2.00	Haplic Chernozem (FAO)	Kandeler et al. (1999)
	0.38–1.15	Phaeocem, Lithosol, Cambisol (FAO)	Kandeler et al. (1999)
	0.24–1.83	Haplic Luvisol, Entisol (FAO)	Stemmer et al. (1999)
Urease (µg N g^{-1} 2 h^{-1})	**0.59–5.48**	Typic Haplorthel (USDA)	Tscherko et al. (2002)
	60–439	Typic Umbrorthel (USDA) ornithog.	Tscherko et al. (2002)
	120–190	Pachic Arguistoll (USDA)	Ajwa et al. (1999)
	10.1–131.5	Argixeroll, Haplohumult (USDA)	Burket and Dick (1988)
	60–170	Hapludalf (USDA)	Deng and Tabatabai (1996)
	100–270	Orthic Luvisol (FAO)	Kandeler and Eder (1993)
Arylsulfatase activity (µg p-nitrophenol g^{-1} h^{-1})	**0.2–1.3**	Typic Haplorthel (USDA)	Tscherko et al. (2002)
	0.9–16.2	Typic Umbrorthel (USDA) ornithogenic	Tscherko et al. (2002)
	30–50	Pachic Arguistoll (USDA)	Ajwa et al. (1999)
	115–340	Hapludoll (USDA)	Klose et al. (1999)
	50–350	Hapludoll (USDA)	Deng and Tabatabai (1997)
	28.1–57.9	Phaeocem, Lithosol, Cambisol (FAO)	Kandeler et al. (1996)
Proteases (complex) (µg N g^{-1} 2 h^{-1})	**16–160**	Typic Umbrorthel (USDA) ornithog.[a]	Pietr et al. (1983)
	133–290	Histosols (USDA)	Pietr et al. (1995)
Acid phosphatase (µg phenol g^{-1} 3 h^{-1})	**78–120**	Typic Umbrorthel (USDA) ornithogenic[a]	Pietr et al. (1983)
	247–343	Histosols (USDA)	Pietr et al. (1995)
Alkaline phosphatase (µg phenol g^{-1} 3 h^{-1})	**1.4–53.2**	Typic Haplorthel (USDA)	Tscherko et al. (2002)
	251–548	Typic Umbrorthel (USDA) ornithog.	Tscherko et al. (2002)
	143–258	Typic Umbrorthel (USDA) ornithog.[a]	Pietr et al. (1983)
	310–800	Haplic Chernozem (FAO)	Kandeler et al. (1999)
	64–187	Haplumbrepts, Hapludalfs (USDA)	Beyer et al. (1999)
	553–992	Phaeocem, Lithosol, Cambisol (FAO)	Kandeler et al. (1996)

[a]Temperature of incubation 10 °C.

activity in Antarctic soils. As such, data from soils of the Casey area (Bölter 1989, 1992a) showed a ranking for MUF-substrate degrading activity with MUF -acetate >-butyrate >-phosphate >-glucose >-galactose. It also became evident during this study that the activity correlated directly to contents of organic matter and has a high response to temperature (Bölter 1989). Results showing comparable trends are found with FDA by Bölter (1992b) and Roser et al. (1993) from soils of the Casey area. Soils from Arctowski did not show such a clear relationship to soil organic matter when analysed with different MUF-substrates (Bölter 1992a).

The comparison of enzyme activities involved in the C-, N-, S-, or P-cycling from Antarctic and temperate ecosystems is given in Table 11.2. Antarctic soils show much lower ranges of xylanase, urease, arysulfatase and alkaline phosphatase than soils from temperate environments. The high input of nutrients into Antarctic soils by birds increases several enzyme activities (alkaline and acid phosphatases, urease and complex of proteases) to the level of temperate soil types.

Fungi (*Verticillium, Aspergillus* and *Phoma*) were investigated from Victoria Land and showed the ability to produce lipases in high quantities, polygalacturonase, amylase and phosphatase were common (Fenice et al. 1997). Experiments with marine psychrophilic bacterial isolates from Dumont d'Urville Station showed that they were able to produce exoenzymes with very low temperature optima (Feller et al. 1994). A lipase from *Moraxella*, α-amylase from *Alteromonas haloplanctis*, β-lactamase from *Psychrobacter immobilis* and protease from *Bacillus* sp. had maximal production rates close to environmental conditions, –2 to 4 °C, but higher temperatures inhibited them, a fact which indicates the potential of cold adapted enzymes and can also be relevant for terrestrial bacteria, although such observations have not been described so far. Only the measurements of enzymatic activity of some soil enzymes (acid and alkaline phosphatases, proteases, uric acid hydrolase and asparagine lyase) done at 10 °C suggest that such cold adaptation could exist (Pietr et al. 1983).

11.5 CO_2 Evolution and O_2 Consumption

The acceptance of CO_2 evolution as a measure of overall activity accepts the model that the majority of metabolic processes are aerobic. As herbivory can be widely neglected in Antarctic soils systems (Vincent 1988), this end product can be taken into consideration as a measure for total microbial activity. Protozoans are mainly restricted to wet moss cushions (Block 1984; Smith and Tearle 1985). The share of protozoans to total respiration was assumed to be about 10 % in a moss cushion at Signy Island (Davis 1981), other animals, including collembola and nematodes accounted less than 1 %.

Both methods of CO_2 evolution and O_2 consumption have thus found application in Antarctic environments. However, results which allow a direct comparison between these methods are still rare (e.g. Wynn-Williams 1984; Bölter 1991, 1992a). It becomes evident from these analyses that these descriptors do not match properly, although stoichiometric rules imply direct relations and have been used as a basis for decomposition models (Davis 1980). Reasons for this must be seen in the different experimental conditions. A fact that also explains mismatches while comparing such data with remineralisation rates obtained from isotopic approaches. The data of Wynn-Williams (1985b) showed that in wet peat samples O_2 uptake refers only partly to respiration and anaerobic processes occur concomitantly.

The application of the dynamic gas exchange measurements via infra-red gas analysis (IRGA) or the static method using gas chromatography allows analyses of metabolic activity with low disturbances of the system. Discrete soil samples can be analysed in gas exchange chambers or gas samples can be obtained from various kinds of enclosures. Measurements of CO_2 provide data of an integral nature across all metabolic process and all organisms involved in it. Simultaneous respiration (by soil heterotrophs) and incorpora-

Table 11.3. CO_2 evolution of soils from Casey and Arctowski (µg CO_2 g^{-1} h^{-1}) in relation to incubation temperature. *Highlighted values* show median values, other values represent maxima

t (°C) Location	−3	0	3	5	6	10	15	20	25	30
Casey 0–2 cm *n*=11	n.d.	n.d.	**5.4** 33.3	n.d.	**5.4** 50.0	**10.7** 66.5	**16.1** 91.5	**21.5** 124.0	**21.3** 158.0	n.d.
Casey >2 cm *n*=12	n.d.	n.d.	**0** 3.5	n.d.	**0** 5.2	**0** 1.7	**0.9** 3.5	**1.2** 5.2	**1.5** 5.2	n.d.
Arctowski 0–2 cm[a] *n*=11	**2.1** 9.3	**6.7** 17.5	n.d.	**11.4** 33.8	n.d.	**15.7** 37.1	**26.2** 61.9	**34.9** 86.6	**36.8** 116.8	**43.6** 122.6
Arctowski 0–2 cm *n*=7	**0** 1.1	**0** 1.1	n.d.	**0** 1.6	n.d.	**0** 1.8	**0** 2.7	**0** 3.6	**0** 2.6	**0** 5.4
Arctowski >2 cm *n*=17	**0** 4.0	**0** 4.0	n.d.	**0** 6.4	n.d.	**1.2** 9.1	**1.2** 9.1	**1.5** 11.1	**0.6** 20.1	**2.7** 7.5

[a] Surface samples from Arctowski soils with significant plant cover of moss/lichen crusts, lichen heaths or meadows formed by *Deschampsia antarctica* and *Colobanthus quitensis*.

tion (by plants) occurs in environments with plant cushions, resulting in a net signal of CO_2 and makes it difficult to elucidate the individual shares of organisms.

As already stated for other parameters of microbial activity, the upper layers show highest data where significant organic matter is present. Actual data for CO_2 evolution show values in ranges from 0 to 43 μg CO_2 g^{-} h^{-1}, depending on incubation temperature (see Table 11.3).

Recalculating these data to area-based values, for a temperature of 10 °C a figure of 100–150 mg CO_2 m^{-2} h^{-1} can be taken into consideration. Including the time spans of appropriate soil temperatures, corresponding values of 2 and 3 g CO_2 m^{-2} d^{-1} for Casey and Arctowski, respectively, can be assumed. Similar values were obtained for tundra biomes in Alaska and Scandinavia during the IPB studies (Bunnell et al. 1975; Skartveit et al. 1975; Billings et al. 1977; Poole and Miller 1982; Moore 1986).

The main problem in comparing these data with other biotopes is the assessment of environmental factors. The shifts of CO_2 evolution are not only connected with temperature, as shown in Table 11.3, but also with soil moisture and organic matter availability. The elevation of moisture can enhance the CO_2-output by a factor of three for individual temperatures and leads to shifts of an activity signal to lower temperatures (Bölter 1992a). Such observations need to be incorporated carefully into balances and extrapolations into larger areas as well as to longer time scales.

11.6 Driving Factors of Microbial Activity

11.6.1 Environmental Conditions

Temperature, soil moisture and substrate availability are obviously the most effective driving forces of microbial activity. This must have consequences on any attempt to balance the systems over larger areas or longer time scales. As there are no direct hints that the microbial community has physiologically adapted to low temperatures or low substrate contents, other facts must be taken into consideration to explain high values of enzymatic activity or low turnover times of organic matter. It is evident from the environment itself and the ranges in which the environmental and biological parameters can be monitored, as well as the time spans which allow an active metabolism and the short periods over which these factors can change, that the microbial community needs several strategies for survival.

Campbell and Claridge (1987) have shown that the main reason for low development on the Antarctic continent are low temperatures and low precipitation, i.e. desert-like constraints. There are several indications that show direct effects between moisture availability and the distribution and abundance of organisms (Kennedy 1993). This fact has been used before by Lewis Smith (1985) to explain low plant productivity by these main external factors with respect to low turnover rates of organic matter. However, detailed studies have shown that this model holds true at the broad scale but also claims many exceptions at lower scales. Both temperature and precipitation can be found valid for developments in space and time.

Local factors like relief and exposure in combination with microclimate show pronounced effects on plant production and subsequent biological patterns (e.g. Kappen 1985; Bölter et al. 1989; Bölter 1992a, b). Although elevated temperatures, as measured at sites near Casey Station with maxima at 34 °C in soils and even 42 °C in moss beds (Bölter et al. 1989; Lewis Smith 1986), are exceptions for only short times, they pose special conditions and force the microbial flora to find survival strategies over extremely wide temperature ranges. On the other hand, there is much evidence that moisture and substrate quality play an even more important role than temperature alone (e.g. Wynn-Williams 1982, 1984) in a multivariate system. Bearing in mind that most bacteria in this system might be defined as oligotrophs, we have to consider that they may also have wide nutritional flexibility with respect to organic substrates (Upton and Nedwell 1989).

However, with regard to the overall process, temperatures in ranges between 5 and 15 °C seem to be of greater importance. Within this range, the maximal Q_{10}-values were observed (Bölter 1992a) for uptake and mineralisation of glucose as well as for CO_2 evolution. Similar observations were made for the substrate induced respiration (SIR) by Roser et al. (1993). It could also be shown that the sum of glucose mineralisation within this window of proper conditions ($0<t<10$ °C) makes up nearly 80 % of total mineralisation; the rest can be found in the next step, $10<t<20$ °C (Bölter 1992a). Kerry (1990a) reported from fungi isolated at Casey that many of them showed lowest Q_{10}-values for the temperature range 4 to 15 °C, which can be interpreted that they are not in real competition with the bacterial community. Their temperature characteristic, however, has been classified as psychrotroph or cold tolerant mesophil and minimum growth temperatures are indicated as zero or below.

The reason for this shift to lower temperatures seems contradictory to the above mentioned fact that higher temperatures create higher metabolic activity. However, the situation is that the organisms have to cope with a strong drought at higher temperatures – similar to freezing conditions at the other end of this small temperature window. Osmotrophic organisms like bacteria and fungi find themselves in a pair of shears represented by water availability on the one hand and low nutrient conditions on the other. Hence, it is not only water stress which hampers potential microbial activity, but increasing concentrations of free dissolved anions and cations in the remaining fluid water increases the osmotic stress factor which needs a concomitant high tolerance of the organisms against it.

Life on this narrow ledge of appropriate environmental conditions for metabolism, awaiting drastic changes within time-spans of minutes seems to be a more important factor than the adaptation to a single variable like low temperatures. Thus, observations on rapid reactions of microbial metabolism to thaw-freeze cycles (Wynn-Williams 1982) or dry-wet conditions (Parker et al. 1983) provide suitable descriptions of the microbial system in Antarctic environments. The main effect of this process, however, is probably the stimulation of microbial activity by sustaining fresh low molecular weight substances, especially carbohydrates (Wynn-Williams 1980; Tearle 1987; Melick and Seppelt 1992).

Consequently, life conditions are more linked to events rather than to long-term, stable conditions. The low productivity of the Antarctic terrestrial system thus cannot be regarded as a permanently dormant or starving community which only needs extra solar energy, but is a complex system of balanced switches at different levels and high stress. This further needs the determination of special constraints and thresholds which have to be regarded as active switches for metabolic processes. The microclimatic conditions with drastic changes at the 0 °C threshold for availability can account for one of these important switches (Walton 1982; Morley et al. 1983).

Ino (1985) supposed that organisms living in harsh environments should have short recovery periods and showed this for respiratory activity of mosses. It has been proposed that because of the low production rates, the complete food web acts at low activity levels (Lewis Smith 1985). This indeed holds especially true for environments which are dominated by a detritus food web, where the link across faunal components is widely blocked. Another crucial ecological feature which is missing in these habitats is the bioturbation as maintained by an active soil fauna in other environments. Small and often isolated patches of soils with plants as servants for organic matter have to stay as their own systems without possibilities for exchanges of matter, energy and information with neighbourhoods.

From the viewpoint of a microbial organism, the most important factor for survival is the accessibility of organic substrates as well as nitrogen and phosphorous. These can be present in organic or inorganic forms. Antarctic soils which are normally deficient in inorganic nutrients – except in the vicinity of coastlines or penguin rookeries – have to cope with the problem that they are also poor in humus and other retardant substances. Also, particulate organic carbon, particulate carbohydrates and dissolved carbohydrates dominate the surface layers (Bölter 1993b). The same study showed a ratio <1 for dissolved carbohydrates vs amino acids, indicating that the latter play an important role in the nutrition of bacteria.

11.6.2 Dissolved Organic Matter

Dominant inputs of dissolved organic matter are leaching products from plants, photosynthetic microorganisms, active free enzymes or products excreted from microfauna. Hence, carbohydrate leaching may constitute a major nutrient input into Antarctic soils in areas away from bird colonies. This is particularly evident in the maritime Antarctic where seasonal increases in soil microbial activity have been linked to flushes of soluble carbohydrates from vegetation in the spring thaw (Wynn-Williams 1980, 1982; Tearle 1987). Tearle (1987) showed that the leaching of dissolved organics from lichens and mosses at Signy Island can account for 62–236 mg g^{-1} sugar. Polyols, of which arabitol had a prominent position with 115 mg g^{-1}. Bölter (1990a) detected free amino acids (24–136 mg g^{-1} soil) and dissolved monocarbohydrates (12–32 mg g^{-1} soil) in Windmill Island soils. Melick and Seppelt (1992) found 5–65 mg g^{-1} in bryophyte samples. Roser et al. (1992a,b, 1994) have reported similar values from the Casey area, which reached values up to 3300 µg g^{-1} in cryptogam dominated soils, Melick and Seppelt (1994) report data with a maximum between 52–273 mg g^{-1} for various cryptogams.

Such organic material is nutritionally of high quality and taken up rapidly by microbial communities (Melick et al. 1994). Free carbohydrates, especially glucose, were shown to be a significant and preferred substrate while comparing bacterial biomass production based on different dissolved compounds (Bölter 1993a). Hence, residual amounts of glucose are low. Estimates of free glucose show only values less than 0.01 % of the particulate carbohydrates in soils from Casey and Arctowski regions (Bölter 1993b). Roser et al. (1994), however, found carbohydrates in algae-dominated soils up to nearly 7 % of organic carbon. Further, high amounts of low weight organic acids were evident contributors to dissolved organic mater.

A diverse microbial flora and fauna has been found to reflect site nutrient status in continental Antarctic soils (Cameron et al. 1970; Bölter 1990a, b, 1992b; Roser et al. 1993; Melick et al. 1994). The carbohydrates leached from Antarctic bryophytes have been shown to be primarily sucrose, fructose and glucose, while arabitol, ribitol and mannitol are the major carbohydrates leached from Antarctic lichens (Chapman et al. 1994; Melick and Seppelt 1992). Non-marine algae reveal a variety of water soluble carbohydrates and polyhydric alcohols (Chapman et al. 1994). In contrast to the maritime Antarctic, the production and metabolism of leachates within soils in the drier, harsher climate of continental Antarctic sites is limited to the short periods when free water is available. Despite this, significant accumulations of short chain alcohols and organic acids may occur as a result of microbial fermentation and may play an important role in controlling the availability of plant micronutrients (Roser et al. 1994). The soil microbiota in the Windmill Islands has been shown to respond rapidly to sudden changes in microclimate and nutrient status, a response which appears essential for the successful survival in the continental Antarctic environment.

11.7 Mineralisation of Guano

Large active and abandoned penguin rookeries substantially drive microbial systems in Antarctic terrestrial systems. This special kind of cycling of nutrients not only within the terrestrial systems but also between the terrestrial and the marine environment (Dawson et al. 1985) is a special feature of Antarctica (see also Sect. 2.1). Hence, this specialised environment needs a separate assessment.

Bacterial, enzymatic and chemical analyses indicate active microbiological mineralisation of penguin excrement in the maritime Antarctic (Pietr et al. 1983; Myrcha et al. 1985; Pietr 1993). The following physiological groups of bacteria were found: proteolytic, ammonifying, nitrifying, lecithin degrading, $Ca_3(PO_4)_2$ dissolving, chitin degrading bacteria. Processes of proteolysis, ammonification and hydrolysis of organophosphate compounds take place at the beginning of excreta decomposition, whereas nitrification proceeds in guano solutions with a delay after microbiological mineralisation of soluble organic compounds and reduction of ammonia ion concentrations, usually in the marginal zone of the penguins' influence or in deeper layers of the soil (Fig. 11.1).

Bacteria are a dominant group of microbiota in ornithogenic soil and they play the main role in the biological mineralisation of guano (Pietr 1986; Ramsay and Stannard 1986; Bölter et al. 1997). The bacterial biomass in the soil of an active penguin rookery accounted for 92 % of total microbiota (Roser et al. 1993). Large, Gram-negative, non-spore-forming parts of the bacterial population, often grouped in tetrads, cocci constitute about 48 % of visible by direct microscopy microbes in ornithogenic soils (Ramsay and Stannard 1986). However, among those isolated on standard agar organic media, strains of the family *Pseudomonaceae* are most frequently observed in ornithogenic and other Antarctic soils. Recent studies of Bowman et al. (1996) show that *Psychrobacter* is the predominant genus in ornithogenic soil in Antarctica. Strains of *Psychrobacter* belong to three distinct groups represented by *P. immobilis* as well as two novel species *P. urativarans* sp. nov. and *P. frigidicola* sp. nov. The genus *Bacillus* dominates amongst spore-forming bacteria. There are other spore-forming bacteria from the genus *Sporosarcina*, which could not be identified to species level (Pietr, unpubl.). Numbers of enumerated fungi (moulds and yeast) on solid agar media and their biomass are very low in soil with a high content of fresh organic matter of animal origin (Fletcher et al. 1985; Pietr 1986; Roser et al. 1993).

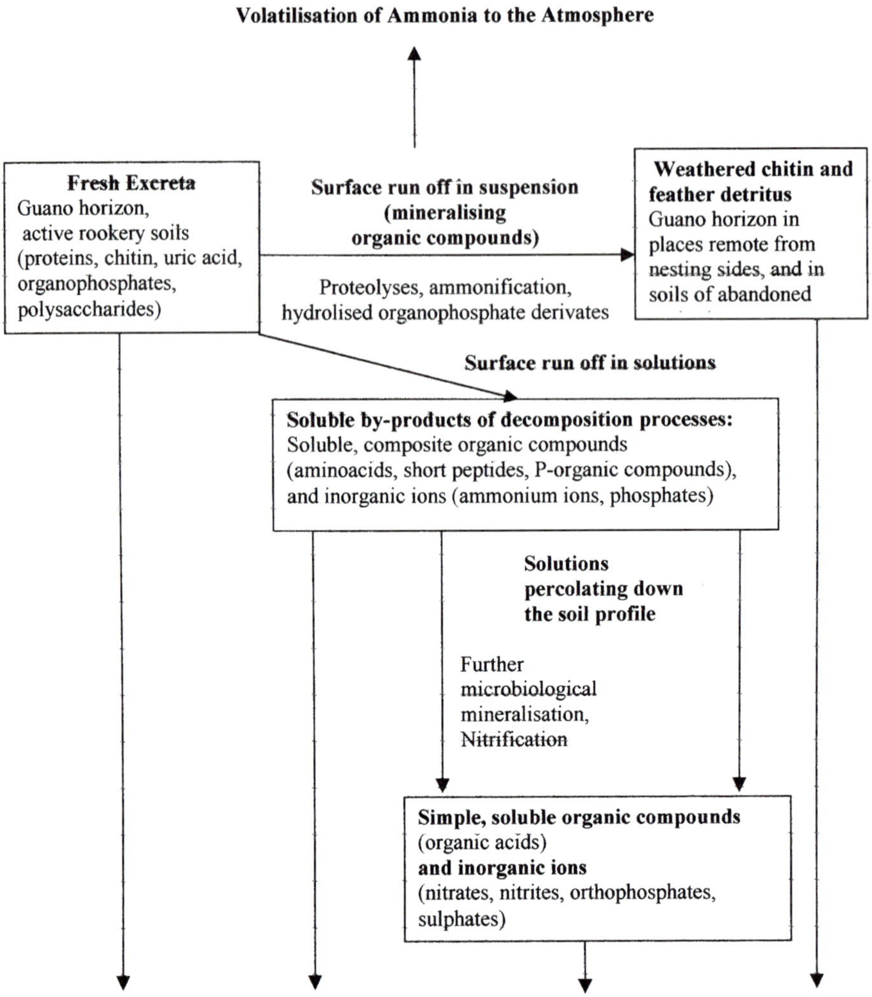

Fig. 11.1. Scheme of the degradation of organic matter in ornithogenic soils

Dominant fungi present in the area manured by guano solutions belong to unidentified *Mycelia sterilia* (Zabawski and Piasecki 1981; Fletcher et al. 1985). Among identified filamentous fungi representative of the genus *Penicillium* (*P. biforme, P. cyclopium, P. notatum, P. commune, P. antarctica* sp. nov.) occur, being the most often isolated fungi from ornithogenic soils. Moreover, *Cladosporium, Hormondedrum* and typical coprophilic *Thelebolus microsporus* are also present (Zabawski and Piasecki 1981; Fletcher et al. 1985; McRae et al. 1999). The presence of keratinolytic fungi from the genus *Chrysosporium* and *Sporotrichum* was typical for the soils at sites of penguin moulting (Zabawski and Piasecki 1981). Moulds take part in the final stage of the decomposing processes of guano remains. The biomass of microfauna is not a significant bulk element of ornithogenic soil (Petz 1997), however, their activity is an important factor controlling organic matter disintegration processes (Myrcha et al. 1991; Petz 1997) during the early stages of the decomposition processes. The rare presence of moulds and yeast in ornithogenic soil could be related to the presence of acrylic acid in penguin excrement (Boyd 1967) and high chitinolytic and antifungal activity of bacteria (Czekanowska and Zabawski 1988; Pietr 1986, 1993).

The nitrate reducers and N_2-fixing bacteria were not detected. High enzymatic activity (protease, asparaginase, uric acid hydrolase) in the surface horizons of these soils were similar or higher than those found in normal topsoil in the temperate zone. microorganisms from fresh penguin excreta, rather than the native soil microflora, appear to be the major source of some enzymes (protease, phosphatases) in the organic layer on active penguin rookeries (Pietr at al. 1983, Speir and Ross 1984). However, the enzymatic breakdown of amino acids, organophosphates and uric acid due to the activity of soil microorganisms takes place mainly after the initial breakdown of excreta.

About 50 % of organic carbon may volatilise to the atmosphere within 1 week from the most susceptible to decomposition compounds of excreta, in extremely warm temperatures for the Antarctic. This process lasted 3 weeks for the average temperatures at Arctowski (Pietr et al. 1983). The C/N ratio did not change significantly during these first steps of mineralisation in the closed system ranging from 2.1 to 2.6. Chitin makes up about a third of fresh excreta. This component has been concentrated during the mineralisation process and represents the main component of the ornithogenic soil horizon mixed with urates and mineral phosphates.

There is much evidence that microbes of different categories have much impact on the soil systems in the Antarctic. Their footprints are small but effective regarding the remineralisation and thus the recycling of freshly produced organic material. Degradation of refractive material, e.g., from mosses or lichens, is hampered by environmental factors as well as by the organic material itself. This, however, can be seen in other environments too where accumulations of non or slowly degradable material occurs. Material fluxes in cold or severe environments like the Antarctic terrestrial zones are thus more limited by physical constraints rather than by the biological properties of their inhabitants.

References

Ajwa HA, Dell CJ, Rice CW (1999) Changes in enzyme activities and microbial biomass of tallgrass prairie soil as related to burning and nitrogen fertilization. Soil Biol Biochem 31:769 – 777

Azmi OR, Seppelt RD (1998) The broad-scale distribution of microfungi in the Windmill Islands region, continental Antarctica. Polar Biol 19:92–100

Bailey AD, Wynn-Williams DD (1982) Soil microbial studies at Signy Island, South Orkney Islands. Br Antarct Surv Bull 51:167–191

Baublis JA, Wharton RA, Volz PA (1991) Diversity of micro-fungi in an Antarctic dry valley. J Basic Microbiol 31:3–12

Beyer L, Sieling K, Pingpank K (1999) The impact of a low humus level in arable soils on microbial properties, soil organic matter quality and crop yield. Biol Fert Soils 28:156–161

Beyer L, Bölter M (2000) Chemical and biological properties, formation, occurrence and classification of spodic Cryosols in a terrestrial ecosystem of East Antarctica (Wilkes Land). Catena 39:95–119

Beyer L, Bölter M, Seppelt RD (2000) Nutrient and thermal regime, microbial biomass, and vegetation of Antarctic soils in the Windmill Islands Region of East Antarctica (Wilkes Land). Arct Antarct Alp Res 32:30–39

Billen G, Joiris C, Wijnant P, Gillian G (1980) Concentration and microbiological utilisation of small organic molecules in the Scheldt Estuary, the Belgian coastal zone of the North Sea and the English Channel. Estuarine Coast Mar Sci 11:279–294

Billings WD, Peterson KM, Shaver GR, Trent AW (1977) Root growth, respiration, and carbon dioxide evolution in an arctic tundra soil. Arct Alp Res 9:129–137

Block W (1984) Terrestrial microbiology, invertebrates and ecosystems. In: Laws RM (ed) Antarctic ecology, vol 1, Academic Press, London, pp 163–236

Bölter, M (1981) DOC-turnover and microbial biomass production. Kieler Meeresforsch Sdh 5: 304–310

Bölter M (1989) Microbial activity in soils from Antarctica (Casey Station, Wilkes Land). Proc NIPR Symp Polar Biol 2:146–153

Bölter M (1990a) Microbial ecology of soils from Wilkes Land, Antarctica. I. The bacterial population and its activity in relation to dissolved organic matter. Proc NIPR Symp Polar Biol 3:104–109

Bölter, M (1990b) Microbial ecology of soils from Wilkes Land (Antarctica). II. Patterns of microbial activity and related organic and inorganic matter. Proc NIPR Symp Polar Biol 3:120–132

Bölter, M (1990 c) Evaluation – by cluster analysis – of descriptors for the establishment of significant subunits in Antarctic soils. Ecol Modell 50:79–94

Bölter M (1991) Microbial mineralization in soils and on plant material from Antarctica. In: Weller G, Wilson CL, Severin BAB (eds) Int Conference on the Role of the polar regions in global change. Proceedings of a Conference held 11–15 June 1990 at the University of Alaska Fairbanks, Vol II, Geophysical Institute University of Alaska, Fairbanks, pp 418–422

Bölter M (1992a) Vergleichende Untersuchungen zur mikrobiellen Aktivität in Böden und an Kryptogamen aus der kontinentalen und maritimen Antarktis (Casey, Wilkes Land, und Arctowski, King George Island). Habilschrift, Univ Kiel, pp 1–203

Bölter M (1992b) Environmental conditions and microbiological properties from soils and lichens from Antarctic (Casey Station, Wilkes land). Polar Biol 11:591–599

Bölter M (1993a) Effects of carbohydrates and leucine on growth of bacteria from Antarctic soils (Casey Station, Wilkes Land). Polar Biol 13:297–306

Bölter M (1993b) Organic matter and its availability to microorganisms in Antarctic soils. In: Gilichinski D (ed) 1st Int Conf Cryopedol (Cryosols: the effects of cryogenesis on the processes and peculiarities of soil formation). Russ Acad Sciences, Pushchino, pp 189–199

Bölter M (1994) Estimations of microbial biomass by direct and indirect methods with special respect to monitoring programs. Proc NIPR Symp Polar Biol 7:198–208

Bölter M (1995) Distribution of bacterial numbers and biomass in soils and on plants from King George Island (Arctowski Station, Maritime Antarctica). Polar Biol 15:115–124

Bölter M (1996) Analysis of soil microbial communities (autotrophs and heterotrophs) from King George Island (Arctowski Station). Proc NIPR Symp Polar Biol 9:283–298

Bölter M (1997a) Soil properties and distributions of invertebrates and bacteria from King George Island (Arctowski Station), maritime Antarctic. Polar Biol 18:295–304

Bölter M (1997b) Microbial communities in soils and on plants from King George Island (Arctowski Station, Maritime Antarctic). In: Battaglia B, Valencia J, Walton DWH (eds) Antarctic communities: species, structure and survival. Cambridge University Press, Cambridge, pp 162–169

Bölter M (1998) Structure of bacterial communities in Arctic permafrost soils (Tajmyr Peninsula, Siberia) In: Glowacki P, Bednarek J (eds) Polish Polar Studies, 25th Intern Polar Symp, Warszawa 1998. Inst Geophys Pol Acad Sci, Warszawa, pp 61–66

Bölter M, Kappen L, Meyer M (1989) The influence of microclimatic conditions on potential photosynthesis of *Usnea sphacelata* – a model. Ecol Res 4:297–307

Bölter M, Möller R, Dzomla W (1993) Determination of bacterial biomass with epifluorescence microscopy: comparison of size distributions from image analysis and size classifications. Micron 24:31–40

Bölter M, Blume H-P, Kappen L (1995) Bodenbiologische Untersuchungen in der maritimen und kontinentalen Antarktis (King George Island und Windmill Islands). I. Umweltparameter und anorganische Nährstoffe. Polarforschung 65:41–61

Bölter M, Blume HP, Schneider D, Beyer L (1997) Soil properties and distribution of invertebrates and bacteria from King George Island (Arctowski Station), maritime Antarctic. Polar Biol 18:295–304

Bölter M, Seppelt RD, Beyer L, Pingpank K (2000) Studies on floristic diversity, soil organic matter, and soil microbes from the Windmill Islands, East Antarctica. Bibl Lichenol 75:421–432

Bowman JP, Cavanagh J, Austin J, Sanderson K (1996) Novel Psychrobacter species from Antarctica ornithogenic soils. Int J Syst Bacteriol 46:841–848

Boyd WL (1967) Ecology and physiology of soil microorganisms in polar regions. JARE Sci Rep, Proc Symp Pacific Antarct Sci, Spec Issue 1:265–275

Broady P (1979) A preliminary survey of the terrestrial algae of the Antarctic Peninsula and South Georgia. Br Antarct Surv Bull 48:47–70

Bunnell FL, MacLean SF Jr, Brown J (1975) Barrow, Alaska, USA. In: Rosswall T, Heal OW (eds) Structure and function of tundra ecosystems. Ecol Bull (Stockholm) 20:73–124

Burket JZ, Dick RP (1998) Microbial and soil parameters in relation to N mineralization in soils of diverse genesis under differing management systems. Biol Fert Soils 27:430–438

Cameron RE, King J, David CN (1970) Microbiology, ecology and microclimatology of soil sites in dry valleys of southern Victoria land, Antarctica In: Holdgate MW (ed) Antarctic ecology, vol 2. Academic Press, London, pp 702–716

Czekanowska E, Zabawski J (1988) Soil actinomycetes of the Admiralty Bay area at King George Island, Antarctic. In: Jahn A, Pereyma J, Szczepankiewicz-Szmyrka A (eds) XVth Polar Symp. Wroclaw Univ Publ House, Wroclaw, pp 279–288

Chapman BE, Roser DJ, Seppelt RD (1994) ^{13}C NMR analysis of Antarctic cryptogam extracts. Antarct Sci 6:295–305
Christie P (1987) Nitrogen in two contrasting Antarctic bryophyte communities. J Ecol 75:73–93
Christie P, Nicolson TH (1983) Are mycorrhizas absent from the Antarctic? Trans Br Mycol Soc 80:557–560
Corpe WA, Winters H (1972) Hydrolytic enzymes of some periphytic bacteria. Can J Microbiol 18:1483–1490
Czarnecki B, Bialasiewicz D (1987) Fungi as a component of the aerosphere in the H. Arctowski Polar Station and its vicinity (King George Island, South Shetland Islands). Pol Polar Res 8: 53–158
Davey A (1983) Effects of abiotic factors on nitrogen fixation by blue-green algae in Antarctica. Polar Biol 2:95–100
Davis RC (1980) Peat respiration and decomposition in Antarctic terrestrial moss communities. Biol J Linnean Soc 14:39–49
Davis RC (1981) Structure and function of two Antarctic terrestrial moss communities. Ecol Monogr 51:125–143
Dawson R, Schramm W, Bölter M (1985) Factors influencing the production, decomposition and distribution of organic and inorganic matter in Admiralty Bay, King George Island. In: Siegfried WR, Condy PR, Laws RM (eds) Antarctic Nutrient Cycles and Food Webs. Springer, Berlin Heidelberg New York, pp 109–114
Del Frate G, Caretta G (1990) Fungi isolated from Antarctic material. Polar Biol 11:1–7
Deng SP, Tabatabai MA (1997) Effect of tillage and residue management on enzyme activities in soils: III. Phosphatases and arylsulfatase. Biol Fert Soils 24:141–146
Feller G, Narinx E, Arpigny JL, Zekhnini Z, Swings J, Gerday C (1994) Temperature dependence of growth, enzyme secretion and activity of psychrophilic Antarctic bacteria. Appl Microbiol Biotechnol 41:477–479
Fenice M, Selbmann L, Zucconi L, Onofri S (1997) Production of extracellular enzymes by Antarctic fungal strains. Polar Biol 17:275–280
Fletcher LD, Kerry EJ, Weste GM (1985) Microfungi of McRobertson and Enderby Lands, Antarctica. Polar Biol 4:81–88
French DD, Smith VR (1986) Bacterial populations in soils of a Subantarctic island. Polar Biol 6:75–82
Fry JC (1988) Determination of biomass. In: Austin, B (ed) Methods in aquatic bacteriology. Wiley, Chichester, pp 27–72
Gallikowski C, Hirsch P (1988) Preliminary characterization and identification of 1984/1985 continental Antarctic soil microorganisms of Linnaeus Terrace (Altitude 1600 m; McMurdo Dry Valleys). Polarforschung 58:93–101
Greenfield LG (1992) Precipitation nitrogen at maritime Signy Island and continental Cape Bird, Antarctica. Polar Biol 11:649–653
Greenfield L, Wilson G (1981) University of Canterbury, Antarctica Expedition, Report No 19. University of Canterbury, Christchurch, pp 1–47
Grobler DC, Torien DF, Smith VR (1987) Bacterial activity in soils of a sub-Antarctic island. Soil Biol Biochem 19:485–490
Hagström A, Larsson U, Hörstedt P, Nordmark S (1979) Frequency of dividing cells, a new approach to the determination of bacterial growth rates in aquatic environments. Appl Environ Microbiol 37:805–812
Heal OW, Flanagan PW, French DD, Maclean SF (1981) Decomposition and accumulation of organic matter. In: Bliss, LC, Heal, OW, Moore, JJ (eds) Tundra ecosystems: a comparative analysis. Cambridge University Press, Cambridge, pp 587–633
Heatwole H, Saenger P, Spain A, Kerry E, Donelan J (1989) Biotic and chemical characteristics of some soils from Wilkes Land, Antarctica. Antarct Sci 1:225–234

Ino Y (1985) Comparative study of the effects of temperature on net photosynthesis and respiration in lichens from the Antarctic and subalpine zones in Japan. Bot Mag Tokyo 98:41–53

Jackson WA, Volk RJ 1970 Photorespiration. Annu Rev Plant Physiol 21:385–432

Jenkinson DS, Powlson DS, Wedderburn RW 1976 The effects of biocidal treatments on metabolism in soil. III. The relationship between soil biovolume, measured by optical microscopy, and the flush of decomposition caused by fumigation. Soil Biol Biochem 8: 89–202

Kandeler E, Eder G (1993) Effect of cattle slurry in grassland on microbial biomass and on activities of various enzymes. Biol Fert Soils 16:249–254

Kandeler E, Kampichler C, Horak O (1996) Influence of heavy metals on the functional diversity of soil microbial communities. Biol Fert Soils 23:299–306

Kandeler E, Tscherko D, Spiegel H (1999) Long-term monitoring of microbial biomass, N-mineralisation and enzyme activities of a Chernozem under different tillage. Biol Fert Soils 28:343 - 351

Kappen L (1985) Lichen-habitats as micro-oases in the Antarctic – the role of temperature. Polarforschung 55:49–54

Kennedy AD (1993) Water as a limiting factor in the Antarctic terrestrial environment: a biogeographical synthesis. Arct Alp Res 25:308–315

Kerry E (1990a) Effects of temperature on growth rates of fungi from sub-Antarctic Macquarie Island and Casey, Antarctica. Polar Biol 10:293–299

Kerry E (1990b) Microorganisms colonizing plants and soil subjected to different degrees of human activity, including petroleum contamination, in the Vestfold Hills and MacRobertson Land, Antarctic. Polar Biol 10:423–430

Klose S, Moore JM, Tabatabai MA (1999) Arylsulfatase activity of microbial biomass in soils as affected by cropping systems. Biol Fert Soils 29:46–54

Ladd JN (1985) Soil enzymes. In: Vaughan, D, Malcolm, RE (eds) Soil organic matter and biological activity, vol 1. Nijhoff/Junk, Dordrecht, pp 175–22

Lewis Smith RI (1985) Nutrient cycling in relation to biological productivity in Antarctic and Sub-Antarctic terrestrial and freshwater ecosystems. In: Siegfried WR, Condy PR, Laws RM (eds) Antarctic nutrient cycles and food webs. Springer, Berlin Heidelberg New York, pp 138–155

Lewis Smith RI (1986) Plant ecological studies in the fellfield ecosystem near Casey Station, Australian Antarctic Territory, 1985–1986. Br Antarct Surv Bull 72:81–91

Line MA (1988) Microbial flora of some soils of Mawson base and the Vestfold Hills, Antarctica. Polar Biol 8:421–427

McRae CF, Hocking AD, Seppelt RD (1999) Penicillium species from terrestrial habitats in the Windmill Islands, East Antarctica, including a new species, *Penicillium antarcticum*. Polar Biol 21:97–111

Melick DR, Seppelt RD (1992) Loss of soluble carbohydrates and changes in freezing point of Antarctic bryophytes after leaching and repeated freeze-thaw cycles. Antarct Sci 4:399–404

Melick DR, Seppelt RD (1994) Seasonal investigation of soluble carbohydrates and pigment levels in Antarctic bryophytes and lichens. Bryologist 97:98–105

Melick DR, Bölter M, Möller R (1994) Rates of soluble carbohydrate utilization in soils from the Windmill Islands Oasis, Wilkes Land, continental Antarctica. Polar Biol 14:59–64

Meyer-Reil L-A 1978 Uptake of glucose by bacteria in the sediment. Mar Biol 44:293–298

Montemartini Corte A, Gestro C (1994) Fungi from mud of Antarctic ponds. In: Battaglia, B, Bisol, PM, Varotto, V (eds) Proceedings of the 2nd Meeting on Antarctic Biology, Padova, 26–28 February 1992. Edizioni Universitairie Patavine, Padova, pp 33–42

Moore TR (1986) Carbon dioxide evolution from sub-Arctic peatlands in eastern Canada. Arct Alp Res 18:189–193

Morita RY, Moyer CL (1989) Bioavailability of energy and the starvation of cells. In: Hattori, T, Ishida, Y, Maruyama, Y, Morita, RY and Uchida, A (eds) Recent advances in microbial ecology. Japan Scientific Societies Press, Tokyo, pp 75–79

Morley CR, Trofymow JA, Coleman DC, Cambardella C (1983) Effects of freeze-thaw stress on bacterial populations in soil microcosms. Microb Ecol 9:329–340

Myrcha A, Pietr SJ, Tatur A (1985) The role of pygoscelid penguin rookeries in nutrient cycles at Admiralty Bay King George Island. In: Siegfried WR, Condy PR, Laws RM (eds) Antarctic nutrient cycle, food webs. Springer, Berlin Heidelberg New York, pp 156–162

Myrcha A, Ochyra R, Tatur A (1991) Site of Special Scientific Interest no. 8, Western shore of Admiralty Bay, King George Island, South Shetland Islands. In: Klekowski, RZ, Opalinski KW (eds) First Polish-Soviet Antarctic Symp 'Arctowski 85'. Institute of Ecology Publ Office, Warsaw, pp 157–168

Nakatsubo T, Ino Y (1986) Nitrogen cycling in an Antarctic ecosystem. 1. Biological nitrogen fixation in the vicinity of Syowa Station. Mem Natl Inst Pol Res Ser E 37:1–10

Onofri S, Tosi S, Persiani AM, Maggi O, Riess S, Zucconi L (1994) Mycological researches in Victoria Land terrestrial ecosystems. In: Battaglia B, Bisol PM, Varotto V (eds) Proceedings of the 2nd Meeting on Antarctic Biology, Padova, 26–28 February 1992. Edizioni Universitairie Patavine, Padova, pp 19–32

Parker LW, Miller J, Steinberger Y, Whitford WC (1983) Soil respiration in a Chihuahuan desert rangeland. Soil Biol Biochem 15:303–309

Paul EA, Clark FE (1989) Soil microbiology and biochemistry. Academic Press, London

Petz W (1997) Ecology of the active soil microfauna (Protozoa, Metazoa) of Wilkes Land, East Antarctica Polar Biol 18:33–44

Pietr SJ (1986) The physiological groups of microorganisms in different soils at Admiralty Bay region (King George Island, South Shetland Islands, Antarctica) Pol Polar Res 7:395–406

Pietr SJ (1993) Soil microorganisms In: Rakusa-Suszczewski S (ed) The maritime Antarctic coastal ecosystem of Admiralty Bay. Dept of Antarctic Biology, PAS, Warsaw, pp 167–172

Pietr SJ, Tatur A, Myrcha A (1983) Mineralization of penguin excrements in the Admiralty Bay region (King George Island, South Shetlands, Antarctica). Pol Polar Res 4:97–112

Pietr SJ, Fischer Z, Tyrawska D, Bajan C (1995): Biological processes of organic matter degradation in Karkonosze soils. In: Fischer Z (ed) Ecological problems of alpine part of Karkonosze. Dzieknów Lesny Institute of Ecology Publ Office, Warsaw, pp 131–168

Poole DK, Miller PC (1982) Carbon dioxide flux from three Arctic tundra types in north-central Alaska, USA. Arct Alp Res 14:27–32

Pugh GJF, Allsopp A (1982) Microfungi on Signy Island, South Orkney Islands. Br Antarct Surv Bull 57:55–67

Ramsay AJ (1983) Bacterial biomass in ornithogenic soils of Antarctica. Polar Biol 1:221–225

Ramsay AJ, Stannard RE (1986) Numbers and viability of bacteria in ornithogenic soils of Antarctica. Polar Biol 5:195–198

Raven JR (1990) Use of isotopes in estimating respiration and photorespiration in microalgae. Mar Microb Food Webs 4:59–86

Ray MK, Shivaji S, Shyamala Rao N, Bhargava PM (1989) Yeast strains from Schirmacher Oasis, Antarctica. Polar Biol 9:305–309

Roser DJ, Melick DR, Seppelt RD (1992a) Reductions in the polyhydric alcohol content of lichens as an indicator of environmental pollution. Antarct Sci 4:185–189

Roser DJ, Melick DR, Ling HU, Seppelt RD (1992b) Polyol and sugar content of terrestrial plants from continental Antarctica. Antarct Sci 4:413–420

Roser DJ, Seppelt RD, Ashbolt N (1993) Microbiology of ornithogenic soils from the Windmill Islands, Budd Coast, Continental Antarctica: microbial biomass distribution. Soil Biol Biochem 25:165–175

Roser DJ, Seppelt RD, Nordstrom O (1994) Soluble carbohydrate and organic acid content of soils and associated microbiota from the Windmill islands, Budd Coast, Antarctica. Antarct Sci 6:53–59

Shivaji S, Shyamala Rao N, Saisree L Reddy GSN, Seshu Kumar G, Bhargava PM (1989) Isolates of Arthrobacter from soils of Schirmacher Oasis, Antarctica. Polar Biol 10:225–229

Siebert J, Hirsch P (1988) Characterisation of 15 selected coccal bacteria isolated from Antarctic and soil samples from McMurdo-Dry Valleys (South Victoria Land). Polar Biol 9:37–44

Skartveit A, Ryden BE, Kärenlampi L (1975) Climate and hydrology of some Fennoscandian tundra ecosystems. In: Wielgolaski FE (ed) Fennoscandian tundra ecosystems. Part 1. Plants and microorganisms. Springer, Berlin Heidelberg New York, pp 41–56

Smith HG, Tearle PV (1985) Aspects of microbial and protozoan abundances in Signy Island fellfields. Br Antarct Surv Bull 68:83–90

Smith KA, Swift RS (1983) General isotope techniques. In: Smith KA (ed) Soil analysis. Marcel Dekker, New York, pp 229–298

Smith MJ, Walton DWH (1985) Statistical analysis of the relationships among viable microbial populations, vegetation, and environment in a sub-Antarctic tundra. Microbiol Ecol 11:245–257

Stemmer M, Gerzabek MH, Kandeler E (1999) Invertase and xylanase activity of bulk soil and particle-size fractions during maize straw decomposition. Soil Biol Biochem 31:9–18

Sun SH, Huppert M, Cameron RE (1978) Identification of some fungi from soil and air of Antarctica. Antarct Res Ser 30, American Geophysical Union, Washington, DC, pp 1–26

Tearle PV (1987) Cryptogamic carbohydrate release and microbial response during spring freeze-thaw cycles in Antarctic fellfield soils. Soil Biol Biochem 19:381–390

Tscherko D, Kandeler E, Beyer L, Bölter M, Blume H-P (2002) Microbial biomass and enzyme activity of soil transects in maritime Antarctica (King-George Island, West Antarctica). Arct Antarct Alp Res (in press)

Tubaki K (1961) On some fungi isolated from Antarctic materials. Biological results of the Japanese Antarctic Research Expedition. Seto Mar Biol Laborat Spec Publ 14:3–9

Tubaki K, Asano I (1965) Additional species of fungi isolated from Antarctic materials. Jap Ant Res Exp Sci Rep, Ser E 27:1–12

Upton AC, Nedwell DB (1989) Nutritional flexibility of oligotrophic and copiotrophic Antarctic bacteria with respect to organic substrates. FEMS Microbiol Ecol 62:1–6

Vestal JR (1988a) Biomass of the cryptoendolithic microbiota from the Antarctic desert. Appl Environ Microbiol 54:957–959

Vestal JR (1988b) Carbon metabolism of the cryptoendolithic microbiota from the Antarctic desert. Appl Environ Microbiol 54:960–965

Vincent WF (1988) Microbial ecosystems of Antarctica. Cambridge University Press, Cambridge

Vishniac H (1993) The microbiology of Antarctic soils In: Friedmann, EI (ed) Antarctic microbiology. Wiley-Liss, New York, pp 297–341

Vishniac HS, Hempfling WP (1979) *Cryptococcus vishniacii* sp. nov., an Antarctic yeast. Intern J Syst Bacteriol 29:153–158

Walton DWH (1982) The Signy Island terrestrial reference sites: XV. Micro-climatic monitoring, 1972–74. Br Antarct Surv Bull 55:111–126

Wicklow, DT (1968) *Aspergillus fumigatus* Fresenius isolated from ornithogenic soil collected at Hallett Station, Antarctica. Can J Microbiol 14:717–719

Wiliams PG, Roser DJ, Seppelt RD (1994) Mycorrhizas of hepatics in continental Antarctica. Mycol Res 98:34–36

Wynn-Williams DD (1980) Seasonal fluctuations in microbial activity in Antarctic moss peat. Biol J Linnean Soc 14:11–28

Wynn-Williams DD (1982) Simulation of seasonal changes in microbial activity of maritime Antarctic peat. Soil Biol Biochem 14:1–12

Wynn-Williams DD (1984) Comparative respirometry of peat decomposition on a latitudinal transect in the maritime Antarctic. Polar Biol 3:173–181

Wynn-Williams DD (1985a) Photofading retardant for epifluorescence microscopy in soil micro-ecological studies. Soil Biol Biochem 17:739–746

Wynn-Williams DD (1985b) The Signy Island terrestrial reference sites: XVII. Peat O_2-uptake in a moss carpet relative to edaphic and microbial factors. Br Antarct Surv Bull 68:61–69

Wynn-Williams DD (1988) Microbial colonization of Antarctic fellfield soils. Proc 4th Int Symp Microb Ecol, Ljubljana, pp 191–200

Wynn-Williams DD (1993) Microbial processes and initial stabilization of fellfield soils In: Miles J, Walton DWH (eds) Primary succession on land. Blackwell, Oxford, pp 17–32

Zabawski, J, Piasecki, J (1981) Studies of soil microflora of King George Island. In: Maliszewska, E (ed) Microbiological transformations of nitrogen compounds in soil under various ecological conditions. Institute of Cultivation Fertilization and Pedology, Pulawy, pp 131–138

12 Plant Communities on King George Island

M. Olech

12.1 Introduction

King George Island – the largest island in the South Shetland Islands archipelago – is included in the phytogeographical province of the maritime Antarctic. The maritime Antarctic also includes the western coast of the Antarctic Peninsula, up to the southern part of the Marquerite Bay, and the archipelagos of the South Orkney Islands, South Sandwich Islands and Bouvet Øya. In phytogeographical terms, the region occupies the middle position between the continental Antarctic and sub-Antarctic, as indicated by the number of species in plant communities and the composition of the latter (Lindsay 1976; Filson 1982; Lewis Smith 1988; Olech 1994).

On King George Island only about 10 % of land – in the coastal zone and on the nunataks further inland – is free of ice, but even these areas are only partly occupied by plants, thus the landscape is of a semi-desert type (Zarzycki 1993).

The vegetation of King George Island shows a marked similarity to other regions in the northern parts of the maritime Antarctic (Allison and Lewis Smith 1973; Lewis Smith 1972, 1979). The tundra there is poorly developed – consisting almost exclusively of cryptogams. In dry and rocky places, lichens predominate, whereas mosses prevail in wetter niches on the ground. The vascular plants have only a very limited role and they occur only locally, often as an insignificant admixture in the communities of cryptogams.

The striking feature of King George Island is that the vegetation is less abundant on the northern part of the coast than on the southern part. The northern coast lacks vascular plants entirely, has only a few mosses and the vegetation is restricted almost exclusively to ornithocoprophilous lichen communities (Lindsay 1971).

The studies on tundra communities completed to date are insufficient, principally because of the poor current knowledge of species and the occurrence of many unsolved taxonomic problems – the latter chiefly regarding the lichens and aerophytic algae. The issue of the classification of vegetation in

Ecological Studies, Vol. 154
L. Beyer and M. Bölter (eds.) Geoecology of Antarctic Ice-Free Coastal Landscapes

the Antarctic biome is also unsolved. The Braun-Blanquet method, used so commonly throughout Europe, is used in the Antarctic only to a very limited extent. The main reason for this is the poor knowledge of lichen species. A few attempts to use this method, where the differentiation of plant associations is based on characteristic species or a characteristic combination of species, were conducted in the Antarctica by Follman (1965), Redon (1969), Furmańczyk and Ochyra (1982), and Olech (1990). These studies pertained to some individual community types.

Another system has been proposed and implemented by British biologists (Holdgate 1964; Longton 1967, 1979, 1988; Gimingham and Lewis Smith 1970; Lewis Smith 1972; Lewis Smith and Gimingham 1976). They distinguished in descriptive terms the main units based on physiognomic criteria; the floristic composition is therefore reflected in lower units, namely associations, characterised by a group of permanent species, and several associations, distinguished on the basis of predominating species. Using this system, several authors have tried to describe the vegetation of King George Island. The first general account was provided by Lindsay (1971). The main types of moss communities on the Fildes Peninsula were presented by Hu (1998). A broad review and description of the vegetation on King George Island, mainly including the participation of liverworts and mosses in the plant communities, were presented by Ochyra and Vana (1989) and Ochyra (1998).

There were also the first attempts to map the distribution of the plant communities of the various regions of King George Island (Furmańczyk and Ochyra 1982; Olech 1994).

12.2 Comments on the Flora

The vegetation of the Antarctic biome is characterised by an extremely low number of species, in particular the vascular plants. There are only two native species of flowering plants: the grass *Deschampsia antarctica* Desv. (Fig. 12.1; Poaceae) and the Antarctic pearlwort *Colobanthus quitensis* (Kunth) Bartl. (Fig. 12.2; Caryophyllaceae). The occurrence of these two species is limited to places having the most favourable climatic conditions within the maritime Antarctic, with the southernmost stations located on Alamode Island 68°42'S, 67°32'W, in the Terra Firma Islands group (Komarkowa et al. 1990). They grow at low altitudes, usually up to 50–60 m a.s.l., but are sometimes found at higher locations, e.g. in the region of Admiralty Bay on King George Island, where *Deschampsia antarctica* was found at the altitude of 330 m on Dutkiewicz Cliff, and *Colobanthus quitensis* at an elevation of 160 m on Krzesanica Cliff (Zarzycki 1993).

Some alien vascular plants have also been introduced into the maritime Antarctic (Holdgate 1964; Longton 1966; Walton 1975). Detailed observations

Fig. 12.1. *Deschampsia antarctica* Desv.

Fig. 12.2. *Colobanthus quitensis* (Kunth) Bartl.

on the population of *Poa annua* L. in the Admiralty Bay area, King George Island were subsequently made by Olech (1996).

On King George Island, the mosses have been described by Ochyra (1998), who reported 61 species from this area (out of 106 species known for the whole of Antarctica). The mosses predominate here in more sheltered, moist places that are stable and have some soil and organic material accumulation (Ochyra 1998). The following could be regarded as the most commonly occurring mosses on King George Island: *Sanionia georgico-uncinata* (Müll. Hal.) Ochyra & Hedenäs, *Bryum pseudotriquetrum* (Hedw.) Gaertn., Meyer & Scherb., *Pohlia nutans* (Hedw.) Lindb., *Syntrichia princeps* (De Not.) Mitt., *S. filaris* (Müll. Hal.) Zand., *Bartramia patens* Brid., *Warnstorfia laculosa* (Müll. Hal.) Ochyra & Matteri, *Brachythecium austrosalebrosum* (Müll. Hal.) Kindb., *Andreaea depressinervis* Card.; *A. gainii* Card. and *Polytrichastrum alpinum* (Hedw.) G.L. Smith.

Liverworts (hepatics) are of minor importance in the flora of Antarctica; usually they are only an admixture in compact swards of mosses. Eleven species of liverworts are reported to occur on King George Island (Ochyra and Vana 1989). The most common include: *Cephaloziella varians* (Gott. in Neum.) Steph.; *Lophozia excisa* (Dicks.) Dum., and *Barbilophozia hatcheri* (Evans) Loeske.

Lichens, as in the remainder of ice-free continental Antarctica, have a dominant role in the flora and plant cover of King George Island, particularly in places featuring the more extreme habitat conditions. The list of lichens on King George Island may be estimated at more than 190 species (Olech, 2000). It is still difficult to determine the precise number of taxa of lichens because of the taxonomic confusion that has prevailed for years. Taxonomic revisions have reduced many species to synonymy (Castello 1995; Castello and Nimis 1994a,b, 1995), although, many new species have also been described (Hertel 1984; Øvstedal 1988; Vezda et al. 1992; Inoue 1993; Olech and Søchting 1993; Orange 1993; Øvstedal et al. 1996).

With regard to habitat preferences, Western Antarctica is dominated by epilithic species. Among the most common species, the following predominate: *Usnea antarctica* Du Rietz, *U. aurantiaco-atra* (Jacq.) Bory (Fig. 12.3), *Caloplaca sublobulata* (Nyl.) Zahlbr., *C. johnstonii* (Dodge) Søchting & Olech, *Lecanora dispersa* (Pers.) Sommerf., *L. polytropa* (Ehrh.) Rabenh., *Carbonea assentiens* (Nyl.) Hertel, *Amandinea coniops* (Wahlenb. in Ach.) Scheid., *Placopsis contortuplicata* Lamb (Fig. 12.4), and *Rhizocarpon geographicum* (L.) DC. The lichens which are specialised as epiphytes on mosses (i.e. epibryophytic) are quite commonly represented in Western Antarctica, the following occurring most commonly: *Ochrolechia frigida* (Sw.) Lynge, *Psoroma hypnorum* (Vahl.) Gray, *Physconia muscigena* (Ach) Poelt, *Rinodina turfacea* (Wahlenb.) Körb., *Caloplaca tiroliensis* Zahlbr. and *Lecanora expectans* Darb. The least numerous ecological group, the terricolous species, are represented principally by the following: *Stereocaulon alpinum* Laur.; *Sphaerophorus glo-*

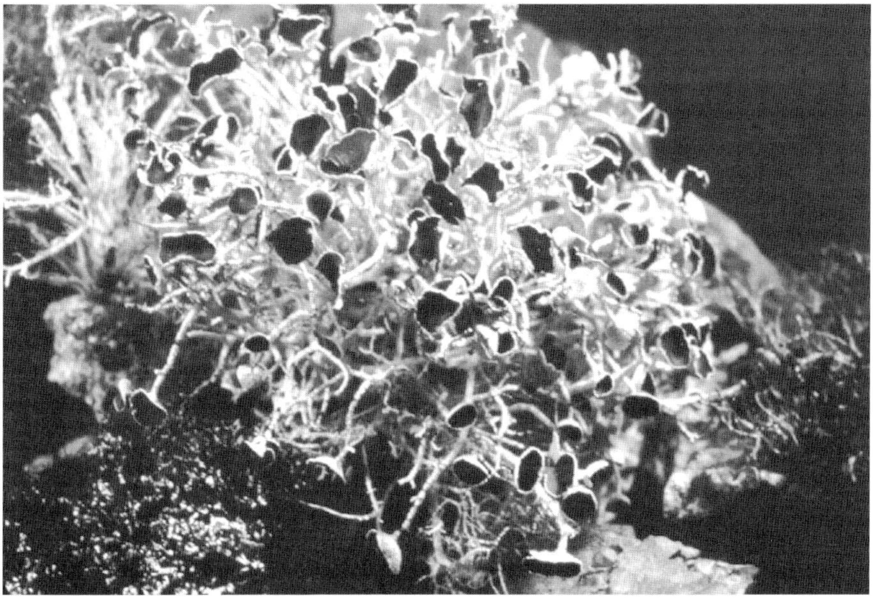

Fig. 12.3. *Usnea aurantiaco-atra* (Jacq.) Bory

Fig. 12.4. *Placopsis contortuplicata* Lamb

bosus (Huds.) Vain.; *Cetraria aculeata* (Schreb.) Fr., and various species of *Cladonia*.

Macromycetes do not play any significant role in the Antarctic tundra. A dozen or so species are known to occur in the Antarctic region. Those most common on King George Island include the Basidiomycetes *Omphalina pyxidata* (Bull.) Quel., *O. antarctica* Singer, *Galerina pseudomycenopsis* Pilát (Guminska et al. 1994), while the most common ascomycete is *Lamprospora miniatopsis* Spooner.

More widely distributed in the polar regions are those microfungi which occur on the thalli of lichens (lichenicolous fungi), on mosses and in the soil. The fungi living on lichens display variable life forms and strategies. Most of these are facultative or obligatory parasites causing necrotic changes of the host's thallus. A much smaller group include saprophytes or parasymbionts. The level of knowledge of lichenicolous fungi is still poor. To date, 54 species have been recorded on King George Island (Olech and Alstrup 1995).

Cyanobacteria and eukaryotic algae play an important part in polar ecosystems, primarily because of their pioneering role in the colonisation of new habitats. They occur in freshwater biotopes and in soil, on mosses, rocks and in snow. These are now under intensive study and their number is tentatively estimated at more than 400 species. The species which may have an important role in forming terrestrial communities should include a common Antarctic nitrophilous green alga, *Prasiola crispa*, (Lightf.) Menegh, and Cyanobacteria or blue-green algae – *Phormidium* spp., *Nostoc punctiforme* (Kützing) Hariot, and *Oscillatoria* spp.

12.3 A Review of Important Plant Communities on King George Island

In the Antarctic tundra, the predominating cryptogams (mainly lichens and mosses) form a mosaic of plants. These communities most often occupy very small areas and the number of species represented is low. More frequent are patches formed by a few crustose lichens covering rock substrates, or by moss cushions in moist depressions. Much rarer are the multi-layer communities composed of a layer of crustose lichens and a layer of mosses or fruticose lichens or of vascular plants.

12.3.1 Communities of Rock Substrates

Saxicolous communities are the most common and widespread on King George Island and in the whole of western and eastern Antarctica. They include many variable lichen communities composed of crustose, squamulose, stipitate, foliose or fruticose thalli. Under the classification made by Lewis Smith and Gimingham (1976), these communities are included in two subformations: (1) the crustose lichen subformation and (2) the fruticose lichen and moss cushion subformation.

Preliminary studies indicate that such communities are often among the richest in terms of floristics. However, these communities are among the least known from the viewpoint of the species which form them; this is a result of poor taxonomic knowledge of the epilithic lichens.

The communities of epilithic lichens occur on coastal cliffs, boulders and pebbles on sea shores, rock faces, ridges and screes further inland.

The composition of the saxicolous communities depends on many factors, such as relief, geological substrate, hydrological regime and microclimate, and particularly the duration of the persistent snow cover and wind. Another major influence is nitrogen enrichment from the large colonies of birds and also sea spray, resulting in the formation, on King George Island, of a variety of communities of a nitrophobic, nitrophilous and halophilous nature.

12.3.2 Nitrophobic Communities

Nitrophobic communities develop on rocky substrates lacking direct influx of nutrients; they occur further inland, and rarely near the coast. Nitrophobic communities are the most widespread in Western Antarctica.

The *Himantormia lugubris–Usnea aurantiaco-atra* community occurs on exposed rock faces, ridges and, more rarely, on stony slopes or plateaux, principally on southern or eastern exposures. It occurs most often in higher locations, above 100 m, and is rarely found lower, e.g. on Fildes Peninsula. Apart from the dominant fruticose species, *Himantormia lugubris* and *Usnea aurantiaco-atra*, this community is formed of lichens, such as *Austrolecia antarctica*, *Carbonea assenties*, *Pertusaria dactylina*, *Rhizocarpon geographicum*, *Ochrolechia frigida* and *Umbilicaria decussata*. The accompanying mosses include *Andreaea gainii*, *A. regularis* and *Distichium capillaceum*.

Umbilicaria decussata dominated communities develop primarily on north and north-westerly exposures in higher zones of rocks, often near or on the summits, on vertical rock faces, and occasionally on moderately inclined slopes in places sheltered from strong winds. This community also includes *Usnea aurantiaco-atra*, *U. antarctica*, *Umbilicaria africana*, and crustose lichens – *Rhizocarpon geminatum*, *R. geographicum* and *Carbonea assentiens*.

On flat protruding rock ledges *Ochrolechia parella* communities develop, which also contain such species as *Pertusaria corallophora*, *Rhizocarpon geographicum*, *Rh. geminatum*, *Ochrolechia frigida*, *Usnea aurantiaco-atra*, *U. antarctica* and, in somewhat lower locations, also *Haematomma erythromma*. In places where some soil has accumulated, the moss *Dicranoweisia brevipes* is also associated with the community.

A community with *Parmelia saxatilis* develops frequently on nearly flat or only slightly inclined rock blocks and rock surfaces. It occupies places well protected from the wind, with a long-persisting snow cover. This community also includes *Coelopogon epiphorellus*, *Hypogymnia lugubris*, *Usnea aurantiaco-atra*, *Ochrolechia frigida*, *Cystocoleus ebeneus*, *Cladonia* spp., as well as *Pseudephebe pubescens*, *Stereocaulon glabrum* and *Sphaerophorus globosus*. They are also accompanied by mosses, such as *Andreaea depressinervis* and *A. gainii*.

On screes and pebbles, particularly at lower locations, a community with *Tremolecia atrata*, *Carbonea assentiens*, *Tephromela atra*, *Candelariella aurella*, and *Usnea antarctica* can be found.

In very moist places e.g. at the bottom of rock faces where water drips, or on pebbles and blocks in seasonal streams, *Staurothele gelida* lichen communities occur, accompanied by such species as *Verrucaria aeleoplaca*, *Aspicilia* spp., *Placopsis contortuplicata*, *Leptogium puberulum. Caloplaca iomma*, *C. executa* and *C. austroshetlandica*. They are all also often found along ephemeral streams.

In the places with a long-persisting snow cover, snow bed communities develop, with a few species only. The species occurring most often are *Caloplaca austroshetlandica*, *Lecidea lapicida*, *Aspicilia* spp. and *Lecanora polytropa*.

12.3.3 Halophilous and Nitrophilous Communities

On the rocks within direct reach of the sea waves, i.e. in the salt-spray zone, hygrohaline mono-specific communities often occur, dominated by species of *Verrucaria*. This community is formed by lichens with dark thalli: *Verrucaria psychrophila*, *V. ceuthocarpa* and *V. tesselatula*. Sometimes they are accompanied by cushions of the moss *Muelleriella crassifolia*, which can tolerate high salt concentrations (Ochyra 1998). In lower rock zones the community is accompanied by filamentose marine green algae.

The *Caloplaca cirrochrooides–Mastodia tesselata* is one of the ornithocoprophilous communities, which also includes halophilous species. The community includes such species as *Xanthoria candelaria*, *Physcia caesia*, *Lecania brialmontii*, *Caloplaca sublobulata*, the alga *Prasiola crispa* and, more rarely, *Caloplaca millegrana*. This community occurs on coastal rocks.

An ornithocoprophilous *Caloplaca regalis–Caloplaca lucens* community (in older references known as *Caloplaca regalis–Xanthoria elegans*) is widespread and often occupies large areas. Apart from the aforementioned species, *Buellia latemarginata*, *Amandinea coniops* and *Physcia caesia* also occur. In crevices and on rock ledges where some soil accumulates, the community is joined by nitrophilous mosses – *Syntrichia princeps* and *Hennediella antarctica*. High fertilisation by bird faeces is the main factor favouring the development of this community. Its physiognomy is determined by the yellow thalli of *Caloplaca regalis* and by the orange thalli of *C. lucens*. It is one of the most beautiful communities of the Antarctic lichen tundra. The community usually occupies places on a northern exposure and seems to be associated with large colonies of penguins (*Pygoscelis adeliae*, *P. antarctica*).

Flat surfaces of manured rocks are occupied by the *Haematomma erythromma* community, with major proportions of *Usnea antarctica*, *Amandinea petermanii*, *Tephromela atra* and *Buellia* spp. This community occupies mainly southern and south-eastern facing exposures (Olech 1990).

Similar exposures but on more steeply inclined rocks up to completely vertical faces are occupied by *Ramalina terebrata* communities where apart from this fruticose lichen, some crustose lichens, such as *Caloplaca lucens*, *Amandinea petermanii* and *Buellia* spp. also occur.

On gravel on seashores and at the bottom of coastal rocks *Acarospora macrocyclos*, *Caloplaca coralligera*, *Rhizoplaca aspidophora* and *Caloplaca sublobulata* occur.

12.4 Terricolous Communities

Terricolous communities of variable structure and floristic composition develop depending on the stability of the substrate, its nitrogen and phosphorus content, and the availability of water. Some other factors can also have modifying effects, such as the duration of snow cover, exposure to wind or the steepness of the slope gradient of the substrate.

12.4.1 Communities with Vascular Plants

Deschampsia antarctica–Colobanthus quitensis communities occur on coastal plains or gentle slopes with a northern exposure, on pebble and moderately moist soil. This community can form widespread swards or small tussocks among mosses such as *Sanionia georgico-uncinata* or *Polytrichastrum alpinum*, *Bryum pseudotriquetrum* and *Polytrichum piliferum*; or as an admixture in lichen communities with *Usnea antarctica*. Near bird colonies, particularly close to penguin colonies where the substrate is fertilised, the proportion of *Colobanthus quitensis* decreases whereas that of nitrophilous algae and cyanobacteria increases.

Deschampsia antarctica–Prasiola crispa communities occur near penguin colonies on gentle slopes of moraines or hills, where water with bird faeces runs off. On flat places enriched by water carrying faecal runoff, e.g. between moraines on which penguins have their colonies, *Deschampsia antarctica* occurs, with a high proportion of cyanobacteria including *Phormidium atten-*

uatum, *Ph. scotii*), the chlorophyte *Prasiola crispa* and the moss *Sanionia georgico-uncianata*.

12.4.2 Communities with a Predominance of Bryophytes

In places with stagnating water, on the fringes of patches of melting snow, in depressions on the terrain, on the banks of streams, lakes and small bodies of water, mossy communities develop. More or less hydrophilous communities of moss carpets are formed by large plagiotropic mosses, in particular *Warnstorfia laculosa*, *W. sarmentosa* and *Sanionia georgico-uncinata* (Ochyra 1998). In individual patches, one species usually predominates, with the remaining species constituting only an insignificant admixture. In permanently waterlogged places communities of *Warnstorfia laculosa* predominate. *Warnstorfia sarmentosa* dominated communities develop in moist places, but those which are only seasonally flooded, usually on a pebble substrate of slopes. Lichens also often occur there (growing on pebbles or on mosses) and include *Placopsis contortuplicata*, *Leptogium puberulum* or *Polyblastia* sp. with the lichenicolous fungus *Thelocarpon cyaneum* (Olech and Alstrup 1990) occurring on its thallus.

In drier sites on the fringes of boggy areas, usually on stone substrates, communities in which *Sanionia georgico-uncinata* predominates can be found. These communities can cover vast areas of flat land, along coasts or in valleys between moraines of glaciers. The community displays a permanent peat forming process (Gimingham and Lewis Smith 1970) and is included in the moss-carpet subformation.

On the fringes of meltwater canals, on wet rocky slopes usually near the mouths of streams, *Brachythecium austrosalebrosum–Syntrichia filaris* communities develop. It consists of mosses growing into large cushions which merge into vast vivid willow-green edges on the stream banks.

On the soil of the stream banks there are also *Bryum pseudotriquetrum-Sanionia georgico-uncinata* communities. They form relatively small patches and often *Syntrichia filaris* can be found there. These communities were classified by Gimingham and Lewis Smith (1970) as a moss hummock subformation.

On scree slopes covered with boulders and sheltered from winds and where some mineral soil accumulates, *Andreaea gainii* and *Andreaea depressinervis* communities develop, sometimes covering large areas. They are accompanied by *Ceratodon purpureus* and *Schistidium antarctici*. Epiphytic lichens, with a predominance of *Ochrolechia frigida* and with *Pertusaria dactylina*, *Massalongia carnosa*, *Pannaria hookeri*, *Japewia tornoensis*, *Caloplaca ammiospila*, *C. siphonospora* and *C. tetraspora*, occur on mosses.

The *Polytrichastrum alpinum–Ochrolechia frigida* community most often occurs on moist pebble soil at the bottom of rock faces or on steeper surfaces, such as near the summits of glacial moraines. It develops in the form of compact swards of *Polytrichastrum alpinum*, often as a result of a confluence between circular colonies. In the later stages of succession, or in drier places the moss is invaded by epibryophytic crustose lichens: *Ochrolechia frigida* and *Rinodina turfacea*, *Psoroma hypnorum*, *Massalongia carnosa*, which accompany the first species.

12.4.3 Communities with a Predominance of Lichens

On the summits of predominantly contemporary moraines, on ridges and hills, in lower locations, on gravel and pebble soils in less sheltered places and those exposed to winds, communities with *Usnea antarctica* occur. In lower locations, a higher proportion of the grass *Deschampsia antarctica* occurs. On the older moraines with a stable substrate and in more sheltered places other fruticose species of lichens occur, such as *Cetraria aculeata*, *Bryoria chalybeiformis* and *Sphaerophorus globosus*.

In places with particularly advantageous conditions, sheltered from the wind, a magnificent tundra develops, formed by fruticose lichens: *Stereocaulon alpinum*, *Usnea antarctica*, *Sphaerophorus globosus*, *Cetraria aculeata*, *Cladonia gracilis*, with a proportion of the mosses *Sanionia georgico-uncianata* and *Polytrichastrum alpinum*. In higher locations, the lichens *Usnea aurantiaco-atra* and *Coelopogon epiphorellus* also occur.

In abandoned penguin colonies, nitrophilous epigeal lichen-moss communities develop, where species such as *Syntrichia princeps*, *Xanthoria candelaria*, *Caloplaca citrina*, *Physconia muscigena*, *Lecanora expectans*, *Caloplaca tiroliensis* predominate, as well as the chlorophyte alga *Prasiola crispa*.

12.5 The Role of Animals in Antarctic Terrestrial Ecosystems

A characteristic feature of the terrestrial ecosystems of Western Antarctica is the major role of vertebrates in both nutrient cycling and energy flow. A special role is played by sea birds, particularly penguins, whose enormous breeding colonies occur on the coasts, in locations not covered by ice. As these birds live in two environments – feeding in the sea and breeding on land – they move huge quantities of organic matter onto the land in the form of guano. Thus they play an important role in nutrient cycling in the Antarctic terrestrial system. This is of particular importance to vegetation because the autochthonous resources of biogenic substances in the tundra – particularly those of nitrogen and phosphorus – are minimal. The faeces of birds, as well as feathers, egg shells and regurgitated stomach pellets are the sources of nutrient enrichment for emerging soils. The ornithogenic soils, a unique form worldwide (Tatur and Myrcha 1993), are colonised by specific nitrophilous communities.

On the most fertilised places, communities with predominantly green algae *Prasiola crispa* occur, and also communities with cyanobacteria such as *Phormidium* spp. and *Nostoc* spp. *Deschampsia antarctica*, a grass which forms communities in places of organic matter influx to the tundra, grows more abundantly and has greater biomass (Fig. 12.5). There is a stimulation of primary production and an emergence of specific ornithocoprophilous communities related to organic matter influx in the form of guano (Eurola and Hakala 1977). Particularly characteristic for Western Antarctica is the development of ornithocoprophilous lichen communities which locally cover large areas of rock faces. The thalli of many ornithocoprophilous lichens have vivid colours and thus give the rock surfaces a unique orange-yellow appearance, breaking the monotony of the Antarctic landscape. The important communities include: *Caloplaca regalis–Caloplaca lucens*; *Caloplaca cirrochrooides–Mastodia tesselata*; *Caloplaca sublobulata*; *Haematomma erythromma* and *Ramalina terebrata*. Many species of lichens have adapted to tolerate high levels of nutrients, particularly nitrogen and phosphorus from the penguin guano. It is worth noting that the composition of carotenoids in the thalli of *Caloplaca regalis* is almost identical to that in the penguin guano, and typical of carotenoids found in krill (Filho et al. 1986). The ornithocoprophilous effect manifests itself principally in the zones of coastal cliffs and boulders on

Fig. 12.5. Patch of *Deschampsia antarctica* with large individual grass tussocks in the vicinity of the penguin rookery.

the coastal plains, near large colonies of penguins (*Pygoscelis adeliae, P. antarctica, P. papua*).

Further inland, nitrophilous epilithic and epibryophytic lichen communities occur sporadically near single nests of birds such as Skuas (*Catharacta antarctica*; *C. maccormicki*). Antarctic tern (*Sterna vittata*) and giant petrel (*Macronectes giganteus*). The more important of these include the communities with *Xanthoria candelaria*; *Physcia muscigena* and *Caloplaca tiroliensis*. On protruding rocks which are vantage points for birds, ornithophilous communities thrive, such as *Haematomma erythromma–Xanthoria elegans*; *Caloplaca isidioclada*.

This effect on the Antarctic tundra, particularly in lower locations, is also exerted by marine mammals. In places occupied by large colonies of seals, particularly elephant seals (*Mirounga leonina*) or fur seals (*Arctocephalus gazella*), the tundra is often destroyed and its substrate enriched with nitrogen. In such places, a community with the moss *Hennediella antarctica* occurs locally (Ochyra 1998).

The contribution to energy flow in the Antarctic tundra ecosystem by invertebrates, the permanent inhabitants of Antarctica, is minimal. There is an absence of herbivorous invertebrates.

12.6 Distribution of the Principal Terrestrial Ecosystems

The distribution of the plant communities on King George Island, as in other parts of Antarctica, shows evidence of zonality, beginning from the coast and continuing further inland. (Fig. 12.6) The zonality can frequently be related to environmental gradients (Fig. 12.7), such as that resulting from variable concentrations of guano around colonies of penguins. With increasing distance from the coast and animal colonies, particularly penguins, and with the simultaneous increase in altitude, the composition and structure of the plant communities changes. At low altitudes, on coastal plains and gentle mountain slopes or on younger glacial moraines, there is more often a marked presence of vascular plants with an increased percentage of cover, resulting from a nutrient gradient effect. It is manifested principally by the occurrence of specific ornithocoprophilous lichen communities, or by large proportions of green algae in developing epigeal plant communities.

On coastal rocks vulnerable to the waves, there are communities of halophilous lichens, formed mostly by crustose *Verrucaria* species. The patches of these communities extend to about 4.5 m, although sometimes they occur below sea level, on rock exposed only during the ebb-tide. Further inland, with increasing altitude and change in trophic gradient, there is a dramatic drop in the proportion of vascular plants and algae in favour of lichens in the communities. In inland locations, nitrophobic communities

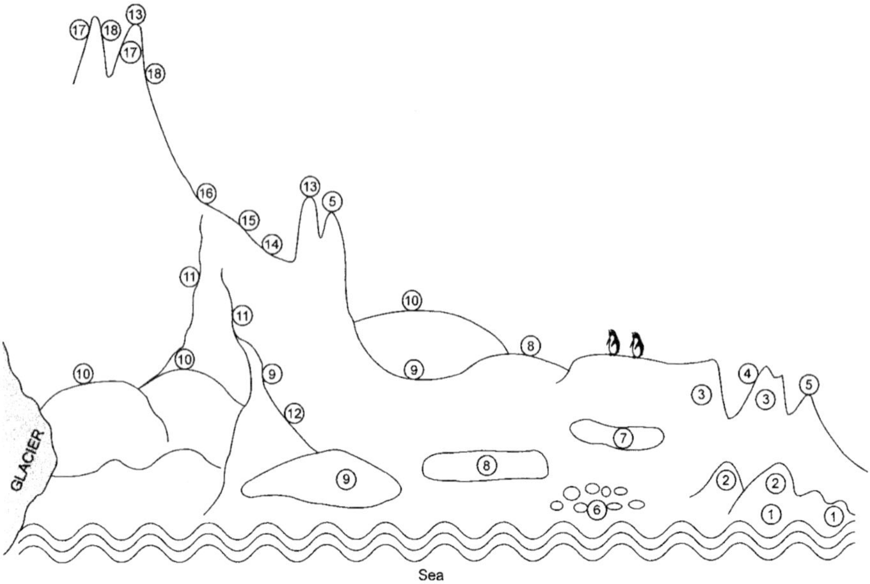

Fig. 12.6. Schematic profile across a coastal area of the Admiralty Bay illustrating the distribution of the principal terrestrial ecosystems. *1 Verrucaria* spp. communities, *2 Caloplaca cirrochrooides–Mastodia tesselata, 3 Caloplaca regalis–C. lucens, 4 Ramalina terebrata, 5 Haematomma erythromma, 6 Acarospora macrocyclos–Caloplaca coralligera, 7 Prasiola crispa, 8 Deschampsia antarctica–Colobanthus quitensis, 9 Sanionia georgico-uncianata–Warnstorfia laculosa–W. sarmentosa, 10 Usnea antarctica–Cetraria aculeata–Bryoria chalybeiformis, 11 Staurothele gelida, Caloplaca iomma–C. exsecuta, 12 Brachythecium salebrosum-Syntrichia filaris, 13 Ochrolechia parella,14 Andreaea gainii–A. depressinervis, 15 Stereocaulon alpinum–Sphaerophorus globosus, 16 Parmelia saxatilis, 17 Himantormia lugubris–Usnea aurantiaco-atra, 18 Umbilicaria decussata*

predominate in general. If nitrophilous communities are found, it is only occasionally and sporadically as small stands among other types of vegetation.

Above the eulitoral '*Verrucaria* zone', a '*Caloplaca* zone' can be distinguished. The communities occurring here will most often include ornithocoprophilous lichen communities, mostly associated with the presence of the colonies of penguins but also of petrels (*Macronectes giganteus, Daption capense*), and of nests of gulls (*Larus dominicanus*). Just above the *Verrucaria* zone, *Caloplaca cirrochrooides–Mastodia tesselata* communities occur. Above these and at greater distances from the sea, *Caloplaca regalis–Caloplaca lucens* communities occur, with the most beautiful and the largest patches occurring in the vicinity of large penguin colonies. In this zone, also *Haematomma erythromma* and *Ramalina terebrata* communities occur.

With increasing distance both vertically and horizontally from the marine and biotic influence, nitrophobic communities appear. In the zone spanning

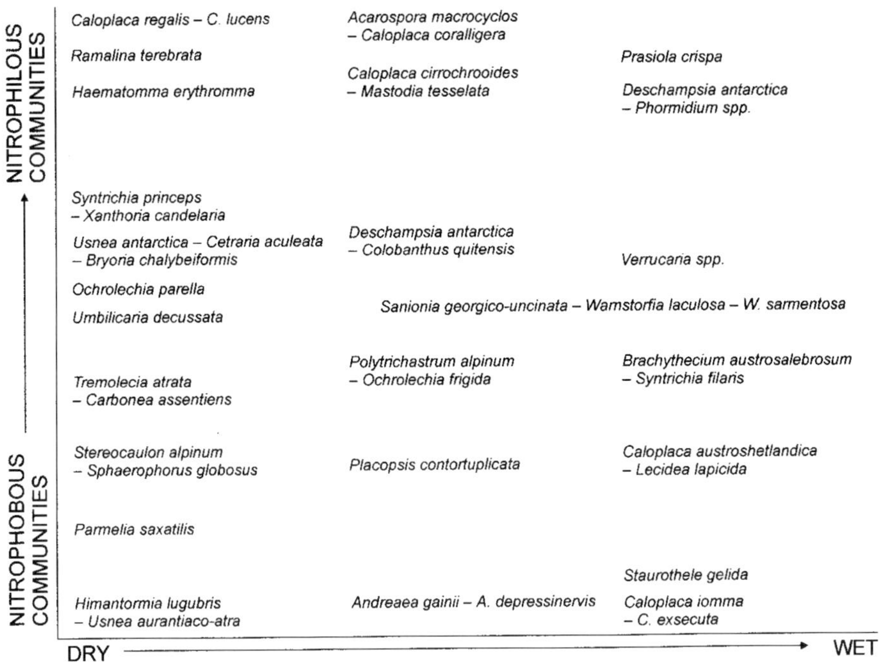

Fig. 12.7. Important plant communities shown against the trophic gradient and the gradient of substrate moisture

from approximately 10–15 m up to approximately 100 m a.s.l., which is defined as the '*Usnea antarctica* zone', a community of fruticose lichens, *Usnea antarctica–Cetraria aculeata–Bryoria chalybeiformis* is found as well as *Deschampsia antarctica–Colobanthus quitensis* community. The communities of epilithic lichens with *Staurothele gelida* and *Brachythecium austrosalebrosum–Synrichia filaris* communities also occur here.

Above a level of 100 m, up to the highest locations, the '*Usnea aurantiaco-atra* zone' occurs. In these elevated places there are communities such as *Usnea aurantiaco-atra–Himantormia lugubris*, *Umbilicaria decussata* communities, or *Andreaea gainii–Andreaea depressinervis* communities and lichen communities with *Parmelia saxatilis*. These communities are often termed 'montane communities' (Lewis Smith 1993).

Acknowledgements. I wish to express my thanks to Dr. R.D. Seppelt, Australian Antarctic Division, for linguistic revision of the English text and I acknowledge the assistance of Malgorzata Matyjaszkiewicz, M.Sc.

References

Allison JS, Lewis Smith RI (1973) The vegetation of Elephant Island, South Shetland Islands. Br Antarct Surv Bull 33–34:185–212

Castello M (1995) The lichen genus *Xanthoria* in Antarctica. Cryptogamic Bryol Lichenol 16:79–87

Castello M, Nimis PL (1994a) Critical notes on Antarctic yellow Acarosporaceae. Lichenologist 26:283–294

Castello M, Nimis PL (1994b) Critical notes on the genus *Candelariella* (Lichenes) in Antarctica. Acta Bot Fenn 150:5–10

Castello M, Nimis PL (1995) A critical revision of Antarctic Lichenes described by C W Dodge. Bibl Lichenol 57: 1–92

Eurola S, Hakala AVK (1997) The bird cliff vegetation of Svalbard. Aquilo Ser Bot 15: –18

Filho LX, Kolakowska A, Vincente C (1986) Carotenoid composition of *Caloplaca regalis* (Vain.) Zahlbr. (Lichenes) and of penguin faeces at King George Island (Antarctica). Pol Polar Res 7:289–294

Filson RB (1982) Lichens of continental Antarctica. J Hattori Bot La 53:357–360

Follmann G (1965) Una asociación nitröfila de liquenes epipétricos de la Antártica occidental con *Ramalina terebrata* Tayl. et Hook. como especie caracterizane. Inst Antarct Chil 4:1–18

Furmańczyk K, Ochyra R (1982) Plant communities of the Admiralty Bay region (King George Island, South Shetland Islands, Antarctic) I. Jasnorzewski Gardens. Pol Polar Res 3:25–39

Gimingham CH, Lewis Smith RI (1970) Bryophyte and Lichen communities in the Maritime Antarctic. In: Holdgate MW (ed) Antarctic ecology, vol 2. Academic Press, London, pp 752–785

Guminska B, Heinrich Z, Olech M (1994) Macromycetes of the South Shetland Islands (Antarctica). Pol Polar Res 15:103–109

Hertel H (1984) Über saxicole, lecideoide Flechten der Subantarctis. Beih Nova Hedwigia 79:399–500

Holdgate MW (1964) Terrestrial ecology in the maritime Antarctic. In: Carrick R, Holdgate MW, Prévost J (eds) Biologie antarctique. Proc 1st SCAR Symposium, Paris, France, 1962. Hermann, Paris, pp 181–194

Hu S-S (1998) Moss community types and species diversity of southern Fildes Peninsula (King George Island, South Shetland Islands) Antarctica. J Hattori Bot Lab 84:187–198

Inoue M (1993) *Buellia subfrigida* sp. nov. (Lichenes, Buelliaceae) from Lützow-Holm Bay area and the Prince Olav Coast, East Antarctica – the asexual sorediate species forming a species pair with *B. frigida* Darb. Antarct Rec 37:19–23

Komárková V, Poncet S, Poncet J (1990). Additional and revisited localities of vascular plants *Deschampsia antarctica* Desv. and *Colobanthus quitensis* (Kunth) Bartl. in the Antarctic Peninsula area. Arct Alp Res 22:108–113

Lewis Smith RI (1972) Vegetation of the South Orkney Islands with particular references to Signy Island. Br Antarct Surv, Sci Rep 68:1–124 + pls 1–6

Lewis Smith RI (1979) The terrestrial vegetation. In: Furse C (ed) Elephant Island; an Antarctic expedition. Anthony Nelson, Shrewsbury, pp 230–236

Lewis Smith RI (1988) Botanical survey of Deception Island. Br Antarct Surv Bull 80:129–136

Lewis Smith RI (1993) Dry coastal ecosystems of Antarctica. In: van der Maarel E (ed) Ecosystems of the world, vol 2A. Dry coastal ecosystems. Polar regions and Europe. Elsevier, Amsterdam, pp 51–71

Lewis Smith RI, Gimingham CH (1976) Classification of cryptogamic communities in the maritime Antarctic. Br Antarct Surv Bull 43:25–47

Lindsay DC (1971) Vegetation of the South Shetland Islands. Br Antarct Surv Bull 25:59–83
Lindsay DC (1976) The lichens of Marion and Prince Edward Islands, Southern Indian Ocean. Nova Hedwigia 28:667–689
Longton RE (1966) Alien vascular plants on Deception Island, South Shetland Islands. Br Antarct Surv Bull 9:55–60
Longton RE (1967) Vegetation in the Maritime Antarctic. In: Smith JE (ed) A discussion of the Antarctic terrestrial ecosystem. Philos Trans R Soc Lond Ser B 252:213–235
Longton RE (1979) Vegetation ecology and classification in the Antarctic zone. Can J Bot 57:2264–2278
Longton RE (1988) The biology of polar bryophytes and lichens. Cambridge University Press, Cambridge
Ochyra R (1998) The moss flora of King George Island Antarctica. Polish Academy of Sciences, W Szafer Institute of Botany, Cracow
Ochyra R, Vana J (1989) The hepatics of King George Island, with particular reference to the Admiralty Bay region (South Shetland Islands, Antarctica). Pol Polar Res 10:183–210
Olech M (1990) Preliminary studies on ornithocoprophilous lichens of the Arctic and Antarctic regions. Proc NIPR Symp Polar Biol 3:218–223
Olech M (1993) Lower plants. In: Rakusa-Suszczewski S (ed) The maritime Antarctic coastal ecosystem of Admiralty Bay. Department of Antarctic Biology, Polish Academy of Sciences, Warsaw, pp 173–179
Olech M (1994) Lichenological assessment of the Cape Lions Rump, King George Island, South Shetland Islands; a baseline for monitoring biological changes. Pol Polar Res 15:111–130
Olech M (1996) Human impact on terrestrial ecosystems in West Antarctica. Proc NIPR Symp Polar Biol 9:299–306
Olech M (2000) Lichens of Antarctica a checklist. The Institute of Botany of the Jagiellonian University, Kraków (in press)
Olech M, Alstrup V (1990) *Thelocarpon cyaneum* sp. nov. Nord J Bot 9: 75–576
Olech M, Alstrup V (1995) Lichenicolus fungi from polar regions. In: Zamek K (ed) Unity of Arctic and Antarctic natural environment. XXII Polar Symposium, 27–28 Oct 1995, pp 111–122
Olech M, Søchting U (1993) Four new species of *Caloplaca* from Antarctica. Lichenologist 25:261–269
Orange A (1993) *Thelidium austroatlanticum* (Verrucariaceae) a new species from the South Orkney Islands. Lichenologist 25:271–277
Øvstedal DO (1988) *Cladonia galindezii* a new Antarctic lichen species. Cryptogamic Bryol Lichenol 9:137–139
Øvstedal DO, Elix JA, Lewis Smith RI (1996) A new species of *Parmelia* (lichenized Ascomycotina) from the Antarctic. Mycotaxon 57:151–153
Redon J (1969) Nueva asociación de liquenes muscicolas de la Antártica Occidental, con Sphaerophorus tener Laur. como especie caracterizante. Bol Inst Antárt Chileno 4:5–11
Tatur A, Myrcha A (1993) Ornithogenic soils. In: Rakusa-Suszczewski S (ed) The maritime Antarctic coastal ecosystem of Admiralty Bay. Dept Antarctic Biology, Polish Academy Science, Warsaw, pp 164–172
Vezda A, Øvstedal DO, Lewis Smith RI (1992) Eine neue Gyalecta-Art aus der Antarktis: *G. pezizoides* sp. n. (lichenisierte Fungi, Gyalectaceae). Nova Hedwigia 55: 27–229
Zarzycki K (1993) Vascular plants and terrestrial biotopes. In: Rakusa-Suszczewski S (ed) The Maritime Antarctic coastal ecosystem of Admiralty Bay. Department of Antarctic Biology, Polish Academy of Sciences, Warsaw, pp 181–187

13 Plant Communities at Wilkes Land

R. Seppelt

13.1 Flora and Vegetation of the Windmill Islands

The Windmill Islands region supports the most extensive vegetation cover of any continental Antarctic locality. Vegetation, dominated principally by the lichen species *Usnea (Neuropogon) sphacelata*, *Umbilicaria decussata*, *Buellia frigida* and *Pseudephebe minuscula* together with mosses, *Grimmia (Schistidium) antarctici*, *Bryum pseudotriquetrum* and *Ceratodon purpureus*, is distributed on ice-free areas to varying degrees, with the richest communities occurring on the three northern peninsulas – Clark, Bailey and Mitchell Peninsulas – and Robinson Ridge (Lewis Smith 1988, 1990b; Kappen et al. 1990; Melick et al. 1994). On Browning Peninsula, in the south, and on the islands, vegetation is generally sparse and confined to isolated communities in sheltered habitats set amongst predominantly barren rocky ground. In eutrophic sites near bird colonies, mosses and lichens are virtually absent and the terrestrial chlorophyte algae *Prasiola crispa*, *Prasiococcus calcareus* and *Desmococcus olivaceus*, dominant.

The flora of the region comprises at least 36 lichens, 4 mosses (Filson 1974; Lewis Smith 1988; Seppelt, unpubl.), 1 hepatic (Seppelt 1983), and 150 non-marine algae including 24 snow algal taxa (Ling 1996; Ling and Seppelt 1990, 1993, 1998a, b). In addition, 120 fungal taxa have recently been isolated (Azmi, McRae and Seppelt, unpublished data; McRae et al. 1999). An ascomycete mycorrhizal fungus has been demonstrated in the rhizoids and invades adjacent cells of the stem of the liverwort *Cephaloziella exiliflora* (Williams et al. 1994). No mycorrhizal fungal associations have been detected in any of the mosses. The fungus is also present in collections of *C. exiliflora* from Botany Bay, Granite Harbour, in southern Victoria Land (77°00' S, 162°35' E).

13.2 Geology, Geomorphology and Vegetation Patterns

Rocks of the Windmill Islands region (Seppelt, Chap. 4) consist of a layered sequence of acid to ultrabasic schists, gneisses and migmatites intruded by

Ecological Studies, Vol. 154
L. Beyer and M. Bölter (eds.) Geoecology of Antarctic Ice-Free Coastal Landscapes

the Ardery Charnockite and the porphyritic Ford Granite (Blight and Oliver 1977, 1982). There are no major-element chemical differences between the northern amphibolite facies and the southern granulite facies rocks (Blight and Oliver 1982). The Ardery Charnokitic Intrusives are magmatic rocks prone to deep weathering related to features of origin rather than to surficial processes (J. Kilpatrick, pers. comm.). Consequently, they tend to crumble readily into their constituent minerals. The metamorphic sequences found in the northerly parts of the region have a much more stable mineral assemblage and crystalline structure. This difference has a marked influence on the suitability of the rock types as a substrate for slow growing lichens and explains, in part, the greater abundance of vegetation on the northerly peninsulas.

Soils of the region are mainly formed by cryoturbation and cryoclastic weathering. Most of the soils are influenced by biotic structure formation, humus accumulation, and chemical weathering, together with acidification and brownification (Blume and Bölter 1993a,b; Beyer at al. 1998a,b; Beyer and Bölter 1999). Polygons 0.3–10 m in diameter have been formed by cryoturbation. The centre of many, perhaps more active, polygons may consist of coarse sand and fines with little vegetation cover, while others may support an almost complete cover of mosses in wetter habitats or lichens in drier habitats. The soils are acid, with a $pH_{(KCl)}$ generally between 4 and 4.5. Organic matter content in the first 1–5 cm of many soils ranges from 0.5–2 %. Shallow peat may accumulate in small depressions but there are no extensive peat deposits. There is a strong correlation between vegetation cover and soil type (Blume and Bölter 1993a,b). Microbial activities within soils also depend significantly on microhabitat moisture and vegetation cover (Bölter 1992, 1993; Melick et al. 1994; for a more detailed description of the soils, see Blume et al., Chap. 7 and 8).

The present surface of the Windmill Islands emerged from beneath a retreating ice plateau during the Holocene period, based on ^{14}C ages of raised marine shorelines, lake bottom sediments, and Adélie penguin remains found in abandoned colonies (Goodwin 1993). Penguin remains have been dated at 4380±250 ^{14}C years B.P. (corrected age of 3289 calendar years B.P.), while a dark brown compacted peat with penguin quills has been dated at 4980±80 ^{14}C years B.P. (corrected age range of 4280–4530 calendar years B.P.). The upper marine limit appears to have reached a level coincident with a present elevation of 31–32 m on Browning Peninsula and Holl Island at the southern end of the region, with upper marine limits being lower towards the northern exposures.

The southern islands were deglaciated by about 8000 years B.P. while the northern islands were deglaciated by about 5500 years B.P. The extent of ice cover at the last glacial maximum and in the period since then has considerable importance for understanding both the potential origin and the abundance of the present day terrestrial biota of the Windmill Islands region. For a discussion of Holocene emergence data, see Seppelt (Chap. 4). Recent studies in the Vestfold Hills (e.g. Adamson and Pickard 1986; Gore and Colhoun 1995, 1997; Fabel et al. 1997; Zwartz et al. 1997), the Larsemann Hills (Gillieson 1991; Burgess et al. 1994, 1997), the Bunger Hills (Colhoun and Adamson 1989, 1992; Colhoun 1991, 1997; Adamson and Colhoun 1992; Colhoun et al. 1992; Melles et al. 1997) and in the Windmill Islands (Hollin and Cameron 1961; Goodwin 1993), have indicated that either the ice sheet was much thinner or some areas of these present day coastal regions may have remained unglaciated during at least the Holocene period. Supporting evidence for the presence of coastal ice-free areas during the last glacial maximum may be found in the radiocarbon dates from moss found in permafrost in the Larsemann Hills (27,950±710 years B.P., Burgess et al. 1994) and for in situ subfossil moss from near Clear Lake in the Vestfold Hills of 7355±150 to 8400±250 years B.P. (Pickard and Seppelt 1984).

Despite the emergence of the Windmill Islands from beneath the retreating ice cap only within the last 5000–8000 years, the most extensive and densest vegetation cover is found on the northernmost Clark and Bailey Peninsulas. From where the vegetation may have been derived cannot be determined but there have probably been repeated arrivals of plant propagules (Melick et al. 1994) over time. Clark Peninsula currently supports Adélie penguin colonies only at two sites, in the northwest (Blakeney Point) and south-west (Whitney Point). Shortly after emergence from the ice during the Holocene, the peninsula, probably initially as a cluster of small islands, supported extensive penguin breeding colonies which have long since been abandoned. Such colonies are a common feature of many localities in the Windmill Islands and most are inland or up-slope of currently occupied colony sites, most of which are now located on islands. There is a substantial nutrient and sea salt input onto the land in the vicinity of extant penguin colonies, which influences microbial numbers and activity (Roser et al. 1993) and vegetation (Melick et al. 1994). Adélie penguins may also provide a significant input of NH_4^+ to the terrestrial environment (Wodehouse and Parker 1978; P. Crittenden, unpublished data). Recent studies of soils in the Windmill Islands have shown that nutrients from these ancient ornithogenic soils may also have a potentially significant role in determining the abundance of the terrestrial vegetation.

13.3 Plants in the Terrestrial Ecosystem

The primary objectives of recent studies on the terrestrial ecosystem of the Windmill Islands have been to determine biodiversity, physiological and biochemical attributes, component interactions, impacts of anthropogenic pollutants, and potential effects of global climate change.

Fruticose and crustose lichens dominate the landscape in the Windmill Islands region, particularly on gneissic substrates. Lewis Smith (1988), in a study of the cryptogamic communities on Bailey Peninsula, recognised two principal subformations – a lichen-dominated association of *Usnea–Pseudephebe–Umbilicaria–Buellia* with five closely related sociations, and a short cushion and turf moss subformation with four moss-dominated sociations within the *Ceratodon–Schistidium–Rinodina* association. *Bryum pseudotriquetrum* (as *B. algens*), while locally dominant in some of the wetter areas, was not considered to merit the status of a *Bryum* sociation. In terms of species composition, many of the sociations within an association were shown to be almost identical, with particular species assuming local dominance depending on the predominance of a particular suite of environmental parameters.

The greatest species diversity and frequency of elements of the macroflora (mosses and lichens) occurs on the four northern peninsulas: Clark (29 spp.), Bailey (26 spp.), Mitchell (17 spp.) and Robinson Ridge (25 spp.). The major distinguishable community types are dominated to varying degrees by bipolar lichens, *Usnea sphacelata*, *Pseudephebe minuscula* and *Umbilicaria decussata* (Melick et al. 1994). The islands are generally poorly vegetated, although Hollin and Beall Islands support relatively high species richness (16 and 12 species respectively), but the vegetation is limited to small communities set amongst generally barren rocky ground. Mosses and lichens are generally absent from eutrophic sites near bird colonies, where the algae *Prasiola crispa*, *Prasiococcus calcareus* and *Desmococcus olivaceus* predominate. The nitrophilous lichen *Xanthoria mawsonii*, together with *Candelariella flava* and *Buellia frigida* are usually found near bird colonies.

Vegetation on the northern peninsulas is best developed on northerly aspects. On exposed sites, such as large boulders or exposed knolls, the windward faces and slopes are generally devoid of lichens (Lewis Smith 1988, 1990b). This partly relates to physical abrasion by wind-borne snow and ice particles, and to availability of moisture for growth and photosynthesis in sheltered habitats (Kappen 1990, 1993; Kappen and Schroeter 1997). Microhabitat or growth conditions may have a significant effect on morphology of lichens (Hancock and Seppelt 1988) and mosses (Seppelt and Selkirk 1984; Seppelt and Hancock 1990) and may lead to confusion in field identification and taxonomic interpretation.

Floristic studies of the Windmill Islands (Lewis Smith 1988, 1990b; Hovenden and Seppelt 1995a) identified small-scale pattern processes and environmental continua correlated with microsite moisture, exposure and nutrient levels. More extensive surveys (Melick et al. 1994) showed that while development is dependent on specific microsite conditions, broad-scale vegetation communities could also be distinguished. Bryophytes appear to be key ecological markers, with health and vigour being indicative of site moisture. This is also particularly evident in extreme habitats, such as the McMurdo Dry Valleys (e.g. Taylor Valley), where the occurrence of mosses on the valley floors is confined almost entirely to the close proximity of meltwater seepage channels from glaciers. The major habitats for lichens are coincident with the commonly developed afternoon cloud band, at around 800 m a.s.l., along the valley sides.

The broad-scale patterns of plant distribution in the Windmill Islands suggest an ongoing expansion of lichen-dominated vegetation and a contraction of bryophyte communities to lower lying areas with reliable moisture supply. These vegetation patterns concur with glaciological and geomorphological evidence indicating a long-term drying of climate and isostatic uplift of the region since the last deglaciation. Analyses of meteorological records show an average lowering of temperature of 0.101 °C $year^{-1}$ in the region over the last 10–15 years following an earlier rise of about 0.086°C $year^{-1}$ since 1960. Similar temperature shifts have been reported for a number of the Subantarctic islands and other continental Antarctic localities (Adamson et al. 1988; Jacka et al. 1984; Jacka and Budd 1991; Selkirk et al. 1990). Fluctuations in interannual climatic parameters in recent years have resulted in estimated times available for plant growth to vary between 29.7 and 9.8 % of the year for mosses and 10.4 and 6.4 % of the year for lichens (Melick and Seppelt 1997). Small shifts in climate may significantly change habitat suitability for animals and plants (Abbott and Benninghoff 1990; Weller 1992). Recent changes in vegetation in the Windmill Islands appear to be much less dramatic than those reported for the Subantarctic islands and some continental Antarctic localities (Allison and Keage 1986; Scott 1990; Lewis Smith 1990a; Smith and Steenkamp 1990). In contrast to the relatively rapid ecological changes being demonstrated for the Subantarctic islands, any change in status of the continental Antarctic terrestrial ecosystem, as in the Arctic, is not yet well understood.

Minor differences in microclimate, particularly moisture availability, may result in significant changes in the composition of plant communities (Selkirk and Seppelt 1987; Hu and Seppelt 1989; Melick and Seppelt 1997). The terrestrial vegetation provides a significant habitat for soil microfauna. What influence these microclimatic differences have on the distribution and abundance of invertebrates, and the role they play in the terrestrial ecosystem, is not known. Little is known of the microfauna of the Windmill Islands, which is limited to a few mites, nematodes, tardigrades, rotifers, and protozoa. The Tardigrada fauna of continental Antarctic localities appears limited to just a few genera and species (Sudzuki 1964; Dastych 1984, 1989; J. D. Miller et al. 1988; W. R. Miller et al. 1994a, b, 1996, 1998). W. R. Miller et al. (1994b, 1996) reported three genera with four species of Tardigrada from the Windmill Islands. Sudzuki (1964), in one of the few comprehensive studies of the zoofauna of continental Antarctic terrestrial habitats found a significant microfauna associated with moss beds in the Langhovde region (69°13'S, 39°45'E). While nematodes have been collected and a new taxon recorded (K. Kito, unpubl. data), no systematic or ecological accounts have yet been published for the Windmill Islands region. Protozoa have been studied by Petz with the first of the results published (Petz 1997; Petz and Foissner 1996, 1997). Although widespread and abundant in some other parts of continental and maritime Antarctica, collembola have not been reported from the Windmill Islands, Bunger Hills, Vestfold Hills, MacRobertson Coast or Sôya Coast. Nothing is known, however, of the interdependence of invertebrates and other components of the biota.

13.4 Ecophysiological Adaptations of the Flora

Estimates of growth rates in the conspicuous flora of the Antarctic vary considerably, depending on species and locality. Lewis Smith (1990a) reported lichen annual growth increments varying from 41 % per annum for colonising thalli of foliose *Umbilicaria antarctica* down to 0.5–0.8 % per annum (equivalent to a radial increase of 5 mm per 100 years) in the crustose *Rhizocarpon geographicum*. In extreme conditions in the Dry Valleys of Southern Victoria Land, growth increments measured for the crustose *Buellia frigida* have been determined as 1 cm per 1000 years (A. Green, unpubl.). Growth rate is dependent on many factors and a paradoxical situation of dry weight loss over the summer in foliose *Umbilicaria decussata* and fruticose *Usnea sphacelata* has been recorded for the Windmill Islands (P. Crittenden and D. Melick, unpubl.).

Melick and Seppelt (1994a,b) demonstrated that in the Windmill Islands mosses and lichens remain physiologically buffered against sudden environmental fluctuations, showing no evidence of the seasonal build-up of carbohydrates and polyhydric alcohols associated with growth flushes and winter hardening seen in cryptogams in the maritime Antarctic region (Tearle 1987), and seasonal nutrient flushes are yet to be demonstrated (Roser et al. 1993). Even a sudden and natural rewetting event in mid winter was shown to have a negligible influence on cation leakage from lichens (Hovenden 1997), except for a significant loss of potassium from *Usnea sphacelata* thalli from the most exposed sites. Jackson and Seppelt (1995) showed that with the onset of winter there was a marked increase in levels of proline concurrent with a decrease in other predominant amino acids in the thallose chlorophyte alga *Prasiola crispa*. Photosynthetic recovery occurs rapidly in this alga when brought from the field to the laboratory in winter, indicating that proline may ameliorate cellular damage due to freezing in this species.

Low productivity of plant communities in polar regions is generally attributed to low summer growth temperatures, the short growing season, nutrient availability and water stress. Maximum irradiance in summer in continental Antarctica may be no different than that received in regions elsewhere. However, that in winter is much less. On an annual basis, this may be compensated for by longer photoperiods during the active spring-summer growing season.

Maximum photosynthetic rates in bryophytes are achieved at comparatively low light intensities (Valanne 1984). It is more likely, therefore, that high light intensities at high latitudes may be growth limiting. Post (1990) has discussed the adaptive and photoprotective nature of anthocyanin pigmentation in *Ceratodon purpureus* in response to high light intensity. H. Adamson et al. (1988), Kappen et al. (1989) and Wilson (1990) have demonstrated photoinhibition in *Grimmia antarctici* in the Windmill Islands region. Recovery of photosynthetic activity may occur under the influence of cloud cover or a light cover of snow (Post et al. 1990). Lovelock et al. (1995a,b) have demonstrated reversible photoinhibition in *Grimmia antarctici* during freezing and thawing, indicating the existence of processes that protect from photoinhibitory damage in such environments, where freezing temperatures occur in conjunction with high solar radiation levels, thus limiting the need for mechanistic repair cycles requiring higher temperatures for enzyme activity.

Davey (1989) showed similar effects on photosynthesis in the green alga *Prasiola crispa* and the cyanobacterium *Phormidium autumnale*, provided that there was an excess of free water. Physiological mechanisms that enable plants to freeze or desiccate without incurring tissue damage have enabled plants to survive in extremely cold and dry environments and will be of particular importance as survival strategies in extreme environments under the influence of global climate change. In plants intolerant of desiccation or freezing, both processes in light cause irreversible reductions in the efficiency of the PSII electron transport mechanism (Somersalo and Krause 1990; Öquist and Huner 1991). In contrast, in plants acclimated to low temperatures or desiccation these processes lead to reversible reductions in electron transport efficiency in PSII (Hällgren et al. 1990; Post et al. 1990; Hurry and Huner 1992; Eickmeier et al. 1993; Muslin and Homann 1992; Seel et al. 1992).

Habitat-related, rather than physiological, causes may give rise to significant differences in apparent photosynthetic responses. Kappen et al. (1989) noted differences in sensitivity to photoinhibition and in photosynthesis in two ‘ecodemes’ of the moss *Grimmia antarctici*. They found that the wetter (turf forming) site was more productive, had a wider temperature range of positive net photosynthesis and was less sensitive to photoinhibition than the dry (cushion-forming) site, and attributed these differences to physiological activities. Tarnawski et al. (1992) demonstrated that these differences could be related to internal turf and cushion CO_2 concentrations arising from differences in the abundance of the rhizo/phylloplane community of protozoa, tardigrades, nematodes, fungi and bacteria involved in the decomposition of organic matter, as well as water supply. Carbon dioxide concentrations in the more open and drier cushions reached only 1.3 times ambient levels and remained relatively constant throughout the summer, while in the wetter turf the internal rhizo/phylloplane CO_2 concentration reached ten times ambient in mid summer.

In polar regions, adaptation of plants to their thermal environment is critical to their survival under the extreme environmental conditions. Kallio and Heinonen (1971) showed that lichens may exhibit a seasonal shift in their optimal temperatures for metabolic processes. Lichens may also show acclimation of photosynthetic processes to changing environmental conditions (Kershaw 1985). Availability of water is the limiting factor for photosynthetic activity in lichens (Harrison et al. 1989; Kappen et al. 1987; Kappen and Breuer 1991; Schroeter et al. 1992, 1997). Optimal temperatures for photosynthesis may never be reached in the field. Of particular significance in understanding the adaptability of lichens to the extreme conditions in polar regions have been the studies by Lange (1965), Lange and Kappen (1972) and Kappen and Lange (1972), demonstrating the capacity of Antarctic lichens to show positive net photosynthesis at temperatures between –10 and –18 °C. Field studies in Southern Victoria Land (Schroeter et al. 1994) demonstrated net photosynthetic activity in *Umbilicaria aprina* at –17 °C with dark respiration ceasing at temperatures below –10 °C, showing that such lichens have an innate ability to be productive at times outside the peak summer season. A capacity to absorb atmospheric water vapour under a covering of snow is a particular advantage since Hovenden et al. (1994) and Hovenden and Seppelt (1995b) have shown the physical impossibility for water vapour uptake from the air in continental Antarctic localities.

Many studies have demonstrated the efficacy of physiological attributes as adaptive strategies to the rigors of the Antarctic environment. Little attention has been given to nutrient availability. Lichens typically grow on substrates deficient in available nitrogen and may thus be largely dependent on atmospheric deposition as a prime source. Lichens appear to have the capacity to actively and efficiently scavenge NO_3^- and NH_4^+ from low solute concentrations (Lang et al. 1976; Rai 1988; Crittenden 1983, 1989; Greenfield 1992). In the field near Casey Station *U. sphacelata* retained 91 % of NO_3^- and 87 % of NH_4^+ in snow meltwater resulting from summer snow showers. The ability to capture NO_3^- from the low concentrations typical of Antarctic snow involves metabolically dependent processes. Crittenden (unpubl.) also found that during one summer melt event there was a significant input of NH_4^+ from dry deposition of NH_3 originating in winds blowing from the west from Adélie penguin colonies, the nearest of which was 2 km distant. The ability to capture inorganic nitrogen in this way at near 0 °C temperatures may be an important growth adaptation. The diversity of the lichen vegetation at another coastal continental Antarctic site, Cape Hallett (72°19'S, 170°16'E) in northern Victoria Land, where there are 75,000 breeding pairs of Adélie penguins close by, is strongly influenced by atmospheric nutrient deposition.

The lichen flora there is dominated by *Xanthoria mawsonii*, *Candelariella flava*, with *Xanthoria elegans*, *Buellia frigida* and *Candelaria murrayi*, species which are either nitrophilous or tolerate higher nitrogen levels in the environment. The moss flora is dominated by *Bryum subrotundifolium*, a species which shows a strong maritime preference and moisture requirement. In the Windmill Islands region Hovenden and Seppelt (1995a) have shown that substrate cationic nutrients, as well as exposure and moisture, may be important delimiters of species distribution patterns in lichen communities.

13.5 Environmental Impacts and Conservation Issues

The extent of the vegetation cover in the Windmill Islands is unique for any continental Antarctic locality. Recognition of this has led to the designation of two Sites of Special Scientific Interest on Bailey Peninsula (SSSI 16) and on Clark Peninsula (SSSI 17). Operation of a rebuilt Casey Station commenced in 1986, with excavation and construction activities commencing in August 1979. The new station site was located in an area of extensive vegetation cover. Building construction called for massive amounts of concrete mixing and emplacement.

By the end of the 1989–1990 austral summer some 1622 m^3 or 3900 tons of concrete mix had been delivered to the station. In the course of the reconstruction, concrete mixing took place at a number of different sites. The cement mixture used (Portland cement type III) and the Darex Set Accelerator used to speed up curing in cold conditions are both alkaline (a cement-water slurry is around pH 11.6). Distribution of wind-borne cement dust, as determined by snow surface pH, soil pH and direct deposition sampling, was extensive, resulting in an elevation of pH above the background snow or soil pH for 300 m or more downwind. At 210 m downwind, surface snow pH was 7.6 and surface soil pH 5.9. A soil surface pH of 4.7 was obtained from a remote site well upwind of the station (E. Adamson and Seppelt 1990).

The effect of cement dust deposition on morphological integrity and photosynthetic physiology in the Casey Station area has been discussed for mosses by E. Adamson et al. (1994) and for lichens by E. Adamson and Seppelt (1990). Examination of affected areas near the station in 1994, 4 years after cessation of concrete mixing, showed that the mosses had substantially recovered under natural conditions, while the lichens showed no apparent recovery, with *Umbilicaria decussata* and *Usnea sphacelata* remaining entirely bleached. Roser et al. (1992) have also shown that an effect of alkaline pollution is to significantly reduce the levels of intracellular polyhydric alcohols and oligosaccharides in *U. decussata* and *U. sphacelata*.

While we have no data on the effect on associated soil- and moss-inhabiting invertebrates, increased alkalinity has been shown to significantly reduce the numbers of soil-dwelling invertebrates in the Arctic (Mackenzie et al. 1990). Meininger and Spatt (1988) showed that increasing alkalinity had a significant effect on the assemblage of tardigrade species in the vicinity of the trans-Alaska pipeline haul road (Dalton Highway), with fungivorous and algal feeding species now found close to the road and omnivorous and carnivorous species found away from the immediate source of calcium-rich road dust.

While there has been a cessation of concrete mixing at Casey Station, the most effected moss beds, those lying to the west or downwind of the station complex, are now being inundated by fine dust from roads. Farmer (1993), in a review of the effects of dust on vegetation, stated that dust may affect photosynthesis, respiration, transpiration, and allow the penetration of phytotoxic gaseous pollutants. Further, he stated: 'The most damaging effects of road dust on any natural community have been found in high arctic systems'. Walker and Everett (1987) found that there was a decline in acidophilous species near to gravel roads in the Arctic, the species gradually being replaced by minerotrophic taxa such as *Ceratodon purpureus* and *Bryum* spp.

References

Abbott SB, Benninghoff WS (1990) Orientation of environmental change studies to the conservation of Antarctic ecosystems. In: Kerry KR, Hempel G (eds) Antarctic ecosystems. ecological change and conservation. Springer, Berlin Heidelberg New York, pp 394–403

Adamson DA, Colhoun EA (1992) The Late Quaternary glaciation and deglaciation of the Bunger Hills, Antarctica. Antarct Sci 4:463–467

Adamson DA, Pickard J (1986) Cainozoic history of the Vestfold Hills. In: Pickard J (ed) Antarctic oasis: terrestrial environments and history of the Vestfold Hills. Academic Press, Sydney, pp 63–98

Adamson DA, Whetton P, Selkirk PM (1988) An analysis of air temperature records for Macquarie Island: decadal warming, EMSO cooling and Southern Hemisphere circulation patterns. Pap Proc R Soc Tasm 122:107–112

Adamson E, Adamson H, Seppelt RD (1994) Cement dust contamination of *Ceratodon purpureus* at Casey, East Antarctica: damage and capacity for recovery. J Bryol 18:127–137

Adamson E, Seppelt RD (1990) A comparison of airborne alkaline pollution damage in selected lichens and mosses at Casey Station, Wilkes Land, Antarctica In: Kerry KR, Hempel G (eds) Antarctic ecosystems. Ecological change and conservation. Springer, Berlin Heidelberg New York, pp 347–353

Adamson H, Wilson M, Selkirk P, Seppelt RD (1988) Photoinhibition in Antarctic mosses. Polarforschung 58:103–112

Allison IF, Keage PL (1986) Recent changes in the glaciers of Heard Island. Polar Rec 23:255–271

Beyer L, Bölter M (1999) Formation, ecology, and geography of cryosols of an ice-free oasis in coastal East Antarctica near Casey Station (Wilkes Land) Aust J Soil Res 37:209–244

Beyer L, Pingpank K, Bölter M, Schneider D, Blume H-P (1998a) Variation of carbon and nitrogen storage in soils of coastal continental Antarctica (Wilkes Land). Euras Soil Sci 31:551–554

Beyer L, Pingpank K, Bölter M, Seppelt RD (1998b) Small-distance variation of carbon and nitrogen in mineral Antarctic Cryosols near Casey Station (Wilkes Land). Z Pflanzenernähr Bodenkd 161:211–220

Blight DL, Oliver RL (1977) The metamorphic geology of the Windmill Islands, Antarctica, a preliminary account. J Geol Soc Aust 24:239–262

Blight DL, Oliver RL (1982) Aspects of the geologic history of the Windmill Islands, Antarctica. In: Craddock C (ed) Antarctic geoscience. Symposium on Antarctic Geology and Geophysics, Madison, Wisconsin, 22–27 Aug 1977, University of Wisconsin Press, Madison, Wisconsin, pp 445–454

Blume H-P, Bölter M (1993a) Podsole, Leptosole und Regosole der Antarktis. Mitteil Dtsch Bodenkd Ges 72:843–846

Blume H-P, Bölter M (1993b) Soils of Casey Station (Wilkes Land, Antarctica). In: Gilichinski D (ed) Joint Russian-American Seminar on Cryopedology and Global Change, Pushchino. Post-Sem Proc 1st Int Conf Cryopedology 1992, Russian Academy of Sciences, Pushchino, pp 96–104

Bölter M (1993) Effects of carbohydratres and leucine on growth of bacteria from Antarctic soils (Casey Station, Wilkes Land). Polar Biol 13:297–306

Burgess J, Carson C, Heat J, Spate A (1997) Larsemann Hills: not heavily glaciated during the last glacial maximum. In: Ricci CA (ed) The Antarctic region: geological evolution and processes. Proc VII Int Symp on Antarctic earth sciences, Siena, 1995. Terra Antarctica Publication, Siena, pp 841–843

Burgess J, Spate A, Shevlin J (1994) The onset of deglaciation in the Larsemann Hills, Eastern Antarctica. Antarct Sci 6:491–495

Colhoun EA (1991) Geological evidence for changes in the East Antarctica ice sheet (60°-120°E) during the last glaciation. Polar Rec 27:345–355

Colhoun EA (1997) A review of geomorphological research in Bunger Hills and expansion of the East Antarctic ice sheet during the last glacial maximum. In: Ricci CA (ed) The Antarctic region: geological evolution and processes. Proc VII Int Symp on Antarctic earth sciences, Siena, 1995. Terra Antarctica Publication, Siena, pp 801–807

Colhoun EA, Adamson DA (1989) Former glacial lakes of the Bunger Hills, Antarctica. Aust Geogr 20:125–136

Colhoun EA, Adamson DA (1992) Raised beaches of the Bunger Hills, Antarctica. Australian National Antarctic Research Expedition Scientific Reports, No 136. Australian Antarctic Division, Kingston

Colhoun EA, Mabin MGC, Adamson DA, Kirk RM (1992) Antarctic ice volume and contribution to sea-level fall at 20,000 yr B.P. from raised beaches. Nature 358:316–319

Crittenden PD (1983) The role of lichens in the nitrogen economy of subarctic woodlands: nitrogen loss from the nitrogen-fixing lichen *Stereocaulon paschale* during rainfall. In: Lee JA, McNeill S, Rorison IH (eds) Nitrogen as an ecological factor. Blackwell, Oxford, pp 43–68

Crittenden PD (1989) Nitrogen relations of mat-forming lichens. In: Boddy L, Marchant R, Read DJ (eds) Nitrogen, phosphorus and sulphur utilization by fungi. Cambridge University Press, Cambridge, pp 243–268

Dastych H (1984) The Tardigrada from the Antarctic with descriptions of several new species. Acta Zool Cracoviensia 27:377–436

Dastych H (1989) An annotated list of Tardigrada from the Antarctic. Entomol Mitt Zool Mus Hamburg 9:249–257

Davey MC (1989) The effects of freezing and desiccation on photosynthesis and survival of terrestrial Antarctic algae and Cyanobacteria. Polar Biol 10:29–36

Eickmeier WG, Casper C, Osmond CB (1993). Chlorophyll fluorescence in the resurrection plant *Selaginella lepidophylla* (Hook. et Grev.) Spring during high-light and desiccation stress, and evidence for zeaxanthin-associated photoprotection. Planta 189:30–38

Fabel D, Stone J, Fifield LK, Cresswell RG (1997) Deglaciation of the Vestfold Hills, East Antarctica: preliminary evidence from exposure dating of three subglacial erratics.

In: Ricci CA (ed) The Antarctic region: geological evolution and processes. Proc VII Int Symp on Antarctic earth sciences, Siena, 1995. Terra Antarctica Publication, Siena, pp 829–834

Farmer AM (1993) The effects of dust on vegetation – a review. Environ Pollut 79:63–75

Filson RB (1974) Studies on Antarctic lichens. II. Lichens from the Windmill Islands, Wilkes Land. Muelleria 3:9–36

Gillieson D (1991) An environmental history of two freshwater lakes in the Larsemann Hills, Antarctica. Hydrobiol 214:327–331

Goodwin ID (1993) Holocene deglaciation, sea level change and the emergence of the Windmill Islands, Budd Coast, Antarctica. Quat Res 40:70–80

Gore DB, Colhoun EA (1995) Regional contrasts in terrain and glacial sediments, Vestfold Hills, East Antarctica. In: Abstracts of the VII Int Symp on Antarctic earth sciences, Siena, 1995, p 163

Gore DB, Colhoun EA (1997) Regional contrasts in weathering and glacial sediments suggests long term subaerial exposure of Vestfold Hills, East Antarctica. In: Ricci CA (ed) The Antarctic region: geological evolution and processes. Proc VII Int Symp on Antarctic earth sciences, Siena, 1995. Terra Antarctica Publication, Siena, pp 835–839

Greenfield LG (1992) Retention of precipitation nitrogen by Antarctic mosses, lichens and fellfield soils. Antarct Sci 4:205–206

Hällgren J-E, Lundmark T, Strand M (1990) Photosynthesis of Scots pine in the field after night frosts during summer. Plant Physiol Biochem 28:437–445

Hancock RJ, Seppelt RD (1988) Habitat specificity and morphological variation in two Antarctic *Usnea* species. Polarforschung 58:285–291

Harrison PM, Walton DWH, Rothery P (1989) The effects of temperature and moisture on CO_2 uptake and total resistance to water loss in the Antarctic foliose lichen *Umbilicaria antarctica*. New Phytol 111:673–682

Hollin JT, Cameron RL (1961) IGY glaciological work at Wilkes Station, Antarctica. J Glaciol 3:833–843

Hovenden M (1997) Effects of a rapid, unseasonal rewetting event on mineral location in Antarctic lichens. New Phytol 137:241–246

Hovenden MJ, Jackson AE, Seppelt RD (1994) Field photosynthetic activity of lichens in the Windmill Islands oasis, Wilkes Land, continental Antarctica. Physiol Plant 90:567–576

Hovenden MJ, Seppelt RD (1995a) Exposure and nutrients as delimiters of lichen communities in continental Antarctica. Lichenologist 27:505–516

Hovenden MJ, Seppelt RD (1995b) Uptake of water from the atmosphere by lichens in continental Antarctica. Symbiosis 18:111–118

Hu S-S, Seppelt RD (1989) Moss communities and their characteristics in ice-free areas of the Windmill Islands, Wilkes land, continental Antarctica. Proceedings of the International Symposium on Antarctic Research, Hangzhou, China, 8–12 May 1989, pp 317–324

Hurry VM, Huner NP (1992) Effect of cold hardening on sensitivity of winter and spring wheat leaves to short-term photoinhibition and recovery of photosynthesis. Plant Physiol 100:1283–1290

Jackson AE, Seppelt RD (1995) The accumulation of proline in *Prasiola crispa* during winter in Antarctica. Physiol Plant 94:25–30

Jacka TH, Budd WF (1991) Detection of temperature and sea-ice extension changes in the Antarctic and Southern Ocean. In: Weller G, Wilson C, Severin B (eds) Proceedings of the International Conference on the Role of the Polar Regions in Global Change. Fairbanks; University of Alaska. University of Alaska, Fairbanks, pp 63–70

Jacka TH, Christou L, Cook BJ (1984) A data bank of mean monthly and annual surface temperatures for Antarctica, the Southern Ocean and South Pacific Ocean. Australian Antarctic Division, ANARE Research Notes 22, pp 1-84

Kallio P, Heinonen S (1971) Influence of short-term low temperature on net photosynthesis in some subarctic lichens. Rep Kevo Subarctic Res Stat 8:63–72

Kappen L (1990) *Usnea sphacelata*, its role in the vegetation and its possible growth capacity on Bailey Peninsula, Wilkes Land. Bibl Lichenol 38:277–289

Kappen L (1993) Plant activity under snow and ice, with particular reference to lichens. Arctic 46:297–302

Kappen L, Breuer M (1991) Ecological and physiological investigations in continental Antarctic cryptogams. II. Moisture relations and photosynthesis of lichens near Casey Station, Wilkes Land. Antarct Sci 3:273–278

Kappen L, Lange OL (1972) Die Kältresistenz einiger Makrolichenen. Flora 161:1–29

Kappen L, Schroeter B (1997) Activity of lichens under the influence of snow and ice. Proc NIPR Symp Polar Biol 10:163–168

Kappen L, Bölter M, Kühn A (1987) Photosynthetic activity of lichens in natural habitats in the maritime Antarctic. Bibl Lichenol 25:297–312

Kappen L, Meyer M, Bölter M (1990) Ecological and physiological investigations in continental Antarctic cryptogams. I. Vegetation pattern and its relation to snow cover on a hill near Casey Station, Wilkes Land. Flora 184:09–220

Kappen L, Lewis Smith RI., Meyer M (1989) Carbon dioxide exchange in two ecodemes of *Schistidium antarctici* in continental Antarctica. Polar Biol 9:415–422

Kershaw KA (1985) Physiological ecology of lichens. Cambridge University Press, Cambridge

Lang GE, Reiners WA, Heier RK (1976) Potential alteration of precipitation chemistry by epiphytic lichens. Oecologia 25:229–241

Lange OL (1965) Der CO_2-Gaswechsel von Flechten bei tiefen Temperaturen. Planta 64:1–19

Lange OL, Kappen L (1972) Photosynthesis of lichens from Antarctica. In: Llano GA (ed) Antarctic terrestrial biology. Antarctic Research Series 20, American Geophysical Union, Washington, pp 83–95

Lewis Smith RI (1988) Classification and ordination of cryptogamic communities in Wilkes Land, Continental Antarctica. Vegetatio 76:155–166

Lewis Smith RI (1990a) Signy Island as a paradigm of biological and environmental change in Antarctic terrestrial ecosystems. In: Kerry KR, Hempel G (eds) Antarctic ecosystems. Ecological change and conservation. Springer, Berlin Heidelberg New York, pp 32–50

Lewis Smith RI (1990b) Plant community dynamics in Wilkes Land, Antarctica. Proc NIPR Symp Polar Biol 3:229–244

Ling HU (1996) Snow algae of the Windmill Islands region, Antarctica. In: Kristiansen J (ed) Biogeography of freshwater algae. Hydrobiologia 336:99–106

Ling HU, Seppelt RD (1990) Snow algae of the Windmill Islands, continental Antarctica. *Mesotaenium berggrenii* (Zygnematales, Chlorophyta) the alga of grey snow. Antarct Sci 2:143–148

Ling HU, Seppelt RD (1993) Snow algae of the Windmill Islands, continental Antarctica. 2. *Chloromonas rubroleosa* sp. nov. (Volvocales, Chlorophyta), an alga of red snow. Eur J Phycol 28:77–84

Ling HU, Seppelt RD (1998a) Non-marine algae and cyanobacteria of the Windmill Islands region, Antarctica with descriptions of two new species. Algol Stud 89:49–62

Ling HU, Seppelt RD (1998b) Snow algae of the Windmill Islands, continental Antarctica. 3. *Chloromonas polyptera* (Volvocales, Chlorophyta) Polar Biol 20:320–324

Llano GA (1962) The terrestrial life of Antarctica. Sci Am 207:213–230

Llano GA (1965) The flora of Antarctica. In: Hatherton T (ed) Antarctica. Reed, Wellington, pp 331–351

Lovelock CE, Osmond CB, Seppelt RD (1995a) Photoinhibition in the Antarctic moss *Grimmia antarctici* Card., when exposed to cycles of freezing and thawing. Plant Cell Environ 18:1395–1402

Lovelock CE, Jackson AE, Melick DR, Seppelt RD (1995b) Reversible photoinhibition in Antarctic moss during freezing and thawing. Plant Physiol 109:955–961

Mackenzie S, Lee JA, Wright JM (1990) Ecological impact of liming blanket bog. CSD Report No 1167 to the Nature Conservancy Council, Peterborough

Meininger CA, Spatt PD (1988) Variations of tardigrade assemblages in dust-impacted Arctic mosses. Arct Alp Res 20:24–30

Melick DR, Seppelt RD (1994b) The effect of hydration on carbohydrate levels, pigment content and freezing point of *Umbilicaria decussata* at a continental Antarctic locality. Crypt Bot 4:212–217

Melick DR, Seppelt RD (1997) Vegetation patterns in relation to climatic and endogenous changes in Wilkes Land, continental Antarctica. J Ecol 85:43–56

Melick DR, Bölter M, Möller R (1994) Rates of soluble carbohydrate utilization in soils from the Windmill Islands oasis, Wilkes Land, continental Antarctica. Polar Biol 14:59–64

Melick DR, Hovenden MJ, Seppelt RD (1994) Phytogeography of bryophyte and lichen vegetation in the Windmill Islands, Wilkes Land, continental Antarctica. Vegetatio 111:71–87

Melick DR, Tarnawski MG, Adam KD, Seppelt RD (1994) Isozyme variation in three mosses from the Windmill Islands oasis, Antarctica: a preliminary study. Biodiv Lett 2:21–27

Melles M, Kulbe T, Verkulich SR, Pushina ZV, Hubberten HW (1997) Late Pleistocene and Holocene environmental history of Bunger Hills, East Antarctica, as revealed by fresh-water and epishelf lake sediments. In: Ricci CA (ed) The Antarctic region: geological evolution and processes. Proc VII Int Symp on Antarctic earth sciences, Siena, 1995. Terra Antarctica Publication, Siena, pp 809–820

Miller JD, Horne P, Heatwole H, Miller WR, Bridges L (1988) A survey of the terrestrial micrometazoa of the Vestfold Hills, Antarctica. I Tardigrada. Hydrobiologia 165:197–208

Miller WR, Miller JD, Heatwole HF (1994a) Tardigrades of the Australian Antarctic Territories: assessing diversity within a sample. Mem Queensl Mus 36:137–145

Miller WR, Heatwole H, Pidgeon RWJ, Gardner GR (1994b) Tardigrades of the Australian Antarctic Territories: the Larsemann Hills, East Antarctica. Trans Am Microsc Soc 113:142–160

Miller WR, Miller JD, Heatwole HF (1996) Tardigrades of the Australian Antarctic Territories: the Windmill Islands, East Antarctica. Zool J Linn Soc 116:175–184

Muslin EH, Homann PH (1992) Light as a hazard for the desiccation-resistant "resurrection" fern *Polypodium polypodioides* L. Plant Cell Environ 15:81–89.

Öquist G, Huner NPA (1991) Effects of cold acclimation on the susceptibility of photosynthesis to photoinhibition in Scots pine and in winter and spring cereals: a fluorescence analysis. Funct Ecol 5:91–100

Petz W (1997) Ecology of the active soil microfauna (Protozoa, Metazoa) of Wilkes Land, East Antarctica. Polar Biol 18:33–44

Petz W, Foissner W (1996) Morphology and morphogenesis of *Lamtostyla edaphoni* Berger and Foissner and *Onychodromopsis flexilis* Stokes, two hypotrichs (Protozoa: Ciliophora) from Antarctic soils. Acta Protozool 35:257–280

Petz W, Foissner W (1997) Morphology and infraciliature of some ciliates (Protozoa, Ciliophora) from continental Antarctica, with notes on the morphogenesis of *Sterkiella histriomuscorum*. Polar Rec 33:307–326

Pickard J, Seppelt RD (1984) Holocene occurrence of the moss *Bryum algens* Card. in the Vestfold Hills, Antarctica. J Bryol 13:209–217
Post A (1990) Photoprotective pigment as an adaptive strategy in the Antarctic moss *Ceratodon purpureus*. Polar Biol 10:241–245
Post A, Adamson E, Adamson H (1990) Photoinhibition and recovery of photosynthesis in Antarctic bryophytes under field conditions. In: Baltscheffsky M (ed) Current research in photosynthesis IV. Kluwer, The Hague, pp 635–638
Rai AN (1988) Nitrogen metabolism. In: Galun M (ed) CRC Handbook of lichenology, vol 1. CRC Press, Boca Raton, pp 201–237
Roser DJ, Melick DR, Seppelt RD (1992) Reductions in the polyhydric alcohol content of lichens as an indicator of environmental pollution. Antarct Sci 4:185–189
Schroeter B, Green TGA, Seppelt RD, Kappen L (1992) Monitoring photosynthetic activity of crustose lichens using a PAM-2000 fluorescence system. Oecologia 92:457–462
Schroeter B, Green TGA, Kappen L, Seppelt RD (1994) Carbon dioxide exchange at subzero temperatures. Field measurements on *Umbilicaria aprina* in Antarctica. Crypt Bot 4:233–241
Schroeter B, Kappen L, Green TGA, Seppelt RD (1997) Lichens and the Antarctic environment: effects of temperature and water availability on photosynthesis. In: Lyons WB, Howard-Williams C, Hawes I (eds) Ecosystem processes in Antarctic ice-free landscapes AA Balkema, Rotterdam, pp 103–118
Scott JJ (1990) Changes in vegetation on Heard Island 1947–1987. In: Kerry KR, Hempel G (eds) Antarctic ecosystems. Ecological change and conservation. Springer, Berlin Heidelberg New York, pp 61–76
Seel WE, Baker NR, Lee JA (1992) Analysis of the decrease in photosynthesis on desiccation of mosses from xeric and hydric environments. Physiol Plant 86:451–458
Selkirk PM, Seppelt RD (1987) Species distribution within a moss bed in greater Antarctica. Symp Biol Hung 35:279–284
Selkirk PM, Seppelt RD, Selkirk DR (1990) Subantarctic Macquarie Island: environment and biology. Cambridge University Press, Cambridge
Seppelt RD (1983) *Cephaloziella exiliflora* (Tayl.) Steph. from the Windmill Islands, continental Antarctica. Lindbergia 9:27–28
Seppelt RD, Hancock RJ (1991) Growth responses of *Ceratodon purpureus* in culture. J Hattori Bot Lab 69:133–145
Seppelt RD, Selkirk PM (1984) Effects of submersion on morphology and the implications of induced environmental modification on the taxonomic interpretation of selected Antarctic moss species. J Hattori Bot Lab 55:273–279
Smith VR, Steenkamp M (1990) Climatic change and its ecological implications at a Subantarctic island. Oecologia 85:14–24
Somersalo S, Krause GH (1990) Reversible photoinhibition of unhardened and cold-acclimated spinach leaves at chilling temperatures. Planta 180:181–187
Sudzuki M (1964) On the microfauna of the Antarctic region. I. Moss-water community at Langhovde. Jpn Antarct Res Exp Sci Rep Ser E 19:1–41
Tarnawski M, Melick D, Roser D, Adamson E, Adamson H, Seppelt R (1992) In situ carbon dioxide levels in cushion and turf forms of *Grimmia antarctici* at Casey Station, East Antarctica. J Bryol 17:241–249
Valanne N (1984) Photosynthesis and photosynthetic products in mosses. In: Dyer AF, Duckett JG (eds) The experimental biology of bryophytes. Academic Press, London, pp 257–273
Walker DA, Everett KR (1987) Road dust and its environmental impact on Alaskan taiga and tundra. Arct Alp Res 19:479–489
Weller G (1992) Antarctica and the detection of environmental change. Philos Trans R Soc Lond 338:201–208

Williams PG, Roser DR, Seppelt RD (1994) Mycorrhizas of hepatics in continental Antarctica. Mycol Res 98:34–36

Wilson ME (1990) Morphology and photosynthetic physiology of *Grimmia antarctici* from wet and dry habitats. Polar Biol 10:337–341

Wodehouse EB, Parker BC (1978) Atmospheric ammonia nitrogen: a potential source of nitrogen eutrophication of freshwater Antarctic ecosystems. In: Parker BC (ed) Terrestrial Biology III. Antarctic Research Series 30, American Geophysical Union, Washington, DC, pp 155–167

Zwartz D, Lambeck K, Bird M, Stone J (1997) Constraints on the former Antarctic ice sheet from sea-level observations and geodynamic modelling. In: Ricci CA (ed) The Antarctic region: geological evolution and processes. Proc VII Int Symp on Antarctic earth sciences, Siena, 1995. Terra Antarctica Publication, Siena, pp 841–843

14 Plant Communities at Dronning Maud Land

H. Kanda, S. Ohtani, and S. Imura

14.1 Introduction

Studies on vegetation in Dronning Maud Land have been performed for many years and most detailed results have been mainly published by Japanese Antarctic Research Expedition (JARE; e.g. Matsuda 1963, 1968; Akiyama 1967; Nakanishi 1977; Kanda 1981, 1986a,b; Newton et al. 1994). This chapter provides a new overview on vegetation research with aspects on their environmental constraints and focuses on moss, lichen and algal communities with different regional aspects in a wider area and their effects on vegetation development.

14.2 Vegetation in Dronning Maud Land

The vegetation is dealt with within the two regions of Antarctic slopes and Antarctic coasts, according to Weyant (1966) and Ohtani and Kanda (Chap. 5).

14.2.1 Vascular Plants

In the Antarctic botanical zone, only two vascular plants, *Deschampsia antarctica* and *Colobanthus quitensis*, are found in the area from the islands of the Scotia Ridge along the west coast of the Antarctic Peninsula south to the current southernmost site at Alamode Island, Terra Firma Islands (68°43' S; Fig. 5.1; Greene 1970; Greene and Holtom 1971; Komárková 1984; Komárková et al. 1985, 1990; Smith and Poncet 1985, 1987). The absence of native vascular plants in the southern Antarctic ('continental' Antarctica) was one of the important reasons for defining the Antarctic botanical zone into the two subzones, the northern and southern Antarctic (Kanda and Komárková 1997). Introduced vascular plants are known, including *Poa annua* at King George Island, South Shetland Islands, *P. annua* and *P. pratensis* growing at Deception Island (Olech 1989).

Ecological Studies, Vol. 154
L. Beyer and M. Bölter (eds.) Geoecology of Antarctic Ice-Free Coastal Landscapes

Fig. 14.1. The Yukidori Valley, Langhovde, East Dronning Maud Land

In Dronning Maud Land, an experimental attempt has been made to grow willow saplings at the Ongul Islands near Syowa Station (Hoshiai 1970; Smith 1996). Plants of *Salix reinii* and *Salix pauciflora* in 1960 and later *Empetrum nigrum* at the same site were transplanted into sandy soil. These species survived one winter and produced new leaves and flowers during the second summer, but failed to survive the following winter. Furthermore, two extraordinary facts were discovered during the three austral summers between 1995 and 1998. Several new shoots of *Salix* species, probably surviving from same site were found in 1998, though it is unknown whether or not similar experiments were performed in this region since 1960 (K. Seto, pers. comm.). However, these were not found in the summer of 2000.

Also, a kind of flowering plant was found in 1995 in a rock fissure about 5 m north of a small prefab house built for a penguin study 20 km south of Syowa Station. The plant grew in a single tuft approximately 20 cm wide and it has a dozen green shoots about 20 cm tall. It had several flowers but no seeds, and has not expanded since this observation in 1995. It was assigned with one of cosmopolitan weedy plants, *Poa* cf. *trivialis* L. This report is still the first record of a flowering plant growing in the southern Antarctic and the species is the first record in the Antarctic botanical zone. Subsequent studies are required to understand how the plant was introduced to Antarctica.

14.2.2 Cryptogamic Organisms

14.2.2.1 West Dronning Maud Land: Heimefrontfjella, Tottanfjella, Vestfjella and Robertskollen

Antarctic Slopes. Very little biological information was available from Heimefrontfjella and Tottanfjella (ca. 75°S, ca. 10°W), until the middle of the twentieth century. During the joint Swedish–British–Norwegian Antarctic Expedition of 1949–1952, O. Wilson was the first who collected mosses, lichens, algae and two species of mites from widely separated inland nunataks in this area, which have been occupied in the recent past by breeding colonies of snow petrels (Dalenius and Wilson 1958). In the austral summer of 1963–1964, lichens, mosses and algae were discovered at many localities and subsequent collections were made both in central Heimefrontfjella and Tottanfjella. Based on these collections, Bowra et al. (1966) recorded 18 species of lichens, two species of mosses, *Sarconeurum glaciale* and *Grimmia* sp. and the green algae *Prasiola* sp.

In the preliminary study of 1987–1988 in Robertskollen, four species of mosses were found, i.e. *Ceratodon purpureus, Bryum pseudotriquetrum, Bryoerythrophyllum recurvirostre*, and *Sarconeurum glaciale* (Ryan et al. 1989). However, no *Grimmia lawiana* was found there, which is the most widespread species in the Gjelsvikfjella and Mühlig-Hofmannfjella to the east. Otherwise, 20 lichen species, including *Usnea antarctica* and *U. sphacelata* and 17 algal species, occur at Robertskollen (Ryan et al. 1989). In this area, four birds were recorded: the Antarctic petrel (*Thalassoica antarctica*), the snow petrel (*Pagodroma nivea*), Wilson's storm petrel (*Oceanites oceanicus*) and the south polar skua (*Catharacta maccormicki*). The snow petrels are the only proven breeding species in this region with an estimated population of 600 breeding pairs (Ryan et al. 1989).

Three species of mites have been recorded, but no springtails (Collembola) are reported from Robertskollen. Sømme (1986) expresses that the apparent absence of springtails is surprising considering the relative abundance at adjacent sites such Heimefrontfjella and Tottanfjella.

Lichens and mosses in Heimefrontfjella and Vestfjella have been recently studied by the Swedish expedition (Thor 1997). Vestfjella is up to 1100 m high and is 120 km from the coast, whereas Heimefrontfjella is up to 2500 m high, and is 300 km from the coast. The Swedish Station Svea is situated at the northernmost end of the nunatak Wasa in the southern part of the nunatak Haldorsentoppen (74°34'36" S, 11°13'24" W, at 1245 m altitude), and the Station Wasa is located in the southern part of the nunatak Basen (73°02'34" S, 13°24'50" W, 450 m altitude). Both stations are located at 135 km long mountain ridges lying parallel to the coast and consisting of nunataks. Thor (1997) established permanent plots with lichens and mosses and recorded 18 species of lichens from Heimefrontfjella and 13 species of lichens from Vestfjella. Ornithocoprophilous species found within permanent plots in Heimefrontfjella are *Caloplaca citrina, C. saxicola, Candelariella flava, Lecanora expectans, Physcia caesia* and *Xanthoria mawsoni.*

14.2.2.2 Central Dronning Maud Land: Gjelsvikfjella, Mühlig-Hofmannfjella and Schirmacher Oasis

Antarctic Slopes. Both mountain areas of Gjelsvikfjella and Mühlig-Hofmannfjella were discovered by air during the Schwabenland Expedition in 1939, but the terrestrial flora was investigated first by the joint Swedish–British–Norwegian Antarctic Research Expedition in 1949–1952 (Engelskjøn 1985). Subsequent lichen and moss samples were collected by the Soviet Antarctic Expedition in 1959–1961 and studied by Golubkova (1969) and Savich-Ljubitskaya and Smirnova (1964). Extensive ecological and geological studies were performed by Engelskjøn who joined the land party of Camp Norway 5 of the Norwegian Antarctic Research Expedition in 1984–1985, working from the eastern outline of the H.U. Sverdrupfjella Mountains through Gjelsvikfjella, and to the western parts of Mühlig-Hofmannfjella. According to his ecological studies, the plant communities of 5 subformations and 11 associations are described (Engelskjøn 1985; Table 14.1).

'Svarthamaren', located in Mühlig-Hofmannfjella, was originally described as SSSI No. 23 in 1987. The area is characterised by a colony of Antarctic petrels, which is the largest known inland seabird colony on the Antarctic continent (SCAR 2000). Flora and vegetation are sparse compared with other areas in Mühlig-Hofmannfjella and Gjelsvikfjella to the west of this site. The foliose green alga, *Prasiola crispa*, lichens, *Candelariella hallettensis* (=*C. antarctica*), *Rhizoplaca* (=*Lecanora*) *melanophthalma*, *Umbilicaria* spp. and *Xanthoria* spp. dominate this area. Areas covered with *Prasiola* show a rich fauna of mites (*Eupodes anghardis*, *Tydeus erebus*), protozoa, nematodes and rotifera.

Antarctic Coasts. The freshwater algae in Schirmacher Oasis near Novolazarevskaya Station were studied by Komárek and Ruzicka (1966). They reported *Scenedesmus armatus* as a common species, almost all of the other species being distributed in the Subantarctic regions. At Schirmacher Oasis, the primary production of lake phytoplankton (Kaup 1986, 1988, 1994) and nitrogen fixation by cyanobacteria associated with moss communities have been measured (Pandey et al. 1992).

Savich-Ljubitskaya and Smirnova (1964) described a new moss species, *Plagiothecium simonovii*, in the same area. This pleurocarpous moss is found at a depth of 32.3 m in Lake Glubokoje. Bednarek-Ochyra et al. (1999) reported that this species should be assigned to *Plagiothecium georgico-antarcticum* occurring in the Subantarctic regions and South America.

Table 14.1. Structure and composition of the vegetation in the two nunataks, Gjeldvikfjella and Mühlig-Hofmannfjella. (After Engelskjøn 1985)

1. Unicellular alga subformation
    - a. *Chlamydomonas* association
    - b. Endolithic alga association
2. Thallose alga subformation
    - a. Encrusted cyanophycean–chlorophycean association
    - b. *Prasiola-Ulothrix* association
3. Epilithic lichen subformation
    - a. *Lecanora* sp.–*Lecidea* spp.–*Umbilicaria decussata* association
    - b. *Xanthoria elegans* association
        - i. *Xanthoria elegans–Candelariella hallettensis–Nostoc* sociation
        - ii. *Xanthoria elegans–Physcia caesia–Umbilicaria aprina* sociation
        - iii. *Xanthoria elegans–Buellia frigida–Acarospora gwynnii* sociation
        - iv. *Xanthoria–Candelaria* sociation
    - c. *Rhizocarpon geographicum* association
4. Fruticose lichen subformation
    - a. *Usnea sphacelata–Pseudephebe minuscula* association
    - b. *Usnea sphacelata–Grimmia lawiana* association
5. Moss cushion subformation
    - a. *Grimmia lawiana* association
    - b. *Sarconeurum glaciale* association

Wafar and Untawale (1983) reported one bryaceous moss and a lichen, *Acarospora* sp. close to the unmanned station (70°45'12" S, 11°38'13" E) of the first Indian Expedition (1981–1982). Tewari and Pant (1996) reported about eight moss samples collected from Zub Lake and Dalini Lake near the Indian station Maitri in the Schirmacher Oasis. These mosses are identified preliminarily as species of *Bryum*, *Grimmia*, *Pohlia* and *Leptobryum*. The last species is interestingly found to be gemmiferous. The pale-green, long, wiry shoots bear reddish-brown, multicellular, globose gemmae characteristically growing on long rhizoidal stalks originating from the leaf axil. Although the present authors have not examined this moss specimen yet, it may coincide with *Leptobryum* sp. (*Leptobryum* cf. *stellatum*) which has been found on some lake bottoms near Syowa Station, East Dronning Maud Land (Imura et al. 1999).

The monograph, *The Schirmacher Oasis, Queen Maud Land, East Antarctica, and Its Surroundings*, was compiled by the Alfred Wegener Institute for Polar and Marine Research (AWI) in 1995. In this work, Richter (1995) presents a checklist with up to 230 species of algae, 26 species of lichens and 7 species of mosses from Schirmacher Oasis and its surroundings, Untersee

Oasis included. Algae are described in the extensive works of Pankow et al. (1997).

14.2.2.3 East Dronning Maud Land: Sør-Rondane, Yamato Mountains (Dronning Fabiolafjella) and the vicinities of Syowa Station

Antarctic Slopes. Sør-Rondane was discovered in 1937 by the Norwegian Expedition led by Lars Christensen. Lichens in these mountains were first reported by Dodge (1962). He reported ten species based on the lichen samples collected by the Belgian Antarctic Expedition in 1959–1961. The presence of nests of snow petrels, Antarctic skuas, Wilson's storm petrels and Antarctic petrels in the Sør-Rondane is mentioned (Loy 1962; Van Autenboer 1964).

The biological survey in the western part of the Sør-Rondane was carried out by the Japanese Antarctic Research Expedition in 1988–1989 (JARE-30 and 31) (Ohyama et al. 1991). In this project, eight nunataks, Brattnipene, Otto Borchgrevinkfjellet, Nils Larsenfjellet, Pingvinane, Lunckeryggen, Vesthaugen, Selungen, Romnanesfjellet were surveyed and nests of snow petrels, communities of lichens, algae and mosses were described. Kanda (1987) reports one moss species, *Grimmia lawiana*, from Tanngarden in the western part of the Sør-Rondane based on the collection by JARE 27 (1985–1986). Later, in JARE 30, *G. lawiana* was observed in association with a rich lichen community (H. Miyawaki, pers. comm.). In the comparative study on invertebrates between the Sør-Rondane (slope) and Sôywa Station area (coast), Ohyama and Hiruta (1995) mention that a collembola, *Cryptopygus sverdrupi*, is present in outcrops of the Sør-Rondane, but surprisingly, this collembola was never found in Syowa Station area, although more prominent vegetation is present there.

Two species of lichens, *Acarospora gwynii* and *Lecidea* sp., have been reported in Yamato mountains (Dronning Fabiolafjella), located ca. 250 km inland from the coast of Lützow-Holm Bay (Nakanishi 1977).

14.2.2.4 Vegetation and Moss Communities in the Vicinities of Syowa Station, East Dronning Maud Land

Antarctic Coasts. The main vegetation formations are found throughout the northern Antarctic, as this region has most subordinated units (Holdgate 1964; Longton 1988):

1. Grass–moss and grass–lichen–moss communities with *Deschampsia antarctica* and *Colobanthus quitensis*, which are best developed on zonal surfaces
2. Lichen–moss–lichen and moss communities, including
   a) Fruticose lichen and moss cushions
   b) Lichen crusts

c) *Polytrichum alpestre–Chorisodontium aciphyllum* moss turf communities in which peat sometimes reaches more than 1 m in thickness
d) *Sanionia* (as *Drepanocladus*)–*Brachythecium–Calliergon* (as *Acrocladium*), usually thin moss carpets
e) *Andreaea–Usnea* rock surface
f) Waterlogged *Calliergidium–Calliergon–Sanionia* moss carpets

In contrast, the terrestrial habitats of the southern Antarctic are dominated by microorganisms, i.e. microalgae and cyanobacteria, as well as lichens, which are the major primary producers (Vishniac 1993). Carpet- and hummock-forming mosses are scarce, liverworts are rare, and vascular plants are absent. The principal communities of the southern Antarctic are as follows (Rudolph 1963; Longton 1973, 1974, 1988; Aleksandrova 1977; Smith 1993; Kobayashi 1974):
1. Moss–lichen and moss
2. Lichen
3. Lichen–algae and lichen–moss–algae which occur near bird colonies
4. Algae–bacteria and algae

Longton (1973) classified the vegetation at coastal sites near McMurdo Station into four subformations, six associations, and 11 sociations of the Antarctic non-vascular cryptogam tundra formation (Table 14.2). In East Dronning Maud Land, including Enderby Land, Kobayashi (1974) classified the vegetation in the vicinities of Syowa Station into moss, moss–lichen, lichen, alga–lichen, and alga. Nakanishi (1977) and Kanda (1981, 1986b, 1987) show

Table 14.2. Antarctic non-vascular cryptogam tundra formation. (After Longton 1973)

1. Fruticose and foliose lichen subformation a. *Usnea* association, incl. 2 sociations
2. Crustaceous lichen subformation
    a. *Caloplaca–Xanthoria* association, incl. 2 sociations
3. Short moss turf and cushion subformation (incl. encrusted moss subformation)
    a. *Bryum* association
        i. *Bryum argenteum* sociation
        ii. *Pottia heimii* sociation
        iii. *Bryum argenteum–Pottia heimii* sociation
        iv. *Bryum* sp. sociation
    b. *Sarconeurum* association
        i. *Sarconeurum glaciale–Caloplaca darbishirei* sociation
4. Alga subformation
    a. *Prasiola* association, incl. 1 sociation
    b. *Nostoc* association, incl. 1 sociation

Table 14.3. Antarctic non-vascular cryptogam tundra formation. (After Nakanishi 1977 and Kanda 1986a,b)

1. Crustaceous lichen subformation
    a. *Buellia* spp.–*Umbilicaria aprina* association
        i. *Buellia frigida* sociation
        ii. *Buellia frigida–Umbilicaria aprina* sociation
        iii. *Xanthoria elegans–Umbilicaria aprina* sociation
2. Encrusted moss subformation
    a. Lichen encrusted *Bryum* spp.–*Ceratodon* spp.–*Pohlia nutans* association
        i. *Rinodina olivaceobrunnea* sociation
3. Short moss turf and cushion subformation
    a. *Bryum pseudotriquetrum* association
        i. *Bryum pseudotriquetrum* sociation
        ii. *Ceratodon purpureus* sociation
        iii. *Ceratodon purpureus–Bryum pseudotriquetrum* sociation
        iv. *Bryum argenteum* sociation
        v. *Bryum argenteum–Bryum pseudotriquetrum* sociation
        vi. *Pottia heimii–Bryum pseudotriquetrum* sociation
        vii. *Pottia austro-georgica–Bryum pseudotriquetrum* sociation
        viii. *Grimmia lawiana* sociation
        ix. *Grimmia lawiana–Ceratodon purpureus* sociation

13 units of communities in the Syowa Station area modifying the systems by Longton (1973) (Table 14.3). The vicinity of Syowa Station implies ice-free areas along the Sôya Coast, Lützow-Holm Bay, and Prince Olav Coast (Fig. 14.1). The Japanese Antarctic Research Expedition (JARE) has studied the community with special reference to mosses.

14.2.3 Moss Habitats and Communities

Nakanishi (1977) classified moss habitats near the Syowa Station area in relation to water availability into six types: small snowdrift type, medium snowdrift type, large snowdrift type, lakeshore type, seepage type and stream type (Fig. 14.2). The moss habitat has water supplies from a small-scale snowdrift which is the most common in the area. In his phytosociological study, Nakanishi (1977) recognises six sociations near Syowa Station. *Ceratodon purpureus* sociation and *C. purpureus–Bryum pseudotriquetrum* sociation are widely found in various habitat types which he has described earlier. In particular, *C. purpureus* sociation is distributed more frequently in rather dry habitats of drift types, and *Bryum argenteum–B. pseudotriquetrum* sociation, *Pottia*

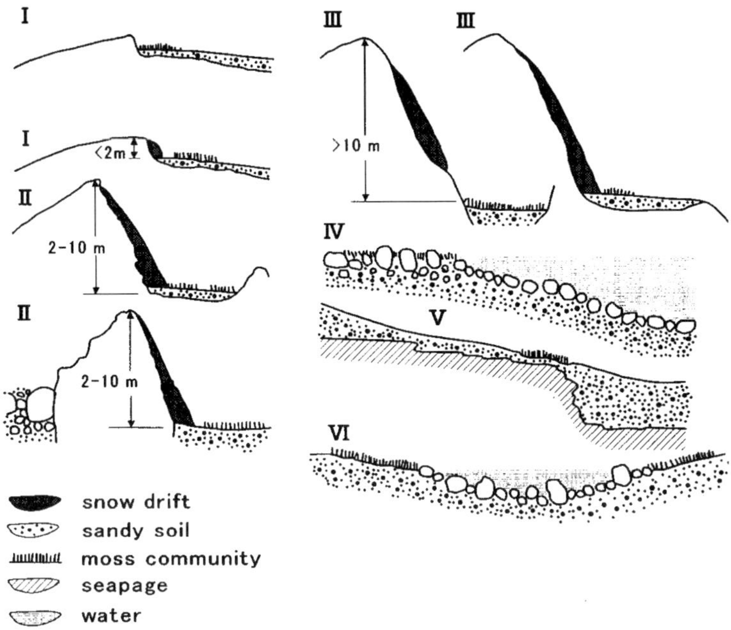

Fig. 14.2. Habitat types classified by the pattern of water supply (after Nakanishi 1977). *I* Small snowdrift type, *II* medium snowdrift type, *III* large snowdrift type, *IV* lake-shore type, *V* seepage type, *VI* stream type

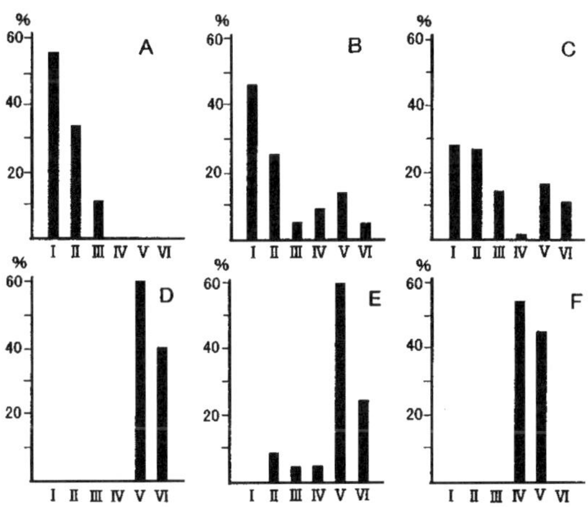

Fig. 14.3. Occurrence frequency of moss sociation in each habitat type (after Nakanishi 1977). *I-VI* Mean habitat types, respectively. **A** *Ceratodon purpureus* sociation. **B** *Bryum pseudotriquetrum* sociation. **C** *Ceratodon purpureus–Bryum pseudotriquetrum*. **D** *Bryum argenteum–B. pseudotriquetrum* sociation. **E** *Pottia heimii–Bryum pseudotriquetrum* sociation. **F** *Pottia austro-georgica–Bryum pseudotriquetrum* sociation

heimii–B. pseudotriquetrum sociation and *Pottia austro-georgica–B. pseudotriquetrum* sociation are developed mainly in the habitats with a constant supply of water (Fig. 14.3). He notes that the most ornithocoprophilic community is the *B. argenteum–B. pseudotriquetrum* sociation, while the *B. pseudotriquetrum* sociation and the *Pottia heimii–B. pseudotriquetrum* sociation both seem to have some relation to the excrement of seabirds. Consequently, he suggests that these sociations could be integrated into the *Bryum* spp.–*Ceratodon* spp.–*Pohlia nutans* association in the northern Antarctic as established by Gimingham and Smith (1970).

Kanda (1981) described three new sociations, i.e. *Grimmia lawiana* sociation and *G. lawiana–Ceratodon purpureus* sociation in Cape Ryûgû, Prince Olav Coast and *Bryum argenteum* sociation in Einstodingen, Sôya Coast. The *Grimmia* type of the *C. purpureus* sociation, which is described by Nakanishi (1977), is closely related to the *G. lawiana–C. purpureus* sociation. Since the *G. lawiana* sociation was described first at Cape Ryûgû in 1981, the typical sociation has been found in some outcrops attached to the continental glacier such as Rundvågshetta and Strandnibba (Kanda 1986b). In these communities, moss colonies usually occur in the moraine, grasping gravel beneath the cushions, attaching to stone surfaces, or sometimes they are turned out and scattered on the moraine. The habitats forming this sociation are strongly exposed to sunshine and often have no water supply available.

The development of the *B. argenteum* sociation described by Kanda (1986b) is influenced not only by calcareous bedrock, but also by nutrient input from seabirds. Small isolated islands near Syowa Station, such as Ongulkalven, Mame-jima Island, Rumpa and Ongul Islands generally have scarce moss vegetation consisting of only one or two species (Matsuda 1964; Kobayashi 1974; Shimizu 1977; Nakanishi 1977). According to Nakanishi (1977), the moss vegetation in the Ongul Islands consists of three sociations, *C. purpureus* sociation, *B. pseudotriquetrum* sociation, and *C. purpureus–B. pseudotriquetrum* sociation. Although these islands have penguin rookeries supplying an effective nutrient for the vegetation, there is no remarkable moss vegetation. In the case of Einstodingen, without penguin rookery, the rich moss vegetation is considered to be due to snow petrel excrement. However, the effects of nutrients supplied by snow petrel and penguin excrement on the moss vegetation have to be studied. The relationships between nutrition at a penguin rookery and soil organisms or vegetation are discussed in Ohtani and Kanda (Chap. 5).

14.2.3.1 Topographic Features Around Communities

In Dronning Maud Land, the ecological investigations of the terrestrial communities were first demonstrated by Matsuda (1963, 1964, 1968) at Syowa Station in East Ongul Island, Sôya Coast. He mentions that the moss habitats are mostly found in the south and west sides of hills according to the topography of East Ongul Island. The environmental factors governing the distribution of moss habitats of the island are considered to be the direction of the north easterly prevailing wind and slope toward the southwest which receives more sunlight and water supply from snow drifts formed on the leeward side. Consequently, the moss communities have been distributed more westerly. Later, Fukushima (1968) and Kobayashi (1974) proved the distribution pattern in West Ongul Island. Kobayashi (1974) daringly fixed the boundary using the existence of moss vegetation in the north and south in the centres of both the islands. Afterward, Shimizu (1977) showed that the southwest slope direction of moss communities on West Ongul Island is the most frequent. He further indicated that the occurrence of plant communities is related to the slope direction and to the solar radiation. However, Yamanaka and Sato (1977) mentioned that the direction of the prevailing wind depends upon ice-free areas and that the moss communities on small islands are distributed generally in the southwest. They are distributed in the west in continental or peninsular ice-free areas, suggesting there is no significant correlation between the slope direction and the plant community on isolated small islands or in small-scale ice-free areas. In the case of the ice-free areas with comparatively gentle slopes or hills, and without wetland such as lakes or streams, the idea of Matsuda (1968) basing relationships on prevailing wind, snowdrift, slope and sunlight may be effective when considering distribution patterns of vegetation. Even in large ice-free areas such as Langhovde, there is hardly any moss vegetation in the vast northern area, but much in the southern part (Kanda and Inoue 1994).

14.2.4 Soil Algal Communities in the Vicinities of Syowa Station, East Dronning Maud Land

Soil algae are widespread and occur in ice-free areas in maritime and continental Antarctica (Broady 1996). Several floristic studies have been carried out in the vicinity of Syowa Station (Akiyama 1967, 1974; Ohtani et al. 1991 2000). Forty-two genera have been identified from ice-free areas in the region. Cyanobacteria, chlorophytes, xanthophytes and diatoms commonly occur in soils (Table 14.4) similar to the Vestfold Hills (Broady 1986).

Macroscopic colonies of *Prasiola crispa* occur at eutrophicated sites such as near penguin rookeries in Ongulkalven and Rumpa (Matsuda 1968; Akiyama 1974; Hoshiai and Matsuda 1979), near nests of snow petrels at Langhovde (Ohtani, unpubl. data) and close to a urinal tank at Syowa Station (Ohtani et al. 2000). Macroscopic colonies of *Nostoc commune* occur widely on wet soil surfaces along a stream in Yukidori Valley, Langhovde (Ohtani, unpubl. data). *Aphanocapsa grevillei*, *Cyanothece aeruginosa* and *Nostoc punctiforme* are commonly found amongst macroscopic colonies on soils in the Ongul Islands (Akiyama 1967).

Table 14.4. Comparison of genera of soil algae found in different studies in the vicinity of Syowa Station, Antarctica. + Present; – absent

Taxa	Akiyama (1967) Ongul Island	Akiyama (1974) LHB[b]	Ohtani et al. (1991) Langhovde[c]	Ohtani et al. (2000) Ongul Island
Cyanophyceae[a]				
Aphanothece	–	+	–	–
Aphanocapsa	+	+	–	+
Gloeothece	–	+	–	–
Synechocystis	+	–	–	–
Synechococcus	+	+	+	–
Gloeocapsa	–	–	+	–
Oscillatoria	+	–	–	–
Phormidium	+	+	–	+
Lyngbya	–	–	+	–
Crinalium	–	–	–	+
Nostoc	+	+	+	+
Anabaena	–	+	–	–
Stigonema	–	+	+	–
Tolypothrix	+	+	–	
Chlorophyta				
Chlamydomonas	+	+	–	+
Carteria	+	–	–	–
Chlorococcum	+	+	–	–
Tetracystis	–	+	–	–
Macrochloris	–	–	–	+
Characium	+	+	–	–
Chlorella	+	–	–	–
Dictyosphaerium	+	–	–	–
Oocystis	–	–	–	+

Most algal species from Syowa Station area are microscopic. They can be isolated using culture methods from sandy soils, and are cosmopolitan or common elsewhere in Antarctica. Akiyama (1967) reported that *Leptolyngbya tenuis*, *Nostoc punctiforme*, *Xanthonema exile*, *Stichococcus bacillaris*, *Raphidonema pyrenoidifera* and *Navicula muticopsis* were the common species in the Ongul Islands. Thirty years later, Ohtani et al. (2000) reported that *Leptolyngbya* cf. *battersii*, *Leptolyngbya tenuis*, *Phormidium autumnale*, *Navicula muticopsis*, *Botrydiopsis* spp., *Xanthonema* spp. and *Macrochloris multinucleata*,were common in East Ongul Island. Although most soil algae found in the Ongul Islands are encountered only from soils, several Cyanobacteria such as *Leptolyngbya tenuis*, *Phormidium autumnale*, and diatoms such as *Navicula muticopsis* and *Hantzschia amphioxys* occur in ponds and streams (Oguni and Takahashi 1989; Hirano 1983). Studies on soil algae at locations other than the Ongul Islands have been made only once at Yukidori Valley, Langhovde. The in situ contact slide method (Ohtani et al. 1991) has shown *Actinotaenium cucurbita* and *Pinnularia borealis* as dominating species of soils

Table 14.4 (*Continued*)

Taxa	Akiyama (1967) Ongul Island	Akiyama (1974) LHB[b]	Ohtani et al. (1991) Langhovde[c]	Ohtani et al. (2000) Ongul Island
Bracteacoccus	+	–	–	–
Kentrosphaera	+	–	–	–
Chlorosphaera	+	–	–	–
Prasiola	–	–	–	+
Chlorosarcina	–	+	–	–
Klebsormidium	+	+	–	+
Stichococcus	+	+	+	+
Raphidonema	+	+	–	+
Actinotaenium	+	–	+	–
Xanthophyceae				
Pleurochloris	–	+	–	–
Monallantos	–	+	–	–
Monodus	+	+	–	–
Botrydiopsis	+	+	–	+
Xanthonema	+	+	–	+
Heterococcus	+	+	–	+
Bacillariophyceae				
Achnanthes	–	–	–	+
Navicula	+	+	+	+
Pinnularia	+	+	+	–
Hantzschia	+	+	+	–
Total	27	26	10	16

[a] Based on Geitler's system.
[b] Soil algae from ice-free areas in Lützow-Holm-Bay (no description of sites).
[c] Observations of contact slide method in situ without culture study.

adjacent to a stream. These species also occur in moss colonies (Ohtani 1986) and in other streams (Hirano 1983). The soil algal flora at ice-free areas in Lützow-Holm Bay remains only poorly known. Further floristic studies are needed, especially in the large ice-free areas such as Langhovde and Skarvsnes.

References

Akiyama M (1967) On some Antarctic terrestrial and subterranean algae. Mem Fac Educ Shimane Univ 1:36–56

Akiyama M (1974) A preliminary note on some algae found in the ice-free area of the coastal region of Lützow-Holm Bay, Antarctica. Mem Fac Educ Shimane Univ 3:37–50 (in Japanese with English summary)

Aleksandrova VD (1977) Geobotanicheskoe raionirovanie Arktiki i Antarctiki (Geobotanical Regions of the Arctic and Antarctic) Komarovskie Chteniya 29, Akademiya

Nauk SSSR Nauka; Leningrad (Translation by D Love). Cambridge University Press, New York, pp 693–696)

Bednarek-Ochyra H, Lewis Smith RI, Ochyra R (1999) The genus *Plagiothecium* (Plagiotheciaceae, Bryopsida) in Antarctica. Bryobrothera 5:55–64

Bowra GT, Holdgate MW, Tilbrook PJ (1966). Biological investigations in Tottanfjella and central Heimefrontfjella. Br Antarct Surv Bull 9:63–70

Broady PA (1986) Ecology and taxonomy of the terrestrial algae of the Vestfold Hills. In: Pickard J (ed) Antarctic oasis, terrestrial environment and history of the Vestfold Hills. Academic Press, New York pp 165–202

Broady PA (1996) Diversity, distribution and dispersal of Antarctic terrestrial algae. Biodiv Conserv 5:1307–1335

Dalenius P, Wilson O (1958) On the soil fauna of the Antarctic and of the sub-Antarctic islands. The Orbatidae (Acari). Arkiv för Zoologi 2:393–425

Dodge CW (1962) Expedition Antarctique Belge Lichens. Bull Jard Bot Etat Brux 32:301–308

Engelskjøn T (1985) Botany of Gjelsvikfjella and Mühlig-Hofmannflella, Dronning Maud Land. In: Orheim O (ed) Report of the Norwegian Antarctic Research Expedition 1984/85, Norsk Polarinst Rapp 22:43–54

Fukushima H (1968) Algal vegetation of the Kasumi Rock ice-free areas, Prince Olav Coast, Antarctica. Antarct Rec 31:73–86

Gimingham CH, Lewis Smith RI (1970) Bryophyte and lichen communities in the maritime Antarctic. In: Holdgate MW (ed) Antarctic ecology, vol 2. Academic Press, New York, pp 752–785

Golubkova NS (1969). Lichens. In: Atlas Antarktiki 2. Hydrometeorological Publishing House, Leningrad. pp 488–492

Greene SW (1970) Studies on *Colobanthus quitensis* (Kunth) Bartl. and *Deschampsia antarctica* Desv.: I. Introduction. Br Antarct Surv Bull 23:19–24

Greene DM, Holtom A (1971) Studies on *Colobanthus quitensis* (Kunth) Bartl. and *Deschampsia antarctica* Desv.: III. Distribution, habitats and performance in the Antarctic botanical zone. Br Antarct Surv Bull 26:1–29

Hirano M (1983) Freshwater algae from Skarvsnes, near Syowa Station, Antarctica. Mem Natl Inst Polar Res Ser E 35:1–31

Holdgate MW (1964) Terrestrial ecology in the maritime Antarctic. In: Carrick R, Holdgate MW, Prevost J (eds) Biologie Antarctique. Proc 1st SCAR Symposium, Paris, France, 1962. Hermann, Paris, pp 181–194

Hoshiai T (1970) Willow cultivation at Syowa Station. Polar News 24(2):16–20

Hoshiai T, Matsuda T (1979) Adelie penguin rookeries in the Lützow-Holm Bay area and relation of rookery to algal biomass in soil. Mem Natl Inst Polar Res Spec Issue 11:140–152

Imura S, Bando T, Saito S, Seto K, Kanda H (1999) Benthic moss pillars in Antarctic lakes. Polar Biol 22:137–140

Kanda H (1981) Flora and vegetation of mosses in ice-free areas of Soya Coast and Prince Olav Coast, East Antarctica. Hikobia Suppl 1:91–100

Kanda H (1986a) Moss communities in some ice-free areas along the Soya Coast, East Antarctica. Mem Natl Inst Polar Res Spec Issue 44:229–240

Kanda H (1986b) Seasonal changes of some environmental factors around the moss vegetation near Syowa Station, East Antarctica. Mem Natl Inst Polar Res Ser E 37:17–26

Kanda H (1987) Ecology of the moss vegetation in the Syowa Station area, Enderby Land, Antarctica. Symp Biol Hung 35:259–267

Kanda H, Inoue M (1994) Ecological monitoring of moss and lichen vegetation in the Syowa Station region, Antarctica. Proc NIPR Symp Polar Biol 7:221–231

Kanda H, Komárková V (1997). Antarctic terrestrial ecosystems. In: Wielgolaski FE (ed) Polar and alpine tundra. Ecosystems of the world, vol 3. Elsevier, Amsterdam, pp 721–761

Kaup E (1986) Die Primärproduktion in Seen der Schirmacher-und Unterseeoase (Ostantarktika) in der Saison 1983/84. Geodät Geophys Veröffentl (Berl) I, 13:56–59

Kaup E (1988) Loads and concentrations of nutrients in the lakes of the Schirmacher Oasis during the season 1983/84. In: Martin J (ed) Limnological studies in Queen Maud Land (East Antarctica), Academy of Sciences of Estonia,Tallin, pp 66–77

Kaup E (1994) Annual primary production of phytoplankton in Lake Verkhneye, Schirmacher Oasis, Antarctica. Polar Biol 14:433–439

Kobayashi, K (1974) A preliminary report on the vegetation of the Prince Olav Coast, Antarctica. Antarct Rec 51:18–28

Komárek J, Ruzicka J (1966) Freshwater algae from a lake in proximity of the Novolazarevskaya Station, Antarctic. Preslia Praha 38:237–244

Komárková V (1984) Studies of plant communities of the Antarctic Peninsula near Palmer Station. Antarct J US 19:180–182

Komárková V, Scott S, Toi M (1985) Summer die-back due to desiccation of *Deschampsia antarctica* Desv. and *Colobanthus quitensis* (Kunth) Bartl. on the largest Stepping Stone Island, Arthur Harbor, Anvers Island, Antarctic Peninsula. Bull Ecol Soc Am 66:211

Komárková V, Poncet S, Poncet J (1990) Additional and revisited localities of vascular plants *Deschampsia antarctica* Desv. and *Colobanthus quitensis* (Kunth) Bartl. in the Antarctic Peninsula area. Arct Alp Res 22:108–113

Lewis Smith RI (1993) Dry coastal ecosystems of Antarctica. In: van del Maarel E (ed) Ecosystems of the world, 2A. Dry coastal ecosystems. Polar regions and Europe. Elsevier, Amsterdam, pp 51–71

Lewis Smith RI (1996) Introduced plants in Antarctica: potential impacts and conservation issues. Biol Conserv 76:135–146

Lewis Smith RI, Poncet S (1985) New southernmost record for Antarctic flowering plants. Polar Rec 22:425–427

Lewis Smith RI, Poncet S (1987) *Deschampsia antarctica* and *Colobanthus quitensis* in the Terra Firma Islands. Br Antarct Surv Bull 74:31–36

Longton RE (1973) A classification of terrestrial vegetation near McMurdo Sound, continental Antarctica. Can J Bot 51:2339–2346

Longton RE (1974) Microclimate and biomass in communities of the Bryum association on Ross Island, Continental Antarctica. Bryologist 77:109–127

Longton RE (1988) Biology of polar bryophytes and lichens. Cambridge University Press, Cambridge

Loy W (1962) An ornithological profile from Iceland to the Antarctic. Gerfaut 52:626–640

Matsuda T (1963) The distribution of mosses on East Ongul Island, Antarctica. Hikobia 3:254–265

Matsuda T (1964) Ecological studies on the community of mosses at Langhovde region, Antarctica. Antarct Rec 21:25–38

Matsuda T (1968) Ecological study of the moss community and microorganisms in the vicinity of Syowa Station, Antarctica. Jpn Antarct Res Exp Sci Rep Ser E 29:1–58

Nakanishi S (1977) Ecological studies of the moss and lichen communities in the ice-free areas near Syowa Station, Antarctica. Antarct Rec 59:69–96

Newton IP, Cooper J, Mehlum F, Thor G (1994) A bibliography of terrestrial biological research in Dronning Maud Land, Antarctica. S Afr J Antarct Res 24 (1/2):111–124

Oguni A, Takahashi E (1989) Floristic studies on algae from inland waters of Antarctica: II. Lake O-ike, West Ongul Island. Proc NIPR Symp Polar Biol 2:154–166

Ohtani S (1986) Epiphytic algae on mosses in the vicinity of Syowa Station, Antarctica. Mem Natl Inst Polar Res Spec Issue 44:209–219

Ohtani S, Akiyama M, Kanda H.(1991) Analysis of Antarctic soil algae by direct observation using the contact slide method. Nankyoku Shiryo (Antarct Rec) 35:285–295

Ohtani S, Suyama K, Yamamoto H, Aridomi Y, Itoh R, Fukuoka Y (2000) Distribution of soil algae at the monitoring sites in the vicinity of Syowa Station between austral summers of 1992/1993 and 1997/1998. Polar Biosci 13:113–132

Ohyama Y, Hiruta S (1995) The terrestrial arthropods of Sor Rondane in eastern Dronning Maud Land, Antarctica, with biogeographical notes. Polar Biol 15:341–347

Ohyama Y, Higashi S, Miyawaki H, Hiruta S, Kanda H (1991) Biological surveys in the Sor Rondane Mountains, Dronning Maud Land, in 1989 and 1990 seasons. Antarct Rec 35:238–246

Olech M (1989) Lichens from the Admiralty Bay region, King George Island (South Shetland Islands, Antarctica). Acta Soc Bot Pol 58:493–512

Pandey KD, Kashyap AK, Gupta RK (1992) Nitrogen fixation by cyanobacteria associated with moss communities in Schirmacher Oasis, Antarctica. Isr J Bot 41:187–198

Pankow H, Haendel D, Richter WE (1991) Die Algenflora in der Schirmacheroase (Ostantarctica). Nova Hedwigia 103:1–150

Richter W (1995) The Schirmacher Oasis, Queen Maud land, East Antarctica, and its surrounding. In: Bormann P, Fritzsche D (ed) Justus Perthes Verlag, Gotha

Rudolph ED (1963) Vegetation of Hallett Station area, Victoria Land, Antarctica. Ecology 44:585–586

Ryan PG, Watkins BP, Lewis Smith RI, Dastych H, Eicker A, Foissner W, Heatwole H, Miller WR, Thompson G (1989) Biological survey of Robertskollen, western Dronning Maud Land: area description and preliminary species lists. S afr T Antarkt. 19:10–20

Savich-Ljubitskaya LI, Smirnova ZN (1964) Glubokovodnyi predstavitel' roda Plagiothecium Br. et Sch. v Antarktide [A deep-water member of the genus *Plagiothecium* Br. et Sch. in Antarctica]. Inf Biul Sov Antarkt Eksped 49:33–39

SCAR (2000) Measures, decisions and resolutions adopted at the XXIII Antarctic Treaty Consultative Meeting, Lima, Peru, 24 May–4 June 1999. SCAR Bull 136:1–5

Shimizu H (1977) Vegetational distribution and habitats on West Ongul and Teoya Islands, Antarctica. Antarct Rec 59:97–107

Sømme L (1986) Ecology of *Cryptopygus sverdrupi* (Insecta: Collembola) from Dronning Maud Land, Antarctica. Polar Biol 6:179–184

Tewari SD, Pant G (1996) Some moss collections from Dakshin Gangotri, Antarctica. Bryol Times 91:7

Thor G (1997) Establishment of permanent plots with lichens and mosses for monitoring local human impact on environment in Heimefrontfjella and Vestfjella, Dronning Maud Land, Antarctica. Antarct Rec 41(3):652–672

Van Autenboer T (1964) Quelques aspects geographiques des Sor-Rondane, Terre de la Reine Maud, Antarctique. La Geogr 3:93–108

Vishniac H S (1993) The microbiology of Antarctic soils. In: Friedmann EI (ed) Antarctic microbiology. Wiley-Liss, New York, pp 297–341

Wafar S, Untawale AG (1983) Flora of Dakshin Gangotri in Antarctica. Scientific Report of First Indian Expedition to Antarctica. Terrestrial Publication 1. Dept. of Ocean Development, South Block, New Delhi, India, pp 182–185

Weyant WS (1966) The Antarctic climate. In: Tedrow JCF (ed.) Antarctic soils and soil forming processes. Antarctic Research Series, vol 8. Am Geophys Union, Washington, DC, pp 47–59

Yamanaka M, Sato K (1977) Distribution of terrestrial plant communities near Syowa Station in Antarctica, with special reference to water supply and soil property. Nankyoku Shiryo (Antarct Rec) 59:54–67 (in Japanese with English sumary)

15 Soil Fauna of Antarctic Coastal Landscapes

I.D. Hogg and M.I. Stevens

15.1 Introduction

The endemic soil fauna of the maritime and continental Antarctic is taxonomically limited and consists almost entirely of arthropods, particularly the springtails (Collembola), and mites (Acari). Chironomid midges (Diptera) and beetles (Coleoptera) have also been recorded from the Antarctic Peninsula and the South Shetland, and Orkney Islands (Gressitt 1964; Balfour-Browne and Tilbrook 1966; Wirth and Gressitt 1967). Notable exceptions to this generalisation are the smaller protozoans, tardigrades, rotifers and nematodes, which are also commonly found in both continental and maritime areas (Gressitt 1965; Utsugi and Ohyama 1989; Bullini et al. 1994). Unfortunately, ecological work on these latter taxa is limited, and our focus here will be primarily on the larger, soil-dwelling arthropods.

Much of the information available on the soil fauna has come from reports documenting the fauna and/or describing new species (e.g. Janetschek 1963; Weiner 1980; Miller et al. 1988), with comparatively few studies examining ecological and physiological aspects (e.g. Matsuda 1977; Block 1985). More recently, interest has been targeted towards predicting the responses of invertebrates to environmental change (e.g. Block and Harrison 1995; Kennedy 1995) as well as evaluating patterns of diversity and dispersal among locations through studies of population genetics (e.g. Valbonesi et al. 1994; Frati et al. 1996). As we will discuss, these latter issues may be inextricably linked.

In this chapter we will discuss the invertebrate soil faunas of the continental and maritime Antarctic. We use the criteria and terminology of Holdgate (1977), referring to the Antarctic Peninsula and South Shetland Islands as 'maritime' and all other regions on the Antarctic coast (and inland) as 'continental'. In keeping with the theme of this volume we pay particular attention to those taxa associated with the continental coastal regions (Windmill Islands and Ongul Island) of East Antarctica and those of the South Shetland Islands. The remainder of the chapter will consist of five sections. Sections 15.2–15.3 will compare and contrast the faunal composition of soil invertebrates from the two regions, and discuss ecological and physiological factors affecting their distribution. Section 15.4 will consider the interhabitat

Ecological Studies, Vol. 154
L. Beyer and M. Bölter (eds.) Geoecology of Antarctic Ice-Free Coastal Landscapes

dispersal of taxa as well as the role of population genetic studies in assessing levels of gene flow among habitats. For this purpose we will also draw on our own and other data from habitats in and around Victoria Land in the Ross Sea sector. Section 15.5 will discuss the potential responses of the soil fauna to environmental changes resulting from both global (e.g. enhanced greenhouse effect) and local disturbances (e.g. habitat destruction), and finally Section 15.6 will provide a summary of further research needs.

15.2 Faunal Composition

From the earliest Antarctic expeditions, several reports have described, added to and revised the inventories of soil invertebrates (e.g. Willem 1902; Salmon 1962; Greenslade and Wise 1984). Perhaps the most comprehensive lists have been provided by Wise (1967, 1971) for the Collembola, and Hunter (1967), Strandtmann (1967) and Wallwork (1967) for the mites. To date, roughly 15 species of springtail and 25 species of mites have been recorded from the Antarctic continent. Of these, only two species of springtail and six species of mite have been recorded from Ongul I. and Windmill Is. (Tables 15.1, 15.2, respectively). The nearby maritime regions (e.g. South Shetland Islands), have recorded higher levels of taxonomic richness with 16 species of springtail and 36 species of mite found on the South Shetland Islands alone (Tables 15.1, 15.2). These lists are undoubtedly incomplete and will continue to be modified, particularly with the advent of molecular taxonomy (e.g. Valbonesi et al. 1994). Such studies have been able to assess diversity to levels previously unattainable, and have frequently revealed the existence of previously unknown, morphologically cryptic species (e.g. Valbonesi et al. 1994; Carapelli et al. 1995a, 1995b).

The taxonomic richness seen throughout maritime and continental Antarctica contrasts with the considerably greater species richness found on the Subantarctic islands (e.g. Crozet, Kerguelen, Heard, Marion, Bird and Macquarie Islands) where between 50–150 species of land arthropods have been recorded. It is evident that Subantarctic islands contain mostly cosmopolitan species (Greenslade 1987; Greenslade and Wise 1984; Greenslade and Wise 1986). The same is also true for maritime regions and Marshall and Pugh (1996) noted that of the total number of mite species found in the maritime Antarctic, 42.5 % are endemic, contrasting with 85.7 % for continental Antarctic. Within the maritime Antarctic, the South Shetland Islands may harbour among the greatest diversities of soil invertebrates. Block and Starý (1996) found that 19 % of all oribatid mite species recorded in the maritime regions were found in the South Shetland Islands, the highest of any of the sites they surveyed. The status of the beetle taxa (endemic or introduced) remains uncertain (Balfour-Browne and Tilbrook 1966), and at least two other species

Table 15.1. List of the collembolan (springtail) fauna found in the maritime Antarctic and continental locations

Taxon	Site[a]	References
Cryptopygus antarcticus antarcticus Willem	S	Willem (1901), Wahlgren (1906), Tilbrook (1967), Wise (1967, 1971), Weiner (1980), Greenslade and Wise (1984), Ohyama and Shimada (1990), Greenslade (1995)
Cryptopygus caecus Wahlgren	S	Wahlgren (1906), Tilbrook (1967), Wise (1967, 1971, 1974)
Cryptopygus sverdrupi Lawrence	O	Sømme (1986), Lawrence (1978), Greenslade (1995)
Cryptopygus badasa n. sp.	S	Greenslade (1995), Convey et al. (1996)
Archisotoma brucei (Carpenter 1907)	S	Tilbrook (1967), Wise (1967, 1971)
Isotoma (Folsomotoma) octooculata (Willem 1901)	S	Willem (1901), Wahlgren (1906), Tilbrook (1967), Wise (1967, 1971), Greenslade (1995)
Isotoma (Folsomotoma) o. kerguelenensis (Enderlein 1903)	S	Gressitt and Weber (1959)
Folsomia candida Willem 1902	S	Greenslade and Wise (1984)
Friesea grisea (Schäeffer 1891)	SO	Wahlgren (1906), Salmon (1962), Tilbrook (1967), Wise (1967, 1971) Ohyama and Sugawara (1989), Greenslade (1995)
Friesea woyciechowskii Weiner	S	Weiner (1980), Greenslade (1995)
Friesea topo n. sp.	S	Greenslade (1995)
Tullbergia mixta Wahlgren	S	Wahlgren (1906), Tilbrook (1967), Wise (1967, 1971), Greenslade (1995)
Onychiurus sp. Strong	S	Greenslade and Wise (1984)
Protophorura sp.	S	Greenslade (1995)
Hypogastrura viatica (Tullberg 1872)	S	Hack (1949), Salmon (1962), Salmon (1964), Tilbrook (1967), Wise (1967, 1971) Greenslade and Wise (1984), Greenslade (1995)
Achorutoides antarcticus n. gen., n. sp.	S	Willem (1901)

[a] S, South Shetland Is.; O, Ongul I.

– one dipteran and one beetle have been confirmed as likely introductions to the Prince Charles Mountains, East Antarctica (Rounsevell 1978, 1979).

Protozoans, tardigrades, rotifers and nematode worms occur widely on the Antarctic continent and in maritime areas (Spain 1971; Jennings 1976a,b; Barker 1977; Dastych 1984; Shishida and Ohyama 1989; Miller et al. 1988; Valbonesi et al. 1994), although these groups appear to be restricted primarily to areas of very high moisture content (Pryor 1962; Tilbrook 1967). Several studies have documented the diversity of these groups, particularly tardigrades

Table 15.2. List of the Acari (free-living mites) found in the maritime Antarctic and continental locations

Taxon	Site[a]	References
Ceratozetes gracilis (Michael)	S	Niedbala (1986)
Liochthonius mollis (Hammer)	S	Tilbrook (1967), Wallwork (1967)
Liochthonius australis Covarrubias	S	Covarrubias (1968)
Oppia loxolineata Wallwork	S	Wallwork (1965, 1967), Tilbrook (1967)
Oppia loxolineata longipilosa Covarrubias	S	Covarrubias (1968), Niedbala (1986)
Oppia pepiptensis brevipectinata	S	Covarrubias (1968)
Austroppia crozetensis (Richters 1908)	S	Block and Starý (1996)
Globoppia loxolineata longipilosa Wallwork 1968	S	Block and Starý (1996)
Globoppia loxolineata (Wallwork)	S	Wallwork (1965), Block and Starý (1996)
Alaskozetes antarcticus antarcticus (Michael)	S	Wallwork (1965, 1967), Tilbrook (1967), Niedbala (1986), Ohyama and Shimada (1990), Block and Convey (1995)
Halozetes antarctica (Michael 1903)	S	Gressit and Weber (1959)
Halozetes belgicae belgicae (Michael 1903)	S	Dalenius and Wilson (1958), Wallwork (1965, 1967), Niedbala (1986)
Halozetes belgicae longisetae Wallwork	S	Wallwork (1967), Niedbala (1986)
Halozetes impeditus sp. nov.	S	Niedbala (1986), Block and Starý (1996)
Antarcticola meyeri n. gen., n. sp.	O	Tilbrook (1967), Wallwork (1967), Rounsevell (1979), Ohyama and Sugawara (1989), Kanda et al. (1990)
Petrozetes oblongus nov. gen., nov. sp.	O	Sitnikova (1969)
Edwardzetes dentifer Hammer 1962	S	Block and Starý (1996)
Magellozetes antarcticus (Michael 1895)	S	Wallwork (1965), Block and Starý (1996)
Magellozetes processus Hammer 1962	S	Block and Starý (1996)
Maculobates nordenskjoeldi (Trägårdh)	S	Wallwork (1965)

(e.g. Barker 1977; Miller et al. 1988; Dastych 1989; Utsugi and Ohyama 1989), and nematodes (Barker 1977; Shishida and Ohyama 1989). To date, over 30 species of tardigrade have been reported from the Antarctic continent, including five confirmed from the areas surrounding Syowa Station (Utsugi and Ohyama 1989), and four species from the Vestfold Hills (Miller et al. 1988). For nematodes, three species (*Plectus antarcticus, P. frigophilus and Scottnema lindsayi*) found by Shishida and Ohyama (1986), were the first records of nematodes from Syowa Station. Yeates (1979) had previously found two of these species (*Plectus antarcticus, P. frigophilus*) from the Bunger Oasis, as well as a species of *Helicotylenchus*.

Table 15.2. (*Continued*)

Taxon	Site[a]	References
Cryptolaelops (Gamasellus) racovitzai (Trouessart)	S	Tilbrook (1967), Jumeau and Usher (1987), Usher et al. (1989), Convey et al. (1996)
Parasitus sp.	S	Tilbrook (1967)
Rhombognathus gressitti Newell	S	Pugh (1993)
Cocceupodes australis	S	Strandtmann and Tilbrook (1968)
Nanorchestes antarcticus Strandtmann 1963	SOW	Tilbrook (1967), Ohyama (1977), Ohyama and Matsuda (1977), Ohyama (1979, 1984), Rounsevell and Greenslade (1988), Heatwole et al. (1989)
Nanorchestes nivalis (Trouessart)	S	Strandtmann (1982), Convey et al. (2000)
Stereotydeus villosus (Trouessart)	S	Tilbrook (1967), Graham (1975)
Stereotydeus meyeri Strandtmann	O	Strandtmann (1967)
Rhagidia gigas gerlachei (Trouessart 1903)	S	Strandtmann (1967), Edwards and Usher (1987), Dalenius (1965)
Rhagidia leechi Strandtmann 1963	S	Tilbrook (1967)
Tydeus tilbrooki Strandtmann 1967	S	Tilbrook (1967)
Tydeus erebus Strandtmann 1967	OW	Johnstone et al. (1973), Ohyama (1977), Ohyama and Matsuda (1977), Rounsevell (1977), Ohyama (1979, 1984)
Eupodes minutus Strandtmann 1967	SO	Tilbrook (1967), Ohyama (1977, 1979), Ohyama and Matsuda (1977), Convey et al. (2000)
Eupodes parvus grahamensis	S	Booth et al. (1985)
Ereynetes macquariensis Fain	S	Strandtmann and Tilbrook (1968)
Bakerdania antarcticus (Mahunka)	S	Pugh (1993), Convey et al. (2000)

[a] S, South Shetland Is.; O, Ongul I.; W, Windmill Is.

15.3 Ecology and Physiology

Moisture has been described as the major factor affecting the distribution and abundance of Antarctic soil invertebrates (Tilbrook 1967; Kennedy 1993). Temperature also plays a critical role in dictating the amount of available water as well as providing taxa with considerable physiological challenges at extreme levels (Strong 1967; Spain 1971; Block 1985; Kennedy 1993). Protozoans, tardigrades, rotifers and nematodes appear particularly dependent on a very high moisture content (Pryor 1962), and thus may have a more limited distribution relative to the arthropod taxa. Several studies examining the effects of moisture have shown that invertebrates are usually associated with areas where relative humidity exceeds 70 %, although arthropods have been collected from areas where values are as low as 30% (Wise and Spain 1967; Rounsevell and Greenslade 1988). This dependence on higher levels of relative humidity is likely related to the cutaneous mode of respiration used by all Antarctic taxa thus far studied – despite having temperate relatives with tracheal respiratory systems (Pryor 1962; Rounsevell and Greenslade 1988). However, cutaneous respiration may offer advantages in coping with the Antarctic environment. For example, Rounsevell and Greenslade (1988), found granulations in the cuticle of *Nanorchestes* spp. and suggest that these help retain a layer of air over the body which facilitates cuticular respiration in soils that are waterlogged (e.g. during the spring thaw). These features may further enhance survival by reducing the chances of freezing when the animal is in direct contact with ice (Rounsevell and Greenslade 1988).

From a feeding perspective, the vast majority of the soil fauna are herbivores and detritivores, and thus contribute directly to the breakdown of organic material and the cycling of nutrients. Studies have reported feeding primarily on algae, fungi, and dead moss shoots (Strong 1967; Fitzsimons 1971a; Lippert 1971), although there is little direct evidence for feeding on live mosses (Fitzsimons 1971a). Instead, mosses appear to serve more as a habitat, as well as maintaining a suitable moisture content (Strong 1967; Convey and Block 1996). The presence of arthropods in association with decaying animals and feather accumulations suggests that scavenging is also possible among these taxa (Strong 1967).

The rate at which the soil fauna feed is strongly temperature-dependent with highest feeding rates and faecal production occurring in the temperature range 5–15 °C (Burn 1981, 1986). Little or no feeding occurs at temperatures less than 0 °C, or greater than 20 °C, where mortality rates, for at least some taxa, can be as high as 80% (Burn 1981). One of the main adaptations of Antarctic taxa to their environment is the ability to maintain metabolic rates equivalent to their temperate relatives at temperatures as low as 5 °C (Burn 1981). The selection of appropriate microhabitats also appears to be important in order to exist within preferred temperature ranges (i.e. 5–15 °C), and as a result most taxa are limited to the upper 3–5 cm of moss and/or soil (Tilbrook 1967). Surprisingly, and despite 24-h daylight and temperatures >0 °C during the summer months, some taxa still maintain diurnal activity patterns. For example, Wise and Spain (1967) found that arthropod activity was limited to the hours between 0600 and 2100 and usually between 0900 and 1600 h. Seasonal activity periods range from mid November to mid February at continental sites (e.g. Wise and Spain 1967), although this period will be extended somewhat in the warmer maritime areas (Convey 1996). Owing to these limited periods of activity, life cycles are typically very long compared with temperate relatives; mature animals have been estimated to live between 3–7 years (Burn 1984), compared with 2 months to a year in temperate habitats (Young 1980).

Species interactions such as competition and predation are limited for the soil fauna, and in particular, the role of predator is greatly under-represented in most Antarctic habitats, although exceptions exist. For example, the largest mite in Victoria Land, *Coccorhagidia gressitti*, preys on other, smaller mite species (Gless 1967). Even in the maritime Antarctic, few predacious species have been found, and these consist almost exclusively of mites (Edwards and Usher 1987), with collembolans serving. In particular, Lister et al. (1988) found predation by the mite *Gamasellus racovitzai* was highly selective for one species of Collembola (*Cryptopygus antarcticus*), and constituted 80 % of all prey traces found during gut content analysis.

The ability of Antarctic taxa to survive the extremely cold temperatures has been studied extensively by W. Block and colleagues (e.g. Block et al. 1978; Young and Block 1980; Sømme 1981; Block and Convey 1995). Sømme (1981) reports that none of the Antarctic springtails and mites thus far examined, are freezing tolerant (i.e. freezing of the body is fatal). Accordingly, their ability to survive low temperatures is dependent on being able to keep their body fluids from freezing (i.e. supercooling). Two main methods appear to be employed, that of avoidance of ice nucleators within their bodies and the use of antifreeze proteins. Surprisingly, these adaptations are not unique to Antarctic taxa and have also been found in temperate relatives (Block and Convey 1995). The production of antifreezes is energetically expensive and occurs only in late autumn and winter. However, the production of antifreezes is not enough to prevent ice nucleators from forming and suppression of feeding is also necessary to avoid 'contaminants' (i.e. food materials) that might initiate ice formation (Young and Block 1980). Accordingly, gut clearing usually occurs when ground temperatures drop below 0 °C and further feeding is restricted as a preparation for winter (e.g. Young and Block 1980). However, not all animals stop feeding simultaneously, and the presence of animals with and without food in their guts explains the bimodal distribution of cold tolerance often reported in animals (Sømme 1981; Pickup 1990).

15.4 Interhabitat Dispersal and Population Genetic Structure

Much of what is known of potential dispersal methods for the invertebrate fauna is based on anecdotal evidence and/or casual observation. For example, air currents are thought to be one of the main agents of dispersal for most Antarctic taxa based on collections of mites in aerial wind traps (Pryor 1962; Strong 1967). However, Marshall and Pugh (1996), suggest that this mode of transport may not be particularly effective for larger invertebrates (e.g. mites) due to the lack of an anhydrobiotic dispersal stage.

Other modes of dispersal are also possible and may include accidental carriage on birds or mammals. Birds are frequently implicated in the transfer of passively dispersed taxa from habitat to habitat, and roosting birds such as skuas (*Catharacta skua*) may provide a suitable dispersal vehicle. Strong

(1967) observed individuals of *Alaskozetes* (Acari) on the feathers of a freshly killed skua, and likewise Tilbrook (1967) collected several species from both bird colonies and nesting areas. At our own study sites in and around Victoria Land, we have noted a strong association between roosting skuas and the presence of both springtails and mites (Hogg and Stevens, unpubl. data). However, Marshall and Pugh (1996) have cast doubt on this mode of transportation, suggesting that the association of arthropods with these areas is more the result of the rich organic food supply, rather than the product of dispersal events per se. We suggest that given the close proximity of soil invertebrates to nesting sites, accidental carriage on birds is at least possible for some Antarctic taxa.

An alternative method of dispersal is that of 'rafting' using sea currents to move among coastal areas. Potential 'rafts' might include ice floes (Strong 1967) or other materials such as feathers from penguins or skuas. Considerable accumulations of moulted feathers from penguins and skuas are present along coastal areas, particularly during January when invertebrates are at their most active (Wise and Spain 1967). Indeed, even in the absence of these materials, Strong (1967) has reported that individuals of *Alaskozetes* were able to survive immersion in salt water for up to 55 days at temperatures of 0 °C. The presence of springtails on the surfaces of melt streams (Pryor 1962; Tilbrook 1967) suggests that access to the sea is even possible for those individuals/populations not directly on the shoreline. Accordingly, movement via sea currents may also provide a viable means of interhabitat dispersal.

A more recent and potentially undesirable method of dispersal may be provided by humans, and there have been several examples of possible species introductions (e.g. Burn 1984; Greenslade and Wise 1984; Niedbala 1986; Rounsevell and Horne 1986). The equally likely, and equally undesirable, transfer of existing species within the Antarctic region is also possible, although this would be difficult to assess on the basis of casual observation only.

Although these observations provide useful information on potential dispersal methods, they rarely provide information on the actual frequency or success of such events. One method to address this issue is to evaluate the population genetic structure of taxa in an effort to quantify rates of dispersal among habitats. By examining the frequencies of alleles at gene or allozyme loci among locations it is possible to evaluate genetic similarities among populations, and thus estimate levels of gene flow (and hence dispersal) among locations.

Despite the usefulness of such studies in determining rates of dispersal among habitats, very few studies have been undertaken to date. Exceptions are the work of Frati and colleagues (e.g. Frati et al. 1992, 1996) on *Isotoma* sp. (Collembola) in northern Victoria Land, and our own work on springtails, predominantly *Gomphiocephalus hodgsoni* (Collembola), in southern Victoria Land and Ross Island (Hogg and Stevens, unpubl. data). Frati et al. (1997b) found fixed allelic differences (non-shared alleles) at sites within 100 km suggesting that no gene flow was occurring among these locations. Our results too, have suggested limited gene flow, particularly between sites on Ross Island and those of southern Victoria Land on the Antarctic continent. For example, in a survey of sites on Ross Island and in the Dry Valleys, we found fixed allelic differences, at two of the ten loci examined between sites on Ross Island and the continent. However, differences on Ross Island were considerably less. Accordingly, these data suggest that gene flow and hence dispersal for Antarctic taxa may be limited to local events.

15.5 Potential Responses to Environmental Change

The ability of the Antarctic soil fauna to respond to environmental changes (e.g. global warming, habitat destruction), will depend to a large extent on the scale and rate at which the change occurs. For rapid, localised changes (e.g. habitat destruction), survival of individuals may be limited due to the relatively low dispersal rates of taxa (see Sect. 15.4) and/or the absence of suitable, alternative habitats. In these instances, local extirpation may be inevitable. However, for slower, more gradual changes (e.g. global warming), dispersal to new areas of suitable habitat may be possible provided that the rate of change does not exceed their ability to find new, alternative habitats. On a larger scale (e.g. hundreds of kilometres), this may occur in conjunction with other changes (e.g. new soil formation/moss growth), although long-distance dispersal among habitats for most taxa may be extremely limited (Balfour and Tilbrook 1966; Convey and Block 1996). However, changes are also likely on relatively smaller spatial scales (e.g. centimetres to metres). For example, considerable temperature differences are known to occur even on relatively small (e.g. 1–5 cm) spatial scales (Wise and Spain 1967), and animals are known to select particular microhabitats that ameliorate ambient air temperatures (Pryor 1962; Tilbrook 1967). Accordingly, and in the case of rising global air temperatures, taxa such as mites and springtails may be able to modify their micro-distribution in the soil/plant profile through behavioural responses to the temperature shift. However, the resulting changes in soil moisture content as well as corresponding changes in feeding, growth rates, and survival may be difficult to predict.

For both large-scale and small-scale changes (although particularly when long-distance dispersal is required) the rate of change must be less than the rate at which individuals are able to find suitable new habitats. Unfortunately, global warming scenarios have suggested that the rate of change may far exceed that previously experienced by the earth's biota (Ojima et al. 1991). Furthermore, several studies have suggested that time since the last glaciation has been insufficient for successful colonisation of some seemingly appropriate habitats by soil taxa (Balfour and Tilbrook 1966; Tilbrook 1967; Convey and Block 1996). Accordingly, the natural migration of animals other than in very localised circumstances appears unlikely to provide a suitable response to most immediate environmental changes.

An alternative is for species to adapt to the changing environmental conditions in situ (i.e. evolutionary response). The ability to evoke an evolutionary response will, in turn, be dictated to a large extent by the underlying genetic variability that is contained within and among populations of a species. Conventional evolutionary theory suggests that species with greater overall levels of genetic variability may be better able to respond to selective pressures such as environmental changes relative to species with comparatively lower levels of variability. For example, if the timing of production of antifreeze proteins was under genetic control, species with the greatest variability for this trait may be more likely to persist following changing environmental temperatures.

However, what is frequently not considered is how this genetic variability may be partitioned among respective populations of a given species (i.e. genetic differentiation among populations). For example, some species with high levels of overall genetic variability may have limited levels of variability within any given population (i.e. have highly differentiated populations, each of limited variability). For species with limited interhabitat dispersal (i.e. Antarctic taxa), variability within populations would be expected to be extremely limited due to the lack of gene flow (and hence genetic exchange) among populations. Accordingly, the ability of these individual populations to respond to environmental change may be extremely limited.

Surprisingly, initial studies had suggested that allelic variability within populations of Collembola was surprisingly high (Frati et al. 1997b; Hogg and Stevens, unpubl. data). Although there are a number of possible explanations, it appears likely that these levels of variability are an artefact of morphologically based classification schemes, which have failed to identify reproductively isolated species or populations. Specifically, currently recognised 'species' may be complexes of morphologically similar, but reproductively isolated taxa. Accordingly, and as initially expected, the allelic variability within true (reproductively isolated) species is likely to be limited. This remains highly speculative, and clearly is an issue that will require further clarification in the immediate future.

In the absence of this information, the preservation and remediation of a range of habitats remains our best option for preserving levels of variability within and among Antarctic taxa.

15.6 Further Research

Although there are a wealth of research opportunities within the maritime and continental Antarctic, it is possible to identify at least three research areas that would benefit from immediate attention to greatly enhance our understanding of these unique systems:

1. Small-scale dispersal experiments, examining vertical and horizontal movements of animals in response to different thermal regimes and humidity gradients would be useful in establishing potential behavioural responses to environmental changes. Furthermore, complementary data on their feeding habits as well as growth rates and survival in these different microhabitats will aid in predicting ecological responses to global atmospheric changes (e.g. global warming).
2. Studies examining the population genetic structure for a range of taxa and throughout their latitudinal distributions will be useful for several reasons. These data will enable accurate determination of true levels of (reproductively isolated) species diversity within and among habitats. They will also provide estimates of gene flow among populations, thus allowing assessment of interhabitat dispersal, as well as providing data on levels of genetic variability within and among populations. This in turn will be important in designating 'evolutionary significant units' which are essential for informed management decisions. Multidisciplinary studies on bird behaviour (e.g. roosting, migration), and invertebrate behaviour (e.g. attachment to feathers, feet etc.), in both continental and maritime locations would be useful for assessing local dispersal dynamics.
3. Ecological work on smaller taxa (e.g. protozoans, nematodes), is woefully lacking and in particular their role in the breakdown of material and cycling of nutrients, as well as their interactions, if any, with the arthropods. This paucity of information may reflect, in part, the challenges of working with these taxa. However, it is essential that their contribution to the biological processes in soils be accurately assessed.

In this chapter, we have presented an overview of soil invertebrates from the South Shetland Is (maritime) and from Ongul I. and Windmill Is (continental). It is clear that knowledge of this important component of the Antarctic soil fauna is incomplete, and in this final section we have tried to outline areas where we feel further research would enhance our understanding of these systems.

Acknowledgements. We are grateful to Manfred Bölter, Allan Green and Rod Seppelt for helpful comments and to Antarctica New Zealand for logistic support. We would also like to thank Penny Greenslade and Bruce Halliday for providing vital references.

References

Balfour-Browne J, Tilbrook PJ (1966) Coleoptera collected in the South Orkney and South Shetland Islands. Br Antarct Surv Bull 9:41–43

Barker RJ (1977) A biological reconnaissance of the Bunger Hills. Antarctic Division Technical Memorandum 67, 55 pp

Block W (1977) Oxygen consumption of the terrestrial mite *Alaskozetes antarcticus* (Acari: Cryptostigmata). J Exp Biol 68:69–87

Block W (1985) Ecological and physiological studies of terrestrial arthropods in the Ross Dependency 1984–85. Br Antarct Surv Bull 68:115–122

Block W, Convey P (1995) The biology, life cycle, and ecophysiology of the Antarctic mite *Alaskozetes antarcticus.* J Zool Lond 236:431–449

Block W, Harrisson PM (1995) Collembolan water relations and environmental change in the maritime Antarctic. Global Change Biol 1:347–359

Block W, Young SR (1978) Metabolic adaptations of Antarctic terrestrial micro-arthropods. Comp Biochem Physiol 61A:363–368

Block W, Young SR, Conradi-Larsen E-M, Somme L (1978) Cold tolerance of two Antarctic terrestrial arthropods. Experimentia 34:1166–1167

Block W, Starý J (1996) Oribatid mites (Acari: Oribatida) of the maritime Antarctic and Antarctic Peninsula. J Nat Hist 30:1059–1067

Booth RG, Edwards M, Usher MB (1985) Mites of the genus *Eupodes* (Acari, Prostigmata) from maritime Antarctica: a biometrical and taxonomic study. J Zool Lond (A) 207:381–406

Bullini L, Arduino P, Cianchi R, Nascetti G, D'Amelio S, Mattiucci S, Paggi L, Orecchia P (1994) Genetic and ecological studies on nematode endoparasites of the genera *Contracaecum* and *Pseudoterranova* in the Antarctic and Arctic-boreal regions. In: Battaglia B, Bisol PM, Varotto V (eds). Proceedings of the 2nd Meeting on Antarctic Biology. Edizioni Universitarie Patavine, Padova, Italy, pp 131–146

Burn AJ (1981) Feeding and growth in the Antarctic collembolan *Cryptopygus antarcticus.* Oikos 36:59–64

Burn AJ (1984) Life cycle strategies in two Antarctic Collembola. Oecologia 64:223–229

Burn AJ (1986) Feeding rates of the cryptostigmatid mites *Alaskozetes antarcticus.* Br Antarct Surv Bull 71:11–17

Carapelli A, Fanciulli PP, Frati F, Dallai R (1995a) The use of genetic markers for the diagnosis of sibling species in the genus *Isotomurus* (Insecta: Collembola). Boll Zool 62:71–76

Carapelli A, Frati F, Fanciulli PP, Dallai R (1995b) Genetic differentiation of six sympatric species of *Isotomurus* (Collembola, Isotomidae); is there any difference in their microhabitat preference? Eur J Soil Biol 31:87–99

Convey P (1996) Overwintering strategies of terrestrial invertebrates in Antarctica – the significance of flexibility in extremely seasonal environments Eur J Entomol 93:489–505

Convey P, Block W (1996) Antarctic Diptera: ecology, physiology and distribution. Eur J Entomol 93:1–13

Convey P, Greenslade P, Pugh PJA (2000) The terrestrial micro-arhtropod fauna of the South Sandwich Islands. J Nat Hist 34:597-609

Convey P, Greenslade P, Richard KJ, Block W (1996) The terrestrial arthropod fauna of the Byers Peninsula, Livingston Island, South Shetland Islands – Collembola. Polar Biol 16:257–259

Covarrubias R (1968) Some observations on Antarctic Orbatei (Acarina) *Liochthonius australia* sp. n., and two *Oppia* spp. n. Acarologia 10:313–356

Dalenius P (1965) The acarology of the Antarctic regions. In: Van Mieghem J, Van Oye P (eds) Biogeography and ecology in Antarctica. Junk, The Hague, pp 414–430
Dalenius P, Wilson O (1958) On the soil fauna of the Antarctic and sub-Antarctic islands. The Oribatidae (Acari). Ark Zool 2:393–425
Dastych H (1984) The Tardigrada from the Antarctic with descriptions of several new species. Acta Zool Cracoviensia 27:377–436
Edwards M, Usher MB (1987) The adult and juvenile stages of two predatory mites, *Rhagidia gerlachei* and *Tuberostoma leechi*, in the maritime Antarctic. Br Antarct Surv Bull 74:37–48
Fitzsimons JM (1971a) On the food habits of certain Antarctic arthropods from coastal Victoria Land and adjacent islands. Pac Insects Monogr 25:121–125
Fitzsimons JM (1971b) Temperature and three species of Antarctic arthropods. Pac Insects Monogr 25:127–135
Frati F, Fanciulli PP, Dallai R (1992) Genetic diversity and taxonomy in soil-dwelling insects: the genus *Orchesella*. J Hered 83:275–281
Frati F, Fanciulli PP, Carapelli A, DeCarlo L, Dallai R (1996) Collembola of northern Victoria Land: distribution, population structure and preliminary molecular data to study origin and evolution of Antarctic Collembola. 3rd Meeting on Antarctic Biology, Santa Margherita Ligure, Italy, Camerino University Press
Frati F, Fanciulli PP, Carapelli A, Dallai R (1997a) The Collembola of northern Victoria Land (Antarctica): distribution and ecological remarks. Pedobiologia 41:50–55
Frati F, Simon C, Sullivan J, Swofford DL (1997b) Evolution of the mitochondrial cytochrome oxidase II gene in Collembola. J Mol Evol 44:145–158
Gless EE (1967) Notes on the biology of *Coccorhagidia gressitti* Womersley and Strandtmann. Antarct Res Ser 10:321–324
Graham WL (1975) Morphology and biology of *Sterotydeus villosus* (Trouessart) (Acarina: Penthalodidae) from Antarctica. MSc Thesis, Texas Technical University, Lubbock, Texas, 84 pp
Greenslade P (1987) Soil fauna studies with particular reference to collembolan faunistics, densities and distributions. In: 1986–87 Australian Antarctic research program. Antarctic Division, Kingston, Tasmania, pp 80–84
Greenslade P (1995) Collembola from the Scotia Arc and Antarctic Peninsula including descriptions of two new species and notes on biogeography. Pol Pismo Entomol 64:305–319
Greenslade P, Wise KAJ (1984) Additions to the collembolan fauna of the Antarctic. Trans R Soc South Aust 108:203–205
Greenslade P, Wise KAJ (1986) Collembola of Macquarie Island. Rec Auckland Inst Mus 23:67–97
Gressitt JL (1964) Ecology and biogeography of land arthropods in Antarctica. In: Carrick R, Holdgate MW, Prévost J (eds) Biologie Antarctique, Hermann, Paris, pp 211–222
Gressitt JL (1965) Terrestrial animals. In: Hatherton T (ed) Antarctica. Methuen, London, pp 351–371
Hack WH (1949) Nota sobre un colémbolo de la Antartida *Argentina Achorutes viaticus* Tullberg. Notas Mus Plata 14:211–212
Heatwole H, Saenger P, Spain A, Kerry E, Donelan J (1989) Biotic and chemical characteristics of some soils from Wilkes Land, Antarctica. Antarct Sci 1:225–234
Holdgate MW (1977) Terrestrial ecosystems in the Antarctic. Philos Trans R Soc Lond, Ser B 279:5–25
Hunter PE (1967) Mesostigmata: Rhodacaridae, Laelapidae (Mesostigmatic mites). Antarct Res Ser 10:35–39
Janetschek H (1963) On the terrestrial fauna of the Ross-Sea area, Antarctica (Preliminary report). Pac Insects 5:305–311

Jennings PG (1976a) The Tardigrada of Signy Island, South Orkney Islands, with a note on the Rotifera. Br Antarct Surv Bull 44:1–25

Jennings PG (1976b) Tardigrada from the Antarctic Peninsula and Scotia Ridge region. Br Antarct Surv Bull 44:77–95

Johnstone GW, Lugg DJ, Brown DA (1973) The biology of the Vestfold Hills, Antarctica. Australian National Antarctic Research Expeditions, Scientific Reports, Series B(1) Zoology 123, pp 1–62

Jumeau PJAM, Usher MB (1987) The Antarctic predatory mite *Gamasellus racovitzai* (Trouessart) (Mesostigmata): a morphometric study of two species. Acarologia 28:15–26

Kanda H, Inoue M, Mochida Y, Sugawara H, Ino Y, Ohtani S, Ohyama Y (1990) Biological studies on ecosystems in the Yukidori Valley, Langhovde, East Antarctica. Antarct Rec 34:76–93

Kennedy AD (1993) Water as a limiting factor in the Antarctic terrestrial environment: a biogeographical synthesis. Arct Alp Res 25:308–315

Kennedy AD (1995) Antarctic terrestrial ecosystem response to global environmental change. Annu Rev Ecol Syst 26:683–704

Lawrence PN (1978) *Cryptopygus sverdrupi* n. sp. A new species of Collembola (Isotomidae) from Sverdrupfjella, Antarctica, with notes on related species in five genera. Nor J Entomol 25:15–20

Lippert G (1971) Occurrence of arthropods in mosses at Anvers Island, Antarctica Peninsula. Pac Insects Monogr 25:137–144

Lister A, Block W, Usher MB (1988) Arthropod predation in an Antarctic terrestrial community. J An Ecol 57:957–971

Marshall DJ, Pugh PJA (1996) Origin of the inland Acari of continental Antarctica with particular reference to Dronning Maud Land. Zool J Linn Soc 118:101–118

Matsuda T (1977) Ecological investigations on free-living mites near Syowa Station, Antarctica. In: Llano GA (ed) Adaptations within Antarctic ecosystems. Smithsonian Institution, Washington, DC, pp 1015–1021

Miller JD, Horne P, Heatwole H, Miller WR, Bridges L (1988) A survey of the terrestrial micrometazoa of the Vestfold Hills. I Tardigrada. Hydrobiologia 165:197–208

Niedbala W (1986) Several species of moss-mites (Acari, Oribatida) from the Antarctic coastal zone. Pol Polar Res 1:119–126

Ohyama Y (1977) Population density of free-living mites in the ice-free areas around Syowa Station, East Antarctic. Antarct Rec 60:47–56

Ohyama Y (1979) Notes on the free-living mites in the ice-free areas around Syowa Station. In: Matsuda T, Hoshiai T (eds) Memoirs of the National Institute of Polar Research, Special issue, no 11. Proc Symp on Terrestrial ecosystem in the Syowa Station area, National Institute of Polar Research, Tokyo, pp 127–129

Ohyama Y (1984) Mites around Syowa Station. Polar News 38:15–19 (In Japanese)

Ohyama Y, Matsuda T (1977) Free-living prostigmatic mites found around Syowa Station, East Antarctica. Antarct Rec 59:172–176

Ohyama Y, Shimada K (1990) Japan-China collaborative research of terrestrial biology at Great Wall Station, King George Island in 1989/90. Antarct Rec 34:235–241

Ohyama Y, Sugawara H (1989) An occurrence of cryptostigmatic mite around Syowa Station area. Proc Int Symp Ant Res China Ocean Press, Tianjin, pp 324–331

Ojima DS, Kittel TGF, Rosswall T, Walker BH (1991) Critical issues for understanding global change effects on terrestrial ecosystems. Ecol Appl 1:316–325

Pickup J (1990) Seasonal variation in the cold-hardiness of three species of free-living Antarctic nematodes. Funct Ecol 4:257–264

Pugh PJA (1993) A synonymic catalogue of the Acari from Antarctica, the sub-Antarctic Islands and the Southern Ocean. J Nat Hist 27:323–421

Pryor ME (1962) Some environmental features of Hallett Station, Antarctica, with special reference to soil arthropods. Pac Insects 4:681–728

Rounsevell DE (1977) Geographic variation in the size of an Antarctic mite *Tydeus erebus* Strandtmann (Acarina: Prostigmata). Acarologia 19:209–219

Rounsevell DE (1978) Populations of introduced arthropods at Australian Antarctic Stations. Pac Insects 18:199–202

Rounsevell DE (1979) Arthropods in a botanical collection from the Prince Charles Mountains, Antarctica. Pac Insects 20:83–85

Rounsevell DE, Greenslade P (1988) Cuticle structure and habitat in the Nanorchestidae (Acari: Prostigmata). Hydrobiologia 165:209–212

Rounsevell DE, Horne PA (1986) Terrestrial, parasitic and introduced invertebrates of the Vestfold Hills. In: Pickard J (ed) Antarctic oasis. Terrestrial environments and history of the Vestfold Hills. Academic Press, Sydney, pp 309–331.

Salmon JT (1962) A new species and redescriptions of Collembola from Antarctica. Pac Insects 4:887–894

Salmon JT (1964) An index to the Collembola. Bull R Soc N Z 7:1–644

Shishida Y, Ohyama Y (1986) A note on the terrestrial nematodes around Syowa Station, Antarctica (extended abstract). Mem Natl Inst Polar Res Spec Issue 44:259–260

Shishida Y, Ohyama Y (1989) A note on the terrestrial nematodes around Palmer Station, Antarctica (extended abstract). Proc NIPR Symp Polar Biol, vol 2, pp 223–224

Sitnikova LG (1969) A new mite, *Petrozetes oblongus* nov. gen., nov. sp. (Acarina: Orbatei) from Antarctica. Sov Antarct Exped Infor Bull 7:38

Sømme L (1981) Cold tolerance of Alpine, Arctic and Antarctic Collembola and mites. Cryobiology 18:212–220

Sømme L (1986) New records of terrestrial arthropods from Dronning Maud Land, Antarctica. Polar Research (NS) 4:225–229

Spain AV (1971) Some aspects of soil conditions and arthropod distribution in Antarctica. Pac Insects Monogr 25:21–26

Strandtmann RW (1967) Terrestrial Prostigmata (Trombidiform mites). Antarct Res Ser 10:51–95

Strandtmann RW (1982) Notes on *Nanorchestes* V. Two new species of Nanorchestes (Acari: Nanorchestidae) from the Antarctic Peninsula and South Atlantic islands. Pac Insects 24:252–258

Strandtmann RW, Tilbrook PJ (1968) Some Prostigmata (Acari) from Signy Island, South Orkney Islands and Deception Island, South Shetland Islands. Br Antarct Surv Bull 17:51–57

Strong J (1967) Ecology of terrestrial arthropods at Palmer station, Antarctic Peninsula. Antarct Res Ser 10:357–371

Tilbrook PJ (1967) Arthropod ecology in the maritime Antarctic. Antarct Res Ser 10:331–356

Usher MB, Block W, Jumeau PJAM (1989) Predation by arthropods in an Antarctic terrestrial community. In: Heywood RB (ed) University research in Antarctica: Proceedings of British Antarctic survey, Antarctic Special Topic Award Scheme Symposium, 9–10 November 1988, British Antarctic Survey, Cambridge, pp 123–129

Utsugi K, Ohyama Y (1989) Antarctic Tardigrada. Proc NIPR Symp Polar Biol 2:190–197

Valbonesi A, Ballarini P, Di Giuseppe G, Miceli C, Felici A, Ortenzi C, Luporini P (1994) Speciation and adaptive biology of Antarctic ciliated protozoa. In: Battaglia B, Bisol PM, Varotto V (eds) Proceedings of the 2nd Meeting on Antarctic Biology. Edizioni Universitarie Patavine, Padova, Italy, pp 111–120

Wahlgren E (1906) Antarktische und Subantarktische Collembolen. Wissenschaftliche Ergebnisse der Schwedischen Südpolar-Expedition, 1901–1903. Zoology 5:1–22

Wallwork JA (1965) A revision of the Family Podacaridae Grandjean (Acari: Oribatei). Acarologia 6:387–399

Wallwork JA (1967) Cryptostigmata (oribatid mites). Antarct Res Ser 10:105–122
Weiner WM (1980) *Friesea woyciechowskii* sp. n. (Collembola, Neanuridae) from King George Island (South Shetland Islands, Antarctica). Pol Pismo Entomol 50:491–495
Willem V (1901) Les Collemboles recueillis par l'Expedition Antarctique Belge. Ann Soc Entomol Belg 45:260–262
Willem V (1902) Collemboles. Expédition Antarctique Belge. Résultants du Voyage du S. Y. Belgica 1897–1899. Rapp Sci Zool 9:1–19
Wirth WW, Gressitt JL (1967) Diptera: Chironomidae (midges). Antarct Res Ser 10:197–203
Wise KAJ (1967) Collembola (Springtails). Antarct Res Ser 10:123–148
Wise KAJ (1971) The Collembola of Antarctica. Pac Insects Monogr 25:57–74
Wise KAJ (1974) A new synonymy in Cryptopygus (Collembola: Isotomidae). Rec Auckland Inst Mus 11:209–212
Wise KAJ, Spain AV (1967) Entomological Investigations in Antarctica, 1963–64 season. Pac Insects 9:271–293
Yeates GW (1979) Terrestrial nematodes from the Bunger Hills and Gaussberg, Antarctica. N Z J Zool 6:641–643
Young SR (1980) Cold hardiness in a temperate cryptostigmatid mite. Oikos 35:350–352
Young SR, Block W (1980) Experimental studies on the cold tolerance of *Alaskozetes antarcticus*. J Ins Physiol 26:189–200

Part IV

Integrating Aspects for Soil Ecology in Antarctic Coastal Landscapes

Preface

Soil ecology is defined as the interaction between the organisms in and on soil as well as their interactions with soil properties (e.g. Working Group Soil Ecology 1995). For this reason, the first four chapters of Part IV point out (1) the significance of soils and their properties for organisms and (2) the opposite point of view, i.e. the influence of organisms on soils. Parts I–III provided detailed information on soil properties, their distribution patterns, and the basic state for life on and in soils. It is now the aim to focus more on interacting items and to see the soil in its role as a host for flora and fauna (Chap. 17). To some extent, new concepts on ecological niches and habitats can be defined (Chap. 18) which can help to classify the environment, and to establish a common view on habitats.

The question about soil formation, as already dealt with in earlier chapters, will be taken into consideration again (Chaps. 19 and 20) and it will be reviewed from the viewpoint of soil formation by biological factors. For temperate climate regimes the importance of the biology factor (ø) in the CLORPT equation (cf. Preface, Part II) is well known for the first steps in soil formation: humus accumulation, bioturbation, root penetration and water and air dynamics etc. (Jenny 1986), but only little is known about the environments of Antarctica. Here, only a very limited number of vascular plants and invertebrates can survive (Part III). This suggests a much greater impact of microbes, algae, lichens and mosses on soil formation. It is obvious that under the climatic regime of Antarctica many processes of soil development work slowly. On the other hand, climatic gradients act on small local scales and thus make processes visible which are often hidden or unrecognisable in temperate regions. More detailed descriptions of various profiles from individual climatic regions as provided in overviews by Claridge and Campbell (1987), Blume et al. (1997), Bölter et al. (1999), or in detail from King George Island (Kuhn 1997) and the Windmill Islands (Beyer and Bölter 1999) provoke discussions about their role in this ecosystem.

As already mentioned in the Preface to Part II, the original CLORPT equation from Jenny (1986) nowadays has to be extended by taking human factors into consideration (Amundsen et al. 1994). Soil formation is influenced to a great extent by humans, up to the total destruction of whole soil landscapes.

Stonehouse (1999) has made an attempt to bring these ideas together with emphasis on disturbance, damage and remediation, i.e. anthropogenic effects on polar environments. Bölter (1996) focused on soil processes which are stimulated by climatic changes in those regions. Although the latter paper refers mainly to the Arctic, parallels to the Antarctic can be drawn, even though many problems act at different time scales (Chap. 21). Growth patterns of microbes, plants, or soil animals are only barely on the scale of climatic change, soil development or man-made disturbance factors. Hence, the authors of this book cannot make a final statement about the state of the art of science, nature or Antarctic politics, but hope to warm people's heart to consider this part of the globe as a mirror for their own efforts in science and politics, and of global history.

References

Amundsen R, Harden J, Singer (1994): Factors of soil formation: a fiftieth anniversary retrospective. SSSA Special Publication No 33, Soil Science Society of America, Madison, Wisconsin

Beyer L, Bölter M (1999) Formation, ecology, and geography of cryosols of an ice-free oasis in coastal East Antarctica near Casey Station (Wilkes Land). Aust J Soil Res 37:209–244

Blume H-P, Beyer L, Bölter M, Erlenkeuser H, Kalk E, Kneesch S, Pfisterer U, Schneider D (1997) Pedogenic zonation in soils of the southern circum-polar region. Adv GeoEcol 30:69–90

Bölter M (1996) Consequences of global warming on soil processes in Arctic Regions. Polarforschung 66:1–10

Bölter M, Blume H-P, Kuhn D (1999) Soils and their microbiological properties from a transect from Cape Horn to the Antarctic Peninsula. Polar Biosci 12:54–67

Campbell IB, Claridge GGC (1987) Antarctica: soils, weathering processes and environment. Elsevier, Amsterdam

Jenny H (1986) The soil resource. Ecological studies 37, 3rd edn. Springer, Berlin Heidelberg New York

Stonehouse B (1999) Biological processes in cold soils. Polar Rec 35:5–10

Working Group Soil Ecology (1995). Workshop on soil ecology. Commision IV. German Association of Soil Science. Mitteil Dtsch Bodenkd Ges (Oldenburg) 78:1–256.

16 Soils as Habitats for Microorganisms

M. Bölter and H.-P. Blume

16.1 Introduction

It is more than a small step into the living sphere of microorganisms. They have special demands for their living space, like plants for their rooting space. The supply of heat, water, soil gases, organic matter and inorganic nutrients presents the general framework. The range of their homes and their activities at the micro-scale is dominated much more directly by physical and chemical laws. Transport mechanisms of dissolved organic and inorganic nutrients as well as gases primarily follow the forces of diffusion rather than active transport mechanisms. Originally, physically driven movements, e.g. cryoturbation, become more important than biomechanical activities, e.g. bioturbation. This holds especially true for the polar soils, where only few micro-arthropods and nematodes can be busy at this job (Block 1984). Thus, we shift from animal-mediated transport of material and breakdown of structural matter and approach an environment of chemical processes driven by physical processes, mainly temperature. This applies to the final degradation of organic matter by enzymes as well as for the chemical weathering processes.

Antarctic coastal terrestrial environments are only small in extent. Broad classifications into 'dry, mesic and wet', as used in the Arctic, can only just be adopted – except for the large areas of the Dry Valleys. Very different areas can be found close to each other, and thus very different conditions for plants and soil-dwelling organisms. We might ask what are the necessary, essential or even indispensable conditions for sustaining microbial life in soils, and what are the conditions to which the organisms have to respond?

16.2 The Microbial Environment

The microbial environment, however, should not only be described as a space which limits life processes, but also as a habitat which promotes different life strategies. Physical and chemical variations lead to many niches and individual habitats. This variability over space has made field measurements extremely variable and has long been recognised as one of the main nuisances

Ecological Studies, Vol. 154
L. Beyer and M. Bölter (eds.) Geoecology of Antarctic Ice-Free Coastal Landscapes

of environmental research (Addicott et al. 1987; Smiles 1988; White 1995). The arrangements in space can be accomplished by strong variability of nutrients and their individual concentrations, leading to numerous gradients. Although this variability might be regarded as a disturbance factor in environmental descriptions, it is the result of the physical, chemical and biological processes.

Nevertheless, the Antarctic soil environment – without bioturbating animals – is basically a rather conservative environment and tends to reduce strong physical shifts. This is, however, a matter of scale and the abundance of organisms which determine its statistical distribution. Much of the pattern of distribution is due to reproductive phenomena, i.e. the formation of micro-colonies and -communities due to apparently good life conditions. A view into the microscope shows such structures. For motile organisms, sensory mechanisms affect the distribution of individuals over space and chemoreception as well as mechanoreception may also operate in this environment. Bacterial chemotaxis for algae has been documented for marine environments for a long time (e.g. Azam and Ammerman 1984); similar processes can be assumed for soil bacteria.

16.2.1 The Stone Sphere

Rock outcrops, block fields and moraine lines are dominated by stones and boulders of various sizes. Such areas are typical of coastal areas of the maritime and continental Antarctic. The fascinating peculiarities of the endolithic communities in the sandstones of the Dry Valleys have been described in detail by Nienow and Friedmann (1993); the weathering process has been reviewed by Chen et al. (2000). In this region, there are several special features, such as shading or provision of substrates for lichens, thus forming a (micro-)milieu with a special character for biological successors and a source of organic debris. Such rocks also show a specially structured environment with respect to their orientation to wind and sun, i.e. forming special micro-climatological niches (Kappen et al. 1987, 1991; Bölter et al. 1989). In terms of environmental impact, rocks seem to be of more importance for structuring the landscape at different scales rather than providing substrates for organisms – except lichens.

Hypolithic communities occur underneath small quartz stones. They can be found frequently along the coast of Admiralty Bay and have been also described for many other regions (Nienow and Friedmann 1993). Although the effect of such miniature sites is minimal on the total environments, they may provide refuges for several algae and cyanobacteria (Broady 1981).

16.2.2 The Aggregate Sphere

From the viewpoint of the microorganisms, the soil environment is primarily determined by inorganic and/or organic particles. The distribution pattern is complicated by the fact that it covers a non-continuous space. The living space, the pores, can be regarded as the frame or shelter of the living environment or as a source of nutrients inside organic aggregates or at their surfaces. For many organisms, e.g. non-motile bacteria or fungi, this living space determines their food supply and their physical conditions. Their distribution is not only due to statistics, but is also determined by food resources and interspecific relations, which leads to a biologically determined structure of the environment. The structure of the microbial sphere is even more strongly affected by nutrient concentrations. Motile microorganisms can optimise their actual position in relation to the field gradients of various nutrients. Azam and Ammerman (1984) showed such a model for the ideal case of water and proposed the formation of 'clusters' 5–100 μm from an appropriate nutrient source.

On the other hand, the partitioning of habitats and resources with subsequent successions of different species of microbes, plants and animals is an important aspect. Monrozier et al. (1991) propose two communities in and on aggregates. The latter community should be more resistant to drying. From observations of small organic aggregates, e.g. diatom shells in samples from Arctowski, colonising is most intensive outside. A reasonable explanation may be a better supply of oxygen or nutrients. Organisms which live inside the aggregate can profit from shelter against grazing (Hattori and Hattori 1976). Foster (1988) could show that for a permanent colonisation of pores a ratio of 3:1 between pore diameter and bacterial diameter is necessary. For soil bacteria of Antarctic environments, which are seldom larger than 3×1 μm, a pore diameter of <10 μm might be well sufficient. Small cocci, which are often less then 0.3 μm in diameter, would only need a pore diameter of 1 μm to have sufficient living space. Much larger pores are necessary for fungi (diameter >10 μm), algae (diameter >100 μm), or nematodes (diameter >30 μm). Flagellates need a pore width of about 5 μm, amoeba at least a width of 10 μm. In terms of water-holding capacity, pore sizes larger than 50 μm are drained at field capacity, pores between 0.2 and 50 μm hold water between the limits of field capacity and the wilting point, pores smaller than 0.2 μm hold water which is unavailable for most plants.

We found two different forms of aggregates in Antarctic soils: those which are formed by soil animals and those which resulted from swell–shrink processes. The first are restricted to small grains (diameter <60 μm) due to mesofauna, the latter can be found in loamy soils where a subangular blocky to angular blocky structure (diameter 2–20 mm) occurs. These aggregates show a higher porosity and have a larger surface due to their blocky structure. Their stability is lower but provides much better habitats for microorganisms.

16.3 Resource Competition

Competition takes place when resources are in short supply. Territorial 'defence' by antibiotics has to be taken into consideration, although proof of this is difficult to obtain from field studies. The action of antibiotics or other growth inhibitors can be described as interference or contest competition and results in unequal access to resources due to action on the reproductive progress of competitors which get better access to nutrients.

It seems more likely that competition is based on different uptake rates of nutrients and access to limiting resources, i.e. exploitation or scramble competition. This has been shown for northern tundra soils by Jonasson et al. (1996a,b). These processes seem to be even more evident in the microenvironment as larger fluctuations in density and population may be built up quickly, based on a temporarily abundant nutrient. Experiments with the development of microcolonies of bacteria by carbohydrates and leucine have demonstrated this clearly (Bölter 1993). However, both strategies may operate simultaneously. The co-existence of different bacterial populations is effective for multiphasic mineralisation (Schmid and Gier 1990). Competition can be controlled by temperature and related substrate affinities of individual populations (Nedwell and Rutter 1994; Rutter and Nedwell 1994).

16.4 Water as a Controlling Factor

As all life functions can be related to water, and keeping in mind that water also controls the diffusion and transport of nutrients and gases, it is necessary to focus on this crucial compound in the soil environment. It has been considered as the primary limiting ecological factor for Antarctica (Kennedy 1993). On the other hand, temperature controls the snow melt, thus affecting the areas with snow cover, which is important for vegetation patterns and therefore related factors on soil developments (Blume and Bölter 1996).

Water availability controls the production of Arctic low tundra ecosystems (Chapin and Shaver 1985). Water stress has been found to be a dominant force in structuring Arctic plant communities (Bliss and Gold 1995), and soil moisture has been recognised as the main parameter explaining vegetation patterns in arctic tundra at Ellesmere Island (Batten and Svoboda 1994). Vegetation patterns, however, reflect the changes on a time scale of years, whereas microorganisms have to cope with short-term changes at weekly, daily or even shorter periods, a fact which is strongly pronounced in unsheltered barren surface horizons. Strong solar irradiation causes high evaporation rates and leads to desertified conditions in dry environments or tropical conditions in wet areas. The content of organic material has an important effect on the water content.

There are direct relationships between these variables verified from field studies at Casey and at Arctowski. Both temperature and moisture were demonstrated to be of high importance for microbial activity (Bölter 1989, 1990a,b).

For most plants, the availability of water is determined by the permanent wilting point, which occurs at about –1.5 MPa water tension, on the one hand, and the field capacity, on the other (–0.01 to –0.05 MPa). Soils with low organic matter content act as hydrophobic bodies (Singer and Ugolini 1976). This is true of sandy and humic soils, like the Podzols at Casey, but cannot be valid for loamy soils, as they have a higher water capacity. It also has to be taken into account that dry structural organic matter, e.g. peat or lichen debris, also has hydrophobic properties. This means that at upper levels (0–2 cm), where the highest numbers of organisms and highest activity levels can be monitored, are the strongest effects on microbes and only those organisms which can produce cysts or spores can use this environment. Deeper horizons are normally not water limited. Gold and Bliss (1995) were able to show that Pergelic Cryaquepts at Ellesmere Island have a matrix potential in the root horizon of >–0.1 MPa; water stress on plants in their study occurred just below 6–8 % water content.

Generally, microbial activity is optimal at field capacity and shifts to lower activities with decreasing matrix potentials. Low matrix potentials where organisms can survive have been described for bacteria (–10 MPa), yeasts (–20 MPa), fungi (–60 MPa) and actinomycetes (–70 MPa; Kilham 1994), the last value corresponds to a pF value of about 5.85. Data for Antarctic soils exist only for the Dry Valley region (Klingler and Vishniac 1988) where values

Table 16.1. Relationships between organisms and/or their activity and water stress. Sources: Paul and Clark (1989); Haider (1996); Schinner and Sonnleitner (1996); Wood (1995); Kilham (1994); Tate (1995)

Organism	Pressure (MPa)
Bacteria, fungi (optimum)	0.01–0.05
Growth of Antarctic yeasts	0.75–3.5
Bacterial respiration	0.8–3.0 (5.0)
Bacterial activity (forests)	0.2
Bacterial activity (deserts)	0.2–2.0
Rhizobium, Nitrosomonas	1.5
Fungi (growth)	4–10
Clostridium	10
Micococcus	25
Mucor	10
Aspergillus in desert soils	14.5–30
Penicillium	25
Xeromyces, Saccharomyces	65

between -6.9 and -0.07 MPa, which relate to 0.25–25% water (v/w), were determined. These data were obtained to describe the environment of Antarctic yeasts. More general data are available for describing limits of microbial abundance and/or activity (Table 16.1).

The data show an extreme span of the growth potential of some microorganisms. Such data are, however, not directly transposable to the soil environment, as water at -20 MPa is not available for microorganisms since it is in small caverns and capillary movement is very slow under such conditions. There are, however, distinctly different problems for plants. Mosses, especially, have to cope with water stress as they do not have internal structures for water transport, in contrast to vascular plants, except the Polytrichaceae. Both dead and living parts take part in the water flow by external capillary transport starting in the waterlogged soil, or from other sources like rain- and meltwater. The widespread of mosses in polar tundras might be regarded as an opportunistic growth on shallow soils or shallow active layers.

16.4.1 Consequences of Changing Water Availability

It could be shown that changes in the matrix potential are followed by changes in the microbial community structure and activity (Lynch 1982; Schinner et al. 1989). Low temperatures shift the metabolic activity optimum to low water contents. A relationship between matrix potential and respiration was described as log-linear for a range from -0.01 to -8.5 MPa (Orchard and Cook 1983). These authors showed that the change of the matrix potential from -0.01 to -0.02 MPa caused a decrease in microbial activity of 10%. Several reasons can be considered for this, such as changes in pH, Eh and sorption of ions to surfaces caused by changing water contents, which definitely alter the microbes' environment, access to nutrients and the confirmation of extracellular enzymes.

Low contents of humic substances and high contents of clay and silt are responsible for the poor buffering capacity of Antarctic soils. Further, soils are sometimes shallow, e.g. Lithic Leptosols. However, biological activity is mostly restricted to the top horizon (~10 cm) and is not very dependent on the deeper layers. During summer, permafrost levels are much deeper, at Casey >0.5 m, at Arctowski >1 m. Ions from weathering processes and, in coastal areas, from the influence of the sea act in more or less free forms in the soils. Hence, they 'compete' for water with living organisms via their osmotic attraction of water. If this attraction of water is strong enough, replacement of it should be by air. On the other hand, at low temperatures, water forms ice crystals and releases its solutes with the effect that the remaining solute becomes stronger and stronger. This effect increases osmotic stress on the organisms.

High, frequent temperature fluctuations just in this range will have considerable ecological consequences. It is not enough to adapt to low water tensions, as observed for cryptobiosis or anhydrobiosis for microarthropods (Block and Convey 1995; Convey and Block 1996). Microbes must be able to react quickly to dramatic changes in their living conditions to avoid dehydration. Strategies to combat this stress are physiological, e.g. increases in cell solutes to maintain a high turgor, or morphological, e.g. reduction in intercellular space, reduction in cell size. The latter seems to be widely used by bacteria when taking into consideration the small mean cell volumes and the high abundances of cocci-shaped cells in polar soils (Bölter 1995; Bölter and Pfeiffer 1997; Bölter et al. 1997).

Although optimal conditions for microbial activity can be found at 50 % saturation of the pore volume with water, which is an optimal value for air humidity in soil pores, lower values can become more favourable for microbial activity (Freytag et al. 1987). Wilson and Griffin (1975) found soil water contents of 3.5 %, equivalent to 12 % water saturation of the pore volume as a lower limit for bacterial respiration. It could also be shown that increasing moisture (50 % water of wet wt.) had drastic positive effects on the CO_2-evolution of soils but also shifted the point of optimal conditions (Bölter 1991). It should, however, be borne in mind that such enhancements of microbial activity occur only under oxygenated conditions, where no drastic shifts between aerobic and anaerobic communities have occurred. Soils have to be regarded as mixtures between aerobic and anaerobic (micro-) sites and not as a partially aerobic system (Tate 1995).

16.4.2 Osmotic Stress

Direct measurements of the osmotic potential become difficult under field conditions. It is possible to measure it by the cryoscopic method. Salinity also affects the temperature at which a solvent has its maximum density. This point drops linearly with increasing salinity. Further, it is possible to recalculate it from conductivity values with regard to the main active salt. Richards (1954) published tables of osmotic pressure values of single salt solutions as related to electrical conductivity. Changes in osmotic potential go along with changes in the thermodynamic activity of the solvent water. The more solutes added, the lower the water activity: non-ionising substances depress it less than salts due to the different chemical potentials of the solutes.

Data from soils of King George Island and the Windmill Islands show moderate osmotic values in these soils as derived from electrical conductivity measurements (Blume and Bölter 1996; Blume et al 1997; Table 16.2). These data are valid for soils which do not contain $CaSO_4$ and where the main salt is NaCl (Blume et al. 1997). It is evident that soils with high sand contents (sites C3, C4) show a higher osmotic water pressure during dry periods than others, due to a stronger influence of evapotranspiration. This also holds true for site C5, a moss bed, where the plant material shrinks and an increasing capillary

Table 16.2. Electrical conductivity (EC) and osmotic water pressure (owp) of representative topsoils at Casey and Arctowski

			C1	C2	C3	C4	C5	A1	A2	A3	A4	A5
Depth		1[a]	0–1	0–.5	0–1	0–8	0–3	0–4	0–1	0–2	0–7	0–4
cm		2	–5	–5	–5	–11	–8	–8	–8	–8	–12	–12
EC_{se}	mS cm^{-1}	1	7.5	0.6	2.1	3.8	3.5	2.9	0.4	0.4	0.7	1.3
		2	0.6	0.1	1.0	1.5	2.8	3.1	0.3	0.4	1.0	1.0
wc_{se}	Mass%	1	22	30	24	30	52	45	16	16	45	20
		2	30	45	20	35	55	36	18	16	36	21
wc_{fc}	vol%	1	13	23	11	6	43	35	12	14	36	14
		2	20	26	13	8	46	26	15	16	31	15
wc_{wp}	vol%	1	4.5	10	1.8	2.5	13	11	5	6	10	5
		2	5.8	15	2	3	14	9	7	6	6	5
owp_{se}	kPa	1	280	20	70	140	130	100	13	12	22	44
		2	20	3	30	50	100	110	9	12	30	30
owp_{fc}	kPa	1	490	25	150	700	150	130	17	14	28	64
		2	29	5	48	220	110	150	11	12	35	42
owp_{wp}	kPa	1	1400	57	940	1700	500	410	40	32	100	180
		2	980	9	300	580	380	430	23	32	180	130

[a] 1, First layer; 2, second layer; wc, water content; se, saturation extract; fc, field capacity (60 hPa); wp, wilting point (1500 kPa). For details of soil descriptions, see Table 16.3.

potential can be anticipated. At sites on King George Island, the osmotic water potential does not become critical due to humid climate conditions (cf. Rakusa-Suszczewski, Chap. 3).

An increase in the concentration of solutes leads to an increase in osmotic pressure and water potential, a decrease in water availability and in the cryoscopic coefficient. Although water stress cannot be separated from osmotic stress, this fact needs special attention in cold environments. The decreasing water availability during the freezing process increases the osmotic stress. Bacteria are known to have a moderately high turgor by accumulating osmotically active solutes. Booth (1999) discussed the possibility that Gram-positive organisms, which have high turgor pressures of about 2 MPa (Gram-negative: 0.5–0.6 MPa), have special advantages in environments with high osmolarity. This tolerance is multi-factorial, i.e. regulation must occur at the level of the cell competitives as well as at the level of membrane. These latter must be of a special nature as they cannot profit from the cytoplasmic solutes.

Amino acids and carbohydrates, e.g. proline, trehalose, and glycerol, are common, but also sucrose, mannitol, glucitol and others can be found. A compilation of various solutes in prokaryotic and eukaryotic cells is given by Borowitzka (1981). Concentrations of several mono- and disaccharides can be found at high levels as leachates from mosses and lichens in Antarctic soils, as well as sugar alcohols (Tearle 1987; Melick and Seppelt 1992, 1994). They are, however, not well used for metabolic processes by bacteria (Bölter 1993; Melick et al. 1994).

The change of the effective concentrations needs to be very quick while living at the 0 °C boundary. Quick intracellular regulation mechanisms for water activity by compatible solutes in response to water stress is important for survival, especially in harsh, cold, rapidly changing environments. So, here we have situations of fluctuating salinity, freezing and dryness. In spring, Antarctic soils are often exposed to short-term flooding due to snowmelt. These areas dry out and leave crusts of algae and cyanobacteria which can be found at many places on King George Island. Soil water not only affects the available moisture but also affects aeration by displacing air and the nature and amount of solutes. Cations become hydrated and change their solubility like small organic molecules which may change their dipole interactions and form new linkages between e.g. oligopeptides.

Drastic changes in water content were monitored within days at Fieldes Peninsula (King George Island, Chen and Blume 1996) and at Arctowski and Casey by R. Möller (pers. comm.). The latter have been measured indirectly via heat conductivity, and varied at Arctowski in meadows of *Deschampsia antarctica* between 13.8 and 49.9 % wet wt. for a period of 2 weeks in January 1992. A barren Regosol in the vicinity showed concomitant values between 9.9 and 20.8 % wet wt. Barren soils at Casey showed a fluctuation during a period of about 4 weeks, in December 1991 and January/March 1992, a variability between 1.7 and 22.4 % wet wt. (Bölter et al. 1995).

16.4.3 Physical Effects on the Habitat While Freezing and Thawing

The expansion of water while changing phase from liquid to ice is one of the most important effects on the microenvironment. It depends, however, on moisture and grain size, as to whether the bulk of matter expands or shrinks. Sand allows space for water to expand, and thus increases the bulk volume, while clay shrinks due to the decrease in temperature on the mineral. A strong destructive component arises from the formation of needle ice close to surface layers (Lawler 1988). This is important for vegetation destruction, soil erosion and frost sorting, and so has great influence on the microhabitat. Thus, organisms are strongly affected by their environment, as well as by ice nucleation inside their bodies. Freezing is thus not only a matter of changing physico-chemical conditions but also involves changing space and matter. The expansion of water needs space and air space is thus reduced.

Raoult's law states that the freezing point is lowered in relation to the concentration of solutes, i.e. its mole fraction. Each solute gives a characteristic value to the extent of the temperature shift, which are expressed by the cryoscopic constants of the solutes. Due to the progression of freezing water, the solute becomes more and more concentrated, which results in a steady drop in the remaining freezing point. This is the reason for the co-existence of liquid and solid phases. The eutectic point determines the end of this process and salt crystals appear. The mixture of salts in Antarctic soils changes from place to place. Often there is dominance of chlorides and sulphates from sea spray. On the other hand, carbonates may be abundant.

An important effect in the system of freezing and thawing of water is supercooling. The formation of ice crystals is strongly mediated by catalysts. The actual freezing thus is an effect not only of temperature but also of other factors. Schnell and Valli (1972) showed that decaying tree leaves could be important sources of ice nucleation particles. Ice nucleation activity can be effectively influenced by bacterial populations (Maki et al. 1974). Epiphytic bacteria have been shown to be nucleators between –2 and –5 °C (Sakai and Larcher 1987).

One physico-chemical action is the possible liberation of occluded organic matter. This is described as a process of disruption of granular structures of particular substances (organic debris) or dead cells (Skogland et al. 1988). Hence, the effect of freeze-thaw cycles can be related to dry–wet cycles going along with changes in the osmotic potentials, acidification, oxygenation, redox potentials and related processes, e.g. the liberation of toxic metals (Marion 1995). This not only affects the activity status of individual organisms at the level of different substrate affinities, but can create changes in overall community structure.

16.5 Ecological Rating of Habitats – An Example from Antarctic Sites

The possible living space for soil-dwelling organisms of continental Antarctic sites (e.g. Casey area) may go down to 10 to 30 cm and deeper (30–60 cm) in the maritime Antarctic (e.g. Arctowski area). It can be assumed to go down to 60 cm, depending on permafrost, nutrient supply and root penetration. Vascular plants and cryptogams depend on these conditions in different ways. Cryptogams depend on the amount of water which arises by capillary forces into the upper 10 cm, whereas vascular plants can profit from deeper horizons, to about 30 cm. This depth can be limited by rocks near the surface and high stone contents. Available water for plants and soil organisms is characterised by the water capacity in the living space (Table 16.3). The availability of oxygen decreases with increasing water content. This is indicated by redoximorphic features, e.g. rusty mottles. Nutrient availability can be characterised by the amount of exchangeable cations in the main rooting zone. Soil organic matter (SOM) contents can be used to characterise the living condi-

Table 16.3. Ecological conditions of representative Antarctic soils for Casey to a depth of 10 cm, and Arctowski 30 cm; Casey data after Blume and Bölter (1996), Arctowski data after Kuhn (1997); for methodological details see Schlichting et al. (1996). *skelet* mass of stones + gravel; *awc* available water capacity; N_{min} NO_3- + NH_4-N; K_a, Ca_a, Mg_a-exchangeable cations; P_a lactate-extractable P (= available); P_v hot HCl-extractable P (= available + weatherable); *org. m.* organic matter (= TOC*2); *feed* relative feeding% over 10 days in relation to offered food

Site	Soil	m a.s.l.[a]	Direction	Months – snow[b]	Temp.[c] (°C)	Oxygen deficieny	bact-C (mg m^{-2})	Plants	Feed%
Casey, Wilkes Land									
C1	Hapli-turbic Cryosol	33	E	0–1	+0.1	No	5.4	Algae	n.d.
C2	Stagni-turbic Cryosol	44	WSW	2–3	+0.2	Subsoil	2.8	Algae	n.d.
C3	Lepti-gelic Podzol	44	WSW	2–3	+0.2	No	19	Lichens	n.d.
C4	Skeleti-gelic Podzol	38	N	2–3	+0.4	Subsoil		Lichens	n.d.
C5	Foli-gelic Histosol	38	N	1–2	+0.4	Subsoil	22	Mosses	n.d.
Arctowski, King George Island									
A1	Molli-gelic Cambisol	38	N	4	+1.7	No	740	Grass	15
A2	Tephri-regic Cryosol	40	N	4	+1.7	No	47	<<Grass	7
A3	Ortithionic-g. Leptosol	55	NO	4	+1.6	No	23	<Grass	11
A4	Dystri-gelic Fluvisol	0.2	flat	5	+2.0	Partly	340	Mosses	15
A5	Tephri-regic Cryosol	122	SO	3	+1.0	No	19	Algae	11

Table 16.3 (*Continued*)

Site	Skelet (%)	awc ($l\ m^{-2}$)	N_{min} ($g\ m^{-2}$)	K_a ($g\ m^{-2}$)	Ca_a ($g\ m^{-2}$)	Mg_a ($g\ m^{-2}$)	P_a ($g\ m^{-2}$)	SOM ($kg\ m^{-2}$)	C/N
Casey, Wilkes Land									
C1	25	16	0.1	2.1	3.1	1.2	19	0.35	6.3
C2	28	14	0.1	19	74	120	19	1.3	10
C3	34	13	0.2	1.8	3.9	2.2	31	6.4	4.6
C4	85	3	0.02	0.8	2.4	0.8	4.9	1.0	12
C5	7	32	0.4	4.9	29	15	7.3	14	12
Arctowski, King-George Island P_v									
A1	35	41	4.2	15	390	190	130	8.3	8.0
A2	45	8	0.2	20	370	210	120	0.87	10
A3	64	22	0.2	16	250	140	90	0.70	n.d.
A4	33	190	0.8	10	89	41	260	7.6	3.8
A5	35	36	0.01	51	250	250	72	0.39	3.1

[a] Meters above sea level.
[b] Months without snow.
[c] Mean temperature of summer months.

tions for heterotrophic organisms. As such, an SOM with low C/N ratios (<20) describe a system of N-remineralisation. The data in Table 16.3 describe the ecological conditions of the (main) living zone of representative soils from the Casey and Arctowski areas. Further site descriptions are presented by Blume et al. (Blume et al., Chap. 7).

Sites C2–C5 are located on a hill which gives a much longer snow-free time than nearby areas in depressions, due to snowdrift during winter. On the one hand, such conditions can favour plant growth because of a longer period of light, on the other hand, such a location suffers from dryness and erosion processes. Locations in depressions or directly leeward of large boulders have much longer periods of snow cover and wet conditions, thus providing habitats for soil algae and cyanobacteria, but only small accumulations of organic matter occurs with subsequently low loads of heterotrophic organisms (sites C1 and C2). Hence, these environmental conditions do not allow the establishment of a developed biocoenosis. Further, they represent Turbic Cryosols (Blume and Bölter 1996), which also prevents the establishment of lichens or mosses. Due to the quartz-rich substrate at site C1 (Beyer et al., Chap. 9), only low amounts of K, Ca, and Mg can be found. Site C2 shows a higher level of Mg because its parent material is schist. Site C3 is a Gelic Podzol with a stabilised surface from fruticose lichens growing on small pebbles. Here, we find elevated bacterial biomass and organic matter with a low C/N ratio. The available water capacity, however, is at a minimum and the top horizon suffers from dry conditions due to its sandy skeletal texture. Site C4 is an ancient penguin rookery and the cover to 15 cm is up to 100 % coarse gravel. Site C5 represents a topsoil of a shallow peat bog influenced by meltwater and dry conditions during late summer in its top layers. Nutrient levels can be described as medium (Table 16.4). All sites at Casey have medium to high amounts of available phosphorus. This can be attributed to recent and ancient bird colonies. Oxygen deficiencies can be found in the subsoils due to waterlogging. Seaborne salts can often be enriched by capillary rise with effects on the osmotic pressure (Blume et al., Chap. 8).

At King George Island, the snow-free period is much longer and extends to 4 to 5 months. The local milder maritime climate allows the establishment of vascular plants. Actual cryoturbation (sites A2 and A5) and patterns of solifluction hamper the establishment of a permanent extended vegetation cover. Sites A1 and A4 show an elevated microbial carbon contribution in relation to high amounts of organic matter. These sites also show the highest results in feeding activity. All sites (except A4, a Fluvisol) show relatively high amounts of available Ca and Mg despite high stone contents due to basaltic and volcanic ashes as parent material.

According to Schlichting et al. (1996), it becomes possible to rank the living conditions provided by these data in the environments of the Admiralty Bay and Windmill Island regions. Main constraints are due to humid climate conditions and a shallow groundwater level. Many sites are further influenced by

Table 16.4. Assessments of ecological site conditions (after Schlichting et al. 1996) in polar regions for 3 dm soil depth for Arctowski area, in brackets for 1 dm for Casey area

Steps	1 very low	2 low	3 medium	4 slightly high	5 high	6 very high	7 extremely high
Sfp[a] (months)		0–1	1.5	3	6	9	12
r.p.[b] (%)		100	85	60	40	15	0
awc[c] ($l\ m^{-2}$)	<	15 (5)	30 (10)	60 (20)	90 (30)	140 (47)	200 (67) >
N_{min} ($g\ m^{-2}$)	<	0.02 (0.01)	0.1 (0.05)	2.0 (1.0)	4.5 (2.3)	12 (6)	25 (13) >
K_a ($g\ m^{-2}$)	<	2 (1)	8 (4)	24 (12)	48 (24)	80 (40)	200 (100) >
Ca_a ($g\ m^{-2}$)	<	4 (2)	15 (7.5)	50 (25)	100 (50)	200 (100)	500 (250) >
Mg_a ($g\ m^{-2}$)	<	2 (1)	5 (2.5)	15 (7.5)	30 (15)	60 (30)	150 (75) >
P_v (P_a) ($g\ m^{-2}$)	<	5 (0.5)	25 (0.5)	125 (5)	175 (13)	250 (20)	400 (50) >
org. m. ($kg\ m^{-2}$)	<	0.2 (0.1)	0.5 (0.25)	1.0 (0.5)	2 (1)	4 (2)	8 (4) >
bac[d] ($g\ m^{-2}$)	<	0.3 (0.2)	0.6 (0.4)	1.2 (0.8)	3 (2)	6 (4)	>
Feed%[e]	<	5	15	40	60	80	95 >

[a] Snow-free period.
[b] Penetrability by roots (mass % of stones + gravels, rock = 100 %; mean of 3 (1) dm.
[c] Available water capacity (+ groundwater if high groundwater table).
[d] Bacterial biomass in $g\ cm^{-2}$ (rating of 1/2 microbial biomass after Machulla 2001), with an adaptation due to mean temperature of the vegetation period after Blume et al. (1991).
[e] Feed activity (relative feeding% during 10 days in relation to offered food stripes after Törne (1990).

groundwater. Increased salt concentrations degrade the water availability by waterlogging and hamper water discharge. According to Richards (1954), a salt concentration of 1 % or 15 mS/cm is equivalent to an osmotic potential of 0.6 MPa. In the Arctowski area, the nutrients are calculated for the first decimetre in full, for the next 2 dm at half values (Table 16.4), due to differences in rooting intensity of the layers

16.6 Adaptation – A Clue to Survival in Harsh Environments?

Most of the effects described above are linked to limitations of survival. However, it has to be questioned whether these facts are actual boundaries of microbial communities or whether their life spans are much longer and not really affected. The fact to be taken into account is that we have to distinguish between potential life spans of individual organisms, i.e. individual physiological properties and biochemical pathways, and those of communities.

The harsh climatic conditions may have led to a reduction of major global gradients of metabolic activity but probably not of viable biomass and genetic potentials (White 1995). The singularity and isolation of individual soil patches and biomes might lead to arguments of genetically poorer populations, but a direct proof on this is missing. Passive transfers of propagules by wind, water or higher organisms may be sufficient to sustain the indispensable exchange of organisms even within the small scale of Antarctic soil habitats.

References

Addicott JF, Aho JM, Antolin MF, Padilla DK, Richardson JS, Soluk DA (1987) Ecological neighbourhoods: scaling environmental patterns. Oikos 49:340–346

Azam F, Ammerman JW (1984) Cycling of organic matter by bacterioplankton in pelagic marine ecosystems: microenvironmental considerations In: Fasham MJR (ed) Flows of energy and materials in marine ecosystems. Plenum Press, New York, pp 345–360

Batten DS, Svoboda J (1994) Plant communities on the uplands in the vicinity of the Alexandra Fiord Lowland. In: Svoboda J, Friedman B (ed) Ecology of a polar oasis, Alexandra Fiord, Ellesmere Island. Captus University Press, Toronto, pp 97–110

Block W (1984) Terrestrial microbiology, invertebrates and ecosystems. In: Laws RM (ed) Antarctic ecology, vol 1. Academic Press, London, pp 163–236

Block W, Convey P (1995) The biology, life cycle and ecophysiology of the Antarctic mite *Alaskozetes antarcticus*. J Zool Lond 236:431–449

Blume H-P, Beyer L, Friedrich F (1991) Correlations between microbial activity, water, air temperature and nutrient status of soils under different land use. In: Esser D, Overdiek D (eds) Modern ecology – basic and applied aspects. Elsevier, Amsterdam, pp 321–346

Blume H-P, Bölter M (1996) Wechselwirkungen zwischen Boden- und Vegetationsentwicklung in der Kontinentalen Antarktis. Verh Ges Ökol 25:25–34

Blume, H-P, Beyer L, Bölter M, Erlenkeuser H, Kalk E, Kneesch S, Pfisterer U, Schneider D (1997) Pedogenic zonation in soils of the southern circumpolar region. Adv GeoEcol 30:69–90

Bölter M (1989) Microbial activity in soils from Antarctica (Casey Station, Wilkes Land). Proc NIPR Symp Polar Biol 2: 46–153

Bölter M (1990a) Microbial ecology of soils from Wilkes Land, Antarctica: II. Patterns of microbial activity and related organic and inorganic matter. Proc NIPR Symp Polar Biol 3:120–132

Bölter M (1990b) Evaluation – by cluster analysis – of descriptors for the establishment of significant subunits in Antarctic soils. Ecol Modell 50:79–94

Bölter M (1991) Microbial mineralization in soils and plant material from Antarctica. In: Weller G, Wilson CL, Severin BAB (eds) International conference on the role of the polar regions in global change. Proc Conf, 11–15 June 1990,University of Alaska Fairbanks, vol II. Geophys Inst Univ Alaska, Fairbanks, pp 418–422

Bölter M (1993) Effects of carbohydrates and leucine on growth of bacteria from Antarctic soils (Casey Station, Wilkes Land). Polar Biol 13:297–306

Bölter M (1995) Distribution of bacterial numbers and biomass in soils and on plants from King George Island (Arctowski Station, maritime Antarctica). Polar Biol 15:115–124

Bölter M, Kappen L, Meyer M (1989) The influence of microclimate conditions on potential photosynthesis of *Usnea sphacelata*: a model. Ecol Res 4:297–307

Bölter M, Blume H-P, Kappen L (1995) Bodenbiologische Untersuchungen in der maritimen und kontinentalen Antarktis (King George Island und Windmill Islands). Teil 1. Umweltparameter und anorganische Nährstoffe. Polarforschung 65:41–61

Bölter M, Pfeiffer E-M (1997) Bacterial biomass and properties of Arctic desert soils. In: Iskandar IK, Wright EA, Radke JK, Sharratt BS, Groenevelt PH, Hinzman LD (eds) Proc Int Symp on Physics, Chemistry, and Ecology of Seasonally Frozen soils. CRREL Spec Rep 97–10,pp 481–487

Bölter M, Blume H-P, Schneider D Beyer L (1997) Soil properties and distributions of invertebrates and bacteria from King George Island (Arctowski Station), maritime Antarctic. Polar Biol 18:295–304

Booth IR (1999) Adaptation to extreme environments. In: Lengeler JW, Drews G Schlegel HG (eds) Biology of the prokaryotes. Thieme, Stuttgart, pp 652–671

Borowitzka LJ (1981) Solute accumulation and regulation of cell water activity. In: Paleg LG, Aspinall D (ed) The physiology and biochemistry of drought resistance in plants. Academic Press, Sydney, pp 652–671

Broady P (1981) The ecology of sublithic terrestrial algae at the Vestfold Hills, Antarctica. Br Phycol J 16:231–240

Chapin FS III, Shaver GR (1985) Arctic. In: Chabot BF, Mooney HA (ed) Physiological ecology of North American plant communities. Chapman and Hall, New York, pp 16–40

Chen J, Blume H-P (1996) Study on the dynamics of soil moisture in an ice-free area of the Fildes Peninsula, King George Island, the maritime Antarctic. Polarforschung 66:11–18

Chen J, Blume H-P, Beyer L (2000) Weathering of rocks induced by lichen colonization – a review. Catena 39:121–146

Convey P, Block W (1996) Antarctic Diptera: ecology, physiology and distribution. Eur J Entomol 93:1–13

Foster RC (1988) Microenvironments of soil microorganisms. Biol Fertil Soils 6:189–203

Freytag HE, Jäger R, Lüttich M (1987) Berechnung des Temperatur- und Feuchteeinflusses auf die Bodenatmung auf zwei verschiedenen Wegen. Arch Acker-Pflanzenbau Bodenkd Berl 31:513–520

Gold WG, Bliss LC (1995) Water limitations and plant community development in a polar desert. Ecology 76:1558–1568

Haider K (1996) Biochemie des Bodens. Enke, Stuttgart

Hattori T, Hattori R (1976) The physical environment in soil microbiology: an attempt to extend principles of microbiology to soil microorganisms. CRC Crit Rev Microbiol 4:423–461

Jonasson S, Vestergaard P, Jensen M, Michelsen A (1996a) Effects of carbohydrate amendments on nutrient partitioning, plant and microbial performance of a grassland-shrub ecosystem. Oikos 75:220–226

Jonasson S, Michelsen A, Schmidt IK, Nielsen EV, Callaghan TV (1996b) Microbial biomass C, N and P in two arctic soils and responses to addition of NPK fertilizer and sugar: implications for plant nutrient uptake. Oecologia 106:507–515

Kappen L, Bölter M, Kühn A (1987) Photosynthetic activity of lichens in natural habitats in the maritime Antarctic Bibl Lichenol 25: 97–312

Kappen L, Breuer M, Bölter M (1991) Ecological and physiological investigations in continental Antarctic cryptogams. 3. Photosynthetic production of *Usnea sphacelata*: diurnal courses, models, and effect of photoinhibition Polar Biol 11:393–401

Kennedy AD (1993) Water as a limiting factor in the Antarctic terrestrial environment: a biogeographical synthesis. Arct Alp Res 25:308–315

Kilham K (1994) Soil ecology. Cambridge University Press, Cambridge

Kuhn D (1997) Genese, Ökologie und Soziologie einer Bodengesellschaft in einem Periglazialgebiet der King George Insel, West-Antarktis. Schriftenr Inst Bodenkd Univ Kiel 40:1–173

Lawler DM (1988) A bibliography of needle ice. Cold Reg Sci Tech 15:295–310

Lynch JM (1982) Limits to microbial growth in soil. J Gen Microbiol 128:405–410

Machulla G, Blume H-P, Jahn R (2001) Schätzung der mikrobiellen Biomasse von Böden aus anthropogenen und natürlichen Substraten – ein Beitrag zur Standortbewertung. J Plant Nutr Soil Sci 164:547–554

Maulla G (1997) Microbial activity. In: Blume H-P, Schleuß U (eds) Assessment of anthropogenic soils in cities. Schriftenr Inst Pflanzenern Bodenkd Univ Kiel 38:176–192

Maki LR, Galyan ME, Chien ME, Caldwell DR (1974) Ice nucleation induced by *Pseudomonas syringae*. Appl Microbiol 28:456–459

Marion GM (1995) Freeze-thaw processes and soil chemistry. CRREL Spec Rep 95–12:1–23

Melick DR, Seppelt RD (1992) Loss of soluble carbohydrates and changes in freezing point of Antarctic bryophytes after leaching and repeated freeze-thaw cycles. Antarct Sci 4:399–404

Melick DR, Seppelt RD (1994) The effect of hydration on carbohydrate levels, pigment content and freezing point of *Umbilicaria decussata* at a continental Antarctic locality. Crypt Bot 4:212–217

Melick DR, Bölter M, Möller R (1994) Rates of soluble carbohydrate utilization in soils from the Windmill Island Oasis, Wilkes Land, continental Antarctica. Polar Biol 14:59–64

Monrozie, LJ, Ladd JN, Fitzpatrick RW, Foster RC, Raupach M (1991) Components and microbial biomass content of size fractions in soils of contrasting aggregation. Geoderma 49:37–62

Nedwell DB, Rutter M (1994) Influence of temperature on growth rate and competition between two psychrotolerant Antarctic bacteria: low temperature diminishes affinity for substrate uptake. Appl Environ Microbiol 60:1984–1992

Nienow JA, Friedmann EI (1993) Terrestrial lithophytic (rock) communities. In: Friedmann, EI (ed) Antarctic microbiology. Wiley-Liss, New York, pp 343–412

Orchard VA, Cook FJ (1983) Relationships between soil respiration and soil moisture. Soil Biol Biochem 15:447–453

Paul EA, Clark FE (1989) Soil microbiology and biochemistry. Academic Press, London

Richards LA (1954) Diagnosis and improvement of saline and alkaline soils. Agric Handbook No 60, US Dept Agric, Washington, DC

Rutter M, Nedwell DB (1994) Influence of changing temperature on growth rate and competition between two psychrotolerant Antarctic bacteria: competition and survival in non-steady-state temperature environments. Appl Environ Microbiol 60:1993–2002

Sakai A, Larcher W (1987) Frost survival of plants. Springer, Berlin Heidelberg New York

Schinner F, Sonnleitner R (1996) Bodenökologie: Mikrobiologie und Bodenenzymatik. Springer, Berlin Heidelberg New York

Schlichting E, Blume H-P, Stahr K(eds) (1995) Pedological traineeship. Blackwell, Berlin

Schmidt SK, Gier MJ (1990) Coexisting bacterial populations responsible for multiphasic mineralization kinetics in soil. Appl Environ Microbiol 56:2692–2697

Schnell RC, Valli G (1972) Atmospheric ice nuclei from decomposing vegetation. Nature 236:163–165

Singer MJ, Ugolini FC (1976) Hydrophobicity in the soils of Findley Lake, Washington. For Sci 22:54–58

Smiles DE (1988) Aspects of the physical environment of soil organisms. Biol Fertil Soils 6:204–215

Tate RL (1995) Soil microbiology. Wiley, New York

Tearle PV (1987) Cryptogamic carbohydrate release and microbial response during spring freeze-thaw cycles in Antarctic fellfield fines. Soil Biol Biochem 19:381–390

Törne Ev (1990) Assessing feeding activities of soil-living animals. I. Bail-lamina-tests (1). Pedologia 34:89–101

White DC (1995) Chemical ecology: possible linkage between macro- and microbial ecology. Oikos 74:177–184

Wilson JM, Griffin DM (1975) Water potential and the respiration of microorganisms in the soil. Soil Biol Biochem 7:199–204

Wood M (1995) Environmental soil biology. Blackie, London

Zak J, Whitford W (1988) Interactions among soil biota in desert ecosystems. Agric Ecosyst Environ 24:87–100

17 Ecological Classification of Terrestrial Algal Communities in Polar Environments

Dedicated to Professor Josef Svoboda on the occasion of his 70th birthday

J. Elster

17.1 Introduction

Cyanobacteria (Cyanoprokaryota, Cyanophyta) and eukaryotic algae (the terms 'algae' and 'algal' in the text always includes Cyanobacteria, unless further specified) play an important role in polar terrestrial ecosystems (e.g. Holdgate 1970; Laws 1984; Chernov 1985; Walton 1987; Aleksandrova 1988; Vincent 1988; Kerry and Hempel 1990; Friedmann 1993; Hempel 1994; Fogg 1998; Priscu 1998). They are widespread in all polar terrestrial environments, including extremes, and frequently produce visible biomass. Their combined biomass represents a sizeable pool of global fixed carbon, influencing mineral cycling and energy flow, and affects the mineral and biological development of polar soils.

Algae possess an oxyphototrophic type of photosynthesis and are one of the main primary colonisers adapted to the extreme natural conditions in polar terrestrial habitats. They are at the very start of nutrient and energy accumulation in the bare polar periglacial landscape. Cyanobacteria are also important for their fixation of elemental nitrogen and its accumulation in nitrogen-limited habitats. Seasonal and diurnal variations of polar terrestrial environments present series of patterns from aquatic (semi aquatic) to dry habitats along a water availability gradient. These patterns differ in periodicity, amplitude, synchrony and regularity, and initiate a number of different ecological and physiological acclimations and adaptations.

The vast barren polar periglacial terrestrial ecosystem, fully or partially covered by ice-caps, is an unstable system because of the impact of global climate change (Chapin et al. 1992). The periglacial terrain, where algae play an especially important role, is designated as the region bordering retreating glaciers. Aquatic and/or semi-aquatic habitats, colonised by algae, function as nourishing veins, initiating the process of primary succession on the newly

Ecological Studies, Vol. 154
L. Beyer and M. Bölter (eds.) Geoecology of Antarctic Ice-Free Coastal Landscapes

ice-free landscape (Lewis Smith 1993; Wynn-Williams 1993; Elster and Svoboda 1995, 1996; Elster et al. 1997).

The present knowledge about the physiology of algae which occupy polar unstable terrestrial environments is poor. A wide spectrum of physiological adaptations and acclimatisation mechanisms, which may protect cells against environmental damage, has been considered. However, due to the extremely high diversity of algal species, life forms and strategies, information concerning algal adaptation and acclimation to polar terrestrial environments is still limited. Elster (1999) summarises this problem under the following points:

1. Changes in species diversity and life strategy in time,
2. Changes in the rate of physiological and biochemical processes,
3. Changes in biomembrane structural-composition and consequently changes in the fluidity of biomembranes,
4. Production of special protective substances.

Much more detailed information concerning algal adaptation and acclimation to stressful conditions – conditions which are common in the polar terrestrial environment – can be found in reviews and original papers (De Winder 1990; Kirst 1990; Vincent and Roy 1993; Peterson 1993; Lawlor and Keys 1993; Kirk 1994; Kirst and Wiencke 1995).

Due to the wide dispersal of algae and their propagules by various vectors, this contribution will take into account:

1. *Overall algal diversity and ecology* in different habitats with respect to other polar ecosystems. The present system of ecological classification for terrestrial polar algae is inconsistent. Various authors studying the terrestrial algae of polar regions used ecological classifications which are common for temperate regions or which were only slightly modified for polar ecological situations. Thus,
2. *Ecological re-classification of polar terrestrial habitats* will be another focal point of this chapter.

17.2 Environmental Constraints

17.2.1 Climate

Climatic conditions are the principal external factors controlling the functioning of terrestrial ecosystems and communities. Near the ground surface (microclimate) is a zone of energy exchange between the substratum and the atmosphere above. Microclimate and environment around individual organisms (microenvironments) are of the highest significance, especially in the latitudes of polar regions. This very narrow zone, just a few centimetres above and below the ground surface, is the environment of life in bare landscapes. It seems like a climatic paradise in cold and frozen ecosystems. However, there are great differences in heat exchange between air and ground because of the presence or absence of water (dry soil, wet soil and/or shallow wetlands). Water acts as an important buffer to temperature change by its large heat storage capacity.

17.2.2 Hydrology and Permafrost

Polar terrestrial ecosystems differ substantially from those of temperate regions because of the presence of permafrost. The thickness of the active layer depends on the solar heat balance and even thin surface layers of algae can influence the exchange of heat at the depths where permafrost occurs. Permafrost limits the subsurface system to the active layer and restricts the migration of water to deeper soil. In an active layer which, in milder localities, is frequently saturated with water, and where, besides the wet soil, various shallow temporary freshwater ecosystems (lotic and lentic wetlands) arise, favourable conditions exist for mass algal microflora development. In such semi-aquatic or aquatic habitats, the leaching of salts and nutrients from the substrate is also more intense.

In terms of environmental priorities in terrestrial polar ecosystems, demands for moisture precede demands for nutrients, which precede demands for higher temperature (Svoboda and Henry 1987; Kennedy 1993; Bliss et al. 1994; Svoboda and Freedman 1994; Lewis Smith 1996, 1997; Lévesque 1997). This statement is convincingly illustrated by patterns of plant distribution along environmental gradients from extreme polar deserts, through semi-deserts, to polar coastal landscapes and/or various types of wetland habitats, and it can also be applied to the microflora.

The last, but by no means least, important factor influencing the distribution, species composition and ecological character of terrestrial polar habitats are the algal aerobiological inputs of viable cells or their propagules. Most of the Arctic periglacial terrain is easily accessible because of the latitudinal distribution of land masses. In contrast, Antarctica is isolated by its surrounding ocean, and even within the continent itself the transportation of cells and propagules is limited.

17.3 Habitat Classification

In this chapter, the ecological classification of the polar terrestrial–periglacial ecosystem is considered with particular attention to algal communities and a new modified structure is proposed. This classification is established with respect to: (1) the water gradient which reflects local climate, the presence or absence of water directly or indirectly determining the properties of the polar soils; and with respect to (2) age which determines the stage of development of algae and, subsequently, vegetation. However, the water gradient and the stage of algal community development can change with climatic conditions.

Three categories of environment are proposed: lacustrine, hydro-terrestrial and terrestrial. This terminology was proposed by Metting (1981) as algae are found along the gradation between planktonic, hydro-terrestrial, and terrestrial microenvironments.

The *lacustrine* (lake) environment has a special limnological character, although the algal benthic communities of lakes overlap, in many cases, with the hydro-terrestrial discussed later. Because lakes are more stable in time, and also because of the indirect connection with the soil environment, the lacustrine algae are not discussed here.

The distinction between aquatic and terrestrial habitats (Vincent and James 1996) is not clear and overlaps between these types of habitats frequently occur. These overlaps have a special ecological importance in polar terrestrial ecosystems because of the presence of permafrost and a wet active ground layer. In the papers reviewed by Broady (1996), terrestrial algae are regarded as those which occur in situations ranging from extremely wet, e.g. on and/or amongst bryophyte cushions bordering streams, to extremely dry, e.g. in soils of the Dry Valleys of Southern Victoria Land.

Hydro-terrestrial environments (wetlands) will be proposed here, to group together all types of habitats where liquid water is available for almost the entire period of summer, for temporary shallow freshwater or brackish, lotic or lentic wetlands and wetted soils – shallow lakes and pools, glacial and snow-fed streams and rivers, including their wet banks and periodically-flooded surroundings, irrigated rock faces or wet walls, saturated mineral soils or wet slope seepages, secondary springs and algal crusts, wet meadows, sea terrace brackish pools and salt marshes. Such types of habitats with flowing and stagnant waters often physically grade into each other. A common distinction between hydro-terrestrial (wetland) and limnetic (lake) areas in polar regions is that wetlands freeze solid during winter, whereas most lakes do not (Hawes et al. 1992). This inevitability is a strong habitat-defining characteristic which will place considerable stress on resident organisms (e.g. Hawes et al. 1992; Vincent et al. 1993b). Algae living in hydro-terrestrial environments have two life strategies which produce macroscopically visible algal growths: algal mats and crusts, and mucilaginous clusters and jelly biomass floating in waters. Under the category of the hydro-terrestrial environment, cryophilic algae also belong, growing between ice crystals in the surface layers of melting snow, and algae occurring in so-called cryoconite holes and/or supraglacial pools on the surface of melting ice caps. All the above categories of algae in the hydro-terrestrial environment directly influence the ecosystem of polar soils.

17.3.1 Hydro-Terrestrial Environments

Polar hydro-terrestrial habitats represent a unique mosaic of unstable extreme environments where large variations in algal species composition and biomass between various types of habitat is evident. Apart from the influence of the many environmental conditions, such as nutrients, salinity, desiccation, pH, and light conditions, as well as natural disturbances – sediment load, and/or water and wind erosion, the overwintering life strategies of organisms can also influence the presence of algal biomass and species composition in a given habitat. Due to the perennial character of many, mainly cyanobacteria-dominated, communities in more extreme situations, a higher biomass of algae can occur although the growth rate remains low. However, between these two algal life strategies which produce visible biomass (i.e. mats or mucilaginous clusters), a series of other transient living strategies also exist.

Hydro-terrestrial algal communities can occur at up to the highest and lowest possible latitudes and altitudes as long as free water is available for some time in the year. Algal distribution, abundance and species diversity vary in accordance with habitat (microhabitat) characteristics. In localities with a steady moisture and nutrient supply, abundance and species diversity of algae is relatively high. However, as the severity of living conditions increases (mainly changes in desiccation–rehydration and subsequent changes in salinity), algal abundance and species diversity decreases.

In the Arctic, hydro-terrestrial algae occur up to the most northern land margins because of easy access and because of less stressful conditions (for example, open seas and a moister environment). Hydro-terrestrial algae have long been studied here (Dickie 1878; Borgesen 1910; Petersen 1924; Ross, 1947; Whelden 1947; Croasdale 1973; Moore 1974a,b; Hamilton and Edlund 1994; Elster and Svoboda 1996; Elster et al. 1997).

Many studies have been published from both continental and the climatically-milder maritime Antarctica. These data cover a wide spectrum of geographical localities from periglacial areas around the continent and the Antarctic Peninsula to sub-antarctic islands. Considerable attention has been given to the climatologically-unique, areas of the Dry Valleys, Victoria Land, where hydro-terrestrial habitats also occur.

17.3.1.1 Algal Mats and Crusts

Algal mats grow on stones (epilithic), on sand (epipsammic), on the uppermost layer of mud (epipelic), on submersed mosses and vascular plants (epiphytic). Algal mats and crusts are cohesive, skin-like, mucilaginous films, flocculi and aggregated microscopic-macroscopic structures produced mainly by prokaryotic (cyanobacteria), and to a lesser extent by eukaryotic, microorganisms (Logan et al. 1964; Golubic 1973; Parker et al. 1981; Wharton et al. 1983; Vincent et al. 1993a). They are well-adapted to the severe conditions of polar hydro-terrestrial environments, such as: changes in desiccation–rehydration or salinity stress, freeze-thaw cycles, and bright, continuous mid summer solar radiation including UV-B (Howard-Williams and Vincent 1989; Vincent and Howard-Williams 1989; Howard-Williams et al. 1989; Davey 1989; Vincent et al. 1993a,b; Vincent and Roy 1993; Vincent and Quesada 1994; Wynn-Williams 1994; Tang et al. 1997).

The ability to recover from desiccation gives a great ecological advantage, and this allows cyanobacteria-dominating mats and crusts to be perennial (Vincent and Howard-Williams 1986; Davey 1989: Hawes 1989; Hawes et al. 1992). The production of algal biomass is higher than subsequent decomposition, which leads to an accumulation of organic carbon in nutrient-poor polar terrestrial ecosystems (Davey 1989; Howard-Williams and Vincent 1989; Vincent and Howard-Williams 1989; Vincent et al. 1993a,b; Tang et al. 1997).

Algal benthic mats and crusts are up to several centimetres thick and may completely cover the bottom substrate (Howard-Williams et al. 1986; Vincent et al. 1993a; Davey and Clarke 1991). Whereas some algal mat types are largely products of single cyanobacterial species, others are complex differentiated microbial communities with several dominant cyanobacteria or eukaryotic algae.

Despite the large variation in mat thickness, surface morphology and colour, most communities have a similar pigment organisation. The carotenoid-rich surface layer serves as a protection against damage from excessive light. Under this layer, a chlorophyll-a maximum enriched in phycocyanin occurs. In several of these mats, deep-living trichomes can migrate to the surface (Vincent and Howard-Williams 1986; Vincent 1988; Howard-Williams and Vincent 1989; Davey and Clarke 1991; Hawes et al. 1992; Vincent et al. 1993a; Elster and Svoboda 1996; Elster et al. 1997). Due to the mobility of filamentous cyanobacteria and mucilage production, sediment particles are trapped and bound together. Sediment is therefore stabilised by mats and crusts. Similar fine structures of cyanobacterial mats and/or crusts were studied by Broady (1979), Davey and Clarke (1991), Elster et al. (1999) from fellfield soils. These types of cyanobacterial mats and crusts occur on soil surfaces, so-called soil crusts, and will be discussed later in more detail. Algal benthic mats and crusts (modern stromatolites) have been studied in the benthos of deep continental Antarctic lakes (Parker et al. 1981; Wharton et al. 1983; Parker and Wharton 1985).

Surprisingly, the mat and crust morphology of cyanobacteria-dominated stromatolites, originating from completely different ecosystems at various tropical localities around the world, coincide in many respects with these mats and crusts formed in polar hydro-terrestrial habitats. (Friedman et al. 1973; Golubic 1973, 1976, 1985; Walter et al. 1976; Krumbein et al. 1977; Bauld

1981; Awramik et al. 1976; Golubic 1985). However, in the case of tropical localities, mats with a single cyanobacterial species, or with only limited numbers of cyanobacteria, frequently achieve their highest natural abundance, but also under extreme conditions of high pH, salinity, or temperature (Golubic 1973) just as in the polar environment.

Cyanobacterial strains isolated from mats and crusts of polar hydro-terrestrial habitats are genetically not adapted to the cold environment. Among 27 isolates of cyanobacteria studied, the temperature optima for growth ranged from 15 to 35 °C, with an average of 19.9 °C (Tang et al. 1997). These laboratory results are in agreement with field measurements of maximum photosynthesis being between 15 °C (Davey 1989) and 25 °C (Vincent and Howard-Williams 1989). Not only are there similarities in the morphology of mats and crusts of both tropical stromatolites and polar hydro-terrestrial habitats, but also their competitive advantages, such as tolerance to extreme conditions, are common for both types of microbial communities. It is known that cyanobacteria which have built up peculiar laminated rocks serve as records for the evolution of photosynthetic microorganisms. However, what we did not know, and thus are interested in, is the role of polar cyanobacterial mats and crusts in the phylogeny of these special life forms. Such research should tell us the evolutionary differences between these two types of 'Precambrian' microbial communities.

17.3.1.2 Mucilaginous Clusters and Jelly Biomass Floating in Waters

Mucilaginous clusters are attached to stones (epilithic) and on submersed mosses or vascular plants (epiphytic). Jelly biomass can occur in the form of metaphyton (algae loosely associated with bottom vegetation). In shallow stagnant habitats, phytoplankton (phytoflagellates) also frequently occur. Mucilaginous clusters and jelly biomass are produced by both prokaryotic and eukaryotic microorganisms, however, eukaryotic microorganisms more frequently produce visible biomass. This group of algae has a different life strategy in comparison with prokaryotic microorganisms dominated by cyanobacteria. Higher rates of photosynthesis (Davey 1989; Hawes 1989) and lower resistance to freeze-thaw cycles predetermine their annual character (Hawes 1989; Hawes et al. 1993, 1997; Davey 1989; Elster and Svoboda 1996; Elster et al. 1997). They are very susceptible to low temperatures and have high mortality on freezing. However, the retention of viability by a very small number of cells, after prolonged exposure to temperatures below 0 °C, will ensure the year-to-year persistence of the algae. Their survival during winter depends on a few highly resistant cells (Hawes et al. 1992).

17.3.1.3 Nutrient Cycling and Matter Flow in Algal Mats

Herbivory is an important factor influencing the species diversity and occurrence of algal biomass in terrestrial habitats. Exclusion experiments in mountain streams at temperate latitudes have shown that grazer activity influences microbial diversity and abundance (Hunter et al. 1992; Lamberti and Feminella 1996). Benthic or wet soil grazers are generally restricted to small organisms – insect larvae, nematodes, rotifers, tardigrades and protozoa (Hawes 1989; Vincent and Howard-Williams 1989). Grazing pressure thus seems unlikely to be a significant influence on Antarctic algal hydro-terrestrial communities.

Grazing pressure increases as environmental conditions become milder, for example, in accessible terrain where transport of larvae of herbivorous animals is easy (Elster et al., unpubl.). However, little information has been available on this topic in polar hydro-terrestrial habitats until recently. Vincent and Howard-Williams (1989) showed that decomposition of algal biomass under natural polar conditions is slow and its rate increases with temperature. Even a small rise in temperature can influence the decomposition processes and herbivore activity remarkably (Chapin et al. 1992). Fungal hyphae are frequent components of algal mats and crusts. They can have various roles from the symbiotic to the parasitic. It has been shown (Elster et al. 1997) that parasitic fungi attack mats of filamentous green algae.

17.3.1.4 Community Structure of Algal Mats

In both macroscopically-visible algal growths discussed above (mats or mucilaginous clusters), similarities have been found in life-strategy and ecological role in the Arctic and the Antarctic hydro-terrestrial environments. In cyanobacteria-dominated mats and crusts similar genera produce visible biomass – Chroococcales – *Aphanocapsa*, *Aphanothece*, *Gloeocapsa*, *Gloeocapsopsis*; Oscillatoriales – *Leptolyngbya*, *Lyngbya*, *Schizothrix*, *Phormidium*, *Microcoleus*, *Oscillatoria;* Nostocales – *Scytonema*, *Tolypothrix*, *Calothrix*, *Dichothrix*, and *Nostoc* have been determined. Benthic mats (cohesive, skin-like, mucilaginous films, flocculi of various colours) are composed mainly of *Phormidium* and *Nostoc*, together with soft *Leptolyngbya* trichomes. In localities where percolating water flows and desiccation often occur, *Gloeocapsa*, *Schizothrix*, *Dichothrix* and *Scytonema*, again together with *Leptolyngbya* trichomes, occur (Croasdale 1973; Broady 1981a; Engelskjøn 1986; Vincent and Howard-Williams 1986; Broady 1989a, b, c; Wynn-Williams 1990; Elster et al. 1997, 1999).

Filamentous forms of eukaryotic algae are the most frequent component of visible algal biomass in hydro-terrestrial habitats. Here again, compositions of

Arctic and Antarctic algal genera coincide. Algal mats and mucilaginous clusters floating in water are mainly composed by the following genera: Chrysophyceae – *Hydrurus;* Xanthophyceae – *Tribonema;* Chlorophyceae – *Microspora, Prasiola*; Charophyceae – *Klebsormidium*; Ulvophyceae – *Ulothrix*; Zygnemaphyceae – *Zygnema, Mougeotia, Spirogyra*. Jelly biomass floating in water is produced by *Cylindrocystis,* and soft yellow–brown mats on alluvial sands and gravels are produced by several pennate diatoms – *Tabellaria, Hannea, Fragilaria, Synedra, Eunotia, Stauroneis, Achnanthes, Hantzschia, Navicula, Pinnularia, Cymbella, Nitzschia* (Croasdale 1973; Moore 1974a,b; Broady 1982, 1986, 1987, 1989a–c; Engelskjøn 1986; Vincent-Howard-Williams 1986; Ohtani 1986; Ohtani and Kanda 1987; Hawes 1989; Davey 1989; Kawecka and Olech 1993; Vinocur and Pizarro 1995; Pizarro et al. 1996; Elster and Svoboda 1996; Elster et al. 1997, 1999; Van de Vijver and Beyens 1997; McKnight et al. 1998).

17.3.1.5 Snow Algae Communities (Cryophilic Algae)

Snow fields, where melt-water percolates through the upper layer, may have their surfaces coloured by algae communities. More than 110 algal species occur in melted snow (Kol 1968). Their distribution is worldwide, in alpine regions as well as in the Arctic and Antarctic, where snow ablation occurs. These algae have also received extensive attention in the literature of both polar regions (Fogg 1967; Akiyama 1977; Ishikawa et al. 1986; Kol 1969, 1971, 1972; Kol and Flint 1968; Broady 1989 c; Pankow et al. 1991; Ling 1996; Ling and Seppelt 1990, 1993, 1998; Gradinger and Nürnberg 1996; Mataloni and Tesolin 1997; Muller et al. 1998).

They can cover white snow with colours ranging from green through yellow and orange to red or brown and grey depending on the species, cell cycle timing and environmental conditions. The colour of snow is induced mainly by algae such as *Chlamydomonas* spp., *Chloromonas* spp. and *Mesotaenium* spp. However, a few studies have noted patterns in species distribution which appear to depend on snow chemistry. On the ice-free area of the Windmill Islands, Antarctica, grey snow was dominated by *Mesotaenium berggrenii* (Ling and Seppelt 1990) and red snow by *Chloromonas rubroleosa* (Ling and Seppelt 1993). Both of them are widespread on mainland ice-free areas but are very rare on offshore islands where green snow occurs around the large penguin rookeries. In contrast, on the snow fields of Arctic Svalbard, Muller et al. (1998) showed that an extensive accumulation of astaxanthin and its esters accompanied the transition from green biflagellated cells to orange spores, hypnozygotes and dark-red cysts. The nutrient content of meltwater and snow algae had no direct correlation with the content of secondary carotenoids.

A wide spectrum of cyanobacteria and algae have been found in melting snow. However, only a limited number of true psychrophiles which complete their life cycles in the snow is known. They assimilate and develop at temperatures in a very narrow range around 0 °C. They have an optimum growth below 10 °C and no growth at temperatures above 10 °C (Hoham 1975; Hoham et al. 1979; Kawecka 1978, 1986; Ling and Seppelt 1993).

Similar cold conditions exist in cryoconite holes and/or supraglacial pools. Cryoconite is the rock debris which is blown onto the ice surface. It absorbs radiation and melts underlying ice to produce a hole or pool (Fogg 1998). Cryoconite holes and pools are also occupied by algae communities (Vincent 1988). However, little has been known about these algae until now.

17.3.2 Terrestrial Environments

In polar terrestrial environments water is available in liquid form for only a short period (e.g. snow melt, summer rain or snow fall) or is available as air humidity (vapour absorbed directly from the air). Algae, in nearly all polar terrestrial environments, are invariably encountered both on and beneath the surface of soils and rocks. They range from communities on periodically-wetted soils and rock, to two clearly terrestrial categories which were established on the basis of habitat types. These three categories have been defined by Petersen (1935) as the hydro-terrestrial, the euterrestrial (including both epiterranean and subterranean soil algae), and the aeroterrestrial (aerial and subaerial). However, the first community of wetted soils and rocks, where algae produce crusts and mats, overlap in many cases with the hydro-terrestrial environment discussed above. Real terrestrial algae, including edaphophytes (*edaphic* – soil algae) and aerophytes (*lithophytic* – aerial algae), play an especially important ecological role in polar desert-semi-desert and/or periglacial ecosystems.

In the following text only edaphic and lithophytic communities will be briefly discussed because of their close connection with the geoecology of terrestrial polar habitats.

17.3.2.1 Soil Algae (Edaphophytes)

Polar substrata (ground–bedrock) vary greatly in their physical and chemical characteristics (e.g. Tedrow 1977, 1991; Walton 1984; Campbell and Claridge 1987). Soil substratum is derived from a range of parent rock types, where concentration of salts differs widely. The organic content is usually extremely low, but the soil can contain high levels. Soils also vary in texture, pH and nutrient content and can differ markedly in moisture from extreme aridity to water saturation. Cold conditions and aridity slow down chemical weathering and the rate of biological activity. Water saturation, in contrast, results in leaching of salts, exclusion of oxygen, and in accumulation of organic matter.

All these soil factors can change remarkably over short distances and with time. However, altitudinal and latitudinal gradients in soil conditions also affect the composition and abundance of algal communities (Broady 1996; Elster et al. 1999) that exist through both polar regions (Longton 1988). It has been proposed that in soils of the colder and more arid Coastal and Slope Provinces of Continental Antarctica, cyanobacteria and algae predominate but, in contrast, in milder and moister maritime Antarctica, bryophytes and lichens are more common (Smith 1984).

The polar soil flora is selected from a limited range of algae within the Cyanobacteria, Chlorophyta, Bacillariophyta and Xanthophyta (Akiyama 1967; Novichkova-Ivanova 1972; Prescott 1979; Engelskjøn 1981; Klaveness and Rueness 1986; Ohtani et al. 1991; Broady 1996; Ling 1996; Elster et al. 1999). Most soil algae are capable of movement, and most of the other groups are slowly motile or have flagellate forms. Many soil algae exist as unicells or in short filaments (Round 1981; Hoffmann 1989; Archibald 1990). Both surface and subsurface growth of algae occurs in soils. Obviously, surface growths can be washed to deeper levels while, on the other hand, water of melted permafrost can heave cells and spores to the surface. As already discussed, for hydro-terrestrial environments, soil algae have the ability to survive freezing and desiccation – it is thus quite important as a selection factor and exceeds in importance the ability to metabolise in dry conditions. However, for soil algae, the physical disruption of the unstable environment, in which organisms may become exchanged between the surface and soil profile, is another crucial environmental stress factor (Chambers 1967; Elster et al. 1999).

Soil algae, especially the filamentous cyanobacteria and some groups of algae form confluent mucilaginous sheets which protect them against damage by desiccation and freeze–thaw cycles, and are frequently recorded as primary stabilisers of bare, eroded soils. The mucilage binds the soil particles, and reduces water loss and removal by wind (Bailey et al. 1973; Davey and Clarke 1991; Wynn-Williams 1992, 1993; Broady 1996).

17.3.2.2 Soil Algal Crusts

In polar desert and semi-desert localities, soil algae can produce distinct visible biotic crust layers on the ground surface which are called soil algal crusts or cryptogamic crusts (Fletcher and Martin 1948; Campbell et al,. 1989; Davey 1991; Davey and Clarke 1991; Johansen 1993; Broady 1996; Elster et al. 1999). They consist of water-stable, surface soil aggregates held together by algae, fungi, lichens, and mosses. Cyanobacteria, together with some green algae, are the most conspicuous elements of algal crusts (Broady 1996; Elster et al. 1999). In periglacial environments, a gradient in the algal crust can occur. In younger soils, close to the glacial front, cryptogamic growths of different colours can exist. With time and distance from the glacier, the cryptogamic spots increase their size and finally produce distinct soil crusts (Elster et al., unpubl.).

Davey and Clarke (1991) studied the fine structure of algal crust from fell-field soils of Signy Island, maritime Antarctica. The crust was up to 5-mm thick and was distinctly layered. The surface of the crust consisted of nonliving material above a zone of *Phormidium* type. It has been suggested that the surface layer protects the cyanobacterium from the effects of desiccation or high irradiance. The lower layers were less structured than the upper layer and included other taxa of cyanobacteria and eukaryotic algae. The lowest layer consisted of dead organic material.

17.3.2.3 Nutrient Cycling and Flow of Matter

Water availability has been considered to be one of the most crucial factors in limiting the distribution and abundance of soil algae in polar regions (Boyd et al. 1966; Cameron 1969; Akiyama et al. 1986b; Ohtani et al. 1991; Elster et al. 1999; Ohtani et al. 2000). At Signy Island, maritime Antarctica, Davey (1988, 1991) observed a clear seasonal periodicity in soil microflora. The peak of the algal biomass coincided with the peak of the spring snowmelt.

Hoshiai and Matsuda (1979) found a high value of chlorophyll *a* content near the presently-used and abandoned rookeries. High values were also found in depressions where meltwater was supplied from snowdrifts near abandoned rookeries. Penguin guano has antibiotic characteristics and Akiyama et al. (1986a) determined two allelopathic substances (acrylic and oxalic acids) in the excrement of Adélie penguins. These two organic acids, either in co-existence or separately, inhibited the growth of soil algae. A great deal of research concerning the mineralisation of excrement and nutrient cycling in soils near penguin rookeries was done by Pietr et al. (1983), Tatur and Myrcha (1984), Myrcha et al. (1985) and Tatur et al. (1997).

At Ellesmere Island, in the High Arctic, Elster et al. (1999) showed that edaphic conditions, soil texture and substratum stability play important roles in the establishment of algal biomass and diversity. Granitic substrata produced a more abundant algal biomass than dolomitic substrata. The large-grained granitic substratum is more porous and allows the light to penetrate deeper into the soil profile. The dolomitic substratum is of fine material with a high content of silt and clay. This seals and cements the ground surface, preventing light from reaching deeper soil strata. But dolomitic ground is more susceptible to soil heave and churning and microorganisms could be dispersed throughout the profile. The analyses of photosynthetic pigments also showed that algae on granitic substratum had a higher total chlorophyll $a + b$ concentration and a lower chlorophyll $a : b$ ratio, green algae predominate. Dolomitic substrata were lower in total chlorophyll $a + b$, with a high chlorophyll *a* : *b* ratio. Mostly motile cyanobacteria succeeded in penetrating into deeper soil layers in this case. However, Ohtani et al. (1991) in continental Antarctica, near Syowa Station, determined that the higher clay content measured in barren soils adjacent to moss vegetation supports the growth of soil algae. They also considered that the clay has a higher water-holding capacity than surrounding fellfields.

In a study of the vertical distribution of algae in soils of central Ellesmere Island, in the High Arctic, Elster et al. (1999) showed that in the granitic substratum cell numbers peaked at a depth of 3–5 cm, and cells were still present in high numbers at 7-cm depth. The dolomitic soil profile was occupied by algae up to 4-cm depth, but these were not observed below 6 cm. Other authors (Davey and Clarke 1991; and Ohtani et al. 1991) describe the maximal abundance of algae in soils of Antarctica as being at a depth of about 3 cm, however, they demonstrate that in undisturbed soils most algae were present at a depth of 0–2 cm.

At the present time, we do not have enough information concerning soil-algae growth strategies. It is not clear, yet, if perennial or annual life strategy is more common in the polar soil ecosystem. Fogg (1998) mentions that rock debris, which originated from glacial action, is occupied by various microbial opportunists. However, the degree of biomass depends on the stability of the substratum. In an unstable substratum, mainly motile cyanobacteria and algae occur.

17.3.2.4 Community Structure of Soil Algae

Soil algae have been widely studied in both circumpolar regions. In the Antarctic, considerable attention has been given to their diversity and abundance by Akiyama (1967), Broady (1979a–c, 1986, 1987, 1989a–c, 1996), Davey (1988, 1991), Davey and Clarke (1991), Ohtani et al. (1991), Davey and Rothery (1992). Curiously, in the more accessible Arctic, studies of soil algae have been more sporadic (Akiyama 1970; Novichkova-Ivanova 1972; Elster et al. 1999). It is probable that the Arctic soil microflora is more species-rich and more abundant than that of Antarctica. In addition, in the Arctic, green algae are probably more diverse than cyanobacteria (Elster et al. 1999). However, at present, the soil algal diversity of polar habitats is poorly known because collections have not yet been made from all areas and, moreover, techniques have been inadequate for the detection and identification of the total microflora (Broady 1996).

The following genera of edaphic algae have been found in Arctic as well as in Antarctic soils (Broady 1996; Elster et al. 1999; Kanda et al., Chap. 14): cyanobacteria: Chroococcales – *Aphanocapsa*, *Gloeocapsa*, *Synechococcus*; *Oscillatoriales* – *Leptolyngbya*, *Microcoleus*, *Oscillatoria*, *Phormidium*, *Schizothrix*; Nostocales – *Nostoc*, *Tolypothrix*; Bacillariophyceae – *Achnanthes*,

Hantzschia, Navicula, Pinnularia; Xanthophyceae – *Botrydiopsis, Heterococcus, Monodus;* Chlamydophyceae – *Chlamydomonas, Chloromonas, Chlorococcum*; Chlorophyceae – *Bracteacoccus, Chlorella, Coccomyxa, Diplosphaera, Pseudococcomyxa*; Charophyceae – *Stichococcus*; Zygnemaphyceae – *Actinotaenium, Cylindrocystis, Mesotaenium.*

17.3.2.5 Aerophytic Algae (Lithophytic)

The aerophytic (lithophytic) algae occur mainly on or within rock substrata which are elevated above the ground (soil). Moreover, they can also cover or grow within wood, bone, and other natural or artificial materials, which can be found in polar regions. A rock surface offers some hospitality to life, those organisms able to take advantage of it being termed *epilithic. Endolithic* microbial life-forms predominate within rock outcrops (grow beneath the rock surface). The endolithic habitat was described originally for porous rocks in hot deserts such as the Negev (Friedmann 1971, 1980). Its communities have been defined by Golubic et al. (1981) as *chasmoendolithic*, if found in fissures, and *cryptoendolithic*, if located in the structural cavities within rock. Additionally, *hypolithic* algae are those which grow on stone undersurfaces at the stone-soil interface.

This section briefly summarises information concerning lithophytic algae. Reviews summarising our present knowledge are: Barkman (1958), Friedmann and Galun (1974), Wynn-Williams (1990), Nienow and Friedmann (1993), Broady (1996), and Nienow (1996).

Cyanobacteria, green algae and Xanthophyceae are the most dominant oxyphototrophic microorganisms of lithophytic habitats (Friedmann 1971, 1980; Golubic et al. 1981). Endolithic communities are the most thoroughly investigated terrestrial algal communities in Antarctica, in regions ranging from southern Victoria Land to Signy Island in maritime Antarctica (Friedmann and Ocampo 1976, 1984; Friedmann 1977, 1978, 1982; Friedmann et al. 1987, 1988; Broady 1981b, c; Friedmann and Weed 1987; Colwell et al. 1989; Feingold et al. 1990). In extreme desert areas, lithophytic algae are the only photoautotrophic organisms (Cameron and Blank 1967).

The most recent ecological review of lithophytic algae was by Nienow and Friedmann (1993). The morphological and physiological characteristics of the algae interact with the spatial and physicochemical properties of the habitat. These algae live in a wide spectrum of all translucent substratum types, including sandstone, dolomite, granite, granodiorite. As conditions become more extreme, minor differences in physical parameters above and below the rock surface become critical. Differences in the millimetre range, the nanoclimate, control the environment in which algae exist. In the most extreme climates, the absolute limit for life is reached in terrestrial lithophytic communities.

As well as algae, the lithophytic community includes filamentous fungi, yeasts, and heterotrophic bacteria, which together form conspicuous multicoloured horizontal and vertical zones on and under the surface. Algae and fungi can be associated as lichens.

Vertical and horizontal zonation occurs within individual communities. Light intensity, moisture availability, pH, rock chemistry and temperature have been proposed as being the most important environmental parameters influencing the development of lithophytic communities. However, the lithophytic microbial communities are composed of physiologically non-adapted organisms which can survive in hostile environments. The conservation of moisture and nutrients, the buffering of temperature extremes, the penetration of light as an energy source to drive anabolism, and protection from abrasion are the main ecological principles of their survival (Kappen and Friedmann 1983; McKay and Friedmann 1985; Meyer et al. 1988; Nienow et al. 1988a,b; Ocampo-Friedmann et al. 1988; Palmer and Friedmann 1990a, b; Johnston and Vestal 1991; Nienow and Friedmann 1993; Konhauser 1994).

17.4 Conclusion

This chapter has re-classified and described the pattern of algal distribution along a water availability gradient: *aquatic (planktonic communities) – semi-aquatic (hydro-terrestrial communities)* and *dry (terrestrial communities).* Such a gradient has already been demonstrated in higher plant distribution from the extreme of polar deserts, via semi-deserts, to polar coasts with various types of wetland habitats (Svoboda and Henry 1987). However, in both the plant distribution and algal distribution, a series of transient ecological situations also exist.

The presence of liquid water acts as a buffer of temperature change by having a large heat storage capacity, and generally stabilises the environment near the ground surface. Overall, the polar terrestrial environment is extreme in temperature, light (including UV radiation), desiccation and freeze-thaw cycles, and in changes in mineral salt concentration. However, if water availability fluctuates frequently, or even should not be available in liquid form, then conditions become even more extreme.

As has been proposed in many papers, e.g. Friedmann (1993), Nienow and Friedman (1993), Vincent et al. (1993a,b), Vincent and Roy (1993), Vincent and James (1996), Broady (1996), Elster et al. (1997), most algal genotypes occurring in the extreme conditions of polar terrestrial ecosystems exhibit a wide ontogenetic ecological plasticity. Most of the algal genotypes, which are common in the polar terrestrial ecosystem, have been dispersed from lower latitudes. This conclusion is in agreement with many plant ecological studies which have been performed in the polar ecosystem recently (Chapin et al.

1992; Bliss et al. 1994; Svoboda and Freedman 1994). However, because algae have a short life- and cell-cycle, the changes and improvements in the genome can be very fast, in accord with the special climatic conditions of the polar terrestrial environment.

The direction of future phycological research of the polar terrestrial ecosystem should be oriented towards the following:

1. *Taxonomic studies.* Many genera are: either morphologically simple, and the description of unrelated genera tend to overlap; or morphologically pleomorphic, i.e. members of the genus can change their appearance with changes in environmental conditions. This makes identification of field material, at the generic level, difficult. The only way to address this problem is to study the organisms in culture and describe their complete life-history and the range of their metabolic processes. This knowledge should be combined with a description of the physiological, biochemical and molecular genetic (DNA sequencing) properties of strains isolated from polar regions, and compared with those originating from lower latitudes. In cyanobacterial research, for example, Anagnostidis and Komárek (1985, 1988, 1990), Komárek (1987) and Castenholz (1992) have used all these approaches.
2. *Ecophysiological studies.* Algae occurring in polar terrestrial habitats have to overcome a wide range of environmental stresses. Algae (either community or population) react to stress and those which improve their own physiology to a new level of homeostasis are the most successful in competition in a new environment. The latest ecophysiological and biochemical studies of algal adaptation have brought a lot of new information. Future research should focus mainly on the processes of algal adaptation to water deficiency (desiccation, freeze-thaw cycle, extremes of salt concentration and salt stress), and to extremes in low temperature, including cryopreservation.
3. *Ecological functioning of habitats.* Algae greatly influence the cycling of mineral and energy flow, and the mineral and biological development of polar terrestrial ecosystems, including ice-free periglacial landscapes. The vast barren polar landscapes are potentially 'explosive' systems if the trend of global warming should continue (Chapin et al. 1992). Studies focusing on the ecological functioning and role of algae in energy and mineral flows will help us to understand the development of the polar terrestrial ecosystem in space and time.

Acknowledgements. This work was funded by the Grant Agency of the Academy of Sciences of the Czech Republic (Grant No. A 600 5002), the Grant Agency of the Czech Republic (Grant No. 05/94/0156 and 204/93/1177) and the Grant Agency of the Ministry of Education of the Czech Republic (MSM 1231 00004)). The author wishes to thank Dr. P. Broady, Dr. J. Kvit, Prof. J. Komárek, Prof. R.D. Seppelt and Prof. M. Bölter for critical comments. I am also grateful to Prof. J. Svoboda, Prof. J. Komárek, Prof. P. Prosek, Prof. H. Kanda and Prof. S. Kojima who have been supporting my research in polar regions for many years.

References

Akiyama M (1967) On some Antarctic terrestrial and subterranean algae from the Ongul Islands, Antarctica. Antarct Rec 32:71–77

Akiyama M (1970) Some soil algae from the Arctic Alaska, Canada and Greenland. Mem Fac Educ Shimane Univ 4:53–75

Akiyama M (1977) Some ecological and taxonomical observations on the coloured snow algae found in Rumpa and Skarvsnes Antarctica. Mem Natl Polar Res Spec Issue 11:27–34

Akiyama M, Kanda H, Ohyama Y (1986a) Allelopathic effect of penguin excrements and guanos on the growth of Antarctic soil algae. Mem Natl Inst Polar Res Ser E 37:11–16

Akiyama M, Ohyama Y, Kanda H (1986b) Soil nutrient condition related to the distribution of terrestrial algae near Syowa Station (extended abstract). Mem Natl Inst Polar Res Spec Issue 44:189–201

Aleksandrova V D (1988) The Arctic and Antarctic: their division into geobotanical areas. Cambridge University Press., Cambridge

Anagnostidis K, Komárek J (1985) Modern approaches to the classification system of cyanophytes. 1. Introduction. Algol Stud 38/39:291–302

Anagnostidis K, Komárek J (1988) Modern approaches to classification system of cyanophytes. 3 – Oscillatoriales. Algol Stud 50–53:327–472

Anagnostidis K, Komárek J (1990) Modern approaches to the classification of cyanophytes. 5. Stigonematales. Algol Stud 59:1–73

Archibald P A (1990) Soil algae. In: Dindal DL (ed) Soil biology guide. Wiley, New York, pp 69–96

Awramik S M, Margulis L, Barghoorn ES (1976) Evolutionary processes in the formation of stromatolites. In: Walter MR (ed) Stromatolites. Elsevier, Amsterdam, pp 149–162

Bailey D, Mazurak AP, Rosowski JR (1973) Aggregation of soil particles by algae. J Phycol 9:99–101

Barkman JJ (1958) Phytosociology and ecology of cryptogamic epiphytes. Van Gorcum & Co. N. V., Assen, The Netherlands

Bauld J (1981) Occurrence of benthic microbial mats in saline lakes. Hydrobiologia 81:87–111

Bliss LC, Henry GHR, Svoboda J (1994) Patterns of plant distribution within two polar desert landscapes. Arct Alp Res 26:46–55

Borgesen F (1910) Freshwater algae from the 'Danmarks-Expedition' to northeast Greenland (N. of 76°lat.). Medd Gronland 43:69–90

Boyd WL, Staley JT, Boyd JW (1966) Ecology of soil microorganisms of Antarctica. Antarctic soils and soil forming processes. Antarct Res 8:125–159

Broady PA (1979a) The terrestrial algae of Signy Island, South Orkney Islands. Br Antarct Surv Sci Rep 98:1–117

Broady PA (1979b) Signy Island terrestrial reference sites: IX. The ecology of the algae of site 2, a moss carpet. Br Antarct Surv Bull 47: 3–30

Broady PA (1979 c) Quantitative studies on the terrestrial algae of Signy Island, South Orkney Islands. Br Antarct Surv Bull 47:31–41

Broady PA (1981a) Ecological and taxonomical observations on subaerial epilithic algae from Princess Elisabeth Land and Mac. Robertson Land, Antarctica. Br Phycol J 16:257–266

Broady PA (1981b) The ecology of sublithic terrestrial algae at the Vestfold Hills, Antarctica. Br Phycol J 16:231–240

Broady PA (1982) Taxonomy and ecology of algae in a freshwater stream in Taylor Valley, Victoria Land, Antarctica. Arch Hydrobiol Suppl Algol Stud 63:331–339

Broady PA (1986) Ecology and taxonomy of the terrestrial algae. In: Pickard J (ed) Antarctic oasis, terrestrial environment and history of the Vestfold Hills. Academic Press, New York pp 165–292

Broady PA (1987) A floristic survey of algae at four locations in northern Victoria land. N Z Antarct Rec 1:8–19

Broady PA (1989a) Broadscale patterns in the distribution of aquatic and terrestrial vegetation at three ice-free regions on Ross Island, Antarctica. Hydrobiologia 172:77–95

Broady PA (1989b) The distribution of *Prasiola calophylla* (Carmich.) Menegh. (Chlorophyta) in Antarctic freshwater and terrestrial habitats. Antarct Sci 1:109–118

Broady PA (1989 c) Survey of algae and other terrestrial biota at Edward VII Peninsula, Marie Byrd Land. Antarct Sci 1:215–224

Broady PA (1996) Diversity, distribution and dispersal of Antarctic algae. Biodiv Conserv 5:1307–1335

Cameron RE, Black GB (1967) Desert soil algae survival at extremely low temperatures. Cryogen Technol 3:151–156

Cameron RE (1969) Abundance of microflora in soils of desert region. JPL Techn Rep 32–1378:1–16

Campbell IB, Claridge GGC (1987) Antarctica: soils, weathering processes and environment. Elsevier, Amsterdam

Campbell SE, Seeler J-S, Golubic S (1989) Desert crust formation and soil stabilization. Arid Soil Res Rehabil 3:217–228

Castenholz RW (1992) Species usage, concept, and evolution in the cyanobacteria (blue-green algae). J Phycol 28:737–745

Chambers MJG (1967) Investigations of patterned ground at Signy Island, South Orkney Islands: III. Miniature patterns, frost heaving and general conclusions. Br Antarct Surv Bull 12:1–22

Chapin FS, Jefferies RL, Reynolds JF, Shaver GR, Svoboda J (eds) (1992) Arctic ecosystems in a changing climate. An ecophysiological perspectives. Academic Press, London

Chernov YI (1985) The living tundra. Cambridge University Press, Cambridge

Colwell RR, MacDonell MT, Swartz D (1989) Identification of an Antarctic endolithic microorganism by 5SrRNA sequence analysis. System Appl Microbiol 11:182–186

De Winder B (1990) Ecophysiological strategies of drought tolerant phototrophic micro-organisms in dune soils. Academisch Profschrift, University of Amsterdam, 95 pp

Croasdale H (1973) Freshwater algae of Ellesmere Island, N. W. T. Publ Nat Mus Nat Sci, Publ Bot, No 3:1–131

Davey MC (1988) Ecology of terrestrial algae of the fellfield ecosystem of Signy Island, South Orkney Islands. Br Antarct Surv Bull 81:69–74

Davey MC (1989): The effects of freezing and desiccation on photosynthesis and survival of terrestrial Antarctic algae and cyanobacteria. Polar Biol 10:29–36

Davey MC (1991) The seasonal periodicity of algae on Antarctic fellfield soils. Holarct Ecol 14:112–120

Davey M C, Clarke KJ (1991) The spatial distribution of microalgae on Antarctic fellfield soils. Antarct Sci 3:257–263

Davey MC, Rothery P (1992) Factors causing the limitation of growth of terrestrial algae in maritime Antarctica during late summer. Polar Biol 12:595–601

Dickie G (1878) On the algae found during the Arctic Expedition. J Linn Soc Bot 17:6–12

Elster J, Svoboda J (1995) In situ simulation and manipulation of a glacial stream ecosystem in the Canadian High Arctic. In: Jenkins A, Ferrier RC, Kirby C (eds.) Ecosystem manipulation experiments: scientific approaches, experimental design and relevant results. Comm Eur Commun, Inst Hydrol UK and Environment Canada. Ecosystem Res Rep No 20. Brussels, Luxemburg

Elster J, Svoboda J (1996) Algal seasonality and abundance in, and along, glacial stream Sverdrup Pass 79 °N, Central Ellesmere Island, Canada. Mem Natl Inst Polar Res, Spec Issue 51:99–118

Elster J, Svoboda J, Komárek J, Marvan P (1997) Algal and cyanoprocaryote communities in a glacial stream, Sverdrup Pass, 79 °N, central Ellesmere Island, Canada. Arch Hydrobiol Suppl Algol Stud 85:57–93

Elster J (1999) Algal versatility in various extreme environments. In: Seckbach J (ed) Enigmatic microorganisms and life in extreme environments, Kluver, Dordrecht, pp 215–227

Elster J, Lukesova A, Svoboda J, Kopecky J, Kanda H (1999) Diversity and abundance of soil algae in the polar desert, Sverdrup Pass, central Ellesmere Island. Polar Rec 35:231–254

Engelskjøn T (1981) Terrestrial vegetation of Bouvetoya. Norsk Polarinst Skrif 175:17–28

Engelskjøn T (1986) Botany of two Antarctic mountain ranges: Gjelsvikfjella and Muhling-Hofmannfjella, Dronning Maud Land. Polar Res 4:295–224

Fletcher JE, Martin WP (1948) Some effects of algae and moulds in the rain crust of desert soils. Ecol 29:95–100

Feingold L, Singer MA, Federle TW, Vestal JR (1990) Composition and thermal properties of membrane lipids in cryptoendolithic lichen microbiota from Antarctica. Appl Environ Microbiol 56:1191–1194

Fogg GE (1967) Observations on the snow algae of the South Orkney Islands. Philos Trans R Soc Lond Ser B 252:279–294

Friedman GM, Amiel AJ, Braun M, Miller DS (1973) Generation of carbonate particles and laminates in algal mats – example from sea-marginal hypersaline pool, Gulf of Aqaba, Red Sea. Am Assoc Petrol Geol Bull 57:541–557

Friedmann EI (1971) Light and scanning electron microscopy of the endolithic desert habitat. Phycol 10:411–428

Friedmann EI, Galun M (1974) Desert algae, lichens and fungi. In: Brown GW (ed) Desert biology, Vol II., Academic Press, New York, pp 165–212

Friedmann EI, Ocampo R (1976) Endolithic blue-green algae in the dry valleys: primary producers in the Antarctic desert ecosystem. Science 193:1247–1249

Friedmann EI (1977) Microorganisms in Antarctic desert rocks from Dry Valley and Dufek Massif. Antarct J US 12:26–29

Friedmann EI (1978) Melting snow in the dry valleys is a source of water for endolithic micro-organisms. Antarct J US 13:162–163

Friedmann EI (1980) Endolithic microbial life in hot and cold deserts. Orig Life 10:233–245

Friedmann EI (1982) Endolithic micro-organisms in the Antarctic cold desert. Science 215:1045–1053

Friedmann EI, Ocampo-Friedmann R (1984) Endolithic micro-organisms in extreme dry environments: analysis of a lithobiontic microbial habitat. In: Klug HJ, Reddy CA (eds) Current perspectives in microbial ecology. Am Soc Microbiol, Washington, DC, pp 177–185

Friedmann EI, McKay CP, Nienow JA (1987) The cryptoendolithic microbial environment in the Ross Desert of Antarctica: satellite-transmitted continuous nanoclimate data, 1984 to 1986. Polar Biol 7:273–287

Friedmann EI, Weed R (1987) Microbial trace-fossil formation, biogenous, and abiotic weathering in the Antarctic cold desert. Science 236:703–705

Friedmann EI, Hua M, Ocampo-Friedmann R (1988) Cryptoendolithic lichen and cyanobacterial communities of the Ross Desert, Antarctica. Polarforschung 58: 251–259

Friedmann EI (ed.) 1993 Antarctic microbiology. Wiley-Liss, New York

Fogg GE (1998) The biology of polar habitats. Oxford University Press, Oxford

Fumanti B, Cavacini P, Alfinito S (1997) Benthic algal mats of some lakes of Inexpressible Island (northern Victoria land, Antarctica). Polar Biol 17:25–30

Golubic S (1973) The relationship between blue-green algae and carbonate deposits. In: Carr NC, Whitton BA (eds) The biology of blue-green algae. University of California Press, Berkley, pp 434–472

Golubic S (1976) Organisms that build stromatolites. In: Walter MR (ed), Stromatolites. Dev Sedimentol 20:113–126

Golubic S, Friedmann EI, Schneider J (1981) Lithobiontic ecological niche, with special reference to micro-organisms. J Sediment Petrol 51:475–478

Golubic S (1985) Microbial mats and modern stromatolites in Shark bay, Western Australia. In: Caldwell DE, Brierley JA, Brierley CL (eds) Planetary ecology. Van Nostrand Reinhold, New York, pp 3–16

Gradinger R, Nürnberg D (1996) Snow algal communities on arctic pack ice floes dominated by *Chlamydomonas nivalis* (Bauer) Will. Proc NIPR Symp Polar Biol 9:35–43

Hamilton PB, Edlund SA (1994) Occurrence of *Prasiola fluviatis* (Chlorophyta) on Ellesmere Island in the Canadian Arctic. J Phycol 30:217–221

Hawes I (1989) Filamentous green algae in freshwater streams on Signy Island, Antarctica. Hydrobiologia 172:1–18

Hawes I, Howard-Williams C, Vincent WF (1992) Desiccation and recovery of Antarctic cyanobacterial mats. Polar Biol 12:587–594

Hawes I, Howard-Williams C, Pridmore RD (1993) Environmental control of microbial communities in the ponds of the McMurdo Ice Shelf, Antarctica. Arch Hydrobiol 127:271–287

Hawes I, Howard-Williams C, Schwarz AM, Downes M (1997) Environment and microbial communities in a tidal lagoon at Bratina Island, McMurdo Ice Shelf, Antarctica. In: Battaglia B, Valencia J, Walton DWH (eds) Antarctic communities: species, structure and survival. Cambridge University Press, Cambridge, pp 170–177

Hempel G (ed.) (1994) Antarctic science: global concerns. Springer, Berlin Heidelberg New York

Hoffmann L (1989) Algae of terrestrial habitats. Bot Rev 55:77–105

Hoham RW (1975) Optimum temperatures and temperature range for growth of snow algae. Arct Alp Res 7:13–24

Hoham RW, Roemer SC, Mullet JE (1979) The life history and ecology of the snow algae *Chloromonas brevispina* comb. nov. (Chlorophyta, Volvocales). Phycologia 18:55–70

Holdgate MW (ed) (1970) Antarctic ecology. Academic Press, London

Hoshiai T, Matsuda T (1979) Adelie penguin rookeries in the Lützow-Holm Bay area and relation of rookery to algal biomass in soil. Mem Natl Inst Polar Res Spec Issue 11:140–152

Howard-Williams C, Vincent L, Broady P A, Vincent WF (1986) Antarctic stream ecosystems: variability in environmental properties and algal community structure. Int Rev Ges Hydrobiol 71:511–544

Howard-Williams C, Vincent WF (1989) Microbial communities in southern Victoria Land stream: I. Photosynthesis. Hydrobiologia 172:27–38

Howard-Williams C, Pridmore R, Downes MT, Vincent WF (1989): Microbial biomass, photosynthesis and chlorophyll a related pigments in the ponds of the McMurdo ice shelf, Antarctica. Antarct Sci 1:125–131

Hunter MD, Ohgushi T, Price PW (1992) Effect of resource distribution on animal-plant interactions. Academic Press, San Diego

Johansen JR (1993) Cryptogamic crusts of semiarid and arid lands of North America. J Phycol 29:140–147

Johnston CG, Vestal JR (1991) Photosynthetic carbon incorporation and turnover in Antarctic cryptoendolithic microbial communities: are they slowest-growing communities on Earth? Appl Environ Microbiol 57:2308–2311

Ishikawa S, Matsuda O, Kawaguchi K (1986) Snow algae blooms and their habitat conditions observed at Syowa Station, Antarctica. Mem Natl Inst Polar Res Spec Issue 44:191–197

Kappen L, Friedmann EI (1983) Ecophysiology of lichens in the dry valleys of Southern Victoria Land, Antarctica. II. CO_2 gas exchange in cryptoendolithic lichens. Polar Biol 1:227–232

Kawecka B (1978) Biology and ecology of snow algae. 1. The sexual reproduction of *Chlamydomonas nivalis* (Bauer) Wille (Chlorophyta, Volvocales). Acta Hydrobiol 20:111–116

Kawecka B (1986) Ecology of snow algae. Pol Polar Res 7:407–415

Kawecka B, Olech M (1993) Diatom communities in the Vanishing and Ornithologist Creek, King George Island, South Shetlands, Antarctica. Hydrobiologia 269/270:327 –333

Kennedy AD (1993) Water as a limiting factor in the Antarctic terrestrial environment: A Biogeographical Synthesis. Arct Alp Res 25:308–315

Kerry KR, Hempel G (eds) (1990) Antarctic ecosystems. Ecological changes and conservation. Springer, Berlin Heidelberg New York

Kirk JTO (1994) Light and photosynthesis in aquatic ecosystems. Cambridge University Press, Cambridge

Kirst GO (1990) Salt tolerance of eukaryotic marine algae. Ann Rev Plant Physiol Plant Mol Biol 41:21–53

Kirst GO, Wiencke C (1995) Ecophysiology of polar algae. J Phycol 31:181–199

Klaveness D, Rueness J (1986) The supralittoral, freshwater and terrestrial algae vegetation of Bouvetoya. Norsk Polarinst Skrift 185:65–69

Kol E (1968) Kryobiologie. Biologie und Limnologie des Schnees und Eises. I. Kryovegetation. In: Elster HJ, Ohle W (eds) Die Binnengewässer, vol 24. Schweizerbart'sche Verlagsbuchhandlung. Stuttgart, pp 1–216

Kol E (1969) The red snow of Greenland II. Acta Bot Acad Sci Hung 15:281–289

Kol E (1971) Green snow and ice from the Antarctic. Ann Hist-Nat Mus Nat Hung 63:51–56

Kol E (1972) Snow algae from Signy Island (South Orkney Islands, Antarctica) Ann Hist-Nat Mus Nat Hung 64:63–70

Kol E, Flint EA (1968) Algae in green ice from the Balleny Islands, Antarctica. N Z J Bot 6:249–261

Komárek J (1987) Species concept in coccal green algae. Algol Stud 45:437–471

Konhauser KO, Fyfe WS, Schultze-Lam S, Ferris FG, Beverige TJ (1994) Iron phosphate precipitation by epilithic microbial biofilms in Arctic Canada. Can J Earth Sci 31:1320–1324

Krumbein WE, Cohen Y, Shilo M (1977) Solar lake (Sinai). 4. Stromatolitic cyanobacterial mats. Limnol Oceanogr 22:635–636

Lamberti GA, Feminella JW (1996) Plant-herbivore interactions. In: Hauer FR, Lamberti GA (eds) Methods in stream ecology. Academic Press, London, pp 409–430

Lawlor DW, Keys AJ (1993) Understanding photosynthetic adaptation to changing climate. In: Fowden L, Mansfield T, Stoddart J (eds) Plant adaptation to environmental stress. Chapman and Hall, London, pp 85–106

Laws RM (ed) (1984) Antarctic ecology. Academic Press, London

Lévesque E (1997) Plant distribution and colonisation in extreme polar deserts, Ellesmere Island, Canada. PhD Thesis. Department of Botany, University of Toronto, Toronto

Lewis Smith RI (1984) Terrestrial plant biology of the sub-Antarctic and Antarctic. In: Laws RM (ed) Antarctic ecology, vol 1. Academic Press, London, pp 63–162

Lewis Smith RI (1993) The role of bryophyte propagule banks in primary succession: case study of an Antarctic fellfield soil. In: Miles J, Walton DWH (eds) Primary succession on land, Blackwell, Oxford, pp 55–78

Lewis Smith RI (1996) Terrestrial and freshwater biotic components of the western Antarctic Peninsula. In: Ross R, Hofmann E, Quetin L (eds) Foundations for ecological research west of the Antarctic Peninsula. Amer Geophys Union, Washington DC, Antarct Res Ser 70:15–59

Lewis Smith RI (1997) Oases as centres of high plant diversity and dispersal in Antarctica. In: Howard-Williams L, Hawes I (eds) Ecosystem processes in Antarctic ice-free landscapes. Balkema, Rotterdam, pp 119–128

Ling HU (1996) Snow algae of the Windmill Islands region, Antarctica. In: Kristiansen J (ed) Biogeography of freshwater algae. Hydrobiologia 336:99–106

Ling HU, Seppelt RD (1990) Snow algae of the Windmill Islands, continental Antarctica. *Mesotaenium berggrenii* (Zygnematales, Chlorophyta) the alga of grey snow. Antarct Sci 2: 43–148

Ling HU, Seppelt RD (1993) Snow algae of the Windmill Islands, continental Antarctica. 2. *Chloromonas rubroleosa* sp. nov. (Volvocales, Chlorophyta) an alga of red snow. Eur J Phycol 28:77–84

Ling HU, Seppelt RD (1998) Snow algae of the Windmill Islands, continental Antarctica. 3. *Chloromonas polyptera* (Volvocales, Chlorophyta) Polar Biol 20:320–324

Longton RE (1988) Biology of polar bryophytes and lichens. Cambridge University Press, Cambridge

Logan BW, Rezek R, Ginsburg RN (1964) Classification and environmental significance of stromatolites. Bull Geol Soc Am 71:1918–1919

Mataloni G, Tesolin G (1997) A preliminary survey of cryobiontic algae communities from Cierva Point (Antarctic Peninsula). Antarct Sci 9:250–258

McKay CP, Friedmann EI (1985) The cryptoendolithic microbial environment in the Antarctic cold desert: temperature variations in nature. Polar Biol 4:19–25

McKnight DM, Alger A, Tate CM, Shupe G, Spaulding S (1998) Longitudinal patterns in algal abundance and species distribution in meltwater streams in Taylor Valley, Southern Victoria land, Antarctica. In: Priscu JC (ed) Ecosystem dynamics in a polar desert. American Geophysical Union, Washington, DC, Antarct Res Ser 72:109–127

McLean AL (1918) Bacteria of ice and snow in Antarctica. Nature 102:35–39

Metting B (1981) The systematics and ecology of soil algae. Botan Rev 47:195–312

Meyer MA, Huang GJ, Morris GJ, Friedmann EI (1988) The effect of low temperatures on Antarctic endolithic green algae. Polarforschung 58:113–119

Moore JW (1974a) Benthic algae of Southern Baffin Island. I. Epilithic communities in rivers. J Phycol 10:50–57

Moore JW (1974b) Benthic algae of Southern Baffin Island. III. Epilithic and epiphytic communities. J Phycol 10:456–462

Muller T, Bleiss W, Martin C-D, Rogaschewski S (1998) Snow algae from northwest Svalbard: their identification, distribution, pigment and nutrient content. Polar Biol 20:14–32

Myrcha A, Pietr SJ, Tatur A (1985) The role of pygoscelid penguin rookeries in nutrient cycles at Admiralty Bay, King George Island. In: Siegfried WR, Condy PR, Laws RM (eds) Antarctic nutrient cycles and food webs. Springer, Berlin Heidelberg New York, pp 156–162

Nienow JA, McKay CP, Friedmann IE (1988a) The cryptoendolithic microbial environment in the Ross Desert of Antarctica: mathematical models of the thermal regime. Microb Ecol 16:253–270

Nienow JA, McKay CP, Friedmann EI (1988b) The cryptoendolithic microbial environment in the Ross Desert of Antarctica: light in the photosynthetically active region. Microb Ecol 16:271–289

Nienow JA, Friedmann EI (1993) Terrestrial lithophytic (rock) communities. In: Friedmann EI (ed) Antarctic microbiology, Wiley-Liss, New York, pp 343–412

Nienow JA (1996) Ecology of subaerial algae. Nova Hedwigia 112:537–552
Novichkova-Ivanova LN (1972) Soil and aerial algae of polar deserts and arctic tundra. In: Wielgolaski FE, Rosswall T (eds) I. B. P. Tundra biome. Proc 4th Int Meeting on the Biological Productivity of Tundra. I.B.P. Tundra Steering Committee, Stockholm, pp 261–265
Ocampo-Friedmann R, Meyer MA, Chen M, Friedmann EI (1988) Temperature response of Antarctic cryptoendolithic photosynthetic micro-organisms. Polarforschung 58:121–124
Ohtani S (1986) Epiphytic algae on mosses in the vicinity of Syowa Station, Antarctica. Mem Natl Inst Polar Res Spec Issue 44:209–219
Ohtani S, Kanda H (1987) Epiphytic algae on the moss community of *Grimmia lawiana* around Syowa Station, Antarctica. Proc NIPR Symp Polar Biol 1:255–264
Ohtani S, Akiyama M, Kanda H (1991) Analysis of Antarctic soil algae by the direct observation using the contact slide method. Nankyoku Shiryo (Antarct Rec) 35:285–295
Ohtani S, Suyama K, Yamamoto H, Aridomi Y, Itoh R, Fukuoka Y (2000) Distribution of soil algae at the monitoring sites in the vicinity of Syowa Station between austral summers of 1992/1993 and 1997/1998. Polar Biosci 13:113–132
Palmer RJ, Friedmann EI (1990a) Water relations and photosynthesis in the cryptoendolithic fungi. Polarforschung 58:189–192
Palmer RJ, Friedmann EI (1990b) Water relations and photosynthesis in the cryptoendolithic microbial habitat of hot and cold deserts. Microb Ecol 19:111–118
Pankow H, Haendel D, Richter W (1991) Die Algenflora in der Schirmacheroase (Ostantarktika). Nova Hedwigia 103:1–150
Parker BC, Simmons Jr GM, Seaburg KG, Cathey DD, Allnutt FCT (1981) Modern stromatolites in Antarctic Dry Valley Lakes. BioScience 31:656–661
Parker BC, Wharton Jr RA (1985) Physiological ecology of bluegreen algal mats (modern stromatolites) in Antarctic oasis lakes. Arch Hydrobiol Suppl 71, 1/2 (Algological Studies 38/39):331–348
Petersen JB (1924) Freshwater algae from the north coast of Greenland collected by the late Dr. Th. Wulff. Medd. Gronland 64:305–319
Petersen JB (1935) Studies on the biology and taxonomy of soil algae. Dansk Bot Arkiv 8:1–180
Peterson PJ (1993) Plant adaptation to environmental stress: metal pollutant tolerance. In: Fowden L, Mansfield T, Stoddart J (eds) Plant adaptation to environmental stress. Chapman and Hall, London, pp 151–171
Pietr SJ, Tatur A, Myrcha A (1983) Mineralization of penguin excrements in the Admiralty Bay region (King George Island, South Shetlands Islands, Antarctica). Pol Polar Res 4:97–112
Pizarro H, Izaguirre I, Tell G (1996) Epilithic algae from a freshwater stream at Hope Bay, Antarctica. Antarct Sci 8: 61–167
Priscu JC (ed) (1998) Ecosystem dynamics in a polar desert. Antarctic Res Ser 72, Amer Geophys Union, Washington DC
Ross R (1947) Freshwater Diatomeae (Bacillariophyta). In: Polunin N (ed) Botany of the Canadian Eastern Arctic, Part II. Thallophyta and Bryophyta. Nat Mus Can Bull 97:178–233
Round FE (1981) The Ecology of algae. Cambridge University Press, Cambridge
Svoboda J, Henry GHR (1987) Succession in marginal arctic environments. Arct Alp Res 4:373–384
Svoboda J, Freedman B (1994) Ecology of a polar oasis, Alexandra Fiord, Ellesmere Island, Canada. Captus University Publications, Toronto
Tang EPY, Tremblay R, Vincent WF (1997) Cyanobacterial dominance of polar freshwater ecosystems: are high-latitude mat-formers adapted to low temperature? J Phycol 33:171–181

Tatur A, Myrcha A. (1984) Ornithogenic soils on King George Island, South Shetlands Islands (maritime Antarctic zone). Pol Polar Res 5:1–60
Tatur A, Myrcha A, Niegodzisz J (1997) Formation of abandoned penguin rookery ecosystems in the maritime Antarctic. Polar Biol 17: 405–417
Tedrow JCF (1977) Soils of the polar landscapes. Rutgers University Press, New Brunswick
Tedrow JCF (1991) Pedogenic linkage between the cold desert of Antarctica and the polar desert of the high Arctic. Contrib Antarct Res II 53:1–17
Van De Vijver B, Beyens L (1997) Freshwater diatoms from some islands in maritime Antarctic region. Antarct Sci 9: 18–425
Vincent WF, Howard-Williams C (1986) Antarctic stream ecosystems: physiological ecology of a blue-green algae epilithon. Freshwater Biol 16:216–233
Vincent WF (1988) Microbial ecosystem of Antarctica. Cambridge University Press, Cambridge
Vincent WF, Howard-Williams C (1989) Microbial communities in southern Victoria Land streams: II. The effect of low temperature. Hydrobiologia 172:39–49
Vincent WF, Howard-Williams C, Broady PA (1993a) Microbial communities and processes in Antarctic flowing waters. In: Friedmann EI (ed) Antarctic microbiology. Wiley-Liss, New York, pp 543–569
Vincent WF, Downes MT, Castenholz RW, Howard-Williams C (1993b) Community structure and pigment organisation of cyanobacteria-dominated microbial mats in Antarctica. Eur J Phycol 28:213–221
Vincent WF, Roy S (1993) Solar ultraviolet-B radiation and aquatic primary production: damage, protection and recovery. Environ Rev 1:1–12
Vincent WF, Quesada A (1994) Ultraviolet radiation effects on cyanobacteria: implication for Antarctic microbial ecosystem. Antarct Res Ser 62:111–124
Vincent WF, James MR (1996) Biodiversity in extreme aquatic environments: lakes, ponds and streams of the Ross Sea Sector, Antarctica. Biodiv Conserv 5:1451–1471
Vinocur A, Pizarro H (1995) Periphyton flora of some lotic and lentic environment of Hope Bay (Antarctic Peninsula). Polar Biol 15:401–414
Walter MR, Bauld J, Brock, TD (1976) Microbiology and morphogenesis of columnar stromatolites (Conophyta, Vacerrilla) from hot springs in Yellowstone National Park. In: Walter MR (ed) Stromatolites. Devt Sedimentol 20:273–310
Wharton Jr RA, Parker BC, Simmons Jr GM (1983) Distribution, Species composition, and morphology of algal mats (stromatolites) in Antarctic Dry Valley Lakes. Phycologia 22:355–365
Walton DWH (1984) The terrestrial environment. In: Laws RM (ed) Antarctic ecology, vol 1. Academic press, London, pp 1–160
Walton, DWH (ed) (1987) Antarctic science. Cambridge University Press, Cambridge
Whelden RM (1947) Algae. In: Polunin N (ed) Botany of the Canadian Eastern Arctic, II. Thallophyta and Pryophyta. Nat Mus Can Bull 97:13–137
Wynn-Williams DD (1990) Ecological aspects of Antarctic microbiology. Adv Microb Ecol 11:71–146
Wynn-Williams DD (1992) Epifluorescence image analysis of the 3-D structure of phototrophic microbial biofilms at Antarctic soil surface. Microbiol 4:53–57
Wynn-Williams DD (1993) Microbial processes and initial stabilisation of fellfield soil. In: Miles J, Walton DWH (eds) Primary succession on land. Blackwell, Oxford, pp 17–32
Wynn-Williams DD (1994) Potential effects of ultraviolet radiation on Antarctic primary terrestrial colonizers: cyanobacteria and cryptogams. Antarct Res Ser 62:243–257

18 Plants and Lichens in the Antarctic, Their Way of Life and Their Relevance to Soil Formation

L. KAPPEN and B. SCHROETER

18.1 Introduction

In classical terms, soil is a product of rock weathering and influences of soil organisms, dead and transformed organic substances, water and air (Scheffer and Schachtschabel 1984). In other words, textbooks define the pedosphere to be the part of the earth's crust in which the lithosphere is altered by atmospheric forces and organisms. These definitions underline the importance of biota in and on the soil. Campbell and Claridge (1987) therefore discuss the question of whether there are real soils in most of Antarctica because of the virtual absence of an effective soil biological component. They therefore conclude that a broader rather than a narrower approach to the definition of soils in Antarctica is necessary.

In the this chapter we investigate what kind of vascular plants, bryophytes and lichens can exist in Antarctica. Their dependence on tolerance, productivity and their immigration capacity into Antarctica is discussed. Properties of these vegetation elements that enable them to interact with the mineral substrate are demonstrated. Of relevance are growth form, extension into the soil, water relations, metabolic activity and biogenic weathering. With respect to a frequently observed strong ion accumulation in the vegetation, it is discussed to what extent this is relevant to plant nutrition and to pedogenesis.

18.2 Vascular Plants

18.2.1 Immigration and Distribution in Antarctica

The paucity of vascular plants and the dominance of bryophytes and lichens or algal mats in the terrestrial habitats may indicate that Antarctica is immediately comparable with deserts elsewhere in the world. However, hot deserts and Arctic deserts show a greater diversity of vascular plant species. Usually,

Ecological Studies, Vol. 154
L. Beyer and M. Bölter (eds.) Geoecology of Antarctic Ice-Free Coastal Landscapes

the desert character is demonstrated by the wide distance between the individual plants. Therefore, the more scattered the plant vegetation, the more desertic is the area. However, in the maritime Antarctic the vegetation can form closed canopies because precipitation allows lush growth, as in temperate regions or in the sub-Arctic. It is still debated whether the presence of only two vascular plant species in Antarctica is due only to migrating difficulties or due to the specific climate or to the lack of particular microorganisms that usually coexist in the rhizosphere of the plants, such as mycorrhizal fungi, sheath-forming bacteria etc.

In contrast to the Arctic where vascular plants, even woody species, reach the northern-most fringe of land around the Pole (84°N), Antarctica forms a strict distributional border for the expansion of vascular plants. No species reach the continental region south of 69°S. Only *Colobanthus quitensis* (Caryophyllaceae) and *Deschampsia antarctica* (Poaceae) were able to establish spontaneously in the Antarctic Peninsula region having their southernmost distribution border at 68°42'S at the west coast (southwest of the Antarctic Peninsula; Lewis Smith and Poncet 1987), and at 65°S at the eastern coast which is exposed to the frigid Weddell Sea region (Lewis Smith 1984a).

The reason why just *Colobanthus quitensis* and *Deschampsia antarctica* are the only vascular plants present in Antarctica is still obscure. The latter is distributed also over the west part of South America. *Colobanthus* even reaches the tropics, growing in high mountains in Ecuador. There is nothing that could explain why the two species could have better migrated into Antarctica than other species, although Convey (1996) concluded from his study on the reproduction of the two species, that the major limiting factor is geographical isolation.

The evidence of a climatic border can be derived from the fact that the two vascular plant species grow only in the mildest part of Antarctica and thin out further north in the eastern Peninsula region rather than in the milder western part. It is still an open question as to why just such mainly mesophytic species are capable to exist in the maritime Antarctic. If the winter period was an obstacle, other species that exist in Siberia or in high alpine regions might persist in the Antarctic as well. Perhaps the low energy input and the briefness of the summer period is the limiting factor. A further indication that the major reason for the distribution of vascular plants is the winter conditions in Antarctica may be illustrated by two other examples. The two indigenous species, *Colobanthus* and *Deschampsia*, were transplanted from their southernmost habitats, Biscoe Point on southern Anvers Island (Lewis Smith and Poncet 1987) to Cape Hallett (72°S) and Granite Harbour (77°S). However, later visits to both sites in the 1990s did not detect any trace of these species (Lewis Smith 1994).

It is still a matter of debate as to why no other species have been able to grow successfully in the Antarctic. Only two *Poa* species have spontaneously settled and survived for a couple of years in the maritime Antarctic. The question is still why these tuft-forming mesophytic species are so successful here and not those many species from high mountain regions, or from the Arctic, that might be physiologically and morphologically much better adapted to harsh climatic conditions than these mesophytes.

The grass *Poa annua* successfully colonises geothermal areas near the shore of Deception Island, maritime Antarctic. Populations of this species have been recorded over a period of 13 years, during which flowering and setting seeds and sprouting from seedlings has been observed. The area occupied is almost 25 m^2 and was destroyed by a mud flow from a volcanic eruption in 1967 (Collins 1969). *Poa annua* transplanted from sub-Antarctic South Georgia was shown to produce viable seeds on Signy Island, maritime Antarctic (Edwards 1980).

Poa annua, in other climatic regions a summer annual, performs in Antarctica more as a perennial hemicryptophyte, producing green tillers every year until single plants become reproductive and then die. *Poa pratensis* is a grass species well known in central Europe and by far the most successful alien species in Antarctica. One population has persisted over about 40 years (1955–1995) near Primavera Station at Ciera Point, Deception Island (Lewis Smith 1996a). The presence of *Poa pratensis* dates back even to as early as 1936 in this area (Lewis Smith 1985).

Attempts to naturalise other vascular plants from South America (Patagonia) have always failed (Lewis Smith 1996a). Introduced species from the Falkland Islands (angiosperms and pteridophytes) are able to grow in eroded *Polytrichum* peat on Signy Island for 1 or 2 years. But obviously, the persistence through one or at most three winters appeared to be the limiting factor (Lewis Smith 1985). The same is the case with Japanese willow species (*Salix reinii*, and *S. pauciflora*) at Syowa Station, Enderby Land, which are able to grow in sandy soil having been cleaned from all native soil before. The plants failed to survive more than one winter (Hoshiai 1970; Lewis Smith 1985; Kanda et al., Chap. 14).

Our knowledge about symbiotic fungi and other microorganisms which are essential for plants but cannot exist in Antarctica is very poor. From South Georgia vascular-arbuscular mycorrhizal associations are known for many species of the vascular plant flora (Christie and Nicolson 1983) and have recently been detected in the rhizosphere of *Deschampsia antarctica* on Deception Island but not under *Colobanthus quitensis* (Cabello and Pollero 1994). Also, nitrifying bacteria have not been detected so far in the more southern regions (Lewis Smith 1985). However, Williams et al. (1994) describe a mycorrhiza of hepatics in continental Antarctica. It is noteworthy that plant species introduced even in imported soil from original sites from the Falkland Islands to an island 50 km north of Argentine Island, maritime Antarctic, did not persist longer than one or two winter periods. Seedlings of unintentionally introduced seeds in the soil have, however, germinated in the summer after transportation of the soil, but have subsequently died (Lewis Smith

1985). The attempt to transplant tree samplings of two *Nothofagus* species in their soil failed as well (Corte 1961).

18.2.2 Physiological Properties Allow Existence in Antarctica

The freezing tolerance of leaves of *D. antarctica* and *C. quitensis* is not extremely high. A freezing tolerance with a lethal temperature (=LT_{50}) of –26 °C was achieved with *Deschampsia antarctica*, whereas *Colobanthus quitensis* was already injured (LT_{50}) at –6 °C (Bravo et al. 2000). As the latter plant is not freezing-tolerant, it is supposed that avoidance mechanisms such as supercooling are responsible for the survival of *Colobanthus*. The wintering swards of *D. antarctica* near the ground remain well preserved under snow in winter. *Deschampsia antarctica*, but also *C. quitensis*, behave like Arctic plant species as they have extended their flowering cycle over more than one season (Edwards 1974a). *Colobanthus quitensis* is a typical hemicryptophytic cushion plant. Overwintering flower buds set flowers early in spring and fertilisation of the cleistogamous flowers may occur after spring melt (Lewis Smith 1984a). *D. antarctica* seeds continue development under snow cover (Edwards 1974a). Germination of *D. antarctica* seeds is initiated if they are exposed to temperatures below 0 and 5 °C in the moist state for 2 weeks, whereas *C. quitentsis* seeds need more than 3 months. Highest rates of germination are then achieved within the temperature range of 5–18 °C (Holtom and Greene 1967) but germination is also possible at 0 °C according to laboratory experiments. In the field (Argentine Island) germination was observed in both species in summer ranging between 8 to –5 °C air temperature and freeze thaw cycles during most days (Corner 1971).

Photosynthetic activity measured under laboratory conditions in *C. quitensis* is found to be below –5 °C (Edwards 1974b), and according to Edwards and Lewis Smith (1988) photosynthesis is active at temperatures not lower than –2 °C in *D. antarctica*. According to the light saturation level (3 µmol m^{-2} s^{-1} PPFD) at 0 °C and 150 µmol m^{-2} s^{-1} at 10 °C both species perform like shade plants (Edwards and Lewis Smith 1988). They reach their photosynthetic optimum value below 12 °C, as respiration increases rapidly with temperature (Day et al. 1997). Apparently, they have a low capacity for temperature acclimation, as cultivation at up to 20 °C could shift the photosynthetic optimum temperature only slightly.

Closed swards of *D. antarctica* mixed with moss on Signy Island may reach a total above-ground phytomass of 520–560 g m^{-2} and below ground 1680–2040 g m^{-2} of which 160–190 g m^{-2} are dead shoots and 1600–1910 g m^{-2} dead roots, respectively (Edwards 1973, 1974b). The energy content of *D. antarctica* is 18.6–20.4 kJ g^{-1} dry wt. with the highest values in the influorescences and the old roots, indicating their role in storing assimilates (Edwards 1974b). Grasses have a homorhizic root system and are able to penetrate into the soil over several decimetres. In the Antarctic they may reach only 5–25 cm depth.

18.2.3 Relevance of Vascular Plants to Soil Formation

Decomposition rates under *D. antarctica* in Signy Island were shown at 60–80 % to be relatively high (cotton strip data, Fenton 1978). Lewis Smith (1996b) reports about decaying *D. antarctica* that was mixed with remnants of moss, producing a kind of a brown earth with a pH range of 4.8–5.5 to a maximum depth of 20 cm.

Soil analyses from sites with these species are documented from various sites (Lewis Smith 1996b; for reviews see Blume et al., Chap. 8; Beyer et al., Chap. 9; Bölter and Blume, Chap. 16; Beyer et al., Chap. 19). Holdgate et al. (1967) demonstrate seasonal changes of the soil ion content with a steady increase of ammonia-N and P and to a lesser extent of Na and K in summer, reaching a peak value in winter, whereas Ca and Mg show little seasonal change. It is assumed that the nutrient regimes of the Signy Island ecosystems are mainly influenced by faunal activity, especially bird excreta. Rainwater is also effective as it contributes to an input of Na, K and Ca to the soils (Allan et al. 1967). Fabiszewski and Wojtuń (2000) give concentrations of ions in plants from King George Island (Table 18.1). Above-ground phytomass contains higher mineral contents than rootmass. Sodium and calcium are prominent. *D. antarctica* widely reflects the degree of soil nutrient enrichment as the highest values are recorded in plants growing at the margin of a penguin rookery. Whereas *C. quitensis* reveals highest values if growing 50 m away from a rookery and not close to an elephant seal colony.

Obviously, due to recent regional warming effects (milder winter temperatures, slightly elevated summer temperatures; Kennedy 1995), the two Antarctic vascular plant species have become rapid colonisers in recently exposed ground adjacent to receding ice-fields and on moraines (Lewis Smith 1996b; Fowbert and Lewis Smith 1994; Lewis Smith 1994; Grobe et al. 1997). This again indicates the presently limiting role of the climatic factor for these species. Since the carbon to nitrogen content ratios of soils where the vascular

Table 18.1. Concentration of ions (mg g^{-1} dry wt. or % dry wt.) in the two vascular plants of King George Island. (Fabiszewski and Wojtún 2000)

		Ca	Mg	K	Na	C (%)	N (%)
Deschampsia antarctica	Shoots	2.98–4.71	1.46–2.21	5.26–9.85	1.02–1.97	31–41	1–3
	Roots	1.1–3.04	0.81–1.37	1–2.54	0.2–1.33	30–38	1–3
Colobanthus quitensis	Shoots	4.11–6.87	2.49–3.45	10.29–22.79	3.93–4.75	33–37	1.5–2
	Roots	2.56–6.95	1.27–4.34	4.76–20.08	0.9–4.75	29–33	1–1.6

plants grow are less than 10 (see also Fabiszewski and Wojtuń 2000) it is suspected that soil elemental cycling might be C-limited at present (Day et al. 1996). However, warming could improve mineral cycling with higher plant activity resulting in carbon addition.

Soils in sites where vascular plants grow in the maritime Antarctic (e.g. Signy Island) may contain all major functional groups typical for a soil ecosystem: including mycorrhiza and a well-developed soil fauna as will be described in the following (Bryophytes) section.

18.3 Bryophytes

18.3.1 Bryophytes as a Component of Antarctic Vegetation

Bryophytes are among the first colonists on barren ground, on rock surfaces in clefts or cracks. Being distributed by spores, and in Antarctica more frequently by vegetative diaspores, they eventually precede other cryptogams such as lichens and fungi (Lewis Smith 1993). This is for instance demonstrated by Lewis Smith (1984b, c) by the observation of the colonisation processes following volcanic activity as happened in the early 1970s on Deception Island, maritime Antarctic. Besides algae, only bryophytes are found as colonisers on heated ground in most volcanic sites in Antarctica (Bargagli et al. 1996; Broady et al. 1987). But their effectiveness in paving the way for colonisation of rock surfaces was also stated for other sites in the maritime and continental Antarctic region (Smith 1982; Longton 1988a). Cushion-forming species of *Ceratodon* are typical primary colonisers of bare ground. Mosses usually grow quicker than lichens and thus moss colonies have a great capacity to trap wind-blown material and to contribute organic matter to developing soil (Smith 1982). They also quickly recolonise disturbed ground near building activities or subsequent to the disturbance of vegetation by fur seals (Lewis Smith 1990a). Mosses also play a part in hydroseres, accelerating the build-up of organic matter and sediments in the bottom of lakes (Priddle 1979; Kanda and Iwatsuki 1989; Imura et al. 1999).

The moss banks on Signy Island were initiated by cushion-forming mosses such as *Andreaea* species on rocky slopes or by *Chorisodontium* invading carpet-forming mosses on moist ground (Imura et al. 1994). Elevated banks eventually become blown clear of winter snow and thus are exposed to erosion. Such places may then become colonised by crustose and fruticose lichens (Lamb 1964; Gimingham and Lewis Smith 1970; Lewis Smith 1972; Fenton 1982; Longton 1988a). Lichen incrustations of mosses or a shift from bryophyte communities to those with macrolichens have been observed in continental habitats (Kappen 1985a; Melick and Seppelt 1997).

Among the 104 species of bryophytes in Antarctica (Ochyra 1998) only 25–30 are liverworts (Seppelt et al. 1998) while Anthocerotales are absent. Most of the Hepaticae are restricted to the maritime Antarctic region except for *Cephaloziella exiliflora* (Lewis Smith 1988a, 1997; Melick et al. 1994). This situation can mainly be explained by different tolerance to the harshness of the climate on the Antarctic continent. The presence of moss species on fumarolic ground (Collins 1969) of the Antarctic volcanoes Mt. Melbourne (Broady et al. 1987), Mt. Rittmann (Bargagli et al. 1996) and on Deception Island (Young and Klay 1971; Lewis Smith 1984b, 1988a) demonstrate that the local climatic conditions are beneficial for bryophytes that otherwise cannot exist in the Antarctic. The presence of exotic species at these sites indicates the distribution of diaspores all over Antarctica. Viable diaspores of many bryophyte species are present in apparently barren ground as was demonstrated by experiments with cloches (Lewis Smith 1987). Within a few weeks a closed vegetation of many species could be achieved because of an ameliorated temperature regime, increased soil moisture and perhaps shielding from short-wave radiation (Lewis Smith 1990a,1993). The deposition of other air spora such as pollen grains and spores from plants of other continents has been repeatedly shown (van Zanten 1978; Salas 1983; Kappen and Straka 1988; Marshall 1996).

Another indication of the climatic effect on the presence of bryophytes in Antarctica results from a particular species richness and abundance in coastal habitats, which are particularly sheltered sites with sufficient moisture supply (Lewis Smith 1988a, Vincent 1997) or beneath stones in otherwise barren soil (Kennedy 1993). Lakes in the vicinity of Syowa Station and Schirmacher Oasis provide also a beneficial environment even for terrestrial bryophytes such as *Bryum pseudotriquetrum* (Imura et al. 1999, Longton 1988a).

18.3.2 Physiological Properties That Enable Bryophytes to Live in Antarctica

As poikilohydrous organisms, e.g. bryophytes, are tolerant to water loss but tolerance differs with species (Kappen and Valladares 1999). Therefore, xeric, mesic and hydric species can be distinguished according to their tolerance range and habitat preference. Accordingly, Antarctic xeric species such as *Bryum pseudotriquetrum, Andreaea spp., Dicranoweisia grimmiaceae, Schistidium antarctici, S. apocarpum, Ceratodon purpureus,* and *Hennediella heimii* are able to grow on rocks with changing water regimes and longer-lasting drought periods, whereas *Sanionia uncinata (=Drepanocladus uncinatus), Calliergidium* ssp. and *Chorisodontium aciphyllum* are restricted to moist depressions or places with waterlogging (Noakes and Longton 1989; Davey and Rothery 1996; Lewis Smith 1999; Schlensog and Schroeter 2000), whereas *Bartramia patens* and *Polytrichum alpinum* represent mesic species. Some species show a wide physiological and ecological range such as *Bryum argenteum*, forming terrestrial strains and aquatic forms (Longton 1988a; Lewis Smith 1999; Green et al. 2000) or appearing with xeric and mesic forms (*Schistidium antarctici*; Kappen et al. 1989).

Obviously, the hydric species are able to persist under longer-lasting snow cover. The freezing tolerance of mosses is only measured in European material. In fresh material it varies between –15 and –25 °C in winter (Dircksen 1964; Rütten and Santarius 1992a,b, 1993). However, tolerance can be substantially increased with the degree of thallus water loss (Kappen and Valladares 1999). Maximal freezing tolerance of a fresh moss plant is higher (–28/–33 °C) if one considers that single cells of highly frost injured plants are able to regenerate new plants (cf. Dircksen 1964: 'Lethal tolerance limits').

Moss temperatures in the frigid Antarctic East Ongul Islands are most frequently between –10 and –20 °C (Matsuda 1968) and temperature minima between –25 and –30 °C are recorded in the maritime Antarctic Signy Island in winter (Walton 1982) but temperatures could also be rather high (around 0 °C) during winter under deep snow cover (Longton and Greene 1967; Davey 1997). Arctic and Antarctic moss species (*Dicranum elongatum*, *Racomitrium lanuginosum*) are able to resume photosynthetic activity after 10 h exposure to –30 °C passing a phase of heavy respiration for 2–5 h after returning to above 0 °C (Kallio and Heinonen 1975). Prolonged freezing (30 weeks at –30 °C) cause the arctic xeric *Andreaea gainii* to commence a high net photosynthetic rate 5 h after thawing, whereas the hydric *Calliergon sarmentosum* cannot resume photosynthesis 70 h after thawing, however, photosynthesis is quickly recovered after 2 weeks at –30 °C (Collins and Callaghan 1977).

Bryophytes seem to be well buffered against freeze-thaw fluctuations (Melick et al. 1994) as they store a high amount of carbohydrates as protective substances. In the maritime Antarctic, freeze-thaw events are more frequent (110/year) in mesic turfs of *Polytrichum alpestre* and *Chorisodontium aciphyllum* (13/year) than in the wet carpets of *Calliergon* and *Sanionia uncinata* (Longton 1970, 1972a) but the temperature minima were not far from 0 °C in each case. The incidence of winter kill of mosses is not clear: Collins (1976) attributes dying of apices of *Polytrichum alpestre* in Signy Island to frost damage. But it could not be decided whether the low capacity of resprouting of *Sanionia uncinata* in Signy Island was due to frost action or to heavy waterlogging (Longton 1988a). Freeze-thaw cycling on the other hand causes a significant input of organic material to the soil, as intracellular sugars are released from bryophytes under the changing water status (Melick and Seppelt 1992; Montiel and Cowan 1993).

Bryophytes mainly depend on liquid-water uptake for being metabolically active (Noakes and Longton 1989; Davey 1997b,c; Kappen and Valladares 1999). Photosynthetic activity of the green shoots – being 1 to 6 mm long – is optimal at 200–300 % dry wt. in xeric species, but can be over 500 % dry wt. in hydric species (Lewis Smith 1984a; Convey 1994). Their photosynthetic rates range between 0.49 and 3.22 mg CO_2 g^{-1} h^{-1} (measured at around 15 °C) in 14 maritime Antarctic species (Convey 1994) and in a continental Antarctic species (Kappen et al. 1989). These rates are low compared with those of the Antarctic vascular plants (Lewis Smith 1984a). The optimum temperatures for net photosynthesis of maritime Antarctic mosses are estimated to be around 15 °C (Convey 1994; Davey and Rothery 1996) and according to results by Kanda (1979) and Longton (1988a). They are similar (*Bryum argenteum*, Green et al. 2000) or lower (10–15 °C; Ino 1990a,b; 4–6 °C; Kappen et al. 1989) for continental species. Net photosynthesis is ceased as soon as the water freezes; however, the lowest temperature for photosynthetic activity is –10 °C in *Racomitrium lanuginosum* (Kallio and Heinonen 1971, 1973) and a continental Antarctic species (Ino 1990b). CO_2 gas exchange is detectable in Antarctic *Polytrichum* and *Sanionia* at –8 °C (Davis and Harrison 1982), whereas it ceases at –4 °C according to field measurements in the Antarctic Peninsula region (Gannutz 1970). Photosynthetic activity under snow is recorded by Collins and Callaghan (1980) for mosses in the maritime Antarctic. Dark respiration at 15 °C ranges between 1.03 and 4.69 mg CO_2 g^{-1} h^{-1} (Convey 1994). Antarctic bryophyte species do not differ in principle from related taxa of temperate climates (Longton 1988b), although acclimatisation of the photosynthetic apparatus was recorded for *Polytrichum commune* by a comparison of a series of provenances from the tropics to a cool Arctic site (Sveinbjörnsson and Oechel 1983).

The photosynthetic performance of bryophytes is usually typical of shade plants having low light compensation points of 75–180 µmol m^{-2} s^{-1} (Rastorfer 1970, 1972; Kallio and Kärenlampi 1975) and light-saturation at PPFD of about 300 µmol m^{-2} s^{-1} (Longton 1988a). However, *Sanionia uncinata* attained maximum net photosynthetic rates above 500 µmol m^{-2} s^{-1} (Davis and Harrison 1982) and *Bryum argenteum* in continental Antarctica even near 1000 µmol m^{-2} s^{-1} (Green et al. 2000).

For these reasons bryophytes can suffer from photoinhibition (Adamson et al. 1988, Kappen et al. 1989, Post 1990, Lovelock et al. 1995a, b, Green et al. 1999, Schlensog and Schroeter 2000). Schlensog (2000) demonstrated the activity of the xanthophyll cycle under strong light in the diurnal course in Antarctic bryophytes. Post (1990) shows that photoprotective pigments are produced when the mosses are longer subjected to strong irradiance, and Green et al. (2000) demonstrate that the sun-adapted form of a continental Antarctic *B. argenteum* is protected by its silvery shoot points and non-photochemical quenching within the photosystem. Particularly stressful is the influence of both bright light and freeze-thaw cycles, that occur frequently (up to 110 times/year in the maritime Antarctic; Longton 1988a). Every subzero phase causes decreased photosynthetic efficiency, but recovery at low light has been proved to be rapid (Lovelock et al. 1995a,b). The maritime Antarctic *Polytrichum alpestre* is most sensitive to freeze-thaw cycles with a reduction of gross photosynthesis proportional to the lowest temperature reached and no recovery after passing –5 °C (Kennedy 1993a). Kennedy (1993b) concluded that stress by freeze-thaw cycles will most likely limit species distribution.

18.3.3 Productivity and Growth

Estimates of annual primary production for bryophytes from the maritime Antarctic region range between 200 and 1000 g m^{-2} $year^{-1}$ (Baker 1972; Longton 1970; Collins 1973, 197; Davis 1983) and for those from continental sites to up to 100 g m^{-2} $year^{-1}$ (Longton 1974; Ino 1983). However, negative balances have been calculated for some years in both Antarctic regions (Collins and Callaghan 1980; Ino 1983).

Growth of bryophytes is small compared with that of vascular plants. It is strongest if water content of the shoots is very high (Noakes and Longton 1989). So, *Sanionia uncinata* grows best at 800 % dry wt. water content but attains high growth rates still at 1800 % dry wt. Growth of shoots may occur at 0 °C (*Ceratodon*; Kanda 1979). Shoot elongation of maritime Antarctic species varies between 2 and 40 mm $year^{-1}$ (Longton 1970; Collins 1973; Lewis Smith 1984a). It is generally much less in continental Antarctic species (0.2–1.3 mm $year^{-1}$; see Green et al. 1999) but may reach more than 10 mm $year^{-1}$ in exceptional cases (Melick and Seppelt 1997). Shoot extension usually begins shortly after snow clearance but premelt growth is observed in *Calliergidium austostraminium* on Signy Island (Collins et al. 1975). The moisture limit for growth was at water contents by 63 % dry wt. in healthy *Grimmia antarctici* shoots (Melick and Seppelt 1997).

18.3.4 Growth Forms and Strategies

Most polar bryophytes are perennials, but duration of active growth of individual shoots varies widely. Shoots of carpet and hummock forming species may grow continuously for 2 or more years or shoot apices become killed in winter and will be replaced by branches of subapical position (e.g. *Sanionia*; Callaghan and Collins 1981). In *Polytrichum*, shoots continue to grow for many years, in some species young shoots arise at the apex of horizontal rhizomes that are growing several centimetres off the parental shoots (Longton 1979).

The extent of green active tissues varies widely with species or according to the environmental conditions. In Continental Antarctic species such as *Bryum* sp. only 1–2 mm is green assimilating shoot, but this part may be several years old (Longton 1974), like the green apical sections of other moss species which may be up to 50 mm long. Below the green section, brown stems of several centimetres length are able to form rhizoids or to regenerate new shoots.

Table 18.2. Growth forms of Antarctic bryophytes. (After Longton 1988a)

Growth form	Typical genus	Description
Small cushion	*Andreae, Grimmia, Dicranoweisia*	Sparingly branched acrocarpous mosses forming dome-shaped compact colonies (<50 mm in diameter)
Large cushion	*Brachytecium, Bryum*	Arocarpus or pleurocarpous mosses radiation in dome-shaped colonies (>50 mm in diameter); looser than small cushions
Short turf branches erect	*Bryum, Pottia, Hennediella*	Sparingly branched acrocarpous mosses; shoots erect and parallel; loose colony (<20 mm tall)
Tall turf (1),	*Polytrichum, Dicranum*	Sparingly branched pleurocarpous mosses, main shoots erect >20 mm tall; intermediate growth
Tall turf (2), branches divergent	*Sanionia*	Freely branched pleurocarpous mosses, main shoots erect >20 mm tall; determinate growth
Carpet	*Calliergidium, Sanionia*	Pleurocarpous mosses with distal portions of shoots parallel and erect ascending; loose or compact colonies
Thread	*Cephaloziella*	Hepatic; single shoots, not colony-forming but within colonies of mosses

Growth strategies of bryophytes vary greatly and are of great interest with respect to their influence on soil penetration and soil formation. According to their ramification and position of sporogons, acrocarpous and pleurocarpous types are distinguished. Since bryophytes form shoot complexes instead of growing with singular shoots, the acrocarpous type is able to form turfs or small cushions whereas pleurocarpous species form carpets, mats, wefts or threads (Table 18.2). In Antarctica, the relation between acrocarpous and pleurocarpous species in the bryoflora is 5.1:1 (Ochyra 1998). Whether this has ecological relevance is not yet known.

The acrocarpous species *Polytrichum commune* and perhaps other species of the *Polytrichaceae* in the maritime Antarctic were reported to form shoots from a single subterranean system of radiating stems and were therefore able to form circular colonies up to 30-cm diameter (Lewis Smith 1972). The genus *Polytrichum* is also representative of the most highly developed anatomy amongst the bryophytes as it contains water-conducting strands in the stem, the hydroid, a solid subterranean axis, the rhizome with rhizoids but not a true root and a cuticula on the leaflets. According to these properties, such bryophytes are called endohydrous (Mägdefrau 1935) whereas the majority of bryophytes, whether acrocarpous or pleurocarpous, are ectohydrous, which means that they are less cutinised, have no central strands in the stem and frequently form rhizoids along the stems up to the tips. This has consequences for the uptake and conduction of water and minerals. Whereas the ectohydrous bryophytes take up water and minerals over the entire surface of the thallus, endohydrous species are partially able to take up water from the ground and conduct it to the apex. Also, translocation of photosynthates into new below-ground shoots of arctic *Polytrichum alpinum* was recorded by Collins and Oechel (1974).

Noakes and Longton (1989) have compared the water relations between two endohydrous and three ectohydrous species (*Andreaea geinii*, *Sanionia uncinata* and *Schistidium antarctici*). Shoots of endohydrous mosses have a higher resistance to water loss and internal water movement is larger by a factor of 13 in the endohydrous than in the ectohydrous species. Endohydrous species may therefore benefit from internal conduction, particularly under relatively dry conditions, but under moist conditions external water movement is always more important.

The aggregation of bryophyte shoots into colonies enhances the capillary uptake of soil water and retention capacity of precipitation as well as protection against quick water loss. Gimmingham and Lewis Smith (1971) have shown the influence of growth form on the bryophyte water content in Signy Island. In comparison with carpets and large cushions in wet habitats the small cushions in dry habitats have the greatest protective effect on water loss and are also able to rehydrate through external capillary water when water is supplied from the base. CO_2 concentrations can rise up to ten times the ambient air concentration, as is demonstrated for wet turfs of *Grimmia antarctici* and *Bryum subrotundifolium* in continental Antarctica whereas in dry turfs ist was elevated by only 30 % of ambient concentration (Tarnawski et al. 1992; Green et al. 2000).

Shoots of bryophytes can extend deeply into the soil. For instance, in the maritime Antarctic, living stems of *Polytrichum alpestre* reach a length of 25 cm with the greater part in the peaty ground (Longton 1972). Like plant roots they and their rhizomes contribute to soil respiration (Lewis Smith 1984a). In many cases it is not easy to distinguish between the living moss plant and the peat as remains of older growth. Peat banks of *Chorisodontium aciphyllum* in Signy Island, maritime Antartic, may be 1–2 m thick but the part below 30 cm from the surface is permanently frozen (Gimingham and Lewis Smith 1970, Longton 1985). Moss hummocks of *Brachythetium austrosalebrosum* can reach 10–20 cm height (Longton 1967). In the continental Antarctic region moss turfs reach up to 10 cm (Melick and Seppelt 1997), on wet ground, and cushions of rock-colonising species only a few centimetres. No tall turf, carpet or hummock-forming mosses are present. In wetter sites,

Bryum spp. forms elongate and contorted cushions which are up to 10 cm wide and 5 cm high with an active layer of 1–2 mm depth. Studies of branching pattern and rhizoid banding suggest ages of 25–100 years of the cushions of *Bryum inconnexum* (Matsuda 1968; Seppelt and Ashton 1978). Whereas in *B. algens*, the annual growth bands could only be traced back over a period of 15 years (Lewis Smith 1984a).

Scattered green shoots of *Bryum algens* that occur to a depth of 3 cm in the fine-grained granite debris and sandy soil (Kappen 1985a) in northern Victoria Land suggest that bryophytes are able to resist covering with soil and sediments as happens in dunes, growing up through 4 cm of sand deposited on them (Birse et al. 1957). Being buried in sand, mosses are able to stabilise the sand, they add organic matter and maintain moist conditions by their water-holding capacity.

In the continental Antarctic region, mobile moss balls are observed lying on the ground and on glaciers. At Mawson Station Seppelt and Ashton (1978) found colonies of *Bryum algens* and *Grimmia lawiana* which originated as small cushions with green shoots all over the surface which then may be blown over the ground. A mineral core of the larger cushions suggests that cryoturbation may have caused the formation of these balls (Longton 1988a). In the maritime Antarctic the balls are generated after cushions of *Bartramia patens* and *Ditrichum strictum* are detached from the ground by frost action and subsequently green shoots were developed on the undersides of the detached cushions (Huntley 1971; Longton 1988a).

18.3.5 Phytomass

Phytomass of bryophyte vegetation may reach enormous values in the maritime Antarctic. The large moss turfs of Signy Island contain up to 40 kg m^{-2} peat in a 1-m-deep bank. However, the green shoots on top may reach up to 1000 g m^{-2}. Living phytomass of *Polytrichum* stands reach about 10 kg m^{-2} (Longton 1972; Lewis Smith 1984a). Moss carpets with *Sanionia* reach 2.1 kg m^{-2} (Lewis Smith 1984a). Usually, lower values are expected on the Antarctic continent. Here, the largest figures are about 1100 g m^{-2} green phytomass if a single colony is considered. Related to an area where the moss community occurs, the values for *Bryum antarcticum* could be 14–938 g m^{-2} and for *Bryum argenteum* 5–397 g m^{-2} green and brown phytomass on Ross Island (Longton 1974) and 197 g m^{-2} green phytomass of *Grimmnia laviana* and *Bryum algens* near Mawson Station, Wilkes Land (Seppelt and Ashton 1978).

According to the only value (16.9 kg g^{-1} for *Brachythecium austrosalebrosum*) from a maritime Antarctic habitat, the energy content of Antarctic bryophytes seems to be lower than in those from lower latitudes and the Arctic (Longton 1988a). Protein content is generally low compared with that of vascular plants but Antarctic readings are not available (Longton 1988a). Soluble sugar contents are determined in Antarctic mosses by Rastorfer (1972) and range between 3 and 5% dry wt. in green shoots. From analyses with Arctic bryophytes two results are remarkable, first that the lignin content is consistently higher in bryophytes than in green parts of vascular plants (Skre et al. 1975), and secondly that the content of Calcium can be typically higher in mosses than in leaves of associated monocots (Staatland et al. 1983). We will refer to mineral nutrients later.

18.3.6 Other Organisms Attached to Bryophytes

Apart from what is produced by bryophytes and which may later serve as a nutrient source in the soil, mosses provide niches for other organisms which may also support the system and particularly the soil with their products. Bryophytes are frequently found to be colonised by unicellular algae. Broady (1977) counted $12 \cdot 10^6$ algal cells per gram moss attached to the surface of leaflets of *Polytrichum* and *Chorisodontium* turfs and *Calliergidium-Sanionia* carpets in the maritime Antarctic. The number of algal cells increase with the age of the leaflets down to the seventh or eighth leaflet and then decrease due to light deficiency. Among the algae, species of Xanthophyceae, Euchlorophyceae, Zygophyceae and Ulotrichophyceae and also cyanobacteria have been identified.

Cyanobacteria are frequent colonisers on moss leaflets in Antarctic habitats. They are less abundant on the turf mosses because the pH is not high enough. However, *Nostoc muscorum* is very abundant on hydric bryophytes (Broady 1979). It is not clear whether these cynobacteria can greatly contribute to an N-input, as temperatures which prevail are too low for nitrogen fixation (Fogg and Stewart 1968; Horne 1972; Broady 1979). However, N-fixation in situ was shown to be possible even below 0 °C (Davey 1983); and significant rates are recorded by Davey and Marchant (1983), Nakatsubo and Ino (1986) and Christie (1987) also for samples in continental Antarctica. Davey (1983) demonstrates for *Nostoc* and *Calothrix* a diurnal course of nitrogen fixation with highest values around noon (up to 211 pmol C_2H_4 μg^{-1} Chl *a* h^{-1}) when ground temperatures were near 10 °C.

The presence of protozoa and invertebrates in bryophyte vegetation has been repeatedly recorded. Tilbrook (1973) refers to acari, collembola, nematodes, tardigrades, rotifers and testate amoebae in various moss formations on Signy Island. Considering a bryophyte community in the maritime Antarctic as an ecosystem, it consists of about 15 functional groups of organisms with a total of about 120 species (see also Davis 1981) if we add epiphytic lichens (see Lewis Smith 1988b), microfungi, filamentous fungi, bacteria and cyanobacteria. For the continental Antarctic region Lewis Smith (1996b) reports the presence of protozoans, tardigrades, nematodes, collembola, and mites in the bryophyte vegetation in sites of the Antarctic Peninsula. It is evident from this that a diversity of matter fluxes are active and significantly influencing soil formation.

18.3.7 Decomposition and Mineralisation

Decomposition of plants is one of the major processes in the development of soil. It depends on the resistance of the plant material to decomposition, the climatic conditions and the chemical environment in which decomposition takes place. In Antarctica, decomposition processes are generally slow, particularly slow with bryophytes, as they produce substances which inhibit bacterial growth, although the chemical nature of these substances has not yet been analysed (Banergee and Sen 1979). Bryophyte cell walls are known to contain high concentrations of lignin-like phenolic components (Hebant 1977). The range of fungi that parasitise mosses is broad (Doebbeler 1978). Fungal infections are frequent in cold climatic regions. For instance, carpets of *Brachythetium*, *Sanionia*, *Ceratodon* and *Dicranoweisia* show whitish patches or white rings up to 20-cm across, marking the fungal infection. Rings may increase by about 25 mm during the 4-month growing period in the maritime Antarctic. With the exception of *Dicranoweisia*, the mosses can regenerate new shoots. One of the fungi has been identified as *Thyronectria antarctica*, Ascomycotina (Longton 1973).

Mosses decompose from the base with surface litter deposition. Studies of decomposition processes in the Antarctic are mainly made on peat-forming mosses and on wet carpets in Signy Island, maritime Antarctica. Wet mixed *Calliergon-Sanionia* carpets achieve a dry weight loss of 8.3 % year^{-1} at a depth of 5 cm but 1.5 % respectively at 10 cm depth (Fenton 1978). The decay rate in a wet *Sanionia* stand at 1–3 cm depth is even higher (25 % year^{-1}). The rate is apparently still larger in *Brachythetium austrosalebrosum* (Collins 1973). In most cases the decomposition rate decreases with depth. Davis (1980) shows a steadily decreasing rate with depth ranging from 8.3 % year^{-1} at 1–3 cm to 0.9 % year^{-1} at 12–24 cm. Moss turf peats are dryer and consequently the decay rate is lower. For instance, in *Polytrichum* banks the dry weight loss was about 1 % year^{-1} in the upper layer, *Chorisodontium* banks may decompose only by a rate of 0.3 % year^{-1}.

Using calculations from data for peat respiration, loss of organic matter in moss turfs from Signy Island was calculated by Davis (1980) to be 571 g m^{-2} year^{-1} (0.017 g g^{-1} year^{-1}). The rate of wet carpet decomposition at 288 g m^{-2} year^{-1} (0.01 g g^{-1} year^{-1}) is perhaps an underestimated value. Fenton (1980) has compared peat accumulation with primary production of the moss cover on Signy Island. At a primary production of 350 g m^{-2} year^{-1} of *Polytrichum* the peat accumulation is 850 g m^{-2} year^{-1} and of *Chorisodontium* the respective values are 162 and 89 g m^{-2} year^{-1} only. The decomposition rate in these cases is less than 1 % year^{-1}. All these data illustrate to what extent mosses are able to contribute to humus formation (cf. Longton 1984).

18.3.8 Catchment of Debris, Detritus, Airborne Particles, Air Spora and Trace Substances

It has already been shown that bryophytes provide humic soils with large amounts of living and decaying tissues. Living shoots can penetrate deeply in aeolic sediments and may influence the chemistry by exudates. The mobile moss balls can enclose rock debris and soil from underground where they were originally attached. The surface roughness and the (loose) structure of bryophyte assemblages such as cushions, turfs, mats etc. provide an ideal trap for all kinds of particles. Like in a sponge or a brush, wind-blown sand grains and particles are captured, frequently filling the interstices between the moss shoots. Therefore, moss cushions of e.g. *Andreaea* or *Grimmia* on bare rock contain a great amount of sand and fine rock debris (see also Bargagli et al. 1998). In areas with active volcanoes of the maritime Antarctic, species of *Brachythecium* and *Ceratodon* are impregnated with volcanic ash. In snow beds the material accumulated in late-lying snow gets into the bryophyte vegetation by melt processes (Longton 1988a).

The role of bryophytes as traps for pollen and spores gives information about possible immigration pathways of airspora from other continents into Antarctica (Kappen and Straka 1988; Lewis Smith 1991; Kappen 1993a), but also provides an input of organic material to the habitat. In northern Victoria Land, the entrapment rate of moss is on average 1200 pollen grains m^{-2} year^{-1} (Linskens et al.1993). This is a fairly low rate compared with rates in temperate regions and may be higher in the maritime Antarctic region. This becomes evident from a comparison of pollen concentrations of samples on King George Island (maritime Antarctic) with those from Casey Station on the continent (Kappen and Straka 1988). The much greater abundance of pollen and spore taxa in the maritime Antarctic region is explained by the close (800 km) vicinity to South America and to more favourable meteorological conditions with northerly winds. Up to 44 grains m^{-2} slice of 1.5-mm thickness are found in *Bryum*, *Sarconeuron* and *Tortula* and other taxa. The vertical distribution is determined along a 14-mm-deep profile of *Bryum pseudotriquetrum* and shows high values in the upper part of the cushion and peak values at about 5–8 mm below the surface. The percentage of damaged pollen generally increases with depth in the cushion. Besides pollen and spores, cysts and diaspores and resting unicellular microorganisms such as bacteria and cyanobacteria, yeasts and unicellular algae must also be deposited in the moss vegetation.

Bryophytes were also indicators of atmospheric contamination as they can accumulate chemicals from organic vapours in air. Organochlorine residues with detectable levels of HCB, HCH, DDT and DDE were analysed from moss samples in the Ross Sea area and from several sites in the maritime Antarctic (Bacci et al. 1986). Hydrocarbons, stereols and hydroxy acids that are accumulated in Antarctic mosses may also stem from nearby sources (Matsumoto and Kanda 1985). Even mercury is significantly enriched by mosses in sites with low levels of Hg in soils and sediments in the McMurdo Dry Valleys and Ross Island (continental Antarctica). A close correlation between Hg-content and total organic carbon in the mosses suggests that they could be good biogeochemical markers (Matsumoto et al. 1983). The sources for these chemical components were not anthropogeneous but could be marine aerosols, seabird guano and volcanic emissions, as demonstrated by Bargagli et al. (1998) from accumulation in mosses from North Victoria Land volcanoes. The accumulation process is explained by long-term deposition and then accumulation in the moss surface layer by evaporative water transport.

18.3.9 Nutrient Cycling

The potential significance of cryptogams in polar nutrient cycles is widely accepted although experimental evidence is scarce (Longton 1988a). In the following, what is known about the role of Antarctic bryophytes in nutrient turnover of their habitat is briefly reviewed and finally there is a discussion as to whether these nutrients are essential for bryophytes.

The absorption of N, P and K is mainly intracellular and is by means of active transport, and this is expected for all other ions that are important for metabolism. However, most of the element content of mosses resides outside the cell membrane, being bound to extracellular exchange sites or as particles on plant surfaces. The greatest concentration of N, P and K occurs in the cytoplasm of the green active tissues of the shoot while Ca and Mg tend to accumulate in the brown, partially senescent region (cf. Chapin et al. 1980; Lewis Smith 1978, 1984a). Analyses of ion contents in Antarctic bryophytes were undertaken for Signy Island (Holdgate et al. 1967; Allen et al. 1967; Allen and Northover 1967; Lewis Smith 1978), for the Argentine Islands (Rastorfer 1972) in soils and peat, and for Wilkes Land, continental Antarctic (Lewis Smith 1990b) and from soils and peat under discrete moss communities.

Turfs of Wilkes Land mosses contain equal or slightly higher Na contents (0.05–0.2% dry wt.) than the soil. Sodium (0.08–0.4% dry wt.), calcium (0.09–0.4%), magnesium (0.09–2.2%), manganese (0.006–0.01%) and phosphorus (0.02–0.34%) are accumulated in proportion to the soil ion content, whereas iron (0.05–0.63%) and aluminium (0.05–0.36%) are less concentrated in the moss turfs than in the soil (Lewis Smith 1990b). The ion concentration is mainly proportional to that of the substratum also in *Pohlia nutans* and *Campylopus pyriformis* from central Antarctic volcanoes (Bargagli et al. 1996) and in bryophytes from South Georgia (Walton and Lewis Smith 1980).

Biotically enriched sites (influenced by bird colonies etc.) cause increased N and P contents (Allen et al. 1967; Holdgate et al. 1967). The concentration of most nutrient ions except N in bryophytes is as high as in polar vascular plants (Longton 1988a).

The influence of the coastline is significant for ion accumulations (Lewis Smith 1978), particularly for Na, K, Ca, Mg, Mn, and partly for P which had higher concentrations in moist or wet sites than in dryer sites. This corroborates data by Rastorfer (1972) for the nitrogen content of bryophyte species from Argentine Island (dry/moist: 1.5/2.95 % dry wt.). Similar relations are found for Na, K and Ca in maritime Antarctic bryophytes (Lewis Smith 1978). The moisture regime of the habitat is therefore important for ion uptake.

Lewis Smith (1978) detects considerable differences between summer and winter in ion concentrations in bryophytes from Signy Island. Marine-derived nutrients such as Na were lower in winter when the sea was frozen. Potassium decreases in summer in species from dry sites most likely as a consequence of lack of water in summer.

Storage of nutrients in living and dead moss phytomass may limit its availability to other organisms. Bryophytes are also likely to retain nutrients dissolved in precipitation either by incorporation into tissues or stored in the capillary space within colonies. For instance, Ca stored in Arctic bryophytes has a turnover time of 22 years (Longton 1988a). Summing up, bryophytes contribute to the total nutrient pool but may retard rates of nutrient cycling.

Finally, we discuss to what extent nutrients are essential for the existence of bryophytes in Antarctica. The preference of certain substrata is repeatedly demonstrated for Antarctic bryophytes: *Andreaea* species mainly occur on acidic rock, scree and soil. The *Tortula-Schistidium* association is characteristic of base-rich habitats such as marble outcrops or young volcanic material. Calcicolous species also include *Bartramia patens*, *Bryum argenteum*, *B. pseudotriquetrum*, *Schistidium antarctici* and *Tortula princeps* which grow e.g. on accumulations of decaying limpet shells (Lewis Smith 1996b). But, it has never been shown whether these bryophytes essentially need high amounts of certain ions in such habitats or whether other related factors are more important, and Longton (1988a) concludes that some mosses are fully independent of their substratum.

Nutrient deficiency in Antarctic bryophytes has very rarely been shown. It is widely accepted that Antarctic inland soils are strongly deficient in certain elements, notably nitrogen and phosphorus (Tedrow 1977). In fact, remote sites on nunataks or inland mountains where vegetation is found are mostly in the vicinity of a bird colony. The sub-Antarctic Marion Island receives extremely low N and K and no detectable P from rainfall so that bryophytes are expected to live under extremely nutrient-limited conditions in mires and bogs (Smith 1988). In experiments with sub-Antarctic mosses, Smith (1993) demonstrated that the net photosynthetic rates could be significantly increased if NO_3 or a combination of nitrogen and phosphorus were supplied.

This corroborates earlier fertilisation experiments in very nutrient-poor Arctic inland habitats (Oechel and Sveinbjörnson 1978; Chapin and Chapin 1980). All other elements are available in sufficient amount and the ion concentration in bryophytes simply reflects the input level from the atmosphere and the substratum, sometimes leading to incrustation as may happen on calcareous substrates or with heavy metals (Richardson 1981). Generally speaking, however, the distribution and vegetation pattern of bryophytes in most sites of Antarctica are mainly determined by local water regime, snowpack, action of wind and thermal conditions and in the second instance by nutrient availability, although species richness and density of the vegetation is the greatest under rich N and P supply (Nakanishi 1977; Lewis Smith 1984a, 1985, 1988a; Melick et al 1994).

18.4 Lichens

18.4.1 Floristic Remarks

Lichens form a heterogeneous group of organisms composed mainly of an ascomyceteous mycobiont and a photobiont colony of either unicellular algae or cyanobacteria. The major biomass of the lichen thallus is formed by the mycobiont. Therefore, lichens can be considered as a specialised group of terrestrial fungi. The obligate symbiosis with the photobionts, however, lets them perform like autotrophous organisms, such as plants. Therefore, lichens form a component of the terrestrial vegetation. In Antarctica they are the most obvious and the most successful among all autotrophs in the landscapes of the ice-free regions (Kappen 1993a). Also, their species diversity is the highest compared to other cryptogamic groups. They are represented by about 360 taxa (R.I. Lewis Smith and D. Øvstedal, pers. comm.). The greater number of species (335) exist in the maritime Antarctic whereas only 125 species (Longton 1988a) may be indigenous to continental Antarctica. These figures are small compared with those from sub-arctic and high-arctic recordings (together totalling about 2000 species: Thomson 1972; Greenland 566 species; north of 70°N in North America and Greenland ca. 470 species: Thomson 1997).

The origin of the present lichen flora in the Antarctic is not clear (Kappen 1993a). A few dozens of lichen species may have survived in refuge habitats throughout glacial periods covering over 30 million years, but the majority of species must have migrated into the isolated Antarctic region during the last 10–15,000 years. Although air spora must be blown all over the Antarctic region (Lewis Smith 1984a), air currents allow a more efficient immigration to the maritime Antarctic region than to the remote continental region (Kappen and Straka 1988; Lewis Smith 1991). Graham Land appears to be a transitional zone between the maritime and the continental region (Øvstedal 1983). The import rate is comparatively low. The eventual import of fungal spores from South America to Signy Island was shown by Marshall (1997), who stated that most air spora are of local origin. According to Lewis Smith (1991) no exotic bryophyte or lichen spores have yet been detected in trapping experiments in the Antarctic region. Perhaps the proportion of endemic species in either region will suggest conclusions about the age of the Antarctic lichen flora.

18.4.2 Physiological Properties That Enable Lichens to Exist in Antarctica

The physiological properties of lichens are less likely to limit existence in the harsh climate of Antarctica. The poikilohydrous character is more expressed in lichens than in bryophytes. Having no water-conducting tissues and not being protected against water loss by a surface layer or tissues, the cells are directly exposed to the environmental moisture conditions and usually change their moisture status accordingly. Most of the lichen species are able to tolerate near-total water loss and persist in the desiccated state for periods of several months. In the dry state, the lichens are insensitive to temperature stresses between liquid-air temperature (–196 °C) and up to around 70 °C (Lange 1953; Kappen 1973).

In Antarctica lichens may never suffer from temperature stress even in the fully hydrated state, as most of the tested Antarctic species are able to tolerate far lower temperatures than ever will occur in the coldest part of Antarctica (Kappen 1973). Even though thallus temperatures may reach temperatures close to the heat tolerance border of metabolically active thalli (Lange 1965) with temperatures up to 45 °C in the maritime Antarctic (Lewis Smith 1988 c), it is unlikely that they are stressed because they quickly become desiccated at this warm temperature.

Many lichens, in particular species from polar regions, are able to quickly resume full metabolic activity subsequent to drought or extreme temperature stress (Lange and Kappen 1972; Larson 1978; Kallio and Heinonen 1971) or an Antarctic winter season (Gannutz 1970). The low temperature range of the photosynthetic apparatus of lichens is conspicuous. The lowest temperature of photosynthetic CO_2 fixation is between –24 and –16 °C according to laboratory measurements (Lange and Metzner 1965; Lange and Kappen 1972). Field measurements of net photosnythesis of lichens in sites of continental Antarctica confirm that the lowest values for *Usnea sphacelata* and *Umbilicaria aprina* are between –10 and –20 °C (Kappen 1989; Schroeter et al. 1994).

It may be considered as an adaptive response to the environment that most of the Antarctic lichens gain highest possible photosynthetic rates (T_{opt}) at temperatures between 1 and 12 °C in continental Antarctic species such as

Usnea sphacelata (Kappen 1993a) and up to 18 °C in maritime Antarctic species such as *Ramalina terebrata* (Kappen 1988; Schroeter 1991; Sancho et al. 1997). The highest temperature which still allows positive rates of net photosynthesis (upper temperature compensation point T_{max}) was between 15 and 29 °C for all investigated species.

An acclimative response can also be derived from a comparison of T_{opt} and T_{max} between continental (Hallett Station, Lange and Kappen 1972) and maritime Antarctic (King George Island, Schroeter et al. 1995) provenances which attained T_{opt} 1–6 vs. 2–13 °C, and T_{max} 22 vs 26 °C, or with a comparison of four different provenances of *Parmelia saxatilis* (Schroeter 1997) showing a difference of T_{max} between a mediterranean and an Antarctic provenance of 4 K. But Antarctic populations could even be superior to Arctic populations of *Usnea sphacelata* (Kappen 1993a). The maximal rates of net photosynthesis ranging between 0.1 and 0.5 mg CO_2 g^{-1} h^{-1} or 0.01 and 0.05 mg CO_2 m^{-2} h^{-1} are low (Kappen 1988; Schroeter 1991; Schroeter et al. 1995; Kappen 2000).

Lichens are typical for sun-plant photosynthesis with a light requirement of up to 1000 µmol m^{-2} s^{-1} (Green et al 1999) but differences can be found by a comparison of sun-exposed and sheltered thalli of *U. sphacelata* (=*U. sulphurea,* Kappen 1983). Although data with lichens from various provenances (Demmig-Adams et al. 1990; Manrique et al. 1993) suggest that photoinhibition with the effect of a depression of the photosynthetic carbon balance is most likely to be expected in Antarctica, recent field experiments in the maritime Antarctic (Schlensog et al 1997a; Schlensog and Schroeter 2000; Schlensog 2000) and in the continental Antarctic region (Kappen et al. 1998a) revealed negative evidence of photoinhibition which contrasts to the performance of bryophytes in Antarctica.

Net photosynthetic rates vary drastically with water content of the thalli. Photosynthetic activity commences at a water content of about 20 % dry wt. (Lange et al. 1990) and is maximal at 80 to 250 % dry wt. depending on species (Kappen 1993a). However, the gelatinous thallus of *Leptogium puberulum* is still optimal at 500 % dry wt. (Schlensog et al. 1997). In contrast to bryophytes, green-algal lichens eventually suffer from being covered by surface water films or oversaturation as this causes an inhibition of CO_2 diffusion through the cortex and therefore a depression of the NP rate. This has been shown experimentally with Antarctic lichens such as *Umbilicaria* and *Usnea* species and *Caloplaca regalis* (Kappen 1988, Harrison et al. 1989) but not in *Usnea antarctica* (Schroeter 1991). It is relevant only at temperatures above 7 °C (Kappen 1985, Kappen and Redon 1987)

Under natural conditions, temperatures for photosynthetically active lichens are usually below 10 °C (Kappen 1985 c; Kappen and Redon 1984; Schroeter et al. 2000). In melt streams lichens fail to grow, except *Leptogium puberulum*, a cyanolichen, that is adapted to high water contents (Schlensog et al. 1997a). Crustose lichens such as *Buellia frigida* are shown to reach temperatures of up to 20 °C even in the metabolically active state (Schroeter et al.

1997b; Kappen et al. 1998b). Oversaturation of lichens may happen after heavy rainfalls, as occur in the maritime Antarctic, or under a snow sheet. Field measurements of water content in *Ramalina terebrata* of the maritime Antarctic region (King George Island) reveal that too great water contents (up to 200 %) occurs only rarely in a rainy period. Also, under snow the water content of *U. sphacelata* at a continental Antarctic site was mostly below optimal for net photosynthesis (Kappen and Breuer 1991).

In contrast to bryophytes, lichens with green algae as photobionts can well benefit from uptake of water vapour by the entire thallus surface. This is one reason why lichens are superior to bryophytes in environments with less moisture availability. Obviously, this enables air-dry lichens to reactivate by means of water vapour near or under ice and snow sheets (Kappen 1989, 1993b; Schroeter and Scheidegger 1995; Schroeter et al. 1997b) as for instance shown for *Umbilicaria aprina* which, at –14 °C gains a water content of 23 % dry wt. which is near the minimum necessary for photosynthetic activity (Schroeter and Scheidegger 1995). Photosynthesis at temperatures below freezing of water (Kappen 1993b) may occur in lichens until an equilibrium to a water potential of nearly 40 MPa (Nash et al. 1990). Apparently, due to the high fraction of the mycobiont biomass, the respiratory rate of lichens is about 20 % of maximum net photosynthetic rate at optimum temperature, but at 19 °C respiratory rates may be higher than the maximum net photosynthetic rates (Kappen 1988; Sancho et al. 1997). Nocturnal respiration may reach up to 173 % of diurnal net photosynthesis in *Umbilicaria nylanderiana* in the maritime Antarctic (Sancho et al. 1997) while in arctic lichen species dark respiration reaches only 4–10 % of diurnal net photosynthesis (Lange et al. 2000). As a consequence, negative production can be expected if temperatures are warmer than usual. Recently, negative long-term balances of up to 186 mmol CO_2 kg^{-1} dry wt. have been recorded by Sancho et al. (1997) for some Antarctic species. According to continuous records over 3 years, Schroeter et al. (2000) have been able to demonstrate that thalli of *U. aurantiaco-atra* were photosynthetically active throughout all seasons and that warm (near 0 °C) winter and early spring periods in the maritime Antarctic lead to a negative annual carbon balance. It is therefore likely that the present climatic trend with regional warming effects may cause a loss of the vitality of lichen species such as *Usnea aurantiaco-atra*. In contrast, vascular plants apparently benefit from a regional warming effect, as they have extended their distribution area over the last three decades (Lewis Smith 1994).

Other calculations reveal that the annual primary production may vary between 3 g m^{-2} $year^{-1}$ for cryptoendolithic lichens in the extremely desertic mountains of the McMurdo Dry Valleys and 250 g m^{-2} $year^{-1}$ for *Usnea aurantiaco-atra* in islands of the maritime Antarctic region, whereas the highest value for a continental Antarctic *Usnea* species was 14 mg g^{-1} dry wt. $year^{-1}$ (see Green et al. 1999). These rates are on average two times lower than those for Antarctic bryophytes (see Lewis Smith 1984a).

18.4.3 Growth and Growth Forms

Although the anatomical composition of a lichen thallus is much simpler than a bryophyte thallus, lichens achieve a greater variation of thallus shape (Table 18.3). The anatomical components of homeomerous lichens are an upper and a lower cortex, formed by a dense fungal pseudoparenchyma. The cortex encloses the medulla, a loosely structured plectenchyma of fungal hyphae that contains colonies of photobiont cells in its upper layer, directed to the light. Fruticose species may be strengthened by prosenchymatic hyphal strands in the centre of the branches. Fruiting bodies - apothecia - emerge apically or subapically or marginally or on top of the upper surface of the thallus. Homeomerous thalli have no stratified structure and some fail to produce a cortex structure. The gelatinous lichen *Leptogium puberulum*, however, has a simple cortex but a mixture of fungal hyphae and cyanobacterial colonies.

Relevant to our consideration is the fact that crustose and placodioid thalli have a more homeomerous structure and are firmly attached to a solid and stable substratum of all kinds of material, whether rock, rusted iron containers, old whale bones, boots, glass or concrete. But moss cushions or carpets are also colonised, particularly if they are eroded or senescent. Crustose lichens form less than 5 mm thick layers over the substratum with areoles, scales, or lobes. The width of the thalli may be up to 30 cm. The cryptoendolithic lichens of the Dry Valleys' mountain region form clusters of fungal hyphae mixed with unicellular algae in the porous space of Beacon Sandstone (Friedmann 1982) although sometimes compact crustose thalli are formed under still attached rock flakes (Kappen 1993a). Some Antarctic crustose lichen taxa have the tendency of thickening the thallus or show a stipitate or frutescent habitat. They are typical of the coastal sites mainly of the maritime Antarctic near bird colonies (Kappen 1993a).

Fruticose lichens have a concentric heteromerous anatomy and grow vertically and form branches like shrubs. Mostly, several individual thalli grow together forming a mat or cushion or a turf. This growth form is principally similar to that of mosses. However, they are not as dense as the complexes of moss shoots and therefore cannot store much interstitial water which eventually retards water loss from the thalli. They can grow as tall as 15 cm (*Usnea aurantiaco-atra*, *U. antarctica*) or several decimetres as pendulous growth forms. Besides vertically growing thalli, several *Usnea* species form long prostrate branches like *Alectoria ochroleuca* when they grow over moss carpets. *Pseudephebe pubescens* and *P. minuscula* form tiny mats of prostrate branches attached to small stones.

The genus *Cladonia* is a special form of a fruticose lichen with a horizontal thallus formed by scales and an erect podetium being up to 5 cm tall and carrying the apothecia. Sometimes podetia with many branches form a kind of a turf.

Table 18.3. Growth forms, morphology, and substratum relations of Antarctic lichens. (Modified after Longton 1988)

Growth form	Typical genus or species	Morphology, anatomy	Relations to the substrates
Cryptoendolithic	*Acarospora, Buellia, Rinodina, Lecidea*	Amorphous or distinct thallus	Growing in the porous space of sandstone, sometimes at the surface (limestone endoliths are not known in Antarctica)
Crustose	*Acarospora, Buellia, Rinodina, Lecidea, Lepraria, Ochrolechia, Psoroma*	Homoeomerous (<10 mm tall), forming radially-growing concentric thalli, some with distinct margin; hypo- and euthallus, areolate or squamulous	Closely attached to solid substrata or mosses
Placodioid	*Caloplaca elegans, Placopsis contortuplicata*	Homoeomerous (<10 mm tall), fine lobes, effigurate margin	Closely attached to the substratum
Frutescent	*Caloplaca regalis, Buellia cladocarpiza, Catillaria corymbosa*	Cushion-like, up to 20 mm tall, erect branches, coralloid outgrowth	Closely attached to the substratum
Fruticose	*Usena* (subgenus *Neuropogon*), *Pseudephebe, Ramalina, Stereocaulon*	Erect >50 mm tall, prostrate <20 mm tall (*Pseudephebe*), pendulous >50 mm tall (*Ramalina*)	Basis attached firmly to rock or wood (*Usnea, Pseudephebe*) or loosely lying over moss carpets or (Pseudephebe) attached to small stones and rock debris
Cladonia-type	*Cladonia*	Consists of a scaly horizontal thallus and a vertical part, the podetium (>20 mm tall)	Attached to soil, moss cushions or wood
Foliose laciniate	*Peltigera, Parmelia, Physcia*	Heteromerous: lobate or disciform thalli (<10 mm tall) with rhizines or papillae;	Solidly or loosely attached to the rock, bone, wood, moss, soil.
Foliose umbilicate	*Umbilicaria decussata, U. aprina*	umbilicate with a central rhizinic strand on the thallus underside	Umbilicate thalli solidly fixed on rock by the thick central rhizinic strand
Gelatinous lichen	*Leptogium puberulum*	Homoeomerous, sacculate decumbent lobes (thallus <20 mm tall)	Attached to rock or small stones in or along occasional melt flurries

Foliose lichens typically have a stratified heteromerous thallus with rhizines on the lower cortex that fix the thallus to the substratum. Thalli grow horizontally and can extend over 1 or 2 dm^2. Lobate species such as *Parmelia* are more or less densely attached to solid substrata, *Peltigera* grows over bryophytes and soil. The genus *Umbilicaria* forms a thick hyphal strand in the centre of the disc-like thallus which solidly fixes the lichen to rock. Single thalli may reach a diameter of up to 25 cm.

The growth of lichens is generally slower than that of the bryophytes. Maritime Antarctic foliose, fruticose and also crustose lichens grow faster than those in the continental Antarctic region. Annual growth rates of the branches of fruticose species on Signy Island were between 0.5 and 2 mm (Lewis Smith 1984a). Crustose thalli gain an increase in width by 0.3–3 mm $year^{-1}$ (see Lewis Smith 1990a, Green et al. 1999). Lewis Smith (1990a) illustrates the relatively quick increment of crustose and umbilicate thalli on rock and concrete in Signy Island. For instance, over a period of 15 years the average annual area increase was 41 % in *Umbilicaria antarctica* and 32 % in *Buellia latemarginata*. Whereas the extremely slow growing *Rhizocarpon geographicum* exhibited an annual area increase of only up to 0.8 %. Growth rates are smaller in continental sites. An estimation by Green gave an increase by 0.01 mm $year^{-1}$ in *Buellia frigida* in the Dry Valleys area (Green et al 1999).

18.4.4 Colonisation and Succession

Lichens are well known as pioneering elements of the vegetation. This can be particularly demonstrated for polar regions (Kappen 1973, 1988). In Antarctica they occur as far south as ice-free rock is available. Twenty-three species, most of them crustose but also two *Umbilicaria* and *Xanthoria elegans*, exist south of 80°S and *Carbonea vorticosa*, *Lecidea cancriformis* and *Sarcogyne privigena* exist in the La Gorce Mountains (86°29' S) a few hundred kilometres away from the Pole (Lewis Smith 2000) at an altitude of 1580 m. The lichens grow in isolated and remote mountain ranges and on nunataks. More than 20 taxa are known from altitudes higher than 2000 m. These include crustose species but also *Pseudephebe minuscula*, *Umbilicaria decussata* and the fruticose *Stereocaulon vesuvianum* (Lewis Smith 2000). The highest record is from 2500 m with four species growing in Gjeslvikfjella, Dronning Maud Land (Engelskjön 1986). The summit regions of the big volcanoes in Antarctica provide, however, unfavourable conditions for lichens, as the ground is unstable and too warm.

Even though lichens can form outposts in open exposed surfaces they are not principally the initiators of a more diverse vegetation later associated by mosses or eventually vascular plants. Lichens tend to be slow colonisers. Freshly opened rock faces as a result of frost cracking or opening of ground due to glacier recession may remain uncolonised for several years or even centuries until lichens get established (Cameron 1971; Gannutz 1971; Lewis Smith 1995). This suggests that a certain degree of surface weathering is necessary to prepare the rock for colonisation. This was demonstrated for a moraine in high-arctic Canada where the boulders stayed bare for 80 years until onset of lichen colonisation (Fahselt et al. 1988). Hooker (1980) reports that concrete surfaces needed an exposure period of 14 to 40 years until *Caloplaca cinericola* was able to establish. However, the appearance of lichens on fresh substrates was also recorded (Lewis Smith 1984a; Sancho and Valladares 1993) and Schroeter and Sancho (1996) demonstrated the colonisation of broken glass by lichens in the maritime Antarctic within 20 years.

Longton (1988a) discusses two forms of colonisation and succession. One is the auto-succession, in which pioneering and climax species remain the same and no further stage is to be expected. Such a situation is typical of lichen communities consisting of less than ten species or an assemblage of individuals of one species only on rock and boulders of continental Antarctic coastlines, mountain ranges and nunataks (Filson 1966; Engelskjön 1986; Seppelt and Broady 1988; Ryan et al. 1989; Kappen 1993a; Kappen et al. 1998b). A special case is the so-called cyclic colonisation, in circular-growing crustose or foliose lichens. After the central part has died and disappeared, thalli of the same or other species get established again. This is easily visible on coastal rocks in Wilkes Land, near Casey Station where thalli of the fruticose *U. sphacelata* and *Pseudephebe minuscula* (Kappen 1993a) grow in the centre of *Buellia frigida*. Such communities may be stable over decades.

On the other hand, directional succession starts with a pioneering community with species different from later-stage communities. This process can last two or more centuries as stated for an *Andreaea-Usnea* association at the South Orkney Islands (Lindsay 1978). Usually, the succession follows the degree of weathering or accumulation of fine material in the community (see Longton 1988a). This is the case in many sites in the maritime Antarctic region (Lindsay 1971; Lewis Smith 1972; Lewis Smith and Corner 1973; Lewis Smith 1996) or at places near the coast of the continent with rich vegetation in the vicinity of bird or seal colonies (Lewis Smith 1990; Castello and Nimis 1995). In such areas both types of succession occur, depending on the quality of the substratum ranging from xeric rock habitats with stable but poor lichen communities to lush bryophyte-lichen-plant communities on moist soil (Seppelt and Ashton 1978; Kappen 1985d; c.f. Longton 1988a; Olech 1993; Castello and Nimis 1995; Hovenden and Seppelt 1995; Sancho et al. 1999). Even on the continent the floristic composition can be very rich with a total of 21 lichen species and 5 moss species to a latitude of 78°S (e.g. Seppelt et al. 1995: Kar Plateau).

The phytomass of lichens in the continental Antarctic region can be as high as 1000 g dry wt. m^{-2} and may reach 2000 g dry wt. m^{-2} in the maritime Antarctic region (see Kappen 1993a). The major component in both regions are fruticose *Usnea* species. In such cases, lichens can be as tall as the moss

Fig. 18.1. Dense mixed vegetation on Livigston Island near Estation Juan Carlos I with *Deschampsia antarctica*, *Stereocaulon alpinum*, *Sphaerophorus globosus*, *Usnea aurantiaco-atra*, and *Polytrichum alpinum*

turfs and *Deschampsia antarctica* forming dense canopies such as on Livingston Island (Fig. 18.1).

The cryptoendolithic community in the porous space of Beacon Sandstone in the Ross Desert, Mc Murdo Dry Valleys, exists mainly where sun radiation provides sufficient heat and meltwater (Kappen et al 1981; Nienow and Friedmann 1993). It may exist for thousands of years with a standing biomass of 0.02 to 214 g m^{-2} in the surface layer of the rock. Whether the different species compositions in the rock samples represent successional stages, is hard to say.

18.4.5 Relations to the Substratum

Melick et al. (1994) have discussed whether substratum and nutrient availability is more decisive for the distribution of lichens than physical and meteorological factors in the habitat. Snow cover and site exposure have several aspects. Lichens with a special habitat preference in Antarctica can be silicophilous, calciphilous or ornithocoprophilous species. Ornithocoprophilous species in the Antarctic are typical of places around penguin and seal colonies. However, inside and very close to the animal colonies lichens are not able to grow. Nutrient supply from sea spray is nearly as effective as animal activity, therefore species composition on coastal rocks is quite similar to that of ornithocoprophilous vegetation (Follmann 1965; Schofield and Ahmadjan 1972; Olech 1990; Kappen 1993a; Hovenden and Seppelt 1995). In such places species diversity can be high and most of the lichens are very apparent by their bright coloration. Frutescent growth of otherwise crustose taxa is mostly related to nutrient enriched sites (Jacobsen and Kappen 1988).

Calcicolous species are also identical with most of the ornithocoprophilous species. However, in Antarctica calcareous rock is rare, such as outcrops of marble on Signy Island (Lewis Smith 1972, 1990a) or in the forelands of the Transantarctic Mountains (Marble Point, Victoria Land). Calcicolous species colonise fresh basaltic rock (Lewis Smith 1984 c; Fabiszewski and Wojtuń 2000), whale bones as remnants of the whaling operations at the beginning of the last century (Lindsay 1978; Kappen 1993a) and concrete of buildings (Hooker 1980; Lewis Smith 1990a).

Colonising eroded or senescent regions of bryophytes may be indicative of a substrate dependency of certain lichens such as *Buellia grimmiae*, *Caloplaca athallina*, *Ochrolechia frigida*, *Psoroma hypnorum* etc. (Lewis Smith 1973, 1996b; Collins 1976b) but most of these species can also be found on soils with high organic matter content (see Kappen 1993a). *Usnea* species that grow successfully on bryophyte turfs are not always anchored in the moss but are prostrate on the surface or protruding through the moss, which suggests that they were overgrown by the quicker growing moss.

The majority of inland habitats are colonised by silicolous species such as the cryptoendolithic and many other *Acarospora* and *Lecidea* species, *Rhizocarpon* and most of the *Umbilicaria* and *Usnea* species (see Kappen 1993a). Typically, these species avoid habitats with manuring influence. Thus, they can be considered as nitrophobous or oligotrophic species. The occurrence of *Usnea antarctica* and *Rhizoplaca antarctica* and others on timber or old constructions or remnants of boats on Deception Island may be due to an impregnation with rock dust, as colonisation was possible after more than 20 years exposure to the substratum (see Lewis Smith 1990a).

18.4.6 Ion Uptake

Little is known about ion uptake and transport mechanisms in lichens (Tuominen and Jaakkola 1973; Nash 1996). It is generally accepted that nutrients and ions are mainly derived from the atmosphere by precipitation, wet and dry deposition on the entire thallus surface (Nieboer et al. 1978; Nash 1989), aerosols, metals, and gases (Garty et al. 1979; Puckett 1988; Winner et al. 1988) and blowing dust. But also uptake of nutrients and other mineral components from soil have been shown for terricolous foliose lichens (Goyal and Seaward 1982). Also, snowmelt was shown to be a nutrient source for the Antarctic species *Usnea sphacelata* (Crittenden 1996).

In general, cation uptake may be a rapid passive chemical process, which occurs extracellularly in lichens (Nieboer et al. 1978). The capacity to retain cations within the cell walls is estimated to vary between 6 and 77 $\mu mol\ g^{-1}$ depending on the species (Nash 1989). Anion uptake, however, is a slower process and may result in less total accumulation (Richardson et al. 1984). Ion translocation – most likely symplastic – in the thallus from decaying basal parts to the tip was demonstrated by Hyvärinen and Crittenden (2000) in a *Cladonia* species.

The capacity of accumulating macro-nutrients, except for Mg, is at least as high as in bryophytes and much greater than in vascular plants (Jones 1988). *Usnea antarctica* and *U. aurantiaco-atra* on Signy Island contain extraordinarily high amounts of Ca^{2+} (up to 1.6 % of the lichen dry wt.) which suggests an active accumulation in the thallus. The same is observed by Fabiszewski and Wojtuń (2000) for lichens from King George Island, and this is attributed to their calcicolous nature. In other non-calcicolous species such as *Himantormia lugubris*, *Ramalina terebrata* and *Umbilicaria antarctica* the Ca content is significantly lower according to the substratum chemistry variation. In Arctic lichens, however, a low calcium content is typical (Longton 1988). Concentrations of Ca, Na and K vary significantly during the annual cycle in nine species from Signy Island (Lewis Smith 1978). Ion content is generally low in winter but high in spring and summer as in the mosses. Ion content varies more with environmental conditions such as moisture content or distance from the sea shore than with species (Fabiszewski and Wojtuń 2000). Therefore, different species tend to have similar concentrations of specific minerals in similar habitats (Lewis Smith 1978). Lichens usually contain more trace elements (such as Zn, Cd, Sn, Pb) than bryophytes and plants. But correlation between trace element content of a lichen with that of its rock substratum is usually poor (Tuominen and Jaakkola 1973). Pucket and Finegan (1980), however, detected cellular uptake of Al, Cr, Co, Fe, Na, Ti and V by lichens in a number of localities of Northern Canada. Although heavy-metal ions (Zn, Cu and Cd) are toxic for most lichens (Beckett and Brown 1984), species of the genera *Acarospora*, *Aspicilia*, *Lecidea*, *Rhizocarpon* and *Stereocaulon* are highly specific for metalliferous rocks (see Nash 1996).

Nitrogen and phosphorus are essential for lichens (Hyvärinen and Crittenden 2000) and may be limiting as demonstrated for *Cladonia* in sub-Antarctic habitats (Lewis Smith 1988). The nitrogen content in lichens from King George Island varies typically with species according to their different tolerance to this biogene (Olech 1990, 1993). The distributional gradient between nitrophilous and nitrophobous species is apparent in a transect from the seaside with high animal activity to the poor habitats of inland rocks showing a range of N-contents between up to 4 % and below 1 % respectively (Fabiszewski and Wojtuń 2000).

Lichens with cyanobionts are able to fix nitrogen from air if *Nostoc* is the main photobiont or in so-called cephalodia with *Nostoc* or *Stigonema*. Lichens with cyanobionts (about ten species) live only in the maritime Antarctic and prefer moist habitat conditions (Kappen 1993a). Nitrogen fixation was highest on Signy Island at sites with alkaline rock or under the influence of alkaline meltwater (Horne 1972). Bulk nitrogen fixation was significant at 10 °C in *Leptogium puberulum* and *Stereocaulon* sp. on Signy Island (Fogg and Stewart 1968) and failed below 0 °C (see above). However, some arctic species were reported to fix N until –5 °C (Englund and Meyerson 1974). On the Antarctic continent, cyanolichens are absent. However, free-living *Nostoc* in the Vestfold Hills, Ingrid Christensen Coast, revealed acetylene reduction still at –7 °C (Davey and Marchant 1983).

Although many findings demonstrated the dependency of the lichens on certain substrate chemistries, there is also evidence that this factor may be of minor importance in the Antarctic. In the coastal sites of the maritime and the continental Antarctic region nutrient availability is always greater than limiting (Collins et al.1975) and therefore the distribution of lichens depends more on their tolerance to high concentrations of nutrients such as N and P and Ca. Excessive nutrient concentrations cause a break-down of the lichen symbiosis in certain species adapted to a generally low nutrient supply from the atmosphere like many epiphytic lichens on trees (van Dobben 1993). However, a positive response to fertilising by birds is apparent mainly in ornithocoprophilous species as they were reported to achieve greater thallus size near bird colonies (Ryan and Watkins 1989). Along a gradient of habitats with different degrees of nutrient (NPK) availability from the coast to inland areas, Valladares and Sancho (2000) detect a proportional relation between nutrients and chlorophyll content and photosynthetic rate in fruticose and foliose species of Livingston Island except in nitrogen fixing lichen species. This confirms earlier experiments (Schofield and Ahmadjan1972) with cultivated lichen bionts from Cape Royds on nutrient-rich substrates. The authors therefore conclude that growth of the lichens is nutrient limited in oligotrophic habitats. There is indirect support of the limiting role of major nutrient supply by Seppelt et al.(1988) for the Vestfold Hills where the vegetation is impoverished in areas which have been deglaciated for 3000–5000 years, perhaps also due to toxic effects, and rich stands with lichens and bryophytes are associated with elevated sites (about 60 m) where birds prefer to nest. However, the artificial application of N, P and K fails to increase element content, or net assimilation rate of the Arctic *Cladonia stellaris* (Carstairs and Oechel 1978). Also, the cryptoendolithic organisms in Beacon Sandstone are not stimulated by the addition of ammonium nitrate or nitrite (Johnson and Vestal 1986; Vestal 1988). Apparently, endolithic communities in the Dry Valleys region are not nitrogen-limited but remain on a low level of metabolism. Friedmann and Kibler (1980) assume a major source being abiotic fixation of N by electrical discharges in the upper atmosphere.

The low demand of lichens on nutrients can be demonstrated for major lichen species in the Windmill Island region (Hovenden and Seppelt 1995) where factors like exposure, water availability, wind effects and snow cover appear more important for the vegetation pattern than nutrient availability. A dominance of these factors was also evident by environmentally induced zonations on rocks near Mawson Station (Seppelt and Ashton 1978) and in a vegetation study near Syowa Station (Nakanishi 1977) or with the restriction of lichens to small coastal landscapes being sites where temperature and water availability is favourable (Kappen 1985 c). Another selective factor for lichens is long duration and thick snow cover (Kappen 1993b, 2000; Winkler et al. 1998, 2000).

18.4.7 Contribution to Pedogenetic Processes

The role of lichens in biogenic weathering and pedogenesis has been discussed in reviews by Syers and Iskandar (1973), Jones (1988) and Chen et al. (2000). Ugolini and Edmonts (1983) state that the pedogenetic significance of lichens is largely restricted to rock faces and cliffs and is most prominent in deserts and in the Antarctic region. Physical weathering can be initiated first by rock-penetrating hyphae of lichens that disintegrate plagioclas feldspar crystals in syenite, secondly, by thallus expansion and contraction (Wierzchos and Ascaso 1996) or by a complexing action of lichen compounds (Syers and Iskandar 1973; Jones 1988). On Signy Island lichens induced weathering of mica schist rock with the effect of considerable disruption of the rock surface (Walton 1985).

According to Ascaso et al. (1990) the presence of crystalline oxalate, imogolite, allophane, calcite in epilithic lichen thalli but not in the parent rock (volcanic andesite) indicates biomineralisation processes in rocks of King George Island. Oxalic acid may increase the mobilisation of phosphorous and metal ions of the rock substratum (Nienow and Friedmann 1993). By means of Raman spectroscopy Holder et al. (2000) have demonstrated that the crustose epilithic *Acarospora* produces large amounts of calcium oxalate which indicates a high capacity to bioweather. In the arid habitats of northern Victoria Land (alkali basalt) the lichens produced more calcium oxalate than an *Acarospora* on sandstone enclosing gypsum in central Spain. This was attributed to the different temperature regimes between the two regions that influenced the biogeochemical process. Also, rhizocarpic acid content in the thalli was higher in Antarctic material (Holder et al. 2000).

The biogenous weathering process is most obvious from the exfoliation pattern of Beacon Sandstone in the mountain region of the McMurdo Dry Valleys (Nienow and Friedmann 1993) showing a patchy pattern of dark brown scales bordering lower-lying yellowish surface plots by colour. A transect shows by colour the way iron is enriched at a deeper layer and in a surface crust, and a leached zone with the biotic material occurs at 1–5 mm depth, looking like a miniature podzol profile. Iron mobilisation has also been demonstrated by Weed and Norton (1991) who concluded that the cryptoendolithic system is extremely old, and that fossilised material is present (Friedmann and Weed 1987). The ion content in walls of the fossilised cells has recently determined by Ascaso (2000). The rate of weathering and that of exfoliation, however, is very slow compared with that in a warm tropical climate (Wessels and Schoeman 1988) and may be counted by centuries or longer (Friedmann et al. 1993). On the other hand, long-living foliose and crustose lichens can also protect rock and soil from weathering as the cover avoids erosion by abiotic forces (Lindsay 1978).

Like bryophytes, fruticose lichens are capable of accumulating wind-blown dust which is a potential source of nutrients that finally get into the soil (Nieboer et al. 1978; Nash 1989). Particles can be trapped within the intercellular space of the thallus, as was observed in *Xanthoria parietina* by Collins and Farrar (1978). The capacity of accumulating atmospheric nutrient deposition has already been described. Therefore, epiphytic species contribute a significant amount of phosphorus, nitrogen and trace elements to the ecosystem, when they are decaying on the soil surface (Syers and Iskandar 1973; Knops et al. 1997). At a reduced rate, this also happens with Antarctic lichens, particularly the large species. Nitrogen will also be imported to the soil by lichens with cyanobacteria as photobionts (Arctic Ecosystems: Crittenden and Kershaw 1978). Horne (1972) calculates an annual increase of 2.3 % N by *Leptogium puberulum* on Signy Island, whereas the amount fixed by free-living *Nostoc* is 52 mg N m^{-2} $year^{-1}$ in the Vestfold Hills (Davey and Marchant 1983). This is the same order of magnitude as cryptogamic N-fixation near Barrow, Alaska (Alexander et al. 1978). In the extremely desertic mountainous region of the McMurdo Dry Valleys the cryptoendolithic biota must provide a significant reservoir of nutrients. If their total biomass is estimated for an area of the Ross Desert (1.057 km^2) the annual carbon fixation would amount to 600–900 kg C $year^{-1}$ (Friedmann et al. 1993).

Nutrients can be released from lichens by leaching which happens after changes of the lichen water status or freeze-thaw cycles as shown for lichens from Signy Island with intracellular cations (Hooker 1977). As soon as lichens die, decomposition sets nutrients free to the soil but the decomposition rate of minerals also has to be taken into account (Greenfield 2000). Besides mineral ions, the diversity of organic compounds from the lichens is beneficial for soil formation as these compounds nourish microorganisms that induce mineralisation processes (Ugolini 1970; Tearle 1984; Melick et al. 1994).

Biogenic weathering processes may also happen in many parts of the Antarctic continent where no macroscopic biota are visible. Simple ecosystems may be formed in the remotest sites, as Cameron (1971) and others have demonstrated. In Long Valley (86°14'S) only bacteria, moulds, unicellular algae and a few lichens are present. The record from Brown Peaks (85°32' S) enumerates, besides lichens on rocks, bacteria, moulds, yeasts, unicellular algae, amoebae and rotifers in Lithosols or rock debris (Campbell and Claridge 1987). Mt Howe (87°22'S, 149°18'W) is one of the southernmost sites where bacteria and cyanobacteria were spotted (Cameron et al. 1971).

18.5 Conclusions

The Antarctic region differs from all other continents by the rarity of vascular plants and their absence over most of the continent. Therefore, in the ice-free areas of Antarctica vegetation is mainly formed by cryptogamic organisms such as bryophytes, lichens, fungi, algae and bacteria. These organisms are not principally different from their relatives in other continents. Many species are cosmopolitans or are at least widely distributed, particularly over mountainous regions of the globe. Except for lichens, the cryptogams are not extremely tolerant to freezing. Small size and slow growth are general features of these organisms. Water availability in the habitat and the capability of the organisms to use water in various ways is decisive for the occurrence of the cryptogams and the formation of a vegetation pattern. Ecological adaptation is a major phenomenon which means that cryptogams exploit the synergism of favourable environmental variables such as heat, light, moisture and shelter. Vegetation units larger than 1 or 2 ha exist only in the mildest areas of the maritime Antarctic region. Elsewhere, the vegetation is very scattered or restricted to small spots. Many sites of the ice-free regions in Antarctica are barren.

Significant soil formation in Antarctica is therefore possible in the mildest and wettest part of Antarctica, which are comparable to cold and cool-temperate regions of the world. In most other cases, Antarctic soil formation is comparable to that in extremely dry deserts and high alpine regions and is only the result of cryptogamic activity. Cryptogams have no roots, but several moss species are able to penetrate deeply into the ground and therefore can influence a soil layer of up to 25-cm depth. Bryophytes can produce peat layers of up to 3-m depth in the maritime Antarctic region and therefore create a similar situation as in the boreal climatic regions. Of the lichens, only a few species are able to grow inside peripheral layers of rock or to stay anchored 1 or 2 cm-deep in soil or debris. With epilithic or other surfaces of stable material, colonising thalli lichens may influence only very shallow surface layers of the substratum. By the production of oxalic acid and other specific substances lichens are able to change the mineral composition and structure and to play an active role in bioweathering processes.

Cryptogamic phytomass contains essential nutrients for soil organisms and succeeding vegetation. It also accumulates mineral components. In various ways lichens and mosses are extremely efficient ion accumulators. In addition, cushions, carpets and fruticose growth forms can store a great amount of particulate material as an important source for pedogenesis. In general, however, the process of mineral cycling is slow because most of the cryptogams are slow-growing, longliving and of small size. Thus, in many cases only initial stages of soil formation can be expected.

Acknowledgements. The senior author wishes to dedicate this article to Professor Imre Friedmann, most stimulating and fruitful Antarctic researcher and dear friend.

References

Adamson H, Wilson M, Selkirk P, Seppelt RD (1988) Photoinhibition in Antarctic mosses. Polarforschung 58:103–112

Alexander V, Billington M, Schell DM (1978) Nitrogen fixation in Arctic and alpine tundra, In: Tieszen LL (ed) Vegetation and production ecology of an Alaskan Arctic tundra. Springer, Berlin HeidelbergNew York, pp 539–558

Allen SE, Grimshaw HM, Holdgate MW (1967) Factors affecting the availability of plant nutrients on an Antarctic island. J Ecol 55:381–396

Allen SE, Northover MJ (1967) Soil types and nutrients on Signy Island. In: (JE Smith, organizer) A discussion on the Antarctic terrestrial ecosystem. Philos Trans R Soc B252, 179–185

Ascaso C, Sancho LG, Rodriguez-Pascual C (1990) The weathering action of saxicolous lichens in maritime Antarctica. Polar Biol 11:33–39

Ascaso C (2000) Lichens on rock substrates: observation of the biomineralization proceses. In: Schroeter B, Schlensog M, Green TGA (eds) New aspects in cryptogamic research. Contributions in Honour of Ludger Kappen. Bibliotheca Lichenologica 75. J Cramer in der Gebr. Bornträger Verlagsbuchhandlung, Berlin, pp 127–135

Bacci EB, Calamari D, Gaggi C, Fanelli R, Focardi S, Morosini M (1986) Chlorinated hydrocarbons in lichens and moss samples from the Antarctic Peninsula. Chemosph 15:747–754

Baker JH (1972) The rate of production and decomposition of *Chorisodontium aciphyllum* (Hook. f. & Wils.) Broth. Br Antarct Surv Bull 27:123–129

Banergee RD, Sen SP (1979) Antibiotic activity of bryophytes. Bryologist 82:141–153

Bargagli R, Broady PA, Walton DWH (1996) Preliminary investigation of the thermal biosystem of Mount Rittmann fumaroles (northern Victoria Land, Antarctica). Antarct Sci 8:121–126

Bargagli R, Sanchez-Hernandez JC, Martella L, Monsaci F (1998) Mercury, cadmium and lead accumulation in Antarctic mosses growing along nutrient and moisture gradients. Polar Biol 19:316–322

Beckett RP, Brown RP (1984) The relationship between cadmium uptake and heavy metal uptake tolerance in the lichen genus *Peltigera*. New Phytol 97:301–311

Biebl R (1962) Protoplasmatische Ökologie der Pflanzen, Wasser und Temperatur. Protoplasmatologia XII, 1. Springer, Wien , 344 pp

Birse EM, Landsberg SY, Gimingham CH (1957) The effects of burrial by sand on dune mosses. Trans Br Bryol Soc 3:285–301

Bravo LA, Ulloa N, Zuniga GE, Casanova A, Corcuera LJ, Alberdi M (2000) Cold resistance in Antarctic angiosperms. Physiol Plant 111:55–65

Broady PA (1977) The Signy Island terrestrial reference sites: VII. The ecology of the algae sites 1. A moss turf. Br Antarct Surv Bull 45:47–62

Broady PA (1979) Quantitative studies on the terrestrial algae of Signy Island, South Orkney Islands. Br Antarct Surv Bull 47:31–41

Broady PA, Given D, Greenfield LG, Thompson K (1987) The biota and environment of fumaroles on Mt. Melbourne, northern Victoria Land. Polar Biol 7:97–113

Cabello MLG, Pollero R (1994) *Glomus antarcticum* sp. nov., a vascular-arbuscular mycorrhizal fungus form Antarctica. Mycotaxon 51:123–128

Callaghan TV, Collins NJ (1981) Life cycles, population dynamics, and the growth of tundra plants. In: Bliss LC, Heal OW, Moore JJ (eds) Tundra ecosystems: a comparative analysis. Cambridge University Press, Cambridge, pp 257–284

Cameron RE (1971) Antarctic soil microbial and ecological investigation. In: Quam LO (ed) Research in the Antarctic. American Association for the Advances of Sciences, Washington, DC, pp 137–289

Campbell IB, Claridge GGC (1987) Antarctica: soils, weathering processes and environment. Elsevier, Amsterdam

Carstairs AG, Oechel WC (1978) Effects of several microclimatic factors and nutrients on net carbon dioxide exchange in *Cladonia alpestris* (L.) Rabh. in the subarctic. Arct Alp Res 10:81–94

Castello M, Nimis PL (1995) The lichen vegetation of Terra Nova Bay (Victoria Land, Continental Antarctica). In: Farkas EE, Lücking R, Wirth V (eds) Lichenological papers dedicated to Antonin Vezda. Bibliotheca Lichenologica 58. J. Cramer in der Gebrüder Bornträger Verlagsbuchhandlung, Berlin, pp 43–55

Chen J, Blume HP. Beyer L (2000) Weathering of rocks induced by lichen colonization – a review. Catena 39:121–146

Chapin FS, Chapin MC (1980) Revegetation of an Arctic disturbed site by native tundra species. J Appl Ecol 17:449–456

Chapin FS, Johnson DA, McKendrick JD (1980) Seasonal movement of nutrients in plants of differing growth form in an Alaskan tundra ecosystem: implications for herbivory. J Ecol 68:189–209

Christie P (1987) Nitrogen in two contrasting Antarctic bryophyte communities. J Ecol 75:73–93

Christie P, Nicolson Th (1983) Are mycorrhizas absent from the Antarctic? Trans Br Mycol Soc 80:557–560

Collins NJ (1969) The effects of volcanic activity on the vegetation of Deception Island. Br Antarct Surv Bull 21:79–94

Collins CR, Farrar JF (1978) Structured resistances to moss transfer in the lichen *Xanthoria parietina*. New Phytol 31:71–78

Collins NJ (1973) Productivity of selected bryophytes in the maritime Antarctic. In: Bliss LC, Wiegolaski FE (eds) Proceedings of the conference on Primary Production and Production Processes. Tundra Biome, IBP Tundra Biome Steering Committee, Edmonton, pp 177–183

Collins NJ (1976a) Growth and population dynamics of the moss *Polytrichum alpestre* in the maritime Antarctic. Oikos 27:389–401

Collins NJ (1976b) The development of moss-peat banks in relation to changing climate and ice cover on Signy Island in the maritime Antarctic. Br Antarct Surv Bull 43:85–102

Collins NJ (1977) The growth of mosses in two contrasting communities in the maritime Antarctic: measurement and prediction of net annual production. In: Llano GA (ed) Adaptations within Antarctic ecosystems. Smithsonian Institution, Washington, pp 921–933

Collins NJ, Baker JH, Tilbrook PJ (1975) Signy Island, maritime Antarctic. In: Rosswall T, Heal OW (eds) Structure and function of tundra ecosystems. Ecological Bulletin 20, Stockholm, pp 345–374

Collins NJ, Callaghan TV (1977) Bryophyte ecophysiology in polar regions. Institute of Terrestrial Ecology, Annual Report 1977, pp 11–16

Collins NJ, Callaghan TV (1980) Predicted patterns of photosynthetic production in maritime Antarctic mosses. Ann Bot 45:601–620

Collins NJ, Oechel WC (1974) The pattern of growth and translocation of photosynthate in a tundra moss, *Polytrichum alpinum*. Can J Bot 52:355–363

Convey P (1994) Photosynthesis and dark respiration in Antarctic mosses – an initial comparative study. Polar Biol 14:65–69

Convey P (1996) Reproduction of Antarctic flowering plants. Antarct Sci 8:127–134

Corner RW, Lewis Smith RI (1973) Botanical evidence of ice recession in the Argentine Islands. Br Antarct Surv Bull 35:83–86

Corner RWM (1971) Studies in *Colobanthus quitensis* (Kunth) Bartl and *Deschampsia antarctica* Desv.: IV Distribution and reproduction performance in the Argentine Islands. Br Antarct Surv Bull 26:41–50

Corte A (1961) La primera fanerogama adventicia hallada en el continente antartico. Contrib Inst Antart Argentino (Buenos Aires) 62:1–14

Crittenden PD, Kershaw KA (1978) Discovering the role of lichens in the nitrogen cycle in boreal-Arctic ecosystems. Bryologist 81:258–267

Crittenden PD (1996) The effect of oxygen deprivation on inorganic nitrogen uptake in an Antarctic macrolichen. Lichenologist 28:347–354

Davey A (1983) Effects of abiotic factors on nitrogen fixation by blue-green algae in Antarctica. Polar Biol 2:95–100

Davey A, Marchant HJ (1983) Seasonal variation in nitrogen fixation by *Nostoc commune*, Vaucher at the Vestfold Hills, Antarctica. Phycologia 22:377–385

Davey MC (1997a) Effects of physical factors on photosynthesis by the Antarctic liverwort *Marchantia berteroana*. Polar Biol 17:219–227

Davey MC (1997b) Effects of short-term dehydration and rehydration on photosynthesis and respiration by Antarctic bryophytes. Environ Exp Bot 37:187–198

Davey MC, Rothery P (1996) Seasonal variation in respiratory and photosynthetic parameters in three mosses from the maritime Antarctic. Ann Bot 78:719–728

Davey MC, Rothery P (1997) Interspecific variation in respiratory and photosynthetic parameters in Antarctic bryophytes. New Phytol 137:1–10

Davis R (1980) Peat respiration and decomposition in Antarctic terrestrial moss communities. Biol J Linn Soc 14:39–49

Davis RC (1981) Structure and function of two Antarctic terrestrial moss communities. Ecol Monogr 51:125–143

Davis RC (1983) Prediction of net primary production in two Antarctic mosses by two models of net CO_2 fixation. Br Antarct Surv Bull 59:47–61

Davis RC, Harrison PM (1982) Prediction of photosynthesis in maritime Antarctic mosses. Com Natl Fr Rech Antarct 51:241–247

Day TA, Grobe CW, Ruhland CT (1995) Impacts of climate change on Antarctic vascular plants: warming and ultraviolet-B radiation. Antarct J US Rev 1996:226

Day TA, Ruhland CT, Xiong F (1997) Impacts of ultraviolet-B radiation and regional warming on Antarctic vascular plants. Antarct J US Rev 1997:155

Demmig-Adams B, Maguas C, Adams WW, Meyer A, Kilian E, Lange OL (1990) Effect of high light on the efficiency of photochemical energy conversion in a variety of lichen species with green and blue-green photobionts. Planta 180:400–409

Dircksen A (1964) Vergleichende Untersuchungen zur Frost-, Hitze- und Austrocknungsresistenz einheimischer Laub- und Lebermoose unter besonderer Berücksichtigung jahreszeitlicher Veränderungen. PhD Thesis, University of Göttingen

Doebbeler P (1978) Ascomycete parasites of moss 1. Pyrenomycete species occurring on gametophytes. Mitt Bot Staatssamml München 14:1–360

Edwards JA (1973) Vascular plant production in the maritime Antarctic. In: Bliss LC, Wiegolaski FE (eds) Primary production and production processes. Tundra Biome, Tundra Biome Steering Committee, Stockholm, pp 169–175

Edwards JA (1974a) Studies in *Colobanthus quitensis* (Kunth) Bartl. and *Deschampsia antarctia* Desv.: VI. Reproductive performance on Signy Island. Brit Antarct Surv Bull 39:67–86

Edwards JA (1974b) Studies in *Colobanthus quitensis* (Kunth) Bartl. and *Deschampsia antarctica* Desv. PhD Thesis, University of Birmingham

Edwards JA (1980) An experimental introduction of vascular plants from South Georgia to the maritime Antarctic. Br Antarct Surv Bull 49:73–80

Edwards JA, Greene DM (1973) The survival of Falkland Islands transplants at South Georgia and Signy Island, South Orkney Islands. Br Antarct Surv Bull 33, 34:33–45

Edwards JA, Lewis Smith RI (1988) Photosynthesis and respiration of *Colobanthus quitensis* and *Deschampsia antarctica* from the maritime Antarctic. Br Antarct Surv Bull 81:43–63

Engelskjøn T (1986) Botany of two Antarctic mountain ranges: Gjelsvikfjella and Mühlig-Hofmannfjella, Dronning Maud Land. Pol Res 4:205–224

Englund B, Meyerson H (1974) In situ measurement of nitrogen fixation at low temperatures. Oikos 25:283–287

Fabiszewski J, Wojtuń B (2000) Chemical composition of some dominating plants in the maritime Antarctic tundra (King George Island). Bibl Lichenol 75:127–135

Fahselt D, Maycock PF, Svoboda J (1988) Initial establishment of saxicolous lichens following recent glacial recession in Sverdrup Pass, Ellesmere Island, Canada. Lichenologist 20:253–268

Fenton JHC (1978) The growth of Antarctic moss-peat banks. PhD Thesis, University of London (Westfield College)

Fenton JHC (1980) The rate of peat accumulation in Antarctic moss banks. J Ecol 68:211–228

Fenton JHC (1982) The formation of vertical edges on Antarctic moss peat banks. Arct Alp Res 14:21–26

Filson RB (1966) The lichens and mosses of Mac Robertson Land. ANARE Sci Rep, Series B(2) Bot, Melbourne, 169 pp

Fogg GE, Stewart WDP (1968) In situ determinations of biological nitrogen fixation in Antarctica. Br Antarct Surv Bull 15:39–46

Follmann G (1965) Una asociación nitrófila de líquenes epipétricos de la Antártida occidental con *Ramalina terebrata* Tayl. et al. Hook como especie caracterizante. Inst Antart Chil 4:1–18

Fowbert JA, Lewis Smith RI (1994) Rapid population increases in native vascular plants in the Argentine Islands, Antarctic Peninsula. Arct Alp Res 26:290–296

Friedmann EI (1982) Endolithic microorganisms in the Antarctic cold desert. Science 215:1045–1053

Friedmann EI, Kibler AP (1980) Nitrogen economy of endolithic microbial communities in hot and cold deserts. Microb Ecol 6:95–108

Friedmann EI, Weed R (1987) Microbial trace-fossil formation, biogenous, and abiotic weathering in the Antarctic cold desert. Science 236:703–705

Friedmann EI, Kappen L, Meyer MA, Nienow JA (1993) Long-term productivity in the cryptoendolithic microbial community of the Ross Desert, Antarctica. Microb Ecol 25:51–69

Gannutz TP (1970) Photosynthesis and respiration of plants in the Antarctic Peninsula area. Antarct J US 5:49–52

Gannutz TP (1971) Ecodynamics of lichen communities in Antarctica. In: Quam LO (ed) Research in the Antarctic. AAAS, Washington, pp 213–326

Garty J, Galun M, Kessel M (1979) Localisation of heavy metals and other elements accumulated in the lichen thallus. New Phytol 82:159–168

Gimingham CH, Lewis Smith RI (1970) Bryophyte and lichen communities in the maritime Antarctic. In: Holdgate MW (ed) Antarctic ecology, vol 2. Academic Press, London, pp 752–785
Gimingham CH, Lewis Smith RI (1971) Growth form and water relations in mosses in the maritime Antarctic. Brit Antarct Surv Bull 25:1–21
Goyal R, Seaward MRD (1982) Metal uptake in terricolous lichens III. Translocation in the thallus of *Peltigera canina* and *P. rufescens*. New Phytol 90:73–84
Green TGA, Lange OL (1994) Photosynthesis in poikilohydric plants: a comparison of lichens and bryophytes. In: Schulze ED, Caldwell MM (eds) Ecological studies, vol 100. Springer, Berlin Heidelberg New York, pp 319–341
Green TGA, Schroeter B, Sancho LG (1999) Plant life in Antarctica. In: Pugnaire FI, Valladares F (eds) Handbook of functional plant ecology. Marcel Dekker, New York, pp 495–542
Green TGA, Maseyk K, Pannewitz S, Schroeter B (2000) Extreme elevated in situ carbon dioxide levels around the moss *Bryum subrotundifolium* JAEG., BER. S. GALL. in Antarctica. Bibl Lichenol 75:397–403
Green TGA, Schroeter B, Seppelt RD (2000) Effect of temperature, light and ambient UV on the photosynthesis of the moss *Bryum argenteum*. Hedw. in continental Antarctica. In: Davison W, Howard-Williams C, Broady P (eds) Antarctic ecosystems: models for wider understanding. New Zealand Natural Science, Christchurch, pp 165–170
Greenfield LG (2000) Some properties of water soluble substances from Antarctic and arctic mosses and lichens. Bibl Lichenol 75:127–135
Grobe CW, Ruhland CT, Day TA (1997) A new population of *Colobanthus quitensis* near Arthur Harbor, Antarctica: correlating recruitment with warmer summer temperatures. Arct Alp Res 29:217–221
Harrison PM, Walton DWH, Rothery P (1989) The effects of temperature and moisture on CO_2-uptake and total resistance to water loss in the Antarctic foliose lichen *Umbilicaria antarctica*. New Phytol 111:673–682
Hébant C (1977) The conducting tissues of bryophytes. Cramer, Vaduz
Holder JM, Wynn Williams DD, Rull Perez F, Edwards HGM (2000) Raman spectroscopy of pigments and oxalates in situ within epilithic lichens: *Acarospora* from the Antarctic and Mediterranean. New Phytol 145:271–280
Holdgate MW, Allen SE, Chambers MJG (1967) A preliminary investigation of the soils of Signy Island, South Orkney Islands. Br Antarct Surv Bull 12:53–71
Holtom A, Greene SW (1967) The growth and reproduction of Antarctic flowering plants. In: Smith JE (ed) A discussion on the terrestrial Antarctic ecosystems. Physiol Trans R Soc B252:323–337
Hooker TN (1977) The growth and physiology of Antarctic lichens. PhD Thesis, University of Bristol
Hooker TN (1980) Factors affecting the growth of Antarctic crustose lichens. Lichenologist 12:313–323
Horne AJ (1972) The ecology of nitrogen fixation on Signy Island, South Orkney Islands. Br Antarct Surv Bull 27:1–18
Hoshiai T (1970) Willow cultivation at Syowa Station. Polar News 5,2:30–31
Hovenden MJ, Seppelt RD (1995) Exposure and nutrients as delimiters of lichen communities in continental Antarctica. Lichenologist 27:505–516
Huntley BJ (1971) Vegetation. In: van Zinderen Bakker JM, Winterbottom JM, Dyer RA (eds) Marion and Prince Edward Islands. Balkema, Capetown, pp 98–160
Hyvärinen M, Crittenden PD (2000) ^{33}P translocation in the thallus of the mat-forming lichen *Cladonia portentosa*. New Phytol 145:281–288
Imura S, Higuchi M, Kanda H, Iwatsuki Z (1994) Structure of moss colonies in the Syowa Station area, Antarctica. Proceedings of the National Institute of Polar Research, Symposium of Polar Biol 7:232–236

Imura S, Bando T, Saito S, Seto K, Kanda H (1999) Benthic moss pillars in Antarctic lakes. Polar Biol 22:137–140

Ino Y (1983) Estimation of primary production in moss community on East Ongul Island, Antarcica. Antarct Rec 80:30–38

Ino Y (1990a) Field measurement of net photosynthesis of mosses at Langhovde, East Antarctica. Ecol Res 5:195–205

Ino Y (1990b) Comparison of net photosynthesis and dark respiration of Antarctic mosses measured in the Antarctic and in Japan. Proc NIPR Symp Polar Biol 3:245–253

Jacobsen P, Kappen L (1988) Lichens from the Admiralty Bay region, King George Island (South Shetland Islands), Antarctica. Nova Hedw 46:503–510

Johnston CG, Vestal JR (1986) Does iron inhibit cryptoendolithic microbial communities? Antarct J US 21:225–226

Jones D (1988) Lichens and pedogenesis. In: Galun M (ed) Handbook of lichenology III. CRC Press, Boca Raton, pp 109–124

Kallio P, Heinonen S (1971) Influence of short-term low temperature on net photosynthesis in some sub-Arctic lichens. Rep Kevo Subarct Res Stat 8:63–72

Kallio P, Heinonen S (1973) Ecology of *Rhacomitrium lanuginosum* (Hedw.) Brid. Rep Kevo Subarct Res Stat 10:43–54

Kallio P, Heinonen S (1975) CO_2 exchange and growth of *Rhacomitrium lanuginosum* and *Dicranum elongatum*. In: Wielgolaski FE (ed) Plants and microorganisms. Fennoscandian Tundra Ecosystems. Part 1. Springer, Berlin Heidelberg New York, pp 138–148

Kanda H (1979) Regenerative development in culture of Antarctic plants of *Ceratodon purpureus* (Hedw.). Brid. Mem Natl Inst Pol Res, Spec Issue 11:58–69

Kanda H, Iwatsuki Z (1989) Two aquatic mosses in the lakes near Syowa Station, Continental Antarctica. Hikobia 10:293–297

Kanda H, Komárková V (1997) Antarctic terrestrial ecosystems. In: Wielgolaski FE (ed) Polar and alpine tundra. Ecosystems of the World, vol 3, Elsevier, Amsterdam, pp 721–761

Kappen L (1973) Responses to extreme environments. In: Ahmadjan V, Hale ME (eds) The lichens. Academic Press New York, pp 310–380

Kappen L (1983) Ecology and physiology of the Antarctic fruticose lichen *Usnea sulphurea* (Koenig) Th Fries. Polar Biol 1:249–255

Kappen L (1985a) Vegetation and ecology of ice-free areas of northern Victoria Land, Antarctica. I: The lichen vegetation of Birthday Ridge and an inland mountain. Polar Biol 4:213–215

Kappen L (1985b) Water relations and net photosynthesis of *Usnea*. A comparison between *Usnea fasciata* (maritime Antarctic) and *Usnea sulphure*a (continental Antarctic). In: Brown DH (ed) Lichen physiology and cell biology. Plenum Press, New York, pp 41–56

Kappen L (1985 c) Lichen-habitats as micro-oases in the Antarctic – the role of temperature. Polarforschung 55:49–54

Kappen L (1985d) Vegetation and ecology of ice-free areas of northern Victoria Land, Antarctica. II. Ecological conditions in typical microhabitats of lichens at Birthday Ridge. Polar Biol 4:227–236

Kappen L (1988) Ecophysiological relationships in different climatic regions. In: Galun M (ed) Handbook of lichenology 2, CRC Press, Boca Raton, pp 37–100

Kappen L (1989) Field measurements of carbon dioxide exchange of the Antarctic lichen *Usnea sphacelata* in the frozen state. Antarct Sci 1:31–34

Kappen L (1993a) Lichens in the Antarctic region. In: Friedmann EI (ed) Antarctic microbiology, Wiley-Liss, New York, pp 433–490

Kappen L (1993b) Plant activity under snow and ice, with particular reference to lichens. Arctic 46:297–302

Kappen L (2000) Some aspects of the great success of lichens in Antarctica. Antarct Sci 12:314–324

Kappen L, Breuer M (1991) Ecological and physiological investigations in continental Antarctic cryptogams. II. Moisture relations and photosynthesis of lichens near Casey Station, Wilkes Land. Antarct Sci 3:273–278

Kappen L, Friedmann EI, Garty J (1981) Ecophysiology of lichens in the Dry Valleys of southern Victoria Land, Antarctica. I. Microclimate of the cryptoendolithic lichen habitat. Flora 171:216–235

Kappen L, Redon J (1984) Microclimate influencing the lichen vegetation on different aspects of a coastal rock in the maritime. Ser Cient INACH 31:53–65

Kappen L, Redon J (1987) Photosynthesis and water relations of three maritime Antarctic lichen species. Flora 179:215–229

Kappen L, Schroeter B, Green TGA, Seppelt RD (1998a) Chlorophyll *a* fluorescence and CO_2 exchange of *Umbilicaria aprina* under extreme light stress in the cold. Oecologia 113:325–331

Kappen L, Schroeter B, Green TGA, Seppelt RD (1998b) Microclimatic conditions, meltwater moistening, and the distributional pattern of *Buellia frigida* on rock in a southern continental Antarctic habitat. Polar Biol 19:101–106

Kappen L, Lewis Smith RI, Meyer, M (1989) Carbon dioxide exchange of two ecodemes of *Schistidium antarctici* in continental Antarctica. Polar Biol 9:415–422

Kappen L, Straka H (1988) Pollen and spores transport into the Antarctic. Polar Biol 8:173–180

Kappen L, Valladares F (1999) Opportunistic growth and desiccation tolerance: the ecological success of poikilohydrous autotrophs. In: Pugnaire FI, Valladares F (eds) Handbook of functional plant ecology. Marcel Dekker, New York, pp 9–80

Kennedy AD (1993a) Photosynthetic response of the Antarctic moss *Polytrichum alpestre* Hoppe to low temperatures and freeze-thaw stress. Polar Biol 13:271–279

Kennedy AD (1993b) Water as a limiting factor in the Antarctic terrestrial environment: a biogeographical synthesis. Arct Alp Res 25:308–315

Kennedy AD (1995) Antarctic terrestrial ecosystem response to global environmental change. Ann Rev Ecol Syst 16:683–704

Knops JMH, Nash TH III, Schlesinger WH (1994) The influence of epiphytic lichens on the nutrient cycling of a blue oak woodland. USDA Forest Service Gen. and Technical Reports PWS-GTR-160:75–82

Lamb IM (1964) Antarctic lichens: I. The genera *Usnea, Ramalina, Himantormia, Alectoria, Cornicularia*. Brit Antarct Surv Sci Rep 38:34 pp

Lange OL (1953) Hitze- und Trockenresistenz der Flechten in Beziehung zu ihrer Verbreitung. Flora 140:39–97

Lange OL (1965) Der CO_2 Gaswechsel von Flechten nach Erwärmung im feuchten Zustand. Ber Dtsch Botan Ges 78:441–454

Lange OL, Büdel B, Meyer A, Zellner H, Zotz G (2000) Lichen carbon gain under tropical conditions: water relations and CO_2 exchange of three *Leptogium* species of a lower montane rainforest in Panama. Flora 195:172–190

Lange OL, Kappen L (1972) Photosynthesis of lichens from Antarctica. In: Llano GA (ed) Antarctic terrestrial biology. Antarct Res Ser 20, Am Geophys Union, Washington, pp 83–95

Lange OL, Metzner H (1965) Lichtabhängiger Kohlenstoffeinbau in Flechten bei tiefen Temperaturen. Naturwissenschaften 52:191–192

Lange OL, Pfanz H, Kilian E, Meyer A (1990) Effect of low water potential on photosynthesis in intact lichens and their liberated algal components. Planta 182:467–472

Larson DW (1978) Patterns of lichen photosynthesis and respiration following prolonged frozen storage. Can J Bot 56:2119–2123

Lewis Smith RI (1972) Vegetation of the South Orkney Islands with particular reference to Signy Island. Ph D Thesis, British Antarctic Survey and Department of Botany, University of Aberdeen, 124 pp
Lewis Smith RI (1978) Summer and winter concentrations of sodium, potassium and calcium in some maritime Antarctic cryptogams. J Ecol 66:891–909
Lewis Smith RI (1982) Plant succession and re-exposed moss banks on a deglaciated headland in Arthur Harbour, Anvers Island. Br Antarct Surv Bull 51:193–199
Lewis Smith RI (1984a) Terrestrial plant biology of the sub-Antarctic and Antarctic. In: Laws RM (ed) Antarctic ecology. Academic Press, London, pp 61–162
Lewis Smith RI (1984b) Colonisation by bryophytes following recent volcanic activity on an Antarctic Island. J Hattori Bot Lab 56:53–63
Lewis Smith RI (1984 c) Colonization and recovery by cryptogams following recent volcanic activity on Deception Island, South Shetland Islands. Br Antarctc Surv Bull 62:25–51
Lewis Smith RI (1985) Nutrient cycling in relation to biological productivity in Antarctic and sub-Antarctic terrestrial and freshwater ecosystems. In: Siegfried WR, Condy PR, Laws RM (eds) Antarctic nutrient cycles and food webs, Springer, Berlin Heidelberg New York, pp 138–155
Lewis Smith RI (1987) The bryophyte propagule bank of Antarctic fellfield soils. Symposia Biologica Hungarica 35. British Antarctic Survey, Natural Environment Research Council, Cambridge, pp 233–245
Lewis Smith RI (1988a) Bryophyte oases in Ablation Valleys on Alexander Island, Antarctica. Bryologist 91:45–50
Lewis Smith RI (1988b) Classification and ordination of cryptogamic communities in Wilkes Land, Continental Antarctica. Vegetatio 76:155–166
Lewis Smith RI (1988 c) Recording bryophyte microclimate in remote and severe environments. In: Glime IM (ed) Methods in bryology. Proc Methodol Worksh Maine, Hattori Botany Laboratory Nichinan
Lewis Smith RI (1990a) Signy Island as a paradigm of biological and environmental change in Antarctic terrestrial ecosystems. In: Kerry KR, Hempel G (eds) Antarctic ecosystems. Springer, Berlin Heidelberg New York, pp 32–50
Lewis Smith RI (1990b) Plant community dynamics in Wilkes Land, Antarctica. Proc NIPR Symp Polar Biol 3:229–244
Lewis Smith RI (1991) Exotic sporomorpha as indicators of potential immigrant colonists in Antarctica. Grana 30:313–324
Lewis Smith RI (1993) The role of bryophyte propagule banks in primary succession: case-study of an Antarctic fellfield soil. In: Miles J, Walton DWH (eds) Primary succession on land. Blackwell, Oxford, pp 55–78
Lewis Smith RI (1994) Vascular plants as bioindicators of regional warming in Antarctica. Oecologia 99:322–328
Lewis Smith RI (1995) Ecosystems and community development: new biological research sites in the southwestern Antarctic Peninsula. Br Antarct Surv Annu Rep 1994–1995:84–86
Lewis Smith RI (1996a) Introduced plants in Antarctica: potential impacts and conservation issues. Biol Conserv 76:135–146
Lewis Smith RI (1996b) Terrestrial and freshwater biotic components of the western Antarctic Peninsula. In: Ross RM, Hofmann EE, Quetin LB (eds) Ecological research west of the Antarctic Peninsula. Am Geophys Union, Antarct Res Ser 70: 15–59
Lewis Smith RI (1997) Oases as centres of high plant diversity and dispersal in Antarctica. In: Howard-Williams, Haws I (eds) Ecosystem processes in Antarctic ice-free landscapes, Lyons, Balkema, Rotterdam, pp 119–128
Lewis Smith RI (1999) Biological and environmental characteristics of three cosmopolitan mosses dominant in continental Antarctica. J Veget Sci 10:231–242

Lewis Smith RI (2000) Plants of extreme habitats in Antarctica. Bibl Lichenol 75:405–419
Lewis Smith RI, Corner RWM (1973) Vegetation of the Arthur Harbour-Argentine Islands region of the Antarctic Peninsula. Br Antarct Surv Bull 33, 34:89–122
Lewis Smith RI, Poncet S (1987) *Deschampsia antarctica* and *Colobanthus quitensis* in the Terra Firma Islands. Br Antarct Surv Bull 74:31–35
Lindsay DC (1971) Vegetation of the South Shetland Islands. Br Antarct Surv Bull 25:59–83
Lindsay DC (1978) The role of lichens in Antarctic ecosystems. Bryologist 81:268–276
Linskens HF, Bargagli R, Cresti M, Focardi S (1993) Entrapment of long-distance transported pollen grains by various moss species in coastal Victoria Land, Antarctica. Polar Biol 13:81–87
Longton RE (1967) Vegetation in the maritime Antarctic. Philos Trans R Soc Lond Ser B 252:213–235
Longton RE (1970) Growth and productivity of the moss *Polytrichum alpestre* Hoppe in Antarctic regions. In: Holdgate MW (ed) Antarctic ecology, vol 2. Academic Press, London, pp 818–837
Longton RE (1972) Growth and reproduction in northern and southern hemisphere populations of the peat-forming moss *Polytrichum alpestre* with reference of the estimation of productivity. Proc 4th Int Peat Congr I-IV, Helsinki, pp 259–275
Longton RE (1973) The occurrence of radial infection patterns in colonies of polar bryophytes. Br Antarct Surv Bull 32:41–49
Longton RE (1974) Microclimate and biomass in communities of the *Bryum* association on Ross Island, continental Antarctica. Bryologist 77:109–127
Longton RE (1979) Studies on growth, reproduction and population ecology in relation to microclimate in the bipolar moss *Polytrichum alpestre* Hoppe. Bryologist 82:325–367
Longton RE (1984) The role of bryophytes in terrestrial ecosystems. J Hattori Bot Lab 55:147–163
Longton RE (1985) Terrestrial habitats – Vegetation. In: Bonner WN, Walton DWH (eds) Key environments: Antarctica. Pergamon Press, Oxford, pp 73–105
Longton RE (1988a) Biology of polar bryophytes and lichens. Cambridge University Press, Cambridge
Longton RE (1988b) Adaptations and strategies of polar bryophytes. Bot J Linn Soc 98:253–268
Longton RE, Greene SW (1967) The growth and reproduction of *Polytrichum alpestre* Hoppe on South Georgia. Philos Trans R Soc B 252:295–322
Lovelock CE, Osmond CB, Seppelt RD (1995a) Photoinhibition in the Antarctic moss *Grimmia antarctici* Card. when exposed to cycles of freezing and thawing. Plant Cell Environ 18:1395–1402
Lovelock CE, Jackson AG, Melick DR, Seppelt RD (1995b) Reversible photoinhibition in Antarctic moss during freezing and thawing. Plant Physiol 109:955–961
Mägdefrau H (1935) Untersuchung über die Wasserversorgung der Gametophyten und Sporophyten der Laubmoose. Z Bot 29:337–375
Manrique E, Balagner L, Barnes J, Davidson AW (1993) Photoinhibition studies in lichens using chlorophyll fluorescence analysis. Bryologist 96:443–449
Marshall WA (1996) Biological particles over Antarctica. Nature 383:680
Marshall WA (1997) Seasonality in Antarctic airborne fungal spores. Appl Environ Microbiol 63:2240–2245
Matsuda T (1968) Ecological study of the moss community and microorganisms in the vicinity of Syowa Station, Antarctica. Jpn Antarct Res Exp Sci Rep E29:1–58
Matsumoto G, Chikazawa K, Murayama H, Torii T, Fukushima H, Hanya T (1983) Distribution and correlation of total organic carbon and mercury in Antarctic dry valley soils, sediments and organisms. Geochem J 17:247–255

Matsumoto G, Kanda H (1985) Hydrocarbons, sterols and hydroxy acids in Antarctic mosses. Antarct Rec 87:23–31

Melick DR, Bölter M, Möller R (1994) Rates of soluble carbohydrate utilization in soils from the Windmill Islands Oasis, Wilkes Land, continental Antarctica. Polar Biol 14:59–64

Melick DR, Hovenden MJ, Seppelt RD (1994) Phytogeography of bryophyte and lichen vegetation in the Windmill Islands, Wilkes Land, Continental Antarctica. Vegetatio 111:71–87

Melick DR, Seppelt RD (1992) Loss of soluble carbohydrates and changes in freezing point of Antarctic bryophytes after leaching and repeated freeze-thaw cycles. Antarct Sci 4:339–404

Melick DR, Seppelt RD (1997) Vegetation patterns in relation to climatic and endogenous changes in Wilkes Land, continental Antarctica. J Ecol 85:43–56

Montiel PO, Cowan DA (1993) The possible role of soluble carbohydrates and polyols as cryoprotectants in Antarctic plants. In: Heywood RB (ed) University Research in Antarctica 1989–1992. British Antarctic Survey, Cambridge, pp 111–125

Nakanishi S (1977) Ecological studies of the moss and lichen communities in the ice-free areas near Syowa Station, Antarctica. Antarct Rec 59:68–96

Nakatsubo T, Ino Y (1986) Nitrogen cycling in an Antarctic ecosystem. 1: Biological nitrogen fixation in the vicinity of Syowa Station. Mem Natl Inst Polar Res E37:1–10

Nash III TH (1989) Metal tolerance in lichens. In: Shaw AJ (ed) Heavy metal tolerance in plants. CRC Press, Boca Raton, pp 119–131

Nash III TH (1996) Lichen biology. Cambridge University Press, Cambridge, 303 pp

Nash III TH, Reiner A, Demmig-Adams B, Kilian E, Kaiser WM, Lange OL (1990) The effect of atmospheric desiccation and osmotic water stress on photosynthesis and dark respiration of lichens. New Phytol 16:269–276

Nieboer E, Richardson DHS Tomassini FD (1978) Mineral uptake and release by lichens. An overview. Bryologist 81:226–246

Nienow JA, Friedmann EI (1993) Terrestrial lithophytic (rock) communities. In: Friedmann EI (ed) Antarctic microbiology. Wiley-Liss, New York, pp 297–341

Noakes TD, Longton RE (1989) Studies on water relations in mosses from the cold-Antarctic. Antarct Spec Top:103–116

Oechel WC, Sveinbjörnsson B (1978) The role of bryophytes in nutrient cycling in the taiga. In: Van Cleve K, Chapin FS, Flanagan PW, Viereck LA, Dyrness CT (eds) Forest ecosystems in the Alaskan Taiga. Springer, Berlin Heidelberg New York, pp 121–137

Ochyra R (1998) The moss flora of King George Island, Antarctica. W. Szafer Institute of Botany, Polish Academy of Sciences, Cracow, Poland

Olech M (1990) Preliminary studies on ornithocoprophilous lichens of the Arctic and Antarctic regions. Proc NIPR Symp Polar Biol 3:218–223

Olech M (1993) Lower plants. In: Rakusa-Suszcewski S (ed) The maritime Antarctic coastal ecosystem of Admiralty Bay. Department of Antarctic Biology, Polish Academy of Sciences, Warsaw, pp 173–179

Øvstedal DO (1983) Some lichens from H.U. Sverdrup Mountains, Dronning Maud Land, Antarctis. Nova Hedw 37:683–690

Post A (1990) Photoprotective pigment as an adaptive strategy in the Antarctic moss *Ceratodon purpureus*. Polar Biol 10:241–245

Priddle J (1979) Morphology and adaptation of aquatic mosses in an Antarctic lake. J Bryol 10:517–529

Puckett KJ (1988) Bryophytes and lichens as monitors of metal deposition. Bibl Lichenol 30:231–267

Puckett KJ, Finegan EJ (1980) An analysis of the element content of lichens from the North West Territories, Canada. Canadian J Bot 58:2073–2089

Rastorfer JR (1970) Effects of light intensity and temperature on photosynthesis and respiration of two East Antarctic mosses *Bryum argenteum* and *B. antarcticum*. Bryologist 73:544–556
Rastorfer JR (1972) Comparative physiology of four west Antarctic mosses. In: Llano GA (ed) Antarctic terrestrial biology. American Geophysical Union, Washington, pp 143–161
Richardson DHS (1981) The biology of mosses. Blackwell, Oxford, 220 pp
Richardson DHS, Nieboer E, Laroie P, Padovan D (1984) Anion accumulation by lichens I The characteristics and kinetics of arsenate uptake by *Umbilicaria mühlenbergii*. New Phytol 96:71–82
Rütten D, Santarius KA (1992) Age-related differences in frost sensitivity of the photosynthetic apparatus of two *Plagiomnium* species. Planta 187:224–229
Rütten D, Santarius KA (1992) Relationship between frost tolerance and sugar concentration of various bryophytes in summer and winter. Oecologia 91:260–265
Rütten D, Santarius KA (1993) Seasonal Variation in frost tolerance and sugar content of two *Plagiomnium* species. Bryologist 96:564–568
Ryan PG, Watkins BP (1989) The influence of physical factors and ornithogenic products on plant and arthropod abundance at an inland nunatak group in Antarctica. Polar Biol 10:151–160
Ryan PG, Watkins BP, Lewis Smith RI, Dastych H, Eicker A, Foissner W, Heatwole H, Miller WR, Thompson G (1989) Biological survey of Robertskollen, western Dronning Maud Land: area description and preliminary species lists. S Afr T Antarkt, Deel 19:10–20
Salas MR (1983) Long-distance transport over the Tasman Sea: evidence from Macquarie Island. N Z J Bot 21:285–292
Sancho LG, Pintado A, Valladares F, Schroeter B, Schlensog M (1997) Photosynthetic performance of cosmopolitan lichens in the maritime Antarctic. Bibl Lichenol 67:197–210
Sancho LG, Schulz F, Schroeter B, Kappen L (1999) Bryophyte and lichenflora of South Bay (Livingston Island: South Shetland Islands, Antarctica). Nova Hedw 68:301–337
Sancho LG, Valladares F (1993) Lichen colonization of recent moraines on Livingston Island (South Shetland I., Antarctica). Polar Biol 13:227–233
Sancho LG, Valladares F, Pintado A, Schlensog M, Schroeter B (1997) Compartimiento fotosintético de líquenes cosmopolitas en la Antártida marítima. Bol R Soc Esp Hist Nat (Sec Biol) 93 (1–4):113–118
Scheffer F, Schachtschabel P (1984) Lehrbuch der Bodenkunde, 11. Aufl. F. Enke, Stuttgart
Schlensog M, Schroeter B (2000) Poikilohydry in Antarctic cryptogams and its role for photosynthetic performance in mesic and xeric habitats. In: Davison W, Howard-Williams C, Broady P (eds) Antarctic ecosystems: models for wider ecological understanding. New Zealand Natural Science, Christchurch, pp 175–182
Schlensog M, Schroeter B, Sancho LG, Pintado A, Kappen L (1997a) Effect of strong irradiance on photosynthetic performance of the melt-water dependent cyanobacterial lichen *Leptogium puberulum* (Collemataceae) Hue from the maritime Antarctic. Bibl Lichenol 75:235–246
Schlensog M, Schroeter B, Sancho LG, Pintado A, Kappen L (1997b) Photosynthetic performance of the cyanobacterial lichen *Leptogium puberulum* Hue (Collemataceae) in the maritime Antarctic. Bol R Soc Hesp Hist Nat (Sec Biol) 93:105–111
Schofield E, Ahmadjan V (1972) Field observations and laboratory studies of some Antarctic cold desert cryptogams. In: Llano GA (ed) Antarctic terrestrial biology. Antarctic Research Series 20American Geophysical Union, Washington, pp 97–142
Schroeter B (1991) Untersuchungen zu Primärproduktion und Wasserhaushalt von Flechten der maritimen Antarktis unter besonderer Berücksichtigung von *Usnea antarctica* Du Rietz. Doctorate Thesis Math-Nat Fakultät, CAU, Kiel, 148 pp

Schroeter B (1997) Grundlagen der Stoffproduktion von Kryptogamen unter besonderer Berücksichtigung der Flechten – Eine Synopse. Habilitation Math-Nat Fakultät, CAU, Kiel, 130 pp

Schroeter B, Green TGA, Kappen L, Seppelt RD (1994) Carbon dioxide exchange at subzero temperatures. Field measurements on *Umbilicaria aprina* in Antarctica. Crypt Bot 4:233–241

Schroeter B, Kappen L, Schulz F, Sancho LG (2000) Seasonal variation in the carbon balance of lichens in the maritime Antarctic: long-term measurements of photosynthetic activity in *Usnea aurantiaco-atra*. In: Davison W, Howard-Williams C, Broady P (eds) VII SCAR Intern Biol Symp Antarctic Ecosystems: Models for a wider ecological understanding. New Zealand Natural Science, Christchurch, pp 258–262

Schroeter B, Olech M, Kappen L, Heitland W (1995) Ecophysiological investigations of *Usnea antarctica* in the maritime Antarctic. I. annual microclimatic conditions and potential primary production. Antarct Sci 7 (3):251–260

Schroeter B, Sancho LG (1996) Lichens growing on glass in Antarctica. Lichenologist 28:385–390

Schroeter B, Kappen L, Green TGA, Seppelt RD (1997a) Lichens and the Antarctic environment: effects of temperature and water availability on photosynthesis. In: Lyons WB, Howard-Williams WB, Hawes I (eds) Ecosystem processes in Antarctic ice-free landscapes. Balkema, Rotterdam, pp 103–118

Schroeter B, Schulz F, Kappen L (1997b) Hydration-related spatial and temporal variation of photosynthetic activity in Antarctic lichens. In: Battaglia B, Valencia J, Walton DWH (eds) Antarctic communities, species, structure and survival. Cambridge University Press, Cambridge, pp 221–225

Schroeter B, Scheidegger C (1995) Water relations in lichens at subzero temperatures: structural changes and carbon dioxide exchange in the lichen *Umbilicaria aprina* from continental Antarctica. New Phytol 131:273–285

Seppelt RD, Ashton DH (1978) Studies on the ecology of the vegetation at Mawson Station, Antarctica. Aust J Ecol 3:373–388

Seppelt RD, Broady PA (1988) Antarctic terrestrial ecosystems: the Vestfold Hills in context. In: Ferris JM, Burton HR, Johnstone GW, Bayly IAE (eds) Biology of the Vestfold Hills, Antarctica., Hydrobiol 165:177–184

Seppelt RD Broady PA, Pickard J, Adamson DA (1988) Plants and landscape in the Vestfold Hills, Antarctia. In: Ferris JM, Burton HR, Johnstone GW, Bayly IAE (eds) Biology of the Vestfold Hills, Antarctica, Kluwer, Dordrecht, pp 185–196

Seppelt RD, Green TGA, Schroeter B (1995) Lichens and mosses from the Kar Plateau, Southern Victoria Land, Antarctica. N Z J Bot 33:203–220

Seppelt RD, Lewis Smith RI, Kanda H (1998) Antarctic bryology: past achievements and new perspectives. J Hattori Bot Lab 84:203–239

Skre O, Berg A, Wielgolaski FE (1975) Organic compounds in alpine plants. In: Wielgolaski FE (ed) Fennoscandian Tundra ecosystems. 1. Plants and microorganisms, Springer, Berlin Heidelberg New York, pp 339–350

Smith AJE (1982) Epiphytes and epiliths. In: Smith AJE (ed) Bryophyte ecology. Chapman & Hall, London, pp 191–227

Smith VR (1988) Production and nutrient dynamics of plant communities on a sub-Antarctic Island. 5. Nutrient budgets and turnover times for mire-grasslands, fjaedmark and fernbrakes. Polar Biol 8:255–269

Smith VR (1993) Effects of nutrients on CO_2 assimilation by mosses on a sub-Antarctic Island. New Phytol 123:693–697

Staaland H, Brattbakk I, Ekern K, Kildemo K (1983) Chemical composition of reindeer forage plants in Svalbard and Norway. Holarct Ecol 6:109–122

Sveinbjörnsson B, Oechel WC (1983) The effect of temperature preconditioning on the temperature sensitivity of net CO_2 flux in geographically diverse populations of the moss *Polytrichum commune* Hedw. Ecology 64:1100–1108

Syers JK, Iskandar IK (1973) Pedogenetic significance of lichens. In: Ahmadjan V, Hale ME (eds) The lichens. Academic Press, New York, pp 225–248

Tarnawski MD, Melick D, Roser D, Adamson E, Seppelt RD (1992) In situ carbon dioxide levels in cushion and turf forms of *Grimmia antarctici* at Casey Station, East Antarctica. J Bryol 17:241–249

Tearle PV (1987) Cryptogamic carbohydrate release and microbial response during spring freeze-thaw cycles in Antarctic fellfield fines. Soil Biol Biochem 19:381–390

Tedrow JCF (1977) Soils of the polar landscapes. Rutgers University Press, New Brunswick

Tilbrook PJ (1973) The Signy Island terrestrial reference sites: I. An introduction. Brit Antarct Surv Bull 33, 34:65–76

Thomson JW (1972) Distribution patterns of American Arctic lichens. Can J Bot 50:1135–1156

Thomson JW (1997) American Arctic lichens. 2. The microlichens. The University of Wisconsin Press, Madison, Wisconsin, 675 pp

Tuominen Y, Jaakkola T (1973) Absorption and accumulation of mineral elements and radioactive nucleides. In: Ahmadjan V, Hale ME (eds) The lichens. Academic Press, New York, pp 185–223

Ugolini FC (1970) Antarctic soils and their ecology. In: Holdgate MW (ed) Antarcic ecology, vol 2. Academic Press, London, pp 673–692

Ugolini FC, Edmonds RL (1983) Soil biology In: Wilding LP, Smeck NE, Hall GF (eds) Pedogenesis and soil taxonomy. I. Concepts and interactions. Elsevier, Amsterdam, pp 193–231

Valladares F, Sancho LG (2000) The relevance of nutrient availability for lichen productivity in the maritime Antarctic. Bibl Lichenol 75:405–419

Van Dobben H (1993) Vegetation as a monitor of deposition of nitrogen and activity. Doctorate Thesis, University of Utrecht

Van Zanten BO (1978) Experimental studies on trans-oceanic long-range dispersal of moss spores in the Southern Hemisphere. J Hatt Bot Lab 44:455–482

Vestal JR (1988) Carbon metabolism of the cryptoendolithic microbiota from the Antarctic Desert. Appl Environ Microbiol 54 (4):960–965

Vincent WF (1997) Polar desert ecosystems in a changing climate: a north-south perspective. In: Howard-Williams, Hawes (eds) Ecosystem process in Antarctic ice-free landscapes. Lyons, Balkema, Rotterdam, pp 3–14

Walton DWH (1982) The Signy Island terrestrial reference sites. XV. Micro-climate monitoring, 1972–74. Br Antarct Surv Bull 55:111–126

Walton DWH (1985) A preliminary study of the action of crustose lichens on rock surfaces in Antarctica. In: Siegfried WR, Condy PR, Laws RM (eds) Antarctic nutrient cycles and food webs. Springer, Berlin Heidelberg New York, pp 180–185

Walton DWH, Lewis Smith RI (1980) Chemical composition of South Georgian plants. Br Antarct Surv Bull 49:117–135

Weed R, Norton SA (1991) Siliceous crusts, quartz rinds and biotic weathering of sandstone in the cold desert of Antarctica. In: Berthelin J (ed) Diversity of environmental biogeochemistry. Developments in geochemistry 6. Elsevier, Amsterdam, pp 327–339

Wessels DC, Schoemann P (1988) Mechanisms and rate of weathering of Clarence sandstone by an endolithic lichen. S Afr J Sci 84:274–277

Wierzchos J, Ascaso C (1996) Morphological and chemical features of bioweathered granitic biotite induced by lichen activity. Clays Clay Miner 44:652–657

Williams PG, Roser DJ, Seppelt RD (1994) Mycorrhizas of hepatics in continental Antarctica. Mycol Res 98:34–36

Winkler JB, Schulz F, Kappen L (1998) Seasonal variation of abiotic factors in terrestrial habitats. Ber Polarforsch (Alfred-Wegener-Institut für Polarforschung, Bremerhaven) 299:28–32

Winkler JB, Kappen L, Schulz F (2000) Snow and ice as an important ecological factor for the cryptogams in the maritime Antarctic. In: Howard-Williams, C, Davison W (eds) Antarctic ecosystems: models for wider ecological understanding. New Zealand Natural Science, Christchurch, pp 220–224

Winner WE, Atkinson CJ, Nash III TH (1988) Comparison of SO_2 absorption capacities of mosses, lichens and vascular plants in diverse habitats. Bibl Lichenol 30:217–230

Young SB, Kläy J-R (1971) Bryophytes in the 1969 crater of Deception Island, Antarctica: an apparent case of rapid long-distance dispersal. Ohio J Sci 71: 358–362

19 Soil Ecology in Relation to Plant Patterns

L. BEYER, H.-P. BLUME, M. BÖLTER, L. KAPPEN, D. KUHN,
and R.D. SEPPELT

19.1 Introduction

The southern polar region belongs to the earth's coldest climate region, primarily because of its high average altitude and high latitude. The amount of solar radiation received varies greatly on a seasonal basis and thus poses special and severe conditions to plants and soil organisms which have to find, keep, and when possible extend their niches (Kennedy 1999). Ice-free areas of the Antarctic continent and its adjacent islands are covered primarily by cryptogamic communities (Olech, Chap. 12; Seppelt, Chap. 13; Kanda et al., Chap. 14) and only the mildest places in the maritime Antarctic show the establishment of vascular plants.

Soil formation occurs at a slow rate and mainly by cryoclastic weathering, stimulated by cryoturbation (Campbell and Claridge 1987; Bockheim and Ugolini 1990; Bockheim 1997; Beyer et al. 1999), but under plant cover and soil crusts, chemical weathering, clay formation, and brownification are significant (Blume and Bölter 1993; Blume et al. 1997). Most Antarctic soils belong to the Gelisol soil order, they are present primarily in ice-free areas, mainly scattered around the outer edge of the Antarctic continent (Beyer et al. 1999).

However, little information on soil ecology of the terrestrial ecosystems in Antarctica is available. Effects of vegetation can be anticipated, such as root penetration and litter deposition in the microhabitats: moss colonised areas may generate the accumulation of peat. Possible depletion of nutrients will show effects, or changes in acidity due to decomposition are likely to occur.

19.2 Sites and Environmental Constraints

A comparison of soils of the South Shetland Islands, maritime Antarctic, near Arctowski Station, at King George Island (KGI) (Rakusa-Suszczewski, Chap. 3) and soils of the continental Antarctic, around Casey Station on Bailey Peninsula in the continental Windmill Islands (WMI) region (Seppelt,

Ecological Studies, Vol. 154
L. Beyer and M. Bölter (eds.) Geoecology
of Antarctic Ice-Free Coastal Landscapes

Chap. 4), has been performed recently to ascertain in what way vegetation patterns are related to principal soil ecological properties such as carbon, nitrogen and nutrient contents. A restriction to mainly these two places became necessary as comparable data from other places are not available.

19.2.1 Soils – Texture and Nutrients

The soils range over a wide variety of suborders in both regions (Blume et al. 1997; Beyer and Bölter 1999). As most soils are permafrost-affected and the active layer can be found mostly above 50 (WMI) and 100 cm (KGI: Blume et al. 1997; Beyer and Bölter 1999; Beyer et al. 1999, 2000a), the classification according to the US Soil Taxonomy (Soil Survey Staff 1998) lead to the order of the Gelisols. However, this taxonomy does not recognise the extent of podzolisation at an excepted level (Beyer and Bölter 2000b), it is more useful for an ecological purpose for organic soils and soils units with no visible cryogenic features (Beyer et al. 2000a). Gelisols with organic horizons, podzolised soils and those which show no cryogenic features cannot be described as Cryosols according to the WRB system (ISSS-WRB 1998; Blume et al., Chap. 7). At places along the old moraine lines on King George Island, permafrost can be found often even below 1 m (Kuhn 1997). Soil formation is limited by the slow vegetation turnover rates and by the reduced availability of free water. Physical and chemical weathering can be traced down to the permafrost layer because of cryoturbation (Blume et al. 1997).

In the region of Admiralty Bay, soils cannot be classified strictly as 'lithic' within the US Soil Taxonomy system (Soil Survey Staff 1998) because of the deep solum without bedrock. In the continental WMI region, soils often have a solum <40 cm, and are therefore equivalent to 'lithic' soils according to the US Soil Taxonomy system. Extremely shallow soils are found in this region with a solum of ≤10 cm (Blume et al. 1997; Beyer et al. 2000a), which corresponds to 'lithic' also according to the WRB system. A further differentiation of the soils is made by the occurrence of a mollic epipedon, which means simply a high content of soil organic matter (SOM) and a high base saturation of >50 % and minimal thickness of the Ah horizon, or umbric epipedon similar to mollic, but low base saturation of <50 %, as well as an eutric or dystric acidification level. For both regions, some organic soils and podzolic soils have been described (Blume et al. 1996; Kuhn 1997; Beyer and Bölter 1999) which are not Cryosols according to the WRB system (ISSS-WRB 1998). In addition, many Gelic Leptosols, Cambisols, Regosols, Umbrisols and Podzols can be described because of lacking cryogenic features.

The soil texture ranges between coarse and gravely sand (KGI 500–910 g kg^{-1}; Table 19.1), WMI 240–970 g kg^{-1}; Table 19.2) and in some cases more silt or clay. The stone content (mineral particles >2 mm), can be regarded by the classification as *skeletal* with a mean >350 g stones kg^{-1}, and

Table 19.1. Principal soil ecological properties in the first 10 cm of the mineral soil (<2 mm) and vegetation cover on King George Island (Admiralty Bay region). Laboratory determinations according to Schlichting et al. (1995) on air-dried soil samples. Soil mineral particle composition was determined after sieving at 2 mm by a combination of sieve and elutriate analysis. pH was measured in 10 mM $CaCl_2$ with a commercial glass electrode. Total organic carbon (TOC) was calculated after dry combustion in a Coulomat 702 (Ströhlein Instruments, Kaarst, Germany; Schlichting et al. 1995). Total nitrogen (N_t) was determined by the classical Kjeldahl (Jones 1991). Na, K, Mg, and Ca were extracted with unbuffered $BaCl_2$. Potential H and Al were extracted with Ca-acetate and measured as pH and converted into H+Al value (Schachtschabel 1951). The sum of the cations and potential H+Al is the potential cation exchange capacity (CEC). The percentage of Na, K, Mg, and Ca from CEC is the base saturation (BS). Plant available potassium (K_l), magnesium (Mg_l), and phosphorus (P_l) were extracted with 0.04 *N* Ca lactate + 0.02 *N* HCl at pH 3.7 according to Egner et al. (1960).nK_l and Mg_l were determined by an atomic absorption spectrometer (Perkin Elmer). P_l was determined colorimetrically as the blue-coloured molybdate-phosphate complex according to Harwood et al. (1969). Fe_o/Fe_d ratio describes the soluble iron in oxalate vs. dithionite. (Schlichting et al. 1995)

Code	Vegetation	Sand	Silt	Clay	pH	BS	Fe_o/Fe_d
			($g\ kg^{-1}$)			%	
P8	Nil	810	100	90	6.40	92	0.07
P10	Nil	520	280	200	6.75	91	0.59
P17	<2% Lichens	620	270	110	6.00	n.d.	n.d.
P16	<2% Lichens	500	340	160	5.25	86	0.62
P6	<2% Lichens	690	220	90	5.25	78	0.20
P2	10% *Deschampsia antarctica*	750	140	110	5.15	85	0.35
P5	50% *D. antarctica*, lichens	690	210	100	5.90	93	0.27
P3[a]	70% *D. antarctica*, mosses, lichens	740	160	110	4.85	82	0.42
P14[b]	80% *D. antarctica*, mosses	910	60	30	4.00	8	0.61
P9	90% *D. antarctica*, lichens, mosses	680	260	60	4.25	38	0.38
P1	90% *D. antarctica*, *Colobanthus quitensis*, mosses	720	190	90	5.50	87	0.32
P4[a]	90% *D. antarctica*, *C. quitensis*	560	170	270	4.25	52	0.35
P11	90% *D. antarctica*, *C. quitensis*, mosses	n.d.	n.d.	n.d.	4.05	36	0.36
P12[b]	90% *D. antarctica*, mosses	890	90	20	4.20	9	0.43
P15	95% *D. antarctica*, mosses, lichens	790	120	90	4.00	21	0.59
P7	95% Mosses	890	60	50	4.25	15	0.53
P13[b]	95% *D. antarctica*, mosses	720	230	50	3.60	7	0.48

n.d., Not determined; *BS* base saturation
[a] Spodic soil.
[b] Ornithogenic soil.

Table 19.2. Principal soil ecological properties in the first 10 cm of the mineral soil and vegetation cover in the Windmill Island area close to Casey Station (for methods see Table 19.1)

Code	Vegetation	Sand	Silt $(g\ kg^{-1})$	Clay	pH	BS %	Fe_o/Fe_d	K	Mg $(mg\ kg^{-1})$	P
D1[a,b]	Nil	786	153	61	4.16	19	0.75	54	91	279
B12	Nil	918	51	31	4.00	17	0.44	29	51	149
D3[a,b]	Nil	839	104	57	3.56	21	0.81	59	96	198
L1	<2 % Mosses	784	199	17	5.42	65	0.56	33	59	77
L2	<2 % Mosses	716	260	24	5.63	83	0.46	47	52	8
B2	<2 % Lichens, mosses	662	181	157	4.16	68	0.30	191	904	219
WP1[b]	<2 % Algae	463	263	274	5.28	68	0.24	173	2490	3590
H2[b]	<2 % Algae	945	37	18	5.70	84	0.63	115	281	901
H3	<2 % Algae	894	72	32	5.73	79	0.64	266	859	123
B13	<2 % Lichens, mosses	636	300	64	4.00	9	0.42	28	24	225
A11	<2 % Algae	617	263	66	4.00	20	0.57	74	38	311
WP2[b]	<2 % Algae	236	250	514	5.30	93	0.68	271	14600	4300
C3	2–5 % Mosses, lichens, algae	453	487	34	4.17	19	0.37	75	52	177
A13[a]	2–5 % Burnt lichens, mosses	811	121	68	4.00	21	0.81	80	83	226
D2	10–20 % Lichens, scarce mosses	897	77	26	4.02	14	0.55	32	45	99
H1	80 % Mosses, algae	869	114	17	5.56	53	0.62	356	340	221
A12[a,b]	80–100 % Lichens	902	52	46	3.99	21	0.47	102	86	158
P1	100 % Algae below gravel	921	45	34	5.36	61	0.47	64	121	502
MG1	100 % Algae	965	26	9	6.05	91	0.17	133	333	32
A2	100 % Mosses, lichens	941	55	5	4.20	18	0.53	75	75	30
B14 A	100 % Mosses	858	92	50	4.09	36	0.48	41	106	109
C4	100 % Mosses	524	442	34	3.76	8	0.32	35	23	191
C1 A	100 % Mosses, lichens	589	352	59	3.70	6	0.48	33	39	331
B4	100 % Mosses, lichens	829	129	42	3.85	11	0.65	44	68	84
B14B	100 % Mosses	769	178	53	4.05	33	0.44	38	97	175

Table 19.2 *(Continued)*

Code	Vegetation	Sand	Silt $(g\ kg^{-1})$	Clay	pH	BS %	Fe_o/Fe_d	K	Mg $(mg\ kg^{-1})$	P
C11[a]	100 % Mosses, lichens	890	71	39	3.78	17	0.53	44	86	101
C1	100 % Mosses, lichens	783	161	56	3.71	7	0.50	40	54	163
WP5[a]	100 % Mosses	891	85	24	3.83	11	0.60	66	132	91
C5	100 % Mosses	702	260	38	3.61	14	0.58	8	204	195
A3[a]	100 % Mosses, lichens	849	106	45	3.95	18	0.59	76	101	45
B11	100 % Mosses	683	200	117	3.95	16	0.73	122	144	173
M2	100 % dense moss cushion	300	473	227	4.17	17	0.76	402	378	625
BL209	100 % Dense algal carpet	n.d.	n.d.	n.d.	4.80	n.d.	n.d.	1378	n.d.	1004
BL208	100 % Dense moss cushion	n.d.	n.d.	n.d.	4.08	n.d.	n.d.	209	n.d.	225
M1	100 % Dense moss cushion	117	668	215	4.71	52	0.87	525	1900	262
WP4	100 % Dense moss cushion	389	394	217	3.74	25	0.59	205	377	273
WP6	100 % Dense moss cushion	250	485	265	3.81	18	0.88	597	541	855

[a] Spodic.
[b] Ornithogenic.

Table 19.3. Relationship between soils, vegetation and organic carbon and total nitrogen storage of the entire soil profile on King George Island (Admiralty Bay region). *T.* Typic, *L.* lithic, *C.* cumulic

Code	Soil name	Vegetation	>2 mm ($g\ kg^{-1}$)	TOC	N_t ($g\ m^{-2}$)	C/N
P8	T. Haplorthel	nil	380	250	85	3
P10	T. Haplorthel	nil	450	340	50	7
P17	L. Haplorthel	<2 % Lichens	730	90	20	4
P16	T. Aquiturbel	<2 % Lichens	500	280	75	4
P6	T. Haplorthel	<2 % Lichens	550	460	65	7
P2	T. Haplorthel	10 % *D. antarctica*	440	710	90	8
P5	L. Haplorthel	50 % *D. antarctica*, lichens	680	210	40	5
P3	T. Haplorthel[a]	70 % *D. antarctica*, mosses, lichens	330	5490	500	11
P14	T. Haploturbel[b]	80 % *D. antarctica*, mosses	210	18750	2150	9
P9	T. Umbrorthel	90 % *D. antarctica*, lichens, mosses	330	2380	570	4
P1	T. Umbrorthel	90 % *D. antarctica*, *C. quitensis*, mosses	360	3390	345	10
P4	T. Haplorthel[a]	90 % *D. antarctica*, *C. quitensis*	400	3660	655	6
P11	L. Haplorthel	90 % *D. antarctica*., *C. quitensis*, mosses	230	6820	740	9
P12	C. Umbrorthel[b]	90 % *D. antarctica*, mosses	340	18180	3400	5
P15	L. Umbrorthel	95 % *D. antarctica*, mosses, lichens	570	1720	260	7
P7	T. Psammorthel	95 % Mosses	290	3650	910	4
P13	C. Umbriturbel[b]	95 % *D. antarctica*, mosses	10	38900	5120	8

[a] Podzol
[b] Ornithogenic soil

varies in both areas within the scale of nearly 0 to 900 g kg^{-1} (Tables 19.3 and 19.4). The WMI organic soils from beneath mosses are rich in mineral particles which is attributed to long lasting eolian input from areas without vegetation (Beyer et al. 2000a). In these mineral soils organic carbon (TOC) locally exceeds 100 g C kg^{-1}, whereas in the WMI so-called organic soils contain between 231–342 g kg^{-1} only (Table 19.5).

The C/N ratio varies considerably within a given soil profile due to the different sources of organic matter in the Antarctic soils (Blume et al. 1997; Kuhn 1997; Beyer et al. 2000a) and the selective preservation of carbohydrates or N-containing moieties (Bölter et al., Chap. 11) as well as the translocation processes during soil formation (Blume et al. 1996; Quass et al. 1998; Beyer et al. 2000b, 2001). The mean C/N ratio of the investigated soil horizons is about 7 for KGI and about 12 for WMI (Tables 19.3 and 19.4). This is somewhat lower than that known from natural soils of temperate climatic regions (Scheffer and Schachtschabel 1998) and from Arctic soils (Ping et al. 1995), because nearly all soils in these coastal areas of Antarctica are affected by seabird droppings (Beyer and Bölter 1999) and the absolute amount of C is comparatively lower in Antarctica than in other regions.

Table 19.4. Relationship between soils, vegetation and organic carbon and total nitrogen storage of the entire soil profile in the Windmill Island area close to Casey Station. *NE* not existing

Code	Soil name	Vegetation	>2 mm (g kg^{-1})	TOC	N_t (g m^{-2})	C/N
D1	Lithic Haplorthel[a,b]	Nil	800	1350	80	17
B12	Lithic Haplorthel	Nil	530	1770	170	10
D3	Typic Haplorthel[a,b]	Nil	830	2200	370	6
L1	Typic Haplorthel	<2 % Mosses	480	110	15	8
L2	Lithic Haplorthel	<2 % Mosses	500	170	20	9
B2	Typic Aquiturbel	<2 % Lichens, mosses	590	770	160	5
WP1	Lithic Haplorthel[b]	<2 % Algae	867	860	100	8
H2	Lithic Haplorthel[b]	<2 % Algae	795	1090	260	4
H3	Lithic Haplorthel	<2 % Algae	758	3300	520	6
B13	Lithic Haploturbel	<2 % Lichens, mosses	360	3450	350	10
A11	Lithic Aquiturbel	<2 % Algae	380	4400	310	14
WP2	Lithic Haplorthel[b]	<2 % Algae	800	7680	1400	6
C3	Lithic Haplorthel	2–5 % Mosses, lichens, algae	550	300	20	15
A13	Typic Haplorthel[a]	2–5 % Burnt lichens, mosses	720	3220	260	13
D2	Lithic Haplorthel	10–20 % Lichens, scarce mosses	530	1370	80	17
H1	Lithic Haplorthel	80 % Mosses, algae	667	3770	90	5
A12	Typic Haplorthel[a,b]	80–100 % Lichens	680	1810	210	9
P1	Lithic Haplorthel	100 % Algae below gravel	763	700	110	6
MG1	Lithic Haplorthel	100 % Algae	610	1010	30	31
A2	Lithic Haplorthel	100 % Mosses, lichens	550	1570	170	9
B14 A	Lithic Haplorthel	100 % Mosses	500	1740	130	14
C4	Lithic Aquiturbel	100 % Mosses	160	1780	120	15
C1 A	Lithic Haplorthel	100 % Mosses, lichens	160	2500	240	10
B4	Lithic Haplorthel	100 % Mosses, lichens	410	3300	340	10
B14B	Lithic Haploturbel	100 % Mosses	370	3420	290	12
C11	Lithic Haplorthel[a]	100 % Mosses, lichens	520	3540	290	12
C1	Lithic Haplorthel	100 % Mosses, lichens	290	3950	290	14
WP5	Typic Haplorthel[a]	100 % Mosses	550	4620	390	12
C5	Lithic Haploturbel	100 % Mosses	800	6130	290	22
A3	Lithic Haplorthel[a]	100 % Mosses, lichens	314	8160	310	26
B11	Lithic Haplorthel	100 % Mosses	410	8300	740	11
M2[c]	Lithic Hemistel	100 % Dense moss cushion	106	14480	960	15
BL209[d]	Lithic Fibristel	100 % Dense algal carpet	NE	5200	240	21
BL208[d]	Lithic Hemistel	100 % Dense moss cushion	NE	27000	2340	12
M1[d]	Lithic Fibristel	100 % Dense moss cushion	NE	28610	1310	22
WP4[d]	Typic Hemistel	100 % Dense moss cushion	NE	32100	1780	18
WP6[d]	Lithic Hemistel	100 % Dense moss cushion	NE	45600	2580	18

[a] Spodic.
[b] Ornithogenic.
[c] Including organic layer above mineral A horizon.
[d] Only organic layers, no occurrence of A horizon.

Table 19.5. Differentiation of the uppermost 10 cm of soils at King George Island of the maritime Antarctica and the Windmill Island region of East Antarctica (key to abbreviations see methods)

Vegetation cover %	n	Depth of A horizon	TOC ($g\ kg^{-1}$)	Mean stone content ($g\ kg^{-1}$)	TOC ($g\ m^{-2}$)	N_t ($g\ m^{-2}$)	C/N_t (%)
King George Island area							
0–2	5	3.4	2.8	440	83	17	5
90–95	8	13.8	63.2	300	3643	265	11
LSD_5		4.0	19.3	180	87	102	3
Windmill Island area							
0–5	14	7.5	27.7	640	2190	286	9
100	15	6.7	28.4	430	4346	308	15
LSD_5		2.6	18.5	120	1878	192	4

The pH($CaCl_2$) values range in both Antarctic areas between 3.6 and 6.8 (Tables 19.1 and 19.2) irrespective of the parent material. Carbonatic parent material is very rare in the investigated regions, an evident indication for the presence of carbonate was found only in one profile next to the Ecology Glacier at KGI. This is why some soils are classified in the families euic (pH ≥4.5) or dysic (pH <4.5), and base saturation ranges from 6 to 93 % (Tables 19.1 and 19.2).

For the WMI soils, the stocks of K, Mg and P were investigated (Table 19.2). Beyer et al. (2000 c) show that the availability of the nutrients K, Mg and P in this region is clearly correlated to TOC and clay. In the WMI soils, K and Mg are mainly derived from weathering of gneiss and schist, whereas P is probably derived from seabirds or from eolian transfer of P-rich fine-textured surface soil material (Blume et al. 1997; Beyer et al. 1999). Most of the WMI soils sampled have high P (and in some instances also Mg and K) levels compared with natural soils from temperate regions (Scheffer and Schachtschabel 1998), and Arctic and Antarctic soils with low seabird influence (Blume et al. 1997; Ping et al. 1995; Beyer et al. 1999, 2000a). On KGI, elevated levels of Mg and Ca have been considered to be due to weathering of pyroxenes (Kuhn 1997).

19.2.2 Vegetation Patterns

In the WMI region the flora consists only of mosses, liverworts, lichens growing on pebbles, stones and rocks, soil algae and cyanobacteria (Seppelt, Chap. 13; Elster, Chap. 17). Vegetation establishment and growth is comparatively slow because of the cold (Lewis Smith 1985; 1990), the limited availability of free water (Kennedy 1993) and, particularly for lichens, too little water for much of the summer for photosynthetic carbon gain (Kappen 1993; Hovenden and Seppelt 1995a; Schroeter et al. 1999).

The current vegetation in the ice-free areas of King George Island are characterised locally by the occurrence of the two vascular plants *Deschampsia antarctica* and *Colobanthus quitensis* (Olech, Chap. 12). According to Komárková (1985), both vascular plants have their ecological optimum in the Admiralty Bay region. Nevertheless, huge parts of the open landscape are non-vegetated and the occurrence of vegetation shows a high variation in cover and patchiness (Kuhn 1997).

In both regions, the vegetation cover ranges between scattered and 95–100 % cover (Tables 19.1 and 19.2). On KGI, strong relations can be established between the ice edge of the Ecology Glacier (Olech, Chap. 12) and vegetation cover. Similar observations can be made in the WMI region, where the Løken Moraine close to the ice plateau is only scarcely covered by plants. At the small scale, a great variability in the vegetation cover can be found, which is mainly related to the effects of wind (Lewis Smith 1985, Blume and Bölter 1996), hence, only very small patches between rocks and boulders are areas of primary soils.

19.3 Relationships Between Soils and Vegetation

Around Casey Station a vegetation cover of lichens or mosses exits on soils without present influence of cryoturbation, i.e. on sandy soils with gravel. Silty to loamy soils in ridges which are still affected by active cryoturbation do not show any plant cover (Blume and Bölter 1996). A cover with living moss turfs can be formed only on soils with a high moisture regime. Hovenden and Seppelt (1995b) show that substratum nutrient levels appear to have little effect on the distribution pattern of the dominant lichen species in the WMI region. However, the effect is not significant in determining the distribution of many of the minor species.

Nitrogen content appears to have only minor effects on distribution patterns of vegetation at WMI (Table 19.4). In areas more distant from active bird rookeries the vegetation pattern is related to the soil texture and the micro relief. At KGI, a threshold for higher abundance of vascular plants seems to act at about 100 g N m^{-2}. Below this, only lichens or low coverage of vascular plants occur. This pattern, however, cannot be valid in zones with ornithocoprophilic lichens, e.g. *Caloplaca* sp., *Xanthoria* sp., next to rookeries or the coastal zone as they have specific nutritional demands which are documented by the fact that other frequently occurring species are missing. On siliceous rocks, *Usnea aurantiaco-atra* forms a stable vegetation, but with increasing influences from the sea, mainly salt spray and bird perching, *U. aurantiaco-atra* disappears and is replaced by *U. antarctica* and particularly more eutrophic components. Similar observations with lichens have been described for northern Victoria Land (Kappen 1985).

Large differences in other elements, particularly P and Ca, at WMI, both largely derive from fossil or abandoned penguin colonies, and are also not significant in determining distribution patterns (Table 19.2). As growth is particularly slow in continental Antarctic sites, present distribution patterns may reflect past nutrient environments.

The KGI soils and the density of the mixed cryptogamic and phanerogamic vegetation are compared in the first instance with the total organic carbon and total nitrogen storage of the entire profile and the mean stone content. Weak relationships between soil characteristics and vegetation cover are typical for a mainly cryptogamic vegetation. Lindsay (1978) points out that biological weathering due to lichens is insignificant in areas of the continent. Even a negative effect on weathering by protecting rock surfaces from wind erosion and precipitation has been discussed (Lewis Smith 1973). A dense cover of lichens may protect rock surfaces from wetting and water does not penetrate into the rock's surface with the result that freeze-thaw cycles do not occur (Collins et al. 1975).

Although some input of organic matter and influences from lichens and mosses to soil organic matter have been demonstrated (Tearle 1987; Roser et al. 1992; Melick and Seppelt 1992), the short time scales over which this happens, e.g. during freeze-thaw cycles, have to be taken into account. Reliable dating of dissolved carbon, also of refractable matter, is not present. The same holds true for litter from cryptogams, except for a few datings from peat banks. Lichens cannot be regarded as main sources for litter, as they are very slow growing organisms. This holds especially true for lichens from dry areas. As such, thalli from KGI of *Usnea aurantiaco-atra*, *U. antarctica*, *Ramalina terebrata* and *Caloplaca regalis* are found to be between 32 and 43 years old, those of *Himantormia lugubris* can be estimated to be about 60 years old by using the ^{14}C method (Willkomm et al. 1991).

Nevertheless, accumulations of humic material can be related to vegetation patterns (Kuhn 1997). Mean contents of humics are estimated in the rooting space (30 cm) at sites on KGI. They show values of 0.6 kg m^{-2} humic substance in non-vegetated Leptosols and Regosols. Areas with sparse cover by lichens contain 5.7 kg m^{-2}, and soils with dense coverages of mosses and vascular plants contain 17.4 kg m^{-2}, with a maximum of 31.8 kg m^{-2} (Kuhn 1997). The latter values are comparable to those from boreal forests (Scheffer and Schachtschabel 1998).

According to studies by Lewis Smith (1985, 1990), plant colonisation and moss rhizoidal growth only affects the upper layers (0–10 cm) of the soil. The ability of soils to provide an appropriate environment for the root system is a precondition for potential colonisation by vascular plants. One of these properties is the amount of fine-grained material. At KGI, the non-vegetated soils have higher stone contents. Bulk densities are generally below 1.4 g cm^{-3} and thus well useable for roots. Only one site shows an elevated value due to sesquioxides which are responsible for a higher bulk density of 1.7 g cm^{-3}. Consequently, this soil shows restricted root penetration (Kuhn 1997). Similarly, but in a different way, high contents of silt are also unfavourable for roots as the soil fabric cannot be stabilised because of intensive cryoturbation.

The TOC concentration in the A horizon of the WMI soils is increased due to the old plant and rhizoidal input by mosses as well as leaching products from the dead plant material. Differences between the scarcely and highly vegetated soils are more pronounced in the KGI region where carbon and nitrogen are more closely related to vegetation density (Table 19.3). The amounts of both elements, carbon and nitrogen on bare or scarcely vegetated soils are more than ten times higher in the WMI than in the KGI region although actual vegetation cannot be related to this (Tables 19.3 and 19.4).

Beyer et al. (1999) notes that the origin of this organic matter in the WMI region is still a matter of speculation. Relic material can be taken into consideration. In both regions the carbon storage in the uppermost 10 cm is significantly different between the sparsely and highly vegetated soils (Table 19.5). But nitrogen did not show this pattern, resulting in an increase of the C/N ratio. For both regions this significant increase in C/N suggests more than a simple accumulation of organic matter, but also a qualitative change of soil organic matter (SOM).

Further, soil texture, nutrient retention and the occurrence of active iron oxides, as indicated by a high Fe_o/Fe_d ratio, as well as base saturation for the cation exchange capacity have to be considered as driving forces for plant growth. The oxidation of Fe^{2+} to Fe^{3+} leads to a sixfold volume increase (Louis and Fischer 1979) as well as to browning reactions in these soils (Kuhn 1997). In the KGI region, the nutrient level, indicated through the current pH level and base saturation, are negatively related to plant cover. Relationships between plant cover and soil texture were not apparent on (relic) ornithogenic soils. It is assumed that the reason for missing direct relationships is a result of exceedingly high nutrient levels, namely for K, Mg, and P (Table 19.2) in places with subfossil and extant penguin colonies and an elevated eolian input of nutrient-rich fine soil material of bird or sea colonies (Beyer and Bölter 1999). Both effects are relevant for soils in the WMI area at different soil stages and the sequences of cryptogamic colonisation (Beyer and Bölter 1999).

19.4 Snow and Water as Structural Components

On KGI water is not a limiting factor for the soil and vegetation during the growing period. For the WMI region, however, water may become a limiting factor. This holds especially true for the distribution of cryptogamic communities, like *Usnea sphacelata* sociations in the WMI region, on rocky and stony surfaces which have to withstand long desiccation times during all seasons (Lewis Smith 1986; Kappen et al. 1990), thus biological weathering is also hampered. Occurrence of vegetation on the ground is more related to snow cover and the melting areas during spring and summer. Gaps and sheltered corners between rocks where fine material accumulates are stands for moss cushions, such as *Ceratodon purpureus*, and *Schistidium antarctici*, which also provide habitats for crustose epibryic lichens, such as *Rinodina olivaceobrunnea*, *Buellia grimmiae*, *Candelariella hallettensis*. Moss turfs are also often covered by prostrately growing *Usnea antarctica* (Kappen et al. 1990). The cover of plots between rocks with moss also consists of green algae and cyanobacteria (Bölter 1992). Only where cryoturbation is absent can mosses or soil-colonising lichens develop. *U. sphacelata*, *Pseudephebe minuscula*, and *Buellia soredians* colonise rocks and even small stones as long as they are not moved by cryoturbation. Killed lichens can be observed on the slopes of Fildes Peninsula, on rocks that were turned upside down (Kappen, pers. comm.).

In the WMI region, large areas of shallow soils are covered by moss cushions, and are dominated by *Ceratodon purpureus* and *Schistidium antarctici* (Kappen et al. 1990; Bölter et al. 2000). These communities become more diverse in the vicinity of melt lakes (Lewis Smith 1986). However, observations during the warm summer of 1992/1993 have shown that communities in such low-level areas can be extremely poor because snow cover is high and long-lasting over many years.

Snow is an important factor for the establishment and spread of plant sociations. On the one hand it prevents erosion during the winter storms and offers protection from low temperatures (Longton 1988; Davey et al. 1992), on the other hand, it hampers light penetration. This may lead to the situation that the temperature is high enough to allow metabolic processes and respiration but no photosynthetic gain (Benedict 1990, 1991; Crawford 1997). Snow heights up to 15 cm can be sufficient for photosynthesis (Kappen and Breuer 1991) and carbon gains have been described as valid (e.g. Collins and Callaghan 1980). Melting snow under a snow cover can be regarded as a main source of free water and stimulation for metabolic processes in general, and prevents strong temperature shifts and thus freeze-thaw cycles.

Duration and depth of snow cover strictly control the presence or absence of cryptogams as was demonstrated by Winkler (2000) and Winkler et al. (2000). Mosses dominate in wetter places with more extended snow cover whereas most macrolichens prefer exposed sites with shallow snow cover. Thin snow cover usually disappears quickly and the substratum dries out (Kappen et al. 1998). Hancock and Seppelt (1988) have shown this also in a study at WMI region by analysing substrate moisture and moss cover near a melt lake.

Hence, vegetation is not only a result of substrate and nutrient quality but depends also on actual and long-term snow regimes (Lewis Smith 1990; Winkler 2000). To prevent lichen growth, a snow cover of more than 15 cm thickness is needed, as light and temperature conditions under such snow fields still allow active photosynthesis (Kappen and Breuer 1991; Kappen et al. 1995), although the gain during such conditions by concurrent respiration is very low or even negative (Davis and Harrison 1981; Schroeter et al. 1997). Thus, a significant input to soils from such stands cannot be anticipated.

19.5 Conclusion and Needs for Future Research

In the Antarctic regions, temperature and soil nutrient levels are obviously not the only limiting factors determining the establishment of vegetation. Because of the extremely high stone content, the soil texture has only a small impact on cryptogamic colonisation. A strong effect must be attributed to exposure, especially in continental Antarctica where strong winds play a major role in erosion processes, but also provide sheltered niches, which – on the other hand – can produce extremes on the temperature scale. Examples can be found in the WMI region (Bölter 1992). Bölter et al. (1989) and Lewis Smith (1986) measured temperatures of 34 °C in dry moss carpets and 43 °C in active moss beds and lichen heaths, respectively. This opens up discussions on stress tolerances of Antarctic plants.

C/N ratios in the vicinity of bird colonies are low due to this animal impact. For other areas, additional N-inputs must be attributed to other effects, such as N-fixation, although the total amount is very low (Alexander 1975; Bölter et al. 1995) Microclimate effects, and particularly the availability of free water and length and depth of snow cover have strong influences on the vegetation patterns and thus soil characteristics.

Only an extant vegetation cover influences total organic carbon of the soils in a way that determines soil types, e.g. brownification, or podzolisation (Blume and Bölter 1993; Blume et al. 1996). The surface vegetation creates a favourable soil environment by positive feedback mechanisms between plants and soil, as has been observed in geo-ecological investigations in hot desert systems (Beyer et al. 1998a). This development of local fertile islands observed in hot deserts (Evenari 1985) is also relevant in cold desert ecosystems and should be an aspect of future soil ecological research in the terrestrial ice-free areas of maritime and continental Antarctica.

The classical view, as relevant for other environments, that soils influence vegetation finds an extension in the Antarctic since plant cover has great impact on soil structure and even soil formation. In this environment, which acts for a great part of the year at the borders of living conditions either due to low temperatures or to dry conditions due to low precipitation, interactions between plants and soils become visible at different local and temporal scales. The balance is often delicate and small changes, e.g. caused by cryoturbation or snow cover, can have large effects.

Observations from single soil profiles do not allow us to put together a general hypothesis on ecosystem processes with regard to plant adaptations and soil formation. The analysis of profiles does, however, provide insights to the interactions of soil and plants. Small-scale variability in Antarctic soil ecosystems (Beyer et al. 1998b; Beyer and Bölter 1999), microclimate (Lewis Smith 1990) and soil microbiology (Bölter 1990) makes it, however, difficult to transfer data from a profile level to a landscape level (Beyer and Bölter 1999).

References

Alexander V (1975) Nitrogen fixation by blue-green algae in polar and sub-polar regions. In: Steward WPD (ed) Nitrogen fixation by free-living micro-organisms. IBP No 6. Academic Press, New York, pp 175–188

Benedict JB (1990) Lichen mortality due to late-lying snow: results of a transplant study. Arct Alp Res 22:81–89

Benedict JB (1991) Experiments on lichen growth. II. Effects of a seasonal snow cover. Arct Alp Res 23:189–199

Beyer L, Bölter M (1999) Formation, ecology and geography of cryosols of an ice-free oasis in coastal East Antarctica near Casey Station (Wilkes Land). Aust J Soil Res 37:209–244

Beyer L, Tielbörger K, Blume H-P, Pingpank K, Podlech D (1998a) Geo-ecological soil features and the vegetation pattern in an arid dune area in the Northern Negev, Israel. Z Pflanzenernähr Bodenkd 161:347–356

Beyer L, Pingpank K, Bölter M, Seppelt R (1998b) Small-distance variation of carbon and nitrogen storage in mineral Antarctic Cryosols near Casey Station. Z Pflanzenernähr Bodenkd 161:211–220

Beyer L, Bockheim JG, Campbell IB, Claridge GGC (1999) Genesis, properties and sensitivity of Antarctic Gelisols. Antarct Sci 11:387–398

Beyer L, Pingpank K, Wriedt G, Bölter M (2000a) Soil formation in coastal continental Antarctica. Geoderma 95:283–304

Beyer L, Bölter M (2000b) Chemical and biological properties, formation and classification of spodic Cryosols in a terrestrial ecosystem of East Antarctica (Wilkes Land). Catena 39:95–119

Beyer L, Bölter M, Seppelt RD (2000c) Nutrient and thermal regime, microbial biomass, and vegetation of Antarctic soils in the Windmill Island region of East Antarctica (Wilkes Land). Arct Antarct Alp Res 32:30–39

Beyer L, White DW, Bölter M (2001) Organic matter composition, transformation, and translocation in Haplocryods (Podzols) of an ice-free terrestrial oasis in East Antarctica. Austr J Soil Res 39:543–563

Blume H-P, Bölter M (1993) Soils of Casey Station (Wilkes Land, Antarctica). In: Gilichinski D (ed) Joint Russian-American seminar on cryopedology and global change. Post-Sem Proc 1st Int Conf Cryopedology 1992, Russian Academy of Sciences, Pushchino, pp 96–103

Blume H-P, Bölter M (1996) Wechselwirkungen zwischen Boden- und Vegetationsentwicklung in der Kontinentalen Antarktis. Verh Ges Ökol 25:25–34

Blume H-P, Schneider D, Bölter M (1996) Organic matter accumulation in and podzolisation of Antarctic soils. Z Pflanzenernähr Bodenkd 159:411–412

Blume H-P, Beyer L, Bölter M, Erlenkeuser H, Kalk E, Kneesch S, Pfisterer U, Schneider D (1997) Pedogenic zonation in soils of southern circumpolar region. Adv GeoEcol 30:69–90

Blume H-P, Beyer L, Schneider D (1998) Soils of the Southern circumpolar region and their classification. Euras Soil Sci 31:477–485

Bockheim J G (1997) Properties and classification of cold desert soil from Antarctica. Soil Sci Soc Am J 61:224–231

Bockheim J G, Ugolini F C (1990) A review of pedogenic zonation in well-drained soils of the Southern circumpolar region. Quat Res 34:47–66

Bölter M (1990) Microbial ecology of soils from Wilkes Land, Antarctica: II. Patterns of microbial activity and related organic and inorganic matter. Proc NIPR Symp Polar Biol 3:120–132

Bölter M (1992) Vergleichende Untersuchungen zur mikrobiellen Aktivität in Böden und an Kryptogamen aus der kontinentalen und maritimen Antarktis (Casey, Wilkes Land, und Arctowski, King George Island). Habilschrift, Univ Kiel, pp 1–203

Bölter M, Kappen L, Meyer M (1989) The influence of microclimate conditions on potential photosynthesis of *Usnea sphacelata*: a model. Ecol Res 4:297–307

Bölter M, Blume H-P, Kappen L (1995) Bodenbiologische Untersuchungen in der maritimen und kontinentalen Antarktis (King George Island und Windmill Islands). 1. Umweltparameter und anorganische Nährstoffe. Polarforschung 65:41–61

Bölter M, Seppelt R, Beyer L, Pingpank C (2000) Studies on floristic diversity, soil organic matter, and soil microbes from the Windmill Islands, East Antarctica. Bibl Lichenol 75:421–432

Campbell IB, Claridge GGC (1987) Antarctica: soils, weathering processes and environment. Elsevier, Amsterdam

Collins NJ, Callaghan TV (1980) Predicted patterns of photosynthetic production in maritime Antarctic mosses. Ann Bot 45:601–620

Crawford RMM (1997) Habitat fragility as an aid to long-term survival in Arctic vegetation. In: Woodin SJ, Marquiss M (eds) Ecology of Arctic environments. Blackwell, Oxford, pp 113–136

Davey MC, Pickup J, Block W (1992) Temperature variation and its biological significance in fellfield habitats on a maritime Antarctic island. Antarct Sci 4:383–388

Davis RC, Harrison PM (1981) Prediction of photosynthesis in maritime Antarctic mosses. Coll Ecosyst Subantarct, Comité National Français des Recherches Antarctique 51:241–247

Egner H, Riehm H, Domingo WR (1960) Untersuchungen über die chemische Bodenanalyse als Grundlage für die Beurteilung des Nährstoffzustandes des Bodens: II. Chemische Extraktionsmethoden zur Phosphor- und Kaliumbestimmung. Koenigl Lant Ann 26:204–209

Evenari M (1985) Adaptations of plants and animals to the desert environment. In: Evenari M, Noy-Meir J, Goodall D (eds) Ecosystems of the world 12A. Hot deserts and arid shrublands. Elsevier, Amsterdam, pp 79–93

Hancock RJ, Seppelt RD (1988) Habitat specificity and morphological variation in two Antarctic Usnea species. Polarforschung 58:285–291

Harwood JE, van Steenderen RE, Kühn AL (1969) A rapid method for orthophosphate analysis at high concentrations in water. Water Res 3:417–423

Hovenden MJ, Seppelt RD (1995a) Uptake of water from the atmosphere by lichens in continental Antarctica. Symbiosis 18:111–118

Hovenden MJ, Seppelt RD (1995b) Exposure and nutrients as delimiters of lichen communities in continental Antarctica. Lichenologist 27:505–516

ISSS-WRB [International Soil Science Society – World Reference Base for Soil Resources] (1998) World Reference base for soil resources. World Soil Resources Reports 84, ISSS-ISRIC-FAO, Rome

Jones JB (1991) Kjeldahl method for nitrogen determination. Micro-Macro Publishing, Athens

Kappen L (1985) Vegetation and ecology of ice-free areas of northern Victoria Land, Antarctica. 2. Ecological conditions in typical microhabitats of lichens in Birthday Ridge. Polar Biol 4:227–236

Kappen L (1993) Lichens in the Antarctic region. In: Friedmann EI (ed) Antarctic microbiology. Wiley-Liss Inc, New York, pp 433–491

Kappen L, Meyer M, Bölter M (1990) Ecological and physiological investigations in continental Antarctic cryptogams. 1. Vegetation pattern and its relation to snow cover on a hill near Casey Station, Wilkes Land. Flora 184:209–220

Kappen L, Breuer M (1991) Ecological and physiological investigations in continental Antarctic cryptogams. II. Moisture relations and photosynthesis of lichens near Casey Station, Wilkes Land. Antarct Sci 3:273–278

Kappen L, Sommerkorn M, Schroeter B (1995) Carbon acquisition and water relations of lichens in polar regions – potentials and limitations. Lichenologist 27:531–545

Kappen L, Schroeter B, Green TGA, Seppelt RD (1998) Microclimate conditions, melt water moistening and the distributional pattern of *Buellia frigida* on rocks in a southern continental Antarctic habitat. Polar Biol 19:101–106

Komárková V (1985) Two native Antarctic vascular plants, *Deschampsia antarctica* and *Colobanthus quitensis*: a new southernmost locality and other localities in the Antarctic Peninsula area. Arct Alp Res 17:401–417

Kennedy AD (1993) Water as a limiting factor in the Antarctic terrestrial environment: a biogeographical synthesis. Arct Alp Res 25:308–315

Kennedy AD (1999) Modelling the determinants of species distributions in Antarctica. Arct Antarct Alp Res 31:230–241

Kuhn D (1997) Genese, Ökologie und Soziologie einer Bodengesellschaft in einem Periglazialgebiet der King George Insel, West-Antarktis. Schriftenr Inst Bodenkd Univ Kiel 40:1–173

Lewis Smith, RI (1985): Nutrient cycling in relation to biological productivity in Antarctic and sub-Antarctic terrestrial and freshwater ecosystems. In: Siegfried WR, Condy PR, Laws RM (eds) Antarctic nutrient cycles and food webs. Springer, Berlin Heidelberg New York, pp 138–155

Lewis Smith, RI (1986) Report on Antarctic fieldwork. plant ecological studies in the fellfield ecosystem near Casey Station, Australian Antarctic Territory 1985–86. Br Antarct Surv Bull 72:81–91

Lewis Smith, RI (1990) Plant community dynamics in Wilkes Land, Antarctica. Proc NIPR Symp Polar Biol 3:229–244

Lindsay DC (1978) The role of lichens in Antarctic ecosystems. Bryologist 81:268–276

Longton RE (1988) The biology of poplar bryophytes and lichens. Cambride University Press, Cambridge

Louis H, Fischer K (1979) Allgemeine Geomorphologie. Walter de Gruyter, Berlin

Melick DR, Seppelt RD (1992) Loss of soluble carbohydrates and changes in freezing point of Antarctic bryophytes after leaching and repeated freeze-thaw cycles. Antarct Sci 4:399–404

Ping C-L, Michaelson CJ, Malcom RL (1995) Fractionation and carbon balance of soil organic matter in selected cryic soils in Alaska. In: Lal R, Kimble J, Levine E, Stewart BA (eds) Soils and global change. Lewis, Boca Raton, pp 307–316

Quass W, Beyer L, Schneider D, Blume, H-P (1998) Podzolisation and soil organic matter transfer in soils of Antarctica. Proc 16th ISSS Congr, Montpellier. Electronic CD version, ISSS, Vienna

Roser DJ, Melick DR, Ling HU, Seppelt RD (1992) Polyol and sugar content of terrestrial plants from continental Antarctica. Antarct Sci 4:413–420

Schachtschabel P (1951) Die Methoden zur Bestimmung des Kalkbedarfes. Z Pflanzenernähr Bodenkd 54:134–145

Scheffer F, Schachtschabel P (1998): Lehrbuch der Bodenkunde. 4th edn, Enke, Stuttgart

Schlichting E, Blume H-P, Stahr K (1995) Bodenkundliches Praktikum. 2. Aufl. Blackwell, Berlin

Schroeter B, Kappen L, Schulz F (1997) Hydration-related spatial and temporal variation of photosynthetic activity in Antarctic lichens. In: Battaglia B, Valencia J, Walton DWH (eds) Antarctic communities. Species, structure and survival. Cambridge University Press, Cambridge, pp 221–225

Schroeter B, Kappen L, Schulz F, Sancho LG (2000) Seasonal variation in the carbon balance of lichens in the maritime Antarctic: long-term measurements of photosynthetic activity in *Usnea aurantiaco-atra*. In: Howard-Williams C, Davison W, Broady P (eds) VII SCAR Intern Biol Symp Antarctic Ecosystems: Models for a wider ecological unerstanding. Caxton Press, Christchurch

Soil Survey Staff (1998): Keys to soil taxonomy. 8th edn. US Depart Agric - Nat Res Counc, Washington, DC

Tearle PV (1987) Cryptogamic carbohydrate release and microbial response during spring freeze-thaw cycles in Antarctic fellfield fines. Soil Biol Biochem 19:381–390

Willkomm H, Bölter M, Kappen L (1991) Age estimation of Antarctic macrolichens by radiocarbon measurements. Polarforschung 61:103–112

Winkler B (2000) Die Rolle der Schneebedeckung für die Kryptogamenvegetation in der maritimen Antarktis. Dissertation, Univ Kiel, 140 pp

Winkler B, Kappen L, Schulz F (2000) Snow and ice as an important ecological factor for the cryptogamy in the Maritime Antarctic. In: Howard-Williams C, Davison W (eds) Antarctic ecosystems: models for wider ecological understanding. Caxton Press, Christchurch, pp 220–224

20 Uses, Preservation, and Protection of Antarctic Coastal Regions

M. BÖLTER and B. STONEHOUSE

20.1 Ecological Background

Well over 95 % of the total land area of Antarctica and adjacent islands is permanently ice-covered and unavailable to colonisation by plants and animals. Ice-free areas attract a biota that includes:

1. A flora severely restricted by harsh climatic conditions, isolation and lack of opportunities for recruitment, in which algae, mosses and lichens predominate, and vascular plants are severely limited to the milder and damper marginal areas, linked with
2. A microfauna of soil organisms restricted almost entirely to microphages, and
3. A macrofauna made up mainly of seabirds and seals that depend on the sea for their livelihood, and use the land only for resting and breeding.

Within the milder maritime Antarctic (as defined by Holdgate 1964), some ice-free areas support local populations of recently-introduced plants. Most periantarctic islands (as defined by Stonehouse 1982, 1989) support indigenous land or freshwater birds, introduced plants, and one or more species of introduced mammals.

While the flora and microfauna form a range of closely integrated communities, the macrofauna of marine animals imposes itself superficially, occupying space for activities ashore, and contributing mainly mechanical and chemical perturbations to the soils and terrestrial biota. Humans, too, impose their effects superficially, with little expectation of integration into a predominantly wilderness environment.

Lewis Smith (1984, 1996), Block (1984) and Wynn-Williams (1990) have provided comprehensive reviews of the origins and characteristics of the Antarctic flora, and of Antarctic soil microbiology and microfauna respectively. Longton (1985) provides an alternative review of macrophytic vegetation in the continental and maritime Antarctic, and Vincent (1988) and Friedmann (1993) have reviewed the microbiology of all Antarctic ecosystems, marine, terrestrial and freshwater. Lewis Smith's simple and generally-accepted categorisation of

Ecological Studies, Vol. 154
L. Beyer and M. Bölter (eds.) Geoecology of Antarctic Ice-Free Coastal Landscapes

Table 20.1. Terrestrial plant communities of the Antarctic region. (After Lewis Smith 1984)

Locality and formation	Subformation, with brief description
Continental and maritime Antarctic region	Crustaceous and foliaceous lichen: mainly saxicolous species with many nitrophilous taxa associated with bird colonies or perches, and halophilous taxa on coastal rocks
	Fruticose and foliose lichen: commonly with an understorey of crustose taxa, mainly saxicolous nitrophobous species
	Short moss-cushion and turf: in terricolous communities, may coalesce to form closed stands with a smooth to highly convoluted surface, often encrusted with lichens or cyanobacteria but with fruticose lichens often associated
	Tall moss turf: terricolous communities, some of which develop peat to a depth of 2 m, usually colonised by epiphytic lichens
	Tall moss cushion: tall cushion and deep undulating carpet growth forms, some representing calcicolous types of otherwise shorter forms; on wet soils and rocks and along melt-streams
	Bryophyte carpet and mat: pleurocarpous mosses, or foliose and thallose liverworts, forming shallow turf-like carpets and prostrate mats on moist to wet soils; lichens sparse or absent
	Alga sheet: mainly foliose algae forming loose convoluted sheets mainly on wet soil, often in biotically disturbed habitats
	(MARITIME ANTARCTIC ONLY) Grass and cushion chamaephyte: assemblages of two native phanerogams with associated cryptogams on moist, sheltered coastal soils: plants of *Deschampsia antarctica* may coalesce to form close swards, those of *Colobanthus quitensis* rarely do

Antarctic plant communities is here summarised in Table 20.1. He distinguishes two major geobotanical categories or 'formations' – Antarctic (including maritime Antarctic), and sub-Antarctic (here corresponding to periantarctic), divided respectively into eight and ten subformations.

20.1.1 Ecosystem Development

In the Antarctic region, as elsewhere, ecosystems take time to develop, and change constantly with time. As Lewis Smith (1984) pointed out, long-term studies that are essential for understanding plant colonisation and temporal succession have been undertaken in only a few Antarctic localities, notably South Georgia and Signy Island, South Orkney Islands. Almost two decades later his comment is still largely true: Antarctic biologists continue to find difficulties in securing funding for projects planned to last more than 2 or 3 years, and longer-term continuity is seldom guaranteed at any stage in a pro-

Table 20.1 (*Continued*)

Locality and formation	Subformation, with brief description
Peri-antarctic (sub-Antarctic) islands	Tall tussock grassland: dense or open vegetation dominated by *Poa* spp. Tussocks showing xeromorphic characters; on wet, biotically disturbed ground and moist drained coastal slopes
	Short tussock grassland: closed or slightly discontinuous vegetation dominated by short, narrow-leaved xerophytic grasses (e.g. *Festuca contracta*), with other graminoids, forbs, abundant mosses and lichens; well-drained, moderately exposed habitats
	Meadow: closed vegetation dominated by short soft-leaved mesophytic grasses, sometimes with herbs and bryophytes; moist flats, valley floors and gentle slopes
	Mire: soligenous vegetation forming shallow peat, in wet areas maintained by lateral seepage, dominated by rushes, mesophytic grasses and bryophytes; gentle slopes and valley sides
	Bog: ombrogenous vegetation watered largely by precipitation and retarded run-off, dominated by helophytic graminoids, especially rushes, and bryophytes, including moss turf banks; depressions, valley floors
	Flush: semi-aquatic vegetation dominated by helophytic bryophytes and herbs; stream margins and close to springs
	Swamp: permanently or seasonally submerged vegetation dominated by hydrophytic and helophytic herbs, mosses and algae; zonal around lake and pool margins
	Herbfield: dense vegetation dominated by forbs including *Acaena* spp.; mainly in moist sheltered habitats
	Fern-brake: vegetation dominated by ferns; restricted dry habitats
	Fellfield (= feldmark): discontinuous xerophytic vegetation dominated by cushion-forming forbs, short grasses, mosses and lichens; dry windswept stony soils, especially on high ground

ject. This presents a serious impediment to understanding, not only of natural changes and development, but also of changes induced by animal or human interference (see below). Nevertheless, Lewis Smith's reviews summarise a wide range of contributions on short-term colonisation and succession from many Antarctic localities, and further studies in this important field are constantly emerging.

Block's review (1984) follows Lewis Smith's biogeographical definitions of Antarctic and sub-Antarctic, and summarises information on microflora, microfauna and macro-invertebrate fauna from a wide range of terrestrial localities. Block draws attention to the rapid poleward decrease of species and taxa from the sub-Antarctic to the South Pole, and extensive gaps in our knowledge of soil biology throughout the region. These are pointed by relatively recent discoveries of such unexpected phenomena as the dominance of yeast populations in bryophyte systems during the growing season in the maritime Antarctic, and the presence in the same region of enchytraeid annelids. Block's comment that '... terrestrial ecosystems in the Antarctic, being devoid of large, above-ground herbivores and containing mainly decomposer-oriented communities, provide ideal test-beds for examining perturbation effects ...', is again timely and important when we consider effects of animal and human perturbations (see below).

Such models have also been used to explain the abundance and distribution of microorganisms by Cameron et al. (1976), Wynn-Williams (1990) and others with respect to various environmental factors. Horowitz et al. (1972) and McKay (1993) have extrapolated from extreme desert to extraterrestrial environments. While such remote places are fascinating areas for studying invasion and related phenomena, the coastal environments considered here are above minimal levels both in species and in opportunities for colonisation by plant and animal propagules.

20.1.2 Colonisation by Seabirds and Seals

Whereas almost every Antarctic coastal landscape has a selection of soils and vegetation that characterises its capacity and limitations for colonisation, distribution of macrofauna is more patchy and less predictable from the intrinsic qualities of the land. Of the seven species of penguins, over 40 species of flying birds, and six species of seals that breed in the Antarctic region, very few breed on the land ice that covers so much of the continent and neighbouring islands. Sea ice is more hospitable: emperor penguins *Aptenodytes forsteri* and four species of seals (crabeater *Lobodon carcinophagus*, Weddell *Leptonychotes weddellii*, Ross *Ommatophoca rossii* and leopard *Hydrurga leptonyx*) breed almost exclusively on fast ice or floes. All other species require land for breeding and resting, making prominent use of coasts. Penguins especially form strikingly large communities, concentrating at high densities in sites, some of which extend over several hectares and include hundreds of thousands of nests. Colonies of flying birds, though far outnumbering those of penguins, are usually less prominent.

Some coastal areas, especially those of the maritime region, include huge numbers of several breeding species. Others with superficially similar characteristics are ignored or only lightly colonised. Individual species of seabirds set particular criteria for breeding areas. King penguins *Aptenodytes patagonica* require gently-sloping beaches in the periantarctic zone, that offer easy landings, backed by extensive flats where their chicks can huddle in winter. Adelie penguins *Pygoscelis adeliae* on the continent and nearby islands, and gentoo penguins *P. papua* in warmer latitudes, again favour sloping beaches, backed by nesting areas that become free of winter snow in time for breeding, with suitable nesting sites and materials close at hand. Rockhopper penguins *Eudyptes crestatus* and macaroni penguins *E. chrysolophus* are more likely to find rocky landings, often with heavy surf, and nest on much steeper slopes, including cliffs and headlands (Stonehouse 1985). Burrow-nesting petrels require stable scree slopes or cavernous rocks capable of excavation. Again, many otherwise-suitable sites are invalidated by late dispersal of winter snow. Surface nesting species including albatrosses, several other species of petrels, shags, gulls, skuas and terns each have their own requirements (see for example Siegfried 1985).

Marine criteria are no less important in determining which species live where, and in what concentrations. Along the continental coast, where conditions tend to be harshest and most critical, large colonies exist only where there is a high likelihood of food being locally abundant, especially in mid-to-late summer when the demands of growing chicks increase dramatically (Stonehouse 1967). They form in localities where, for example, currents provide the mixing that brings rich zooplankton to the surface, and sea ice disperses reliably each year by mid-summer, when the chicks are in the nest.

Major breeding colonies of penguins, gulls, skuas, shags and other high-density breeding species are long-term features. Some we now know to have occupied the same sites for well over a century, and probably very much longer. A few have arisen or disappeared within the past few decades of observation. Some disappearances are clearly due to human interference (see below), while others are clearly not. All tend to vary in size from year to year.

Plant life is strongly inhibited close to colonies, due to trampling and high concentrations of guano. New colonies quickly destroy existing vegetation: abandoned colonies or areas within colonies remain abiotic for several seasons, probably until seasonal melt water flushes surface guano into the substrate, when nitrophilous algae become the pioneer colonists.

The criteria by which land-breeding Antarctic fur seals (*Arctocephalus gazella*) and southern elephant seals (*Mirounga leonina*) select their particular breeding and resting sites are less clear, but are likely to be based on similar factors. These species form large breeding colonies, some numbering several hundreds or thousands of animals, particularly in the maritime Antarctic and on periantarctic islands. Stocks of both that were severely reduced by hunting during the late eighteenth to twentieth centuries are currently recovering and expanding.

Southern fur seals (*Arctocephalus gazella*) are invading not only their former breeding haunts, but also southern areas of the maritime region where they were not previously recorded. Their return during recent years has provided opportunities for biologists to study their dramatic effects on plant communities that for many decades have developed without them (see below).

20.2 Occupation, Uses and Impacts by Man

The Antarctic region shows no evidence of ever having supported indigenous human populations. Recorded human visits began in the late eighteenth century, when the first landings were made in coastal areas of the periantarctic and maritime areas, probably by sealers hunting for fur seals and elephant seals. Hunting in these areas persisted through the early decades of the nineteenth century, always by transients whose visits were usually seasonal, only occasionally prolonged by overwintering. Their efforts drastically reduced stocks of both species of seals.

Beltramino (1993) has chronicled changes in the human population of Antarctica in historic times. From the late nineteenth century and throughout the twentieth century, the region attracted explorers and scientists, in numbers that peaked during the International Geophysical Year 1957–1959, and have since fluctuated annually from hundreds overwintering to thousands on summer-only expeditions.

Scientists have regarded the continent and its environs as a unique observatory for studying a wide range of disciplines from atmospheric physics to zoology. Most recent studies have tended to emphasise global change phenomena at a variety of scales and across a variety of disciplines. Those who fund or support their activities, especially governments of claimant states, have in many instances been politically motivated, seeking through physical occupation to advance claims to sovereignty over territory within the region.

A more recent development, from the 1950s onward, has been tourism, which currently brings over 10,000 recreational visitors to the region each year. Practically all tourists are ship-based transients, living on their ship, requiring no infrastructure ashore, and staying for only a few days in the area, within a season that is concentrated in the four summer months. Like the hunters before them, both scientists and tourists congregate in the small coastal areas rather than the much larger expanses of land ice, making inevitable some encroachments on flora and fauna.

20.2.1 Research Installations

Only a few Antarctic scientific stations have been built on the continental ice sheet, where they are maintained by constant renewal, or quickly disappear under the snow. Most stations, whether currently in use or abandoned, are sited in coastal regions, usually in areas of local scientific interest, some in areas of outstanding biological interest. A few are sited primarily for logistic purposes, for example to facilitate passage to the continental interior.

The first scientific party to overwinter on land was based at Cape Adare, Victoria Land, in the closing years of the nineteenth century. More huts were built as accommodation for temporary exploring expeditions, mainly on continental coasts, up to 1939. The first long-term meteorological observatory was built on the South Orkney Islands in 1904, where observations have been maintained ever since. The first whaling station appeared on South Georgia about the same time. In 1939, and sporadically during the 1940s, appeared the first stations designed specifically for long-term occupation. During the 1950s and 1960s, a rash of scientific stations and refuges appeared, mainly in the maritime region, erected both for scientific and political objectives.

Once built, stations tend to remain on site, though not always in useable condition. Historic huts associated with early twentieth century expeditions are venerated and preserved. The huts of Scott's and Shackleton's expeditions, sited in cold, chronically dry McMurdo Sound, remain in a better state of preservation than the near-contemporary huts of Mawson and Nordenskjöld, respectively in the milder, damper ambience of (continental) Commonwealth Bay and (maritime) Snow Hill Island. Mawson's hut, built in what is reputedly one of the windiest places on earth, has been primarily effected by strong winds. It experiences persistent katabatic winds of mean annual force over 61 km/h, with gusts to 320 km/h, and is severely eroded.

Since the mid-twentieth century between 1000 and 5000 scientists are estimated to work each summer in the Antarctic region, based at more than 40 permanent scientific stations, refuges (emergency shelters) and temporary field camps (SCAR 1994). Much of this work, especially biological research, has been conducted in coastal areas. Scientific stations and refuges proliferated from the mid-twentieth century onward. Some remain small: others have developed to the size of small towns, with separate scientific complexes, living quarters, hospital, supermarket, church, bank and attached airstrip. Expensive to build in a hostile environment, they have also proved difficult and expensive to maintain. Many early stations quickly reached the end of their useful life. Difficult and expensive to remove, they remain as graceless relics ringed by litter, to the lasting detriment of their formerly-pristine coastal site.

There can be no doubt that man has profited scientifically from long-term research in the Antarctic laboratory, enhancing knowledge not only of the Antarctic region but of many fundamental geological, atmospheric and oceanic processes that affect the world generally. His impacts on the region, compared with human impacts on any other continent, remain slight. However, the human influences that are currently most apparent in Antarctic coastal region, in environments that were until recently relatively pristine, arise from the erection, use, management, and (often) subsequent dereliction of buildings dedicated for scientific purposes.

Studies of local pollution, e.g. contamination by oil and fuel (Kerry 1990), or more rarely by metals and organochlorides derived from long-term station activities or accidents (Lukowski 1993; Deprez et al. 1999) are well documented. A detailed study of the area surrounding the Polish research station Henryk Arctowski (Rakusa-Suszczewski and Kryszowska 1991), on King

George Island, South Shetland Islands, ends with recommendations for higher standards of station management and conservation behaviour in accordance with recommendations by SCAR and CCAMLR. Other station studies are known to have been made, but appear to be available only as internal reports. Chen and Blume (1995) focus on the more general effects of man's activities, including scientific expeditions and tourism and also on disturbance factors for wildlife: see also Thomson (1977) and Culik et al. (1990), who focus on wildlife.

Despite many such studies, it is still difficult to establish verified monocausal relationships due to direct impacts of men. No comprehensive summaries, comparable with several available for the Arctic (e.g. AMAP 1998; Crane and Galasso 1999) have yet been prepared for the Antarctic. However, the smaller number of Antarctic settlements and more limited scale of human activities makes southern problems less urgent or compelling than those in the north.

20.2.2 Tourism

Antarctic tourism began in the late 1950s with overflights from South America and ship-based visits to a small number of sites in the maritime Antarctic. Since then the ship-based industry has grown substantially. By the mid-1990s annual numbers of visitors exceeded 9000, making use of over 70 landing sites (Stonehouse and Crosbie 1995). Currently, some 15,000 ship-based tourists visit the region annually, making use of over 150 landing sites (Stonehouse 2000). While some locations receive only two or three visits per decade, the more popular ones may be visited two or three times weekly throughout a 4-month summer season, which increases local impacts (Stonehouse 1992; Donachie 1994).

Currently, some 40 areas are protected from public intrusion as scientific reserves, formerly designated Sites of Special Scientific Interest (SSSIs) or Specially Protected Areas (SPAs), now Antarctic Specially Managed Areas (ASMAs) and Antarctic Specially Protected Areas (ASPAs). Changes in their status are determined at annual Antarctic Treaty Consultative Meetings (ATCMs) and promulgated through national governments to tour operators. Responsible tour operators who follow guidelines drawn up jointly between the International Association of Antarctica Tour Operators (IAATO) and the Antarctic Treaty System (ATS) find little reason and less excuse for landing passengers in such scheduled areas: intrusions are indeed rare. Others might argue with some justification that such areas are seldom marked or identified in the field, and therefore vulnerable to misuse.

A useful semi-official document, the *Visitor's Introduction to Antarctica and its Environment*, is available from (http://www.icair.iac.org.nz/) with detailed guidelines to various environments. So long as tour operators continue to accept responsibility for maintaining these largely self-imposed rules, impacts from ship-based visitors remain low. However, Antarctic tourism is following a predictable path of development and diversification, which may or may not be containable within these or similar guidelines.

20.3 Disturbance and Remediation

Polar soils and ecosystems are frequently described as 'fragile' (e.g. Billings 1974), suggesting greater susceptibility to disturbance than, for example, their counterparts in temperate or tropical regions, with less capability of self-regeneration once damage has been done. This concept, bearing strongly on issues of polar wilderness management, has recently been examined and to some degree challenged. Though much of the relevant study is based on Arctic tundra habitats, and gross disturbances in that region arising from industrial and military activities, there are clear parallels with the relatively uncomplicated disturbances arising from human and animal activities in Antarctic landscapes.

In Antarctica, as elsewhere, humans disturb terrestrial ecosystems either (1) mechanically, for example by walking, excavating or moving loads across them, or (2) chemically, by introducing alien reactants. Neither category of disturbance is entirely alien to polar ecosystems. Frost-heave is a violent disturbance to which polar plants must be adapted. Birds and mammals disturb soils and vegetation systems by walking, rolling, scratching and fighting over them, and by concentrations of chemicals contained in their excretions and faeces. An Antarctic moss bed that superficially appears undisturbed and static is in fact responding to constant disturbances, all forms of which to some degree alter the course of soil processes and vegetal succession.

20.3.1 A Terminology for Disturbance and Damage

The term 'damage' in an environmental context carries emotional overtones and needs to be used objectively. Table 20.2 summarises a terminology, drawn from an extensive literature by Walker et al. (1987), Strandberg (1997) and Stonehouse (1999), that is useful in considering the effects of human activities on polar ecosystems, including those of Antarctic coasts.

Table 20.2 re-asserts premises familiar to ecologists concerning ecosystem stability, but often forgotten in debate – that ecosystems are normally subject to perturbation by natural causes, and carry within them substantial capacities for variation and self-recovery.

Disturbance may be intentional or incidental. Intentional disturbance includes damage required by structural engineering, e.g. building scientific

Table 20.2. A vocabulary of terms relating to disturbance and recovery of soil ecosystems, based on Stonehouse (1999) and concepts by Margalef (1968), Strandberg (1997), van Andel and van den Bergh (1987), Webber and Walker (1987), Westman (1978) and White (1979)

Disturbance	A change in conditions that displaces a soil or ecosystem beyond its normal limits of variation, or interferes with its normal functioning
Damage	A change in conditions that results in lasting ecological dysfunction
Resistance	The ability of a soil ecosystem to withstand displacement from a given state
Resilience	A measure of the degree, manner and pace of recovery of a soil or ecosystem to its original state after disturbance
Resilience threshold	The level of disturbance beyond which recovery to an original state will not occur
High and low resilience	A soil or ecosystem that returns rapidly to an original state after disturbance has high resilience: one that returns slowly or not at all has low resilience
Recovery	The processes by which a soil or ecosystem achieves biological and physical stability following disturbance
Recovery potential	A measure of the capacity of a soil or ecosystem to recover after disturbance
Recovery: complete and functional	Recovery is *complete* if the ecosystem is re-established in its original state, or *functional* if it achieves stability at a level different from the original
Remediation	Human-induced processes that encourage recovery of soils and ecosystems after disturbance

stations, roads and airstrips. Incidental disturbance includes upheavals that accompany engineering, e.g. creation of temporary campsites, tracks, rubbish dumps, spills of oil and other industrial liquids and windborne dust, all of which produce long-term scarring. Both forms of disturbance produce secondary effects. Walker et al. (1987) have shown, for example, that bulldozing vegetation and soils over permafrost on Arctic tundra allows unwelcome melting of subsoil and formation of erosion channels. Dust from denuded ground affects melt patterns of snow, neutralises acid soils and lakes, and may inhibit growth of vegetation.

Intentional disturbance, though often extensive, is purpose-oriented and sanctioned on grounds of social or scientific expediency. Incidental disturbance has hitherto been acceptable as a necessary evil, though often accounting for a higher proportion of total damage, but is currently considered both avoidable and antisocial, and increasingly subject both to limiting legislation and remedial action.

Disturbance is countered by two inherent functions of ecosystems, *resistance* (mainly static) and *resilience* (mainly dynamic), that vary widely from one system to another. These determine the nature and degree of response and initiating *recovery*, for which different ecosystems have different capacities. Recovery may be *complete* – a return to an original state, or *functional* – a more probable reversion to a different but no less viable state. Disturbance becomes damage when induced changes overcome resistance and resilience, inhibiting or preventing recovery and resulting in ecological dysfunction.

20.3.2 Remediation

The subject of a substantial and growing literature, remediation is the application of human-induced processes to restore ecosystems after disturbance. Bioremediation seeks restoration through enhancement of natural processes, usually by provision of missing elements, e.g. heat, nutrients or water. See for example Walker et al. (1987), Crawford (1997), Shaver et al. (1986), Jonasson et al. (1996), whose studies focus also on practicalities of restoring northern tundra after severe disturbance and damage, and Kerry (1990), Rakusa-Suszczewski and Krzyszowska (1991), Chen and Blume (1995) for similar studies in the Antarctic region. AMAP (1998) provides a comprehensive review of Arctic pollution issues, and AEON (1998) a more limited overview of Antarctic issues.

Remediation and ultimate restoration of Antarctic landscapes to functional levels of recovery are entirely possible, so long as expectations remain modest and patience is exercised. It may be entirely unrealistic, for example, to expect complete recovery of a site on the continental Antarctic coast after removal of a long-term scientific station. Except in areas where the substrate has been impregnated with oil, vegetation may redevelop within a few years. Even concrete foundations acquire lichens and are favoured as windbreaks by seals or breeding birds. But neighbouring bird colonies that disappeared early in the occupation may never return, and other, more trivial changes will remain: recovery will be functional rather than complete. Recovery that, in temperate regions might be expected within a human life span may take two or three times longer in Antarctica. Remediation is costly, competing with scientific research for time and funding, and governments have so far proved reluctant to spend lavishly in restoring disturbed or damaged sites.

20.4 Coastal Landscapes and Global Change

Possible effects of global environmental changes on Antarctica have been considered for a long time, for example by Smith and Steenkamp (1990), Harris and Stonehouse (1991), Weller (1992) and Kennedy (1993). Most are concerned with warming, but changes in levels of airborne nutrients, pollutants, and propagules and of increasing UV radiation have also been considered.

The role of plants as indicators of changing conditions in terrestrial habitats has been stressed by various authors (e.g. Lewis Smith 1994; Fowbert and Lewis Smith 1994; Green et al. 1999). Plants react conservatively to changing environmental conditions, on time scales that, though largely unknown, are likely to be long-term. Neither they, nor animals show evidence of critical sensitivity to changing temperature. While shifts in their distribution indicates some kind of environmental change, the factors involved may be aridity, length of season, soil quality or other factors, rather than small, long-term temperature changes alone (Stonehouse 1991).

Soils are even more conservative in their responses, and may be used as indicators of environmental changes on a broader scale (Bölter 1996). Thus coastal areas are of critical importance in monitoring global changes, as several other papers in this symposium testify.

20.5 Conservation of Coastal Landscapes

The unique status of Antarctica for both science and tourism makes it essential that their value be recognised, and their conservation – specifically the maintenance of their ecological integrity – be given the highest priority. Many areas have already been substantially altered by human usage, mostly for scientific or political ends. Tourism has hitherto maintained a better record, but is expanding, developing and changing, and may soon be promoting activities that are no less invasive than those of science. Under Antarctic Treaty regulation, specified in the Protocol on Environmental Protection to the Antarctic Treaty, monitoring is required for every human activity permitted. However, standards of monitoring have not been specified, and careful, selective and well-considered monitoring cannot be undertaken casually.

There is a substantial literature on schemes for detecting environmental changes, both on a world scale and specifically in Antarctica: see for example Bockheim et al. (1993), Campbell et al. (1993), Lyons (1993). Guidelines and compilations of monitoring programmes are in use for many individual projects (e.g. COMNAP 1998). Baseline studies and development of regulations for national research programmes are duties imposed on all signatories to the Antarctic Treaty (e.g. Machowski 1992; Carstens et al. 1998; Juterzenka et al. 1998). No single nation is currently responsible for developing monitoring programmes to cover tourist activities, which are different in kind from those of scientists, and involve different monitoring criteria and techniques.

The role of the Antarctic as a natural laboratory continues. Though the questions addressed, the perceived research priorities and the funding available differ from year to year, the region still provides opportunities for both short-term and long-term research, centred mainly on the coastal places that provide the easiest access to its secrets. That the coasts are subject to natural changes, as well as anthropogenic, complicates the work of those who seek their conservation, maintaining interest and providing an additional and welcome challenge within a wide range of disciplines.

References

AEON (1998) Summary of environmental monitoring activities in Antarctica. COMNAP, Washington, DC

AMAP (1998) Assessment report: Arctic pollution issues. Arctic Monitoring and Assessment Program, Olso

Beltramino JCM (1993) The structure and dynamics of Antarctic population. Vantage Press, New York

Billings WD (1974) Arctic and alpine vegetation: plant adaptations to cold summer climates. In: Ives JD, Barry RG (eds) Arctic and alpine environments, Methuen, London, pp 403–443

Block W (1984) Terrestrial microbiology, invertebrates and ecosystems. In: Laws RM (ed) Antarctic ecology, vol 1. Academic Press, London, pp 163–236

Bockheim J, Bölter M, Campbell IB (1993) Design of an experiment to detect the effects of global change on soil development in the Southern circumpolar region. In: Gilichinski DA (ed) Joint Russian-American seminar on cryopedology and global change. Russ Acad Sci, Pushchino, pp 146–164

Bölter M (1996) Consequences of global warming on soil processes in Arctic regions. Polarforschung 66:1–10

Cameron RE, Honour RC, Morelli FA (1976) Antarctic microbiology – preparation for Mars life detection, quarantine, and back contamination. In: Heinrich MR (ed) Extreme environments: mechanisms of microbial adaptation, Academic Press, New York, pp 57–82

Campbell IB, Balks MR, Claridge GGC (1993) A simple technique for estimating the impact of fieldwork on the terrestrial environment in ice-free areas of Antarctica. Polar Rec 29:321–328

Carstens M, Süfke L, Hanssen H, Borkowitz B, Juterzenka Kv, Spindler M (1999) Auswirkungen menschlicher Tätigkeiten auf die Meeresumwelt – erste Ansätze für eine Bewertung nach dem Umweltschutzprotokoll-Ausführungsgesetz zum Antarktis-Vertrag (AUG). Dtsch Hydrogr Z (Suppl) 8:195–202

Chen J, Blume H-P (1995) Impact of human activities on the terrestrial ecosystem of Antarctica: a review. Polarforschung 65:83–92

Claridge GGC, Campbell IB (1985) Physical geography – soils. In: Bonner WN, Walton DWH (eds) Key environments: Antarctica. Pergamon Press, Oxford, pp 62–70

Crane K, Galasso JL (1999) Arctic environmental atlas. Office of Naval Research, Naval Research Laboratories, Hunter College, Washington, DC

Crawford RMM (1997) Disturbance and recovery in Arctic lands. Kluwer, Dordrecht

Culik B, Adelung D, Woakes AJ (1990) The effects of disturbance on the heart rate and behaviour of Adélie penguins (*Pygoscelis adeliae*) during the breeding season. In: Kerry KR, Hempel G (eds) Antarctic ecosystems. Ecological change and conservation. Springer, Berlin Heidelberg New York, pp 178–182

Deprez PP, Arens M, Locher H (1999) Identification and assessment of contaminated sites at Casey Station, Wilkes Land, Antarctica. Polar Rec 35:299–316

Donachie SP (1994) Henryk Arctowski Station – mixing science and tourism. Ann Tourism Res 21:333–343

Fowbert JA, Lewis Smith RI (1994) Rapid population increases in native vascular plants in the Argentine Islands, Antarctic Peninsula. Arct Alp Res 26:290–296

Friedmann EI (1993) Antarctic microbiology. Wiley-Liss, New York

Green TG, Schroeter B, Sancho LG (1999) Plant life in Antarctica. In: Pugnaire FI, Valladares F (eds) Handbook of functional plant ecology. Marcel Dekker, New York, pp 495–543

Harris CM and Stonehouse B, eds. (1991). Holdgate MW (1964) Terrestrial ecology in the maritime Antarctic. In: Carrick R, Holdgate M, Prévost J (eds) Biologie antarctique. Hermann, Paris, pp 181–94

Holdgate MW (1964) Terrestrial ecology in the maritime Antarctic. In: Carrick R, Holdgate MW, Prévost J (eds) Biologie Antarctique. Proc 1st SCAR Symposium, Paris, France, 1962. Hermann, Paris, pp 181–194

Horowitz NH, Cameron RE, Hubbard JS (1972) Microbiology of the Dry Valleys of Antarctica. Science176:242–245

Jonasson S, Lee JA, Callaghan TV, Havström M, Parsons AN (1996) Direct and indirect effects of increasing temperatures on sub-Arctic ecosystems. Ecol Bull 45:45–52

Juterzenka Kv, Hanssen H, Carstens M, Süfke L, Borkowitz B, Spindler M (1998) Bedeutung des Umweltschutzprotokoll-Ausführungsgesetzes zum Antarktis-Vertrag (AUG) für die wissenschaftliche Forschung. Dtsch Hydrogr Z (Suppl) 8:203–210

Kennedy AD (1995) Antarctic terrestrial ecosystem response to global environmental change. Annu Rev Ecol Syst 26:683–704

Kerry E (1990) Microorganisms colonizing plants and soil subjected to different degrees of human activity, including petroleum contamination, in the Vestfold Hills and MacRobertson Land, Antarctica. Polar Biol 10:423–430

Lewis Smith RI (1984) Terrestrial plant biology of the sub-Antarctic and Antarctic. In: Laws RM (ed) Antarctic ecology, vol 1. Academic Press, London, pp 61–162

Lewis Smith RI (1994) Vascular plants as bioindicators of regional warming in Antarctica. Oeocologia 99:322–328

Lewis Smith RI (1996) Terrestrial and freshwater biotic components of the western Antarctic Peninsula. In: Ross RM, Hofmann EE, Quetin LB (eds) Ecological research west of the Antarctic Peninsula. American Geophysical Union, Antarct Res Ser 70, pp 15–59

Longton RE (1985) Terrestrial habitats – vegetation. In: Bonner WN, Walton DWH (eds) Key environments: Antarctica. Pergamon Press, Oxford, pp 73–105

Lukowski AB (1993) Chlorinated hydrocarbons. In: Rakusa-Suszczewski (ed) The maritime Antarctic coastal ecosystem of Admiralty Bay. Dept Antarct Biol, Pol Acad Sci, Warsaw, pp 197–198

Lyons D (1993) Environmental impact assessment in Antarctica under the Protocol on Environmental Protection. Polar Rec 29:11–120

Machowski J (1992) The Antarctic environmental legal regime. Pol Polar Res 13:183–214

McKay CP (1993) Relevance of Antarctic microbial ecosystems to exobiology. In: Firedmann EI (ed) Antarctic microbiology, Wiley-Liss, New York, pp 593–601

Rakusa-Suszcsewski S, Krzyszowska A (1991) Assessment of the environmental impact of the 'H. Arctowski' Polish Antarctic Station (Admiralty Bay, King George Island, South Shetland Islands). Pol Polar Res 12:105–121

SCAR (1994) Stations of SCAR nations operating in the Antarctic, Winter 1993. SCAR Bull 112:4–6

Shaver GR, Chapin FS III, Gartner BL (1986) Factors limiting seasonal growth and peak biomass accumulation in *Eriophorum vaginatum* in Alaskan tussock tundra. J Ecol 74:257–278

Siegfried RW (1985) Birds and mammals – oceanic birds of the Antarctic. In: Bonner WN, Walton DWH (eds) Key environments: Antarctica. Pergamon Press, Oxford, pp 242–65

Smith VR, Steenkamp M (1990) Climatic change and its ecological implications at a sub-Antarctic island. Oecologia 85:14–24

Stonehouse B (1967) Occurrence and effects of open water in McMurdo Sound during winter and early spring. Polar Rec 13:775–78

Stonehouse B (1982) La zonation écologique sous les hautes latitudes australes. In Jouventin P, Massé L, Tréhen P (eds) Colloque sur les ecosystèmes subantarctiques, Comité national français des recherches antarctiques Paris, pp 531–536

Stonehouse B (1985) Birds and mammals – penguins. In: Bonner WN, Walton DWH (eds) Key environments: Antarctica. Pergamon Press, Oxford, pp 266–92

Stonehouse B (1989) Polar ecology. Blackie, Glasgow

Stonehouse B (1991) Polar ecosystems, management and climate modelling. In Harris CM and Stonehouse B (eds) Antarctica and global climate change. Belhaven Press, London

Stonehouse B (1992) Monitoring shipborne visitors in Antarctica: a preliminary field study. Polar Rec 28:213–218

Stonehouse B (1999) Biological processes in cold soils. Polar Rec 35:5–10

Stonehouse B (2000) The last continent: discovering Antarctica. SCP Books, Burgh Castle

Stonehouse B, Crosbie K (1995) Tourist impacts and management in the Antarctic Peninsula area. In: Hall CM, Johnston ME (eds) Polar tourism: tourism in the Arctic and Antarctic regions. Wiley, Chichester, pp 217–33

Strandberg B (1997) Vegetation recovery in Greenland. In: Crawford RMM (ed) Disturbance and recovery in Arctic lands. Proceedings of the NATO Advanced Research Workshop on Disturbance and Recovery of Arctic Terrestrial Ecosystems, Rovaniemi, Finland, 24–30 September 1995. NATO Advanced Science Institutes, Series 2, vol 25. Kluwer, Dordrecht

Thomson RB (1977) Effects of human disturbances on an Adelie penguin rookery and measures of control. In: Llano GA (ed) Adaptations within Antarctic ecosystems. Proc 3rd SCAR Symp on Antarctic Biology. Smithsonian Institution, Washington, DC, pp 1177–1180

Vincent WF (1988) Microbial ecosystems of Antarctica. Cambridge University Press, Cambridge

Walker DA, Cate D, Brown J, Racine C (1987) Disturbance and recovery of arctic Alaskan tundra terrain. CRREL Report 87–11

Weller G (1992) Antarctica and the detection of environmental change. Philos Trans R Soc Lond B 338:201.208

Wynn-Williams DD (1990) Ecological aspects of Antarctic microbiology. Adv Microb Ecol 11:71–146

Concluding Remarks

L. Beyer and M. Bölter

The objective of this book was to present a most recent compilation of our knowledge on soils of Antarctic coastal landscapes. Further, we have tried to compare soils and their biological features from two different climatic regions. These regions have been under investigation for different scientific aspects for many years. Despite national focal points of research, several projects show overlap, e.g. studies on plants, animals or meteorological observations.

A central point for our project was the interaction within the soil sphere, combining data from the soil's biotic and its abiotic world. This close interaction is important for all kind of soils – although working at different scales of space and time. It became possible to show that both plant colonisation and soil formation act in close relationship – when the activity and inputs of microbes, i.e. bacteria, cyanobacteria and algae, are considered along with soil chemical and soil physical processes. As such, the availability of water and its local action becomes a prominent feature of structure soil patterns and the main reason for wide ranges in parameters used to describe its condition in physical, chemical or biological ways.

Soil ecology is defined as the interactions between the organisms in and on the soil as well as their interactions with soil properties. In our geoecological understanding we have used soil ecology in its widest sense and used it comparably to geoecology (Blumenstein et al. 2000). Hence, soils are regarded as the centre of processes and the geoecology describes the interactions between soils and biota, microclimate, topography, parent materials.

Soil formation in Antarctica cannot be entirely characterised by the knowledge from the Polar desert of the Dry Valley region. In contrast to the very old soils of this region, which can be dated back to 10–15 million years, the story of the Antarctic coastal environments is much shorter and much more variable due to evident biological influences. General processes of soil formation in Antarctica are similar to those of other areas of the world but frequently found at a lower level. Most soils have lithic contact within the first 40 cm and show mostly patchy occurrences. Soil formation in the Antarctic coastal landscapes is characterised by soil organic matter accumulation, chemical weathering and brownification, which means iron release, podzolisation, frequently

Ecological Studies, Vol. 154
L. Beyer and M. Bölter (eds.) Geoecology of Antarctic Ice-Free Coastal Landscapes

very high nutrient levels (N, P, K) and redoximorphic features. The latter suggests the occurrence of free liquid water during the thawing season.

Jenny's model, expressed by the 'CLORPT' equation [soil formation=f (cl: climate, ø: biotic factor, r: topography, p: parent materials, t: time)] is – without doubt – valid for soil formation process in Antarctica too, it is not restricted to the pure development of desert soils due to high aridity as has been suggested earlier (e.g. Tedrow 1977; Campbell and Claridge 1987; Bockheim and Ugolini 1990; Priscu 1998). An additional prominent soil formation in the coastal areas is due to seals and birds, i.e. their resting or breeding places. The CLORPT equation from Jenny (1941, 1986) nowadays has to be widened by adding the human factor. The described geoecological interactions are affected strongly by human activities in Antarctica. Even in this region, soil formation is influenced to a great extent by man up to the total destruction of whole soil landscape; examples in the Antarctic can be found around some stations due to high traffic or general station activity. In other parts of the world, soils are becoming more recognised as a third medium besides air and water, but currently very little is done worldwide to regulate soil protection (Tutzing Project 1998). Soil remediation can be a local solution, but the limitation of soil disturbances should be a major concern in order to protect the unique Antarctic environments.

The Antarctic obviously loses a bit of interest for the public, a matter of non-understood reasons and consequences (Walton 2000). Its role in biogenic and mineral resources has declined. Nevertheless, the Antarctic has kept its scientific role as a big natural laboratory and as a basic record of environmental phenomena, and several countries have set up monitoring programmes. For instance, analyses on UV effects on photosynthetic activities increases scientific interest. Plants are used as indicators for shifting margins of growth patterns, similar to glacier retreats. Subsequent ecological processes need detailed monitoring for proofs of changes at different time scales. As the Antarctic loses its remoteness, the direct and indirect effects of man can be observed at many places and on biotic and abiotic processes. As results from research programmes gradually come into the public domain, its effects on decisions cannot be foreseen. We hope that summaries like this have positive feedback on a great environment.

References

Blumenstein O, Schachtzabel H, Barsch H, Bork H-R, Küppers U (2000) Grundlagen der Geoökologie. Erscheinungen und Prozesse in unserer Umwelt. Springer, Berlin Heidelberg New York

Bockheim JG, Ugolini FC (1990) A review of pedogenic zonation in well-drained soils of the Southern circumpolar region. Quat Res 34:47–66

Campbell IB, Claridge GGC (1987) Antarctica: soils, weathering processes and environment. Elsevier, Amsterdam
Jenny H (1941) Factors of soil formation. McGraw-Hill, New York
Jenny H (1986) The soil resource. Ecological studies 37, 3rd edn. Springer, Berlin Heidelberg New York
Priscu JC (ed) (1998) Ecosystem dynamics in a polar desert. Antarctic Res Ser 72, Am Geophys Union, Washington, DC
Tedrow JCF (1977) Soils of Antarctica. In: Tedrow JCF (ed) Soils of the polar landscapes. Rutgers University Press, New Brunswick, New Jersey. pp 518–544
Tutzing Project (1998) Proposal for a 'Convention on sustainable use of soils'. http://www.zukunft.de/soil_convention
Walton DWH (2000) What is the state of the Antarctic environment? Antarct Sci 12: 397

Subject Index

Taxonomic Index

Geographical Index

Ecological Studies

Volumes published since 1996

Volume 114
Crassulacean Acid Metabolism. Biochemistry, Ecophysiology and Evolution (1996)
K. Winter and J.A.C. Smith (Eds.)

Volume 115
Islands. Biological Diversity and Ecosystem Function (1995)
P.M. Vitousek, L.L. Loope, and H. Adsersen (Eds.)

Volume 116
High Latitude Rainforests and Associated Ecosystems of the West Coast of the Americas: Climate, Hydrology, Ecology and Conservation (1996)
R.G. Lawford, P. Alaback, and E. Fuentes (Eds.)

Volume 117
Global Change and Mediterranean-Type Ecosystems (1995)
J. Moreno and W.C. Oechel (Eds.)

Volume 118
Impact of Air Pollutants on Southern Pine Forests (1996)
S. Fox and R.A. Mickler (Eds.)

Volume 119
Freshwater Ecosystems of Alaska. Ecological Syntheses (1997)
A.M. Milner and M.W. Oswood (Eds.)

Volume 120
Landscape Function and Disturbance in Arctic Tundra (1996)
J.F. Reynolds and J.D. Tenhunen (Eds.)

Volume 121
Biodiversity and Savanna Ecosystem Processes. A Global Perspective (1996)
O.T. Solbrig, E. Medina, and J.F. Silva (Eds.)

Volume 122
Biodiversity and Ecosystem Processes in Tropical Forests (1996)
G.H. Orians, R. Dirzo, and J.H. Cushman (Eds.)

Volume 123
Marine Benthic Vegetation. Recent Changes and the Effects of Eutrophication (1996)
W. Schramm and P.H. Nienhuis (Eds.)

Volume 124
Global Change and Arctic Terrestrial Ecosystems (1997)
W.C. Oechel et al. (Eds.)

Volume 125
Ecology and Conservation of Great Plains Vertebrates (1997)
F.L. Knopf and F.B. Samson (Eds.)

Volume 126
The Central Amazon Floodplain: Ecology of a Pulsing System (1997)
W.J. Junk (Ed.)

Volume 127
Forest Decline and Ozone: A Comparison of Controlled Chamber and Field Experiments (1997)
H. Sandermann, A.R. Wellburn, and R.L. Heath (Eds.)

Volume 128
The Productivity and Sustainability of Southern Forest Ecosystems in a Changing Environment (1998)
R.A. Mickler and S. Fox (Eds.)

Volume 129
Pelagic Nutrient Cycles: Herbivores as Sources and Sinks (1997)
T. Andersen

Volume 130
Vertical Food Web Interactions: Evolutionary Patterns and Driving Forces (1997)
K. Dettner, G. Bauer, and W. Völkl (Eds.)

Volume 131
The Structuring Role of Submerged Macrophytes in Lakes (1998)
E. Jeppesen et al. (Eds.)

Volume 132
Vegetation of the Tropical Pacific Islands (1998)
D. Mueller-Dombois and F.R. Fosberg

Volume 133
Aquatic Humic Substances: Ecology and Biogeochemistry (1998)
D.O. Hessen and L.J. Tranvik (Eds.)

Volume 134
Oxidant Air Pollution Impacts in the Montane Forests of Southern California (1999)
P.R. Miller and J.R. McBride (Eds.)

Volume 135
Predation in Vertebrate Communities: The Białowieża Primeval Forest as a Case Study (1998)
B. Jędrzejewska and W. Jędrzejewski

Volume 136
Landscape Disturbance and Biodiversity in Mediterranean-Type Ecosystems (1998)
P.W. Rundel, G. Montenegro, and F.M. Jaksic (Eds.)

Volume 137
Ecology of Mediterranean Evergreen Oak Forests (1999)
F. Rodà et al. (Eds.)

Volume 138
Fire, Climate Change and Carbon Cycling in the North American Boreal Forest (2000)
E.S. Kasischke and B. Stocks (Eds.)

Volume 139
Responses of Northern U.S. Forests to Environmental Change (2000)
R. Mickler, R.A. Birdsey, and J. Hom (Eds.)

Volume 140
Rainforest Ecosystems of East Kalimantan: El Niño, Drought, Fire and Human Impacts (2000)
E. Guhardja et al. (Eds.)

Volume 141
Activity Patterns in Small Mammals: An Ecological Approach (2000)
S. Halle and N.C. Stenseth (Eds.)

Volume 142
Carbon and Nitrogen Cycling in European Forest Ecosystems (2000)
E.-D. Schulze (Ed.)

Volume 143
Global Climate Change and Human Impacts on Forest Ecosystems: Postglacial Development, Present Situation and Future Trends in Central Europe (2001)
J. Puhe and B. Ulrich

Volume 144
Coastal Marine Ecosystems of Latin America (2001)
U. Seeliger and B. Kjerfve (Eds.)

Volume 145
Ecology and Evolution of the Freshwater Mussels Unionoida (2001)
G. Bauer and K. Wächtler (Eds.)

Volume 146
Inselbergs: Biotic Diversity of Isolated Rock Outcrops in Tropical and Temperate Regions (2000)
S. Porembski and W. Barthlott (Eds.)

Volume 147
Ecosystem Approaches to Landscape Management in Central Europe (2001)
J.D. Tenhunen, R. Lenz, and R. Hantschel (Eds.)

Volume 148
A Systems Analysis of the Baltic Sea (2001)
F.V. Wulff, L.A. Rahm, and P. Larsson (Eds.)

Volume 149
Banded Vegetation Patterning in Arid and Semiarid Environments (2001)
D. Tongway and J. Seghieri (Eds.)

Volume 150
Biological Soil Crusts: Structure, Function, and Management (2001)
J. Belnap and O.L. Lange (Eds.)

Volume 151
Ecological Comparisons of Sedimentary Shores (2001)
K. Reise (Ed.)

Volume 152
Global Biodiversity in a Changing Environment: Scenarios for the 21st Century (2001)
F.S. Chapin, O. Sala, and E. Huber-Sannwald (Eds.)

Volume 153
UV Radiation and Arctic Ecosystems (2002)
D.O. Hessen (Ed.)

Volume 154
Geoecology of Antarctic Ice-Free Coastal Landscapes (2002)
L. Beyer and M. Bölter (Eds.)

Volume 155
Conserving Biodiversity in East African Forests: A Study of the Eastern Arc Mountains (2002)
W.D. Newmark

Printing: Saladruck, Berlin
Binding: Stürtz, Würzburg